A C S S Y M P O S I U M S E R I E S **591**

Molecular Action of Insecticides on Ion Channels

J. Marshall Clark, EDITOR

University of Massachusetts

Developed from a symposium sponsored
by the Division of Agrochemicals
at the 207th National Meeting
of the American Chemical Society,
San Diego, California,
March 13–17, 1994

American Chemical Society, Washington, DC 1995

Seplae
Chem

Library of Congress Cataloging-in-Publication Data

Molecular action of insecticides on ion channels / J. Marshall Clark, editor.

  p.   cm.—(ACS symposium series; 591)

"Developed from a symposium sponsored by the Division of Agro-chemicals at the 207th National Meeting of the American Chemical Society, San Diego, California, March 13–17, 1994."

Includes bibliographical references and index.

ISBN 0–8412–3165–6

1. Insecticides—Physiological effect—Congresses. 2. Ion channels—Congresses.   I. Clark, J. Marshall (John Marshall), 1949 –   .
II. American Chemical Society. Division of Agrochemicals. III. American Chemical Society. Meeting (207th: 1994: San Diego, Calif.)
IV. Series.

QP801.I48M65   1995
574.87'5—dc20   95–6612
                                                                CIP

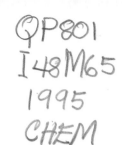

QP801
I48M65
1995
CHEM

1995 Advisory Board

ACS Symposium Series

M. Joan Comstock, *Series Editor*

Foreword

THE ACS SYMPOSIUM SERIES was first published in 1974 to provide a mechanism for publishing symposia quickly in book form. The purpose of this series is to publish comprehensive books developed from symposia, which are usually "snapshots in time" of the current research being done on a topic, plus some review material on the topic. For this reason, it is necessary that the papers be published as quickly as possible.

Before a symposium-based book is put under contract, the proposed table of contents is reviewed for appropriateness to the topic and for comprehensiveness of the collection. Some papers are excluded at this point, and others are added to round out the scope of the volume. In addition, a draft of each paper is peer-reviewed prior to final acceptance or rejection. This anonymous review process is supervised by the organizer(s) of the symposium, who become the editor(s) of the book. The authors then revise their papers according to the recommendations of both the reviewers and the editors, prepare camera-ready copy, and submit the final papers to the editors, who check that all necessary revisions have been made.

As a rule, only original research papers and original review papers are included in the volumes. Verbatim reproductions of previously published papers are not accepted.

M. Joan Comstock
Series Editor

Contents

LIGAND-GATED CHLORIDE CHANNELS

Preface

INSECTICIDES HAVE BEEN THE BOON and the bane of human societies. Their use has resulted in the most plentiful, least expensive, and most secure food source thus far produced by our agrarian efforts. As with all simplistic, directed approaches, the use of insecticides has had far-reaching and, in most cases, unforeseen environmental impacts. Nevertheless, it is unlikely that in the near future the quality or quantity of foodstuff can be maintained in the absence of pesticidal chemicals, particularly insecticides. Thus, our efforts as pesticide scientists should be focused on research approaches that will result in the availability of novel pesticide products that have been evaluated in terms of what we have learned from past faults in pesticide design and unmanaged overuse.

Molecular biological approaches have provided the means to fundamentally change the way in which pesticidal chemicals are discovered and used. Technical knowledge for the production of genetically engineered crop plants and biotechnological pest control strategies already exist and soon will become economically feasible. Because the large majority of currently used insecticides are neurotoxic and many are directed toxicologically to ion-channel disruption, an extensive database is currently available for assessment. Thus, the most enduring contribution of insecticides such as DDT and the cyclodienes may be in providing well-studied models to assess the advantages and disadvantages of new molecular technologies. It is an appropriate time to assess the impact that these molecular approaches have had on pest control and to evaluate the future of molecular pesticide science.

This scenario provided the idealistic focus for the organization of an ACS symposium entitled "Molecular Action and Pharmacology of Insecticides on Ion Channels", upon which this book is based. Topics were chosen principally in terms of their contemporary relevance to insect pest control and concentrated on those aspects in which a molecular understanding of insecticide action was being actively researched. Chapters were then requested from researchers and scholars drawn equally from academia and industry. These researchers are leading experts in their chosen areas of study and present a truly international group from Canada, Israel, Japan, the United Kingdom, and the United States.

This book is divided into three sections, each concerned with the action of insecticides on a major ion-channel family: voltage-sensitive channels, ligand-gated channels, and biopesticide-induced channels.

Presented in this volume are recent advances in how we envision molecular binding sites of insecticides on ion channels and molecular alterations of these sites that result in site-insensitive resistance mechanisms. Additionally, new receptor–ion channels and mosaic receptor organizations are described as novel sites of insecticide action and perhaps as means of genetically managing the development of insecticide resistances in the future.

I thank the authors for their presentations in the symposium and for their contributed chapters that encompass this volume. In particular, I extend my deepest appreciation to the many expert colleagues who provided helpful and necessary critical reviews. I thank Anne Wilson of ACS Books for all her help, suggestions, and encouragement; and Margaret Malone, Amity Lee-Bradley, and Julia Connelly of the Department of Entomology, University of Massachusetts, for endless organizational and editorial concerns. Their efforts and the generous financial support of the Division of Agrochemicals of the American Chemical Society; FMC Corporation; Ciba-Geigy Corporation; Miles, Inc.; Mycogen Corporation; NOR-AM Chemical Corporation; Nissan Chemicals of America Corporation; and Rollins, Inc., made this book possible.

J. MARSHALL CLARK
Department of Entomology
University of Massachusetts
Amherst, MA 01003

January 17, 1995

Chapter 1

Ion-Channel Diversity

Present and Future Roles in Insecticide Action

J. Marshall Clark

Department of Entomology, University of Massachusetts, Amherst, MA 01003

Insecticides have played an enormous role in the development of pesticide and environmental sciences. In our attempts to understand the toxic action of insecticides on target organisms, we have elucidated many fundamental physiological processes and recently have began the study of these interactions at the molecular level. Given the effectiveness and wide-spread use of insecticides, it was probably naive not to have envisioned the environmental concerns that the use of these chemicals has caused. Nevertheless, the negative aspects of insecticides and their use have resulted in a much more complete and rigorous understanding of the toxicokinetics and toxicodynamics of environmental contaminants, their environmental fate and degradation, and problems associated with extensive overuse, such as resistance.

Because the large majority of insecticides are neurotoxic and many interact directly with ion channels in neurons and elsewhere, an extensive research data base now exists that is available for assessment. Thus, the most enduring contribution of insecticides such as DDT, the cyclodienes, etc., may be in providing well-studied models to assess the advantages or disadvantages of new molecular biotechnologies as they become available for use in pesticide science. This aspect, of course, was the focus of the ACS symposium presentations that served as the template for the following chapters which comprise this volume.

The past fifty years of agricultural productivity have been largely the result of the widespread availability of synthetic chemicals, such as fertilizers and pesticides. These compounds have allowed the farmer to protect the high quality

0097–6156/95/0591–0001$12.75/0

food and fiber produced with synthetic fertilizers that are necessary to feed and clothe an ever increasing human population. Currently, the world market value for pesticides is approximately $23 billion with herbicides, insecticides and fungicides accounting for 47, 28 and 22% of this total cost, respectively. (1). Even with a projected growth rate of 10% per annum, the market share for biologicals has been estimated at 2 to 10% of the total pesticide market share by the year 2000 (2). Thus, synthetic chemical pesticides will remain the primary method of efficacious pest control into the next century and, at least in some fashion, well into the future.

As impressive as our progress has been in agriculture, we now face enormous challenges as pesticide scientists. In five years, we will begin the 21st century with an estimated world-wide population of 6 billion people (3). This population is likely to double in only a few decades resulting in tremendous pressure for agriculture to be even more productive. The past excessive and indiscriminate use of nonselective pesticides has made these challenges even more difficult (4). We now are all aware and concerned about the short and long term effects of pesticides in the environment. Societal response to these concerns has resulted in increasingly difficult and costly development and registration processes for new and novel pesticides. More disconcerting, perhaps, than even the environmental impact that certain pesticides have produced, is the ever increasing incidence of pests that are resistant to pesticides and the loss of these chemicals as effective pest control agents. Thus, at precisely the time we should be developing more selective, biodegradable, efficacious, and environmentally-benign chemical pesticides, we are not even keeping pace with those that are being lost through regulatory bannings or pest resistance problems.

From the above scenario, it follows that the pesticide scientists' most serious task at hand is to begin anew the development of innovative and rational means for crop protection (3). These advances must avoid previous faults in pesticide design, lack of selectivity, and over-use. More emphasis needs to be placed on the genetic, biochemical, and pharmacological differences of target versus nontarget organisms. Additionally, more natural products should serve as molecular models for the design of new synthetic but biodegradable pesticides.

Recent advances in molecular biology have provided a fundamentally novel manner in which we analyze the genetic, physiological, and developmental properties of organisms. Molecular approaches have revolutionized the study of pesticide mode of action, the determination of selective toxicities of target versus nontarget organisms, and the elucidation of mechanisms of pest resistance (4). The availability of such molecular biological tools has greatly increased our basic understanding of allied sciences including pharmacology, neurobiology, ion channel and receptor biology, signal transduction, and developmental biology (5). From these basic studies, we now have a wealth of new information applicable to many applied aspects of pesticide science. The technical know-how for the production of genetically-engineered crop plants and biotechnological pest control strategies already exists and will soon be economically feasible (4). Obviously, it is an appropriate time to assess the impact that these molecular approaches have had on pest control and to evaluate the future of molecular pesticide science.

Molecular Basis of Toxicity of Insecticides on Ion Channels

Investigation into the toxic action of insecticidal chemicals has provided much of the leadership on how we conduct, evaluate, and use mode of action studies on pesticides and on their environmental impact. Over the past 25 years, these studies have elucidated new and novel sites of action in insects. Knowledge pertaining to insecticide receptors and xenobiotic metabolism have allowed the more efficient design of efficacious analogues. These findings, together with pharmacological and genetic information, have lead to a fundamentally new way in which insecticides are being used to suppress the selection of resistant insects. Such approaches are now well known as resistance management strategies.

The reasons for these early advancements in the insecticide aspects of pesticide science are many. Included in these are that insect pests are particularly numerous and potentially devastating to many essential crop systems. Therefore, we allotted a large portion of our effort to find ways to control them. Also, insects are advanced physiologically and present complex developmental systems, such as the nervous and endocrine systems, that are easily targeted for chemical disruption. Of particular importance is the role that *Drosophila melanogaster* has played in cytological, developmental, and molecular genetic studies in insect and pesticide science (5).

Pesticide chemists were enormously successful in producing a variety of extremely cheap and effective insecticides by applying the wealth of information obtained from relevant animal studies in neurobiology, pharmacology and developmental biology. Of these early products, the huge majority were found to be active by disrupting the insects' nervous system and in particular, by acting as modulators of various ion channel functions. Because of this, there now exists a robust toxicological data base on the structure-activity relationships, toxicokinetics, metabolism, receptor biology, environmental fate and impacts, and resistance development of these ion channel-directed insecticides. Because some of these neurotoxic insecticides are no longer widely used due to environmental concerns (e.g., DDT, cyclodienes, etc.), their most enduring contribution will most likely be in providing a historical data base to review and to assess the advantages of various molecular biological tools as they become available for use in pesticide science.

Ion Channels Perturbed by Insecticides

Neuronal ion channels certainly have been the most widely exploited group of ion channels in terms of providing a site of action for existing insecticidal chemicals, however, they are far from the only type of ion channel available for such use. As more ion channel pharmacology and molecular sequence data is analyzed, it has become more and more evident that many ion channels, regardless of their tissue location, share many topological and functional similarities. Such similarities can even extend to considerable amino acid and nucleic acid homologies (6). Our understanding of the basic structure and function of ion channels has been revolutionized certainly by the use of molecular biology. Specifically, molecular cloning and sequence analysis of complementary DNAs (cDNAs) have provided primary structures of ion channels which have allowed the first biochemical-based interpretation of the tertiary structure of the encoded proteins. Cloned DNAs can be overexpressed in various cells. Function can then be assigned directly to specific proteins or polypeptides or indirectly by the use of specific antibodies. Additionally, oligonucleotide-directed mutagenesis can be used to site-specifically mutate cDNAs which allows us to test biophysical interpretations concerning channel kinetics, ion-selectivity, ligand binding domains, etc (6).

Thus, our studies on the molecular action of insecticides on ion channels is a start to our understanding of these processes and not an end. Such investigations will inevitably elucidate the nature of vulnerable sites for pest insect control which will be of value in parallel research for the control of weeds and microbial pathogens and vice versa. Molecular knowledge of the target sites of insecticides, natural products, and pharmaceuticals, will be used in the design of new pesticides and in the search of new and novel sites of action.

Voltage-Sensitive Sodium Channels. The voltage-sensitive (or voltage-gated) sodium channel is a transmembrane protein complex that is essential in the generation of action potentials in excitable cells, in particular those associated with the all-or-nothing electrical impulse carried by nerve cell axons (7). On receiving a depolarisation signal, the permeability of the sodium channel rapidly increases for approximately 1 millisecond. The rising phase of the action potential is due to this increased Na^+ permeability and the process that causes it is termed *activation*. The activation process also determines the rate and voltage dependence of Na^+ permeability upon depolarisation. If depolarisation persists, the permeability of the channel will subsequently decrease in a less dramatic fashion and return to a nonconducting state over the next 2-3 milliseconds. This process is termed *inactivation* and determines the rate and voltage dependence of Na^+ permeability as it returns to a nonconducting state. These two processes, activation and inactivation, allow the voltage-sensitive sodium channel to exist in any one of three distinct functional states; active, inactive, or resting. Although both inactive and resting channels are nonconducting, they are very different in how they enter the "active" conducting state upon depolarisation. An inactivated channel is recalcitrant to depolarisation and must first return to the resting state by repolarization prior to being activated.

The primary structures of at least five distinct sodium channels have been reported (8). These are: 1) the sodium channel from the electric organ of the eel *Electrophorus electricus*, 2-4) three separate sodium channels from rat brain (Types I, II & III); and 5) the μI sodium channel from rat skeletal muscle. All sodium channels possess an α subunit that is approximately 1,800-2,000 amino acids in length. The eel electroplax sodium channel is comprised only of an α subunit whereas the rat skeletal channel has both α and β-1 subunits. The rat brain sodium channels all have a heterotrimeric composition with α, β-1 and β-2 subunits. Sodium channel α subunits have been expressed in both *Xenopus* oocytes and in Chinese hamster ovary cells, both resulting in functional channels (9-14). These results and other corroborative experiments have established that the protein structure encoded in the α subunit is sufficient to form a Na^+-specific transmembrane pore that undergoes voltage-dependent activation and inactivation gating processes. Additionally, α subunits expressed in these cells have a number of pharmacologically-specific receptor sites indicative of native sodium channels including a high affinity inhibition by tetrodotoxin and saxitoxin, a prolonged activation by veratridine, a slowed inactivation by α-scorpion toxins, and a frequency- and voltage-dependent inhibition by local anesthetic, antiarrhythmic, and anticonvulsant drugs (8,9). Evidently, the α subunit is of primary importance to the overall functioning of the intact sodium channel. However, in cases where other subunits are present, the function of the α subunit seems to be modified by the presence of these additional subunits and leads to a more native functioning of the channel. This is particularly true in the case of rat brain sodium channels expressed in *Xenopus* where the co-expression of the β-1 subunit greatly increases the decay rate of the sodium current during inactivation (15).

The deduced amino acid sequence of the α subunit of the eel electroplax presents a primary structure with four internal repeats (I-IV) that exhibit a high degree of sequence homology (13,14). Each of these four homologous repeats, or domains, potentially form multiple (S1-S6) α-helical transmembrane segments (Figure 1). To date, all other sodium channels have shown close structural similarities to the electroplax channel. Overall, amino acid homology is highly conserved in the transmembrane repeats S1 to S6 but the cytoplasmic connecting loops are less well conserved except for a short region between repeats III and IV(8,9).

By applying a coordinated experimental approach including polypeptide specific antibodies, site-directed mutagenesis, and consensus sequence determinations, specific structural components of the sodium channel have been identified as essential to specific channel functions. A number of these components are of importance to our current investigations into the molecular basis of insecticide toxicity at the sodium channel.

One, it is now widely held that the voltage sensor which provides the positive gating charges necessary for voltage-dependent activation is associated with specific repeated motifs of positively-charged amino acids (e.g., arginine) followed by two hydrophobic residues within each of the S4 transmembrane segments (16-20).

Two, the fast gating aspect of inactivation appears to be highly dependent on a presence of three contiguous hydrophobic amino acid residues (i.e., isoleucine-1488, phenylalanine-1489, and methionine-1490) that occur within the cytoplasmic linker between homologous repeats III-IV. The phenylalinine-1489 residue appears the most critical and the hydrophobic cluster of the surrounding 10 amino acids probably functions in large part as the inactivation gate for the sodium channel in a

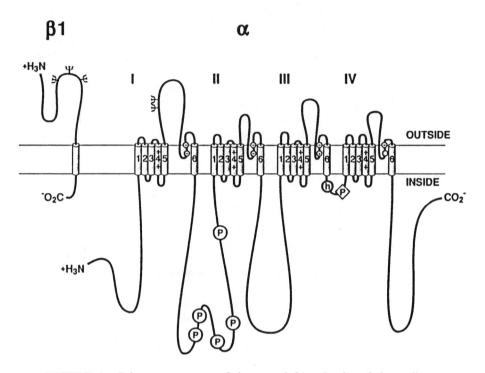

FIGURE 1. Primary structures of the α and β1 subunits of the sodium channel illustrated as transmembrane folding diagrams. The bold line represents the polypeptide chains of the α and β1 subunits with the length of each segment approximately proportional to its true length in the rat brain sodium channel. Cylinders represent probable transmembrane α-helices. Other probable membrane-associated segments are drawn as loops in extended conformation like the remainder of the sequence. Sites of experimentally demonstrated glycosylation (Ψ), cAMP-dependent phosphorylation (P in a circle), protein kinase C phosphorylation (P in a diamond), amino acid residues required for tetrodotoxin binding (small circles with +, -, or open fields depict positively charged, negatively charged, or neutral residues, respectively), and amino acid residues that form the inactivation particle (h in a circle). (Reproduced with permission from ref. 8. Copyright 1993 Annals of New York Acad. of Sci.).

manner similar to the "hinged lids" associated with many allosteric enzymes (8,9,21).

Three, the tetrodotoxin/saxitoxin receptor site of the α subunit of the rat brain sodium channel II has been localized to a set of two clusters of primarily negatively-charged residues (D384, E942, K1422, A1714 and E387, E945, M1425, D1717, respectively). All these amino acids exist as paired residues, one from each cluster (e.g., D384 and E384, etc.), and are associated with the short segment SS2 region between the S5 and S6 hydrophobic segments in each of the four repeating domains I-IV (6,8,9). Mutations of these residues that result in a net decrease in negative charge drastically reduce single channel conductance and overall toxin sensitivity. Thus, it seems likely that these negatively-charged residues in analogous locations in each repeating domain form a ring-like structure that serves as the tetrodotoxin/saxitoxin binding site at the extracellular surface of the transmembrane sodium pore (6,8,9).

Four, at least five distinct sites on the α subunit are phosphorylated by cAMP-dependent protein kinase. The phosphorylations all occur on residues associated with the large cytoplasmic loop connecting homologous domain I and II and all cAMP-dependent phosphorylations result in a reduced sodium ion flux (6,8,9).

Five, α subunits also are phosphorylated by protein kinase C which results in a slowing of sodium inactivation and in a reduced peak sodium current. Apparently, phosphorylation by protein kinase C results in the phosphorylation of serine-1506 which occurs in the intracellular loop between domains III and IV. Because of its close proximity to the inactivation gate, this phosphorylation process likely results in a direct effect on the inactivation gate itself (6,8,9).

Voltage-Sensitive Calcium Channels. Voltage-sensitive (or voltage-gated) calcium channels are transmembrane proteins that allow Ca^{2+} influx to occur when the open configuration is produced during a depolarisation signal (7,22). By allowing Ca^{2+} flux, calcium channels play pivotal roles in the regulation of a vast array of cellular processes, including axonal growth, enzyme modulation, membrane excitability, muscle contraction, and neurotransmitter release (23). Their existence and most early research involved excitable cells such as neuron, muscle and heart. However, voltage-sensitive calcium channels now have been found in most cell types. Recently, it has become apparent that many cell types not only have voltage-sensitive calcium channels but most cell membranes have several types of calcium channels associated with them (23).

At least four separate types of voltage-sensitive calcium channels have been classified based on the electrophysiological and pharmacological properties of each. These four classes have been designated as L-, N-, P-, and T-type calcium channels (Table I). Examples of each of these physiological classes have been described in a variety of cell types and all are expressed in neurons.

L-type voltage-sensitive calcium channels. L-type calcium channels are associated with practically all excitable tissues, are found in many nonexcitable cells, and are the most abundant calcium channel in muscle. L-type channels are

TABLE I. Functional Properties of Voltage-Sensitive Calcium Channels

Type	L	N	P	T
Function				
Activation Range	High voltage	High voltage	High voltage	Low voltage
Single-Channel Kinetics	Hardly any inactivation	Long burst, inactivation	Long burst, inactivation	Late opening, brief burst, inactivation
Physiology	Excitation contraction-, excitation secretion - coupling in endocrine cells and some neurons	Triggers neurotrans- mitter release	Triggers neurotransmitter release, Ca^{2+} spike in some neurons, induces long-term depression	Pacemaker activity and repetitive firing in heart and neurons
Cd^{2+} Block	Yes	Yes	(?)	No
Co^{2+} Block	No	No	(?)	Yes
Ni^{2+} Block	No	No	(No)	Yes
Dihydropyridine (DHP) Block	Yes	No	No	No
ω-conotoxin (ω-CgTx) Block	No	Yes	No	No
ω-agatoxin-IVA (ω-Agal IVA) Block	No	No	Yes	No
Funnel-web spider toxin	No	No	Yes	No
Octanol Block	No	No	No	Yes

SOURCE: Adapted from ref. 24.

distinguished from the other classes by being sensitive to 1,4-dihydropyridine antagonists (DHP-sensitive), eliciting high voltage activation, and having a single-channel conductance of 22-27 pS (24). L-type channels are sensitive to Cd^{2+} block at low concentrations but are less sensitive to Co^{2+}, Ni^{2+}, and omega-conotoxin GVIA (ω-CgTx) block (24, Table I).

Calcium ion flux via L-type channels results in long-lasting calcium currents that inactivate only relatively slowly and then only if Ca^{2+} is not the charge carrier. These currents activate excitation-contraction coupling in heart, skeletal and smooth muscle, and result in hormone and /or neurotransmitter release from endocrine cells and from some neurons (8,24).

The most common conformation for the skeletal muscle L-type calcium channel is a complex of five separate subunits; $\alpha 1$ (165-190 kDa), $\alpha 2$ (143 kDa), β (55 kDa), γ (30 kD), and δ (24-27 kDa) (8). The DHP-sensitive L-type channels of brain are similar to skeletal muscle channels and are comprised of multisubunit complexes of analogous $\alpha 1$, $\alpha 2\delta$, and β subunits (8, Figure 2). Interestingly, the $\alpha 1$ subunit of L-type channels has a high level of homology to the α subunit of the voltage-sensitive sodium channel (24). When expressed in *Xenopus* oocytes or mammalian cells, the $\alpha 1$ subunit can function alone as a voltage-sensitive calcium channel. The cDNA sequence of the $\alpha 1$ subunit encodes a protein of 1,873 amino acids and its hydropathy profile suggests that this subunit consists of four repeating homologous transmembrane domains (24). Each repeating domain has one positive charged segment, S4, that is widely believed to be the voltage sensing region and five hydrophobic segments,S1, S2. S3, S5, and S6). As discussed below for the sodium channel, the glutamic acid residues in the SS1-SS2 region, that are likely responsible for the ion selectivity of calcium channels, are conserved in all $\alpha 1$ subunits to date (8,24). The $\alpha 1$ subunit also contains the pharmacologically-relevant DHP and phenylalkylamine receptors. The phenylalkylamine receptor, an intracellular pore blocker of calcium channels, is associated with residues on the cytoplasmic side of the hydrophobic segment IVS6 (24).

The DHP-sensitive L-type channels are modulated by similar mechanisms as are voltage-sensitive sodium channels including cAMP-dependent kinases (8,23,24), Ca^{2+}/calmodulin-dependent kinases (23), protein kinase C (8), GTP-binding proteins (23), and inositol polyphosphates (23). Given such overall similarities, it is most intriguing that recent mutagenesis experiments have established that mutation of lysine-1422 and alanine-1714 to negatively-charged glutamic acid residues results in an alteration in the ion selectivity of the sodium channel from Na^+-selective to Ca^{2+}-selective (8,24).

Molecular biological approaches have established that at least six distinct genes encode $\alpha 1$ subunits (Table II): rabbit skeletal muscle DHP-sensitive calcium channel (Class Sk); rabbit cardiac muscle DHP-sensitive calcium channel (Class C); rat brain D class or neuroendocrine-type DHP-sensitive calcium channels from brain, human neuroblastoma IMR32 cells, pancreatic islet cells, hamster pancreatic B-cells, and ovarian cells (Class BIV); rat brain A class calcium channels that are widely distributed in the brain but most abundant in the cerebellum, Purkinje cells, and granule cells (Class BI); rat brain E class calcium channels that have a different

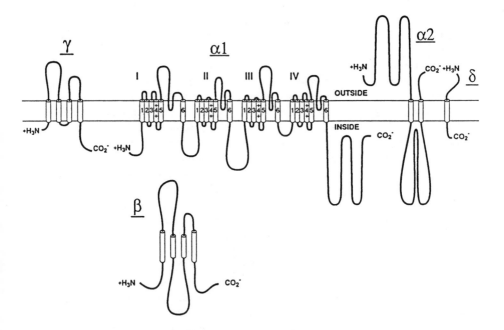

FIGURE 2.   Subunit structure of skeletal muscle calcium channels. Transmembrane folding models of the calcium channel subunits derived from primary structure determination and analysis.   Cylinders represent predicted α-helical segments in the transmembrane regions of the $\alpha 1$, $\alpha 2\delta$, and γ subunits and in the peripherally associated β subunit.   The transmembrane folding patterns are derived only from hydropathy analysis for $\alpha 2\delta$ and γ and from a combination of hydropathy analysis and analogy with the current models for the structures of sodium and potassium channels for $\alpha 1$.   The transmembrane arrangement of $\alpha 2\delta$ is not well-defined by hydropathy analysis and the indicated structure should be taken as tentative. (Reproduced with permission from ref. 8.   Copyright 1993 Annals of New York Acad. of Sci.)

TABLE II. Molecular Classification of Voltage-Sensitive Calcium Channels

Numa Class	Snutch Class	Perez-Reyes Class	Primary Tissue Location	Functional Class
Sk	—	1	skeletal muscle	L type
C	C	1	heart, smooth muscle	L type
BIV	D	3	brain pancreas	L type
BI	A	4	brain	P type
BII	E	—	brain	?
BIII	B	5	brain	N type

SOURCE: Reprinted with permission from ref. 24. Copyright 1993.

distribution than the BI type being most abundant in cerebral cortex, hippocampus, and corpus striatum (Class BII); and rat brain B class ω-CgTx-sensitive calcium channels from cerebral cortex, hippocampus, corpus striatum, midbrain, and cerebellum (Class BIII). Of these six calcium channel isoforms, Class Sk, C, and BIV types represent L-type calcium channels (Table II).

N-type voltage-sensitive calcium channels. N-type calcium channels are largely restricted to neurons and may exist only in neurons (8,23,24). They are pharmacologically distinct from L-type calcium channels due to their lack of DHP-sensitivity and being specifically blocked by the peptide neurotoxin, omega-conotoxin GVIA (ω-CgTx). Like L-type channels, N-type channels are blocked by low concentrations of Cd^{2+} but are resistant to Co^{2+} and Ni^{2+}. Also, N-type channels are high voltage-activated channels similar to L-type channels but have a single channel conductance of 11-15 pS and undergo inactivation much more rapidly than do L-type channels (Table I).

It is widely held that N-type calcium channels are critical in providing the Ca^{2+} which triggers neurotransmitter release at presynaptic nerve terminals (23). The presence of N-type channels at high density in the synaptolemma membrane and the inhibition of omega neurotransmitter release by ω-CgTx supports this contention (24). Additionally, N-type channels are also believed to function in some aspects in the directed migration of immature neurons (24).

The N-type ω-CgTx-sensitive calcium channel has been purified from rat brain and consists of $\alpha 1$, $\alpha 2\delta$, and β subunits that are analogous to those of the neuronal L-type channels. These subunit proteins are likewise phosphorylated by cAMP-dependent kinases and protein kinase C and result in the modulation of Ca^{2+} flux (23). The most striking modulation of N-type channels is their interaction with neurotransmitters (e.g., noradrenaline, acetylcholine, GABA, serotonin, etc.) and neuropeptides (e.g., opioids) (23). Interestingly, in all cases where neurotransmitters have resulted in the inhibition of N-type channels (i.e., down regulation), GTP-binding proteins have been involved, in particular the G_o type (23,24).

The gene encoding N-type ω-CgTx-sensitive calcium channels (Class BIII or class rbB, Table II) has been cloned from rat brain and human neuroblastoma IMR32 cells (8,24). This gene encodes an $\alpha 1$ subunit 2,336 amino acids. Two isoforms were identified by differences in their carboxy-terminal (C-terminal) sequences which probably were the result of alternative RNA splicing. Polyclonal antiserum generated against a peptide from a unique sequence of the Class BIII gene immunoprecipitated high-affinity ω-CgTx binding sites associated with N-type channels isolated from brain membranes. When expressed in mammalian cells, the resulting $\alpha 1$ subunit functions as a voltage-sensitive ω-CgTx-sensitive N-type calcium channel. Additionally, a northern blot analysis using a Class BIII cDNA probe resulted in cross-hybridization only with a specific 9,500 nucleotide-long species of mRNA isolated from brain tissue. The mRNAs from other non-neural tissues did not cross hybridize.

P-type voltage-sensitive calcium channels. P-type calcium channels were first identified in Purkinje cells and have now been found in various types of neuronal cells (24). P-type channels are high voltage-activated calcium channels with single-channel conductance of 16 pS that are insensitive to Ni^{2+}, DHP antagonists (i.e., nifedipine), and ω-CgTx blockage (Table I). They are, however, selectively blocked by funnel-web spider toxin (FTX), ω-agatoxin-IVA (ω-AgaIVA), and low concentrations of Cd^{2+} (e.g., 50% block at 0.5μM). P-type channels function in the Ca^{2+} spike generation in some neurons, have been shown to be involved in neurotransmitter release, and are critically involved in the induction of long-term depression (23,24).

The gene encoding P-type calcium channels (Class BI or class A, Table II) has been cloned and consists of at least two isoforms (BI-1 and BI-2) which apparently have arisen due to alternative RNA splicing (24). Most sequence variation occurs in the cytoplasmic loop between domains II-III, in the transmembrane segment IVS6, and in the C-terminal region. Both isoforms, BI-1 and BI-2, have been expressed in oocytes and result in single-channel kinetics and pharmacology representative of a P-type calcium channel (24).

T-type voltage-sensitive calcium channels. T-type calcium channels were originally identified in vertebrate sensory neurons but have now been recognized in a number of excitable and nonexcitable cells from a variety of organisms (8,23,24). They have been referred to in earlier literature by a number of alternative names including; low threshold activated, I_{fast}, and slowly deactivating channels (23).

T-type channels are low voltage-activated calcium channels with single channel conductance of 8-10 pS and show steady inactivation even for small depolarization (Table I). They are insensitive to DHP antagonists, ω-CgTx, and Cd^{2+} but sensitive to Ni^{2+}, Co^{2+}, octanol, amirolide, and tetramethrin (23) . However, none of these ligands which the T-type channels are sensitive to are particularly specific. The lack of specific ligands has hindered studies on the biochemical and molecular composition of these calcium channels. Nonetheless, they appear to be separate molecular entities due to unique single-channel kinetics, their Ni^{2+} sensitivity, their selective inhibition during oncogene transformations that leave L-type channels unperturbed, and their relatively unique functional role in repetitive firing and pacemaker activity in neurons and heart (23).

T-type calcium channels have not been shown to be modulated by cAMP-dependent kinases and there are conflicting reports of modulation by angiotensin II (23). Recently, however, an inhibitory action by a diacylglycerol analog has been reported, possibly implicating an involvement of protein kinase C (23).

Ligand-Gated Receptors / Ion Channels. Given the enormous amount of information available on this subject, the discussion of ligand-gated receptors/ion channels will be limited to those receptors/ion channels that have been implicated as possible sites of insecticidal action. These are as follows; the $GABA_A$-gated chloride channel, the glycine-gated chloride channel, the glutamate-gated cationic channel, the glutamate-gated chloride channel, and the *Bacillus thuringiensis* (*Bt*) toxin-induced cation-specific channel.

$GABA_A$- and glycine-gated chloride channels. Gamma-aminobutyric acid (GABA) and glycine are major neurotransmitters in the nervous systems of both

vertebrates and invertebrates   (25). Both neurotransmitters mediate synaptic inhibition by opposing excitatory inputs into postsynaptic membranes. This is accomplished by the binding of each neurotransmitter ligand to its specific postsynaptic receptor (GABA$_A$-receptor or glycine receptor, respectively,) which activates chloride ion (Cl$^-$)-selective channels. The increased permeability to Cl$^-$ shifts the postsynaptic potential towards E$_{Cl}$, the equilibrium potential for chloride. Because in most cells the E$_{Cl}$ is close to the resting membrane potential, the increase in Cl$^-$ flux results in membrane hyperpolarization. Thus, chloride channels stabilize the membrane potential by opposing deviations from rest (26).

In vertebrates, GABA$_A$-gated chloride channels are widely distributed in the brain and central nervous system whereas the glycine-gated chloride channels are primarily restricted to the brain stem and spinal cord (27). The two receptors are usually separated by the selective sensitivities to bicuculline (i.e., GABA$_A$-receptor inhibition) and strychnine (i.e., glycine-receptor inhibition) (27). Additionally, these receptors have been implicated as sites of action to a wide variety of pharmaceuticals, toxicants, and insecticides (27-29). Specifically, the vertebrate GABA$_A$-gated chloride channel has at least five types of binding sites: 1) GABA binding site (including binding of agonists such as muscimol and antagonists such as bicuculline); 2) the benzodiazepine binding site (including diazepam and flunitrazepam); 3) the convulsant binding site (including picrotoxins, PTX, and *t*-butylbicyclophosphorothionates,TBPS) within the Cl$^-$ ionophore, itself, 4) the depressant site (including the CNS-depressant barbiturates and barbiturate-acting steroids); 5) externally-facing anionic binding sites for Cl$^-$ ions (30). It is particularly intriguing that each of these types of ligand binding sites interact allosterically with one or more of the other sites. This results in extremely complex pharmaco- and toxicodynamics that may be due to the structural diversity of these receptor/ion channel complexes as evidenced in their heteromultimeric structure and variable subunit make-up (27).

Using molecular biological approaches, it is now apparent that the genes encoding the GABA$_A$- and glycine-gated chloride channel comprise a family of homologous ligand-gated chloride channel genes that is part of a larger superfamily of ligand-gated ion channel genes. Molecular cloning and expression techniques have identified five classes of GABA$_A$-receptor subunits (α, β, γ, δ, and ρ) associated with vertebrate GABA-gated chloride channels and two classes of glycine-receptor subunits (α and β) associated with the glycine-gated chloride channel (28). In the case of the GABA$_A$-receptor, (Figure 3) multiple subunit isoforms have been determined for some of these classes including six α subunits (α1-α6); four β subunits (β1-β4); three γ subunits (γ1-γ3); a δ subunit, and ρ1 subunit that may substitute for an α subunit in the retina. Isoforms of the same class of subunits have a high level of amino acid homology (60-80%) and lower levels of homology between classes (20-40%) (28). Apparently, each subunit isoform is encoded by a separate gene, resulting in more than 15-20 separate genes being characterized as members of this ligand-gated chloride ion channel gene family. Given the multiplicity of various classes of subunits and subunit isoforms within each class, it is finally becoming evident how a single inhibitory neurotransmitter

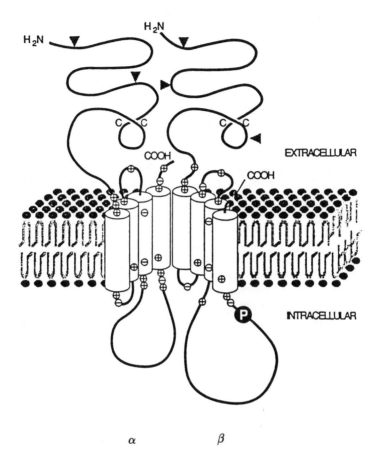

FIGURE 3. A schematic model for the topology of the GABA$_A$ receptor in the membrane. Four membrane-spanning α-helices in each subunit are shown as cylinders. The structures in the extracellular domain are drawn in an arbitrary manner, but the presumed β-loop formed by the disulphide bond predicted at cysteines 139 and 153 (α-subunit numbering) is shown. Potential extracellular sites for N-glycosylation are indicated by triangles, and a possible site for cAMP-dependent serine phosphorylation, present only in the β subunit, is denoted by an encircled P. Those charged residues which are located within or close to the ends of the membrane-spanning domains are shown as small, charged-marked circles. It is proposed that two such structures are complexed in the receptor molecule so as to align the membrane spanning domains, only some of which will form the inner wall of a central ion channel. (Reproduced with permission from ref. 31. Copyright 1987 Nature).

such as GABA can be so widely employed by various neuronal circuits and yield so varied a response.

Sequence analysis of the gene products of the ligand-gated chloride channel gene family has revealed a number of conserved, channel-specific, features (28,30,31). First, all have four regions of highest homology that make up the hydrophobic, transmembrane-helices, M1-M4, that function in the formation of the chloride ionophore (see Knipple *et al.*, this volume). Second, a highly conserved sequence of approximately 12 hydroxy-containing amino acids is observed in the M2 transmembrane domain and contributes to the ion selectivity of the channel. Third, a relatively high degree of homology is associated with the amino terminal (N-terminal) extracellular domain, particularly the 15-residue "cysteine loop". Forth, two homologous segments of the extracellular loop exist between the cysteine loop and the M1 transmembrane domain and are implicated as functioning in the formation of neurotransmitter-binding site.

Vertebrate $GABA_A$-and glycine-gated chloride channels have been determined to be heteromultimeric complexes and probably exist as pentameric arrangements of glycoprotein subunits (27,28,30). Using protein purification and photoaffinity labelling techniques, a 53 kDa α subunit and a 57 kDa β subunit has been identified (28). [^{3}H]flunitrazepam binding to the α subunit has indicated that the α subunit serves as the benzodiazepine binding site whereas [^{3}H]muscimol binding to the β subunit suggests that this subunit serves as the agonist binding site (28).

Glutamate-gated channels. Glutamate receptor-gated ion channels function in several important roles in signal transmission at excitatory synapses in the central nervous systems of vertebrates (32). They also are found in the central nervous systems of invertebrates where they are involved in inhibitory as well as excitatory synaptic transmissions (32). Indeed, the best pharmacologically and physiologically characterized glutamate receptor-gated ion channels are those present postjunctionally at excitatory junctions on arthropod skeletal and visceral muscles (33).

In vertebrates, glutamate is a major excitatory neurotransmitter and besides functioning in synaptic transmission, it is also involved in the modification of synaptic connections during development, and in the modulation of transmission efficacy during plastic changes in the adult brain (34). In such roles, glutamate receptors are believed to be involved in many higher brain functions such as learning and memory. Additionally, overreaction of glutamate receptors results in pathological conditions such as neurodegeneration and cell death. Progressive neurodegenerative disorders such as Huntington's disease and Alzheimer's disease are believed to be the result of such overreaction (34,35).

On vertebrate neurons, glutamate receptors gate integral cation-selective channels and are categorized into two distinct groups, the ionotropic glutamate receptors and the metabotropic glutamate receptors. The ionotropic receptors can be further divided into pharmacologically distinct quisqualate-sensitive receptors, N-methyl-D-aspartate (NMDA)-sensitive receptors, and S-α-amino-3-hydroxy-5-methyl-4-isoxazolepropionate (AMPA)/kainate-sensitive receptors. The metabotrophic receptors modulate intracellular second messengers via GTP-binding proteins (34,35).

The NMDA-sensitive glutamate receptor functions in long-term potentiation, a long-lasting and activity-dependent enhancement of neural transmission that is believed to be critical in the processes of memory and learning (35). Additionally, this receptor is involved in neurodegeneration and cell death under pathological conditions such as epilepsy and cerebral ischemia (35). The AMPA/kainate-sensitive glutamate receptors are most likely involved in nerve rectification and Ca^{2+} permeability (34).

Molecular cloning and expression experiments have established that the NMDA-sensitive glutamate receptors are assembled from NMDAR1 and NMDAR2A-NMDAR2D subunits in mosaics of heteromeric channels (35). All NMDA subunit types are believed to possess four transmembrane segments and belong to the ligand-gated ion channel gene family. The NMDAR1 subunit is a single protein of approximately 940 amino acids, has seven major isoforms, and forms a functional channel as a homomeric complex. However, the NMDAR2A-D subunits are involved in enhancement of the glutamate response and controls functional variability by forming heteromeric native channel complexes (35).

The AMPA/kainate-sensitive glutamate receptors are assembled from four subunits, GluR-1 to GluR-4 (or GluR-A to GluR-D) and, similar to the NMDA-type receptors, are believed to form mosaics of heteromeric channels (34). In heteromeric combinations, the GluR-2 (or -B) subunit dominates the steady-state current-voltage relationship and channel permeability to Ca^{2+} (34).

Pharmacologically and electrophysiologically distinct groups of glutamate-gated receptors/ion channels have also been identified on arthropod muscle (33,36). At the excitatory neuromuscular junctions on locust extensor tibiae muscle bundles, depolarising glutamate receptors gate a mixture of cation-selective channels that are either quisqualate-sensitive, ibotenate-sensitive, or aspartate-sensitive (33). At this site, the quisqulate-sensitive glutamate receptor predominates. Glutamate receptors are also distributed across the extrajunctional sarcolemma of this muscle bundle and are comprised of two distinct types. The D-type glutamate receptor which is quisqulate-sensitive and gates depolarising cation-selective channels and the H-type which is ibotenate-sensitive and gates hyperpolarising chloride channels (33,37).

Bt toxin-induced cation-specific channels. Bacillus thuringiensis (Bt) is a gram-positive, entomopathogenic bacterium that forms a proteinaceous parasporal crystalline inclusion during sporulation (38). The proteins that make up the crystalline inclusion are solubilized in the insect midgut upon ingestion. These protoxins are called δ-endotoxins and range in molecular weight from 140 to 27 kDa. They are activated to the ultimate toxic configuration by midgut proteases and pH by specific C-terminal proteolytic cleavages, releasing a toxic fragment encompassing the N-terminal domain of the protoxin. Once activated, the toxin must pass the insect peritrophic membrane, interact with the larval midgut epithelium, cause a disruption in apical membrane integrity, and induce a leakage of cations (e.g., K^+, etc.) and water into the midgut epithelium cells. Ultimately, the insect will die from complications produced by extensive osmotic swelling and midgut epithelium cell lysis (39,41).

The *Bt* δ-endotoxin proteins are toxic to select larval Lepidoptera, Diptera and Coleoptera. A variety of *Bt* δ-endotoxins and the genes that encode them have been identified. These protein toxins generally can be classified into three size classes; 125-138 kDa (encoded by *cry*I and *cry*IVA-B genes), 65-75 kDa (encoded by *cry*II, III, and IVC-D genes) and 25-28 kDa (encoded by the *cyt* genes)(39). The *cry* toxin genes have high amino acid homology and are believed to share a common evolutionary history. Toxins encoded from *cry*I genes are toxic to lepidopterans, *cry*II gene products to dipterans and lepidopterans, *cry*III to coleopterans, and *cry*IV to dipterans. The *cyt* genes and their encoded protein products are not well related to the *cry* genes or their protein products and have been shown to be toxic only to dipterans (39,40).

The three-dimensional structure of the toxic fragment of the CryIII crystal δ-endotoxin from *B. thuringiensis* subsp. *tenebrionis* has recently been determined by X-ray crystallography (42). The activated toxin consists of three structural domains; a N-terminal hydrophobic and amphipathic domain that consists of 6-7 α-helices (amino acids 1-279), a middle variable region that consists of a three-sheet assembly (amino acids 280-460), and a conserved C-terminal domain that forms a β sandwich (amino acids 461-695). The first domain is apparently involved in pore formation with the hydrophobic and amphipathic α-helics playing a critical role in forming the transmembrane spanning regions of the proposed nonselective cation-specific channel. The three sheet second domain, particularly the region between amino acids 307-382, is probably responsible for glycoprotein receptor binding. The β sandwich third domain has been implicated in the protection of the toxic fragment from nonspecific degradation during proteolytic processing.

At this time, it is widely held that *Bt* toxins form nonselective cation-specific channels in the apical membranes of the epitheliar columnar cells of the insect midgut (41, Figure 4). Although both the Cry and Cyt toxins may share this pore forming process, they do so by interaction with very different cell-membrane "receptors". Cry toxins are believed to bind initially to glycoprotein receptors whereas Cyt toxins bind to *syn*2-unsaturated phospholipids (39,40). However, once the toxins insert, they form pores that are permeant to small cations such as K^+. Ultimately, the leakage of cations and water will destroy the basal-side to apical-side transepithelial potential difference which in turn disrupts pH regulation and nutrient uptake. These physiological events culminate in the cytolysis of midgut epithelium cells and cause the death of the insect (41).

Conclusions

This book presents a summary of our current state of knowledge concerning the molecular action of insecticides on ion channels. The information has been presented in three separate sections: 1) action of insecticides on voltage-gated channels; 2) action of insecticides on ligand-gated channels and 3) action of

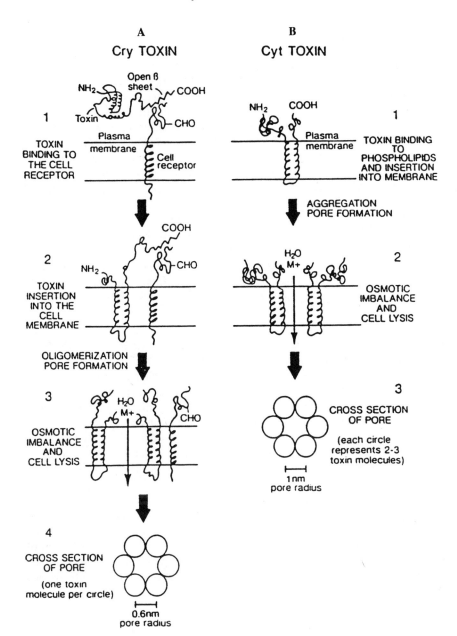

FIGURE 4. Proposed model for the formation of the *Bt* toxin-induced pores. (A) Putative mechanism by which the Cry toxins form a pore. 1. The cell membrane-binding domain of the toxin binds to a high affinity receptor on the apical membrane of the insect midgut columnar cells. The carbohydrate moiety of the receptor may not be involved in binding toxin, as e.g. in *H.*

Continued on next page

FIGURE 4. Continued.

virescens. 2. The toxic domain of the toxin inserts into the cell membrane after a change in toxin conformation. 3. Oligomerization of toxin molecules in the cell membrane, resulting in the formation of a pore, which leads to osmotic imbalance resulting from the influx of water and actions (M^+) or other small molecules. 4. Cross-section of the resulting pore; each circle represents one toxin molecule. (B) Putative mechanism for pore formation by Cyt toxins. 1. The C-terminal hydrophobic region of the Cyt toxin inserts into the cell membrane following hydrophobic interactions between the toxin and the membrane phospholipids. 2. Following this initial binding, aggregation of the Cyt toxin molecules occurs in the cell membrane, resulting in the formation of a pore. Cell lysis occurs because of osmotic imbalance resulting from the influx of water and actions (M^+) and other small molecules. 3. Cross-section of the Cyt toxin pore; each circle represents 2-3 molecules of toxin with 12-18 toxin molecules per pore. Whether the Cry and Cyt toxin pores are formed from the α-helices or β-sheets is not known, although α-helices are depicted in the formation of pores in both *A* and *B*. (Reproduced with permission from ref. 39. Copyright 1992 Annu. Rev. Entomol.)

biopesticide-induced channels. In each section, a variety of channels and affects are examined from both an industrial and an academic point of view.

The overwhelming consensus of this group of scientists was that in their endeavors to determine the molecular site of insecticide action, a wealth of information has been made available concerning their particular ion channel gene and encoded protein. Certainly, as more ion channel sequence is determined and analyzed, additional channel binding motifs will be recognized and utilized in the development of new and novel insecticidal compounds. In cases where insecticide resistant genes have already been selected, information is already available on the mutational basis of resistance and schemes have been designed to use this information in genetically-based resistance management strategies.

Finally, these systems will very likely provide fundamental information on how these insecticide receptor genes are regulated. The roles of alternative RNA splicing, post-transcriptional and -translational processes, etc., in the selection of receptor subunit isoforms are just beginning to be understood. Such mechanistic insights will surely provide a vast variety of selective means to effectively manage pest insects in the future.

Acknowledgments

The symposium that provided the chapters for this volume was made possible by the financial and logistical support of the Agrochemistry Division of the American Chemical Society. Additional support that was critical for the travel of the non-USA speakers was generously provided by FMC Corp., Giba-Geigy Crop., Miles Inc., Mycogen Corp., NOR-AM Chem. Co., Nissan Chem. Amer. Corp., and Rollins Inc.

Literature Cited

1. Anon. Agrow World Crop Protection News,1991,135,18.
2. Duke, S.O.; Menn, J.J.; Plimmer, J.R. In *Pest Control with Enhanced
    Environmental Safety*, Duke, S.O., Menn, J.J., Plimmer, J.R., Ed.; ACS
    Symposium Series ....; American Chemical Society: Washington, DC, 1993; pp
    1013.
3. Kim, L. In *Pesticide/Environment: Molecular Biological Approaches*; Mitsui,
    T., Matsumura, F., Yamaguchi, I. Eds.; Proc. of the First International Sym. on
    Pestic. Sci.; Pestic. Sci. Soc. of Japan: Tokyo, Japan, 1993, pp 265-271.
4. Mitsui, T.; Matsumura, F.; Yamaguchi, I. In *Pesticide/Environment:
    Molecular Biological Approaches*; Mitsui, T., Matsumura, F., Yamaguchi, I.
    Eds.; Proc. of the First International Sym. on Pestic. Sci.; Pestic. Sci. Soc. of
    Japan: Tokyo, Japan, 1993.
5. Matsumura, F.; Kaku, K.; Ewan, E.; Charalambous, P.; Miyazaki, M.; Inagaki,
    M.; Inagaki, S. In *Pesticide/Environment: Molecular Biological Approaches;*
    Mitsui, T., Matsumura, F., Yamaguchi, I., Eds., Proc. of The First International
    Sym. on Pestic. Sci., Pestic. Sci. Soc. of Japan: Tokyo, Japan, 1993, pp 3-15.
6. Imoto, K. In *Molecular Basis of Ion Channels Receptors*; Higashida, H.,
    Yoshioka, T., Mikoshiba, K., Eds.; Annals New York Acad. Sci.; New York
    Acad. Sci.: New York, NY, Vol 707; pp 38-50.
7. Kuftler, S.W.; Nicholls, J.G.; Martin, A.R. *From Neuron to Brain*; Sinauer
    Assoc. Inc. Sunderland, MA, 1984, pp 111-164.
8. Catterall, W.A. In *Molecular Basis of Ion Channels Receptors*; Higashida, H.,
    Yoshioka, T., Mikoshiba, K., Eds.; Annals New York Acad. Sci.; New York
    Acad. Sci.: New York, NY, Vol 707; pp 1-19.
9. Noda, M. In *Molecular Basis of Ion Channels Receptors*; Higashida, H.,
    Yoshioka, T., Mikoshiba, K., Eds.; Annals New York Acad. Sci.; New York
    Acad. Sci.: New York, NY, Vol 707; pp 20-37.
10. Goldin, A.L.; Snutch, T.P.; Lubbert, H.; Dowsett, A.; Marshall, J.; Auld, V.;
    Downey, W.; Fritz, L.C.; Lester, H.A.; Dunn, R.; Catterall, W.A.; Davidson, N.
    In *Messenger RNA coding for only the α subunit of the rat brain Na channel is
    sufficient for expression of functional channels in Xenopus oocytes;* Proc. Natl.
    Acad. Sci. USA, 1983, Vol. 83; pp 7503-7507.
11. Auld, V.J.; Goldin, A.L.; Krafte, D.S.; Marshall, J.; Dunn, J.M.; Catterall,
    W.A.; Lester, H.A.; Davidson, N.; Dunn, R.J. *Neuron.* 1988. Vol. 1, pp. 449-
    461.
12. Auld, V.J.; Goldin, A.L.; Krafte, J.M.; Catterall, W.A.; Lester, H.A.; Davidson,
    N.; Dunn, R.J. In *A neutral amino acid change in segment IIS4 dramatically
    alters the gating properties of the voltage-dependent sodium channel;* Proc.
    Natl. Acad. Sci. USA, 1990, Vol 87, pp 323-327.
13. Noda, M.; Ikeda, T.; Suzuki, T.; Takeshima, H.; Takahashi, T.; Kuno, M.;
    Numa, S. *Nature.* 1986, Vol. 322, pp. 826-828.
14. Suzuki, T.; Beckh, S.; Kubo, H.; Yahagi, N.; Ishida, H.; Kayano, T.; Noda, M.;
    Numa, S. *FEBS Lett.* 1988, Vol 228, pp. 195-200.

15. Isom, L.L.; De Jongh, K.S.; Reber, B.F.X.; Offord, J.; Charbonneau, H.; Walsh, K.; Goldin, A.L.; Catterall, W.A. *Science*. 1992, Vol. 256, pp. 839-842.
16. Catterall, W.A. *Ann. Rev. Biochem.* 1986, Vol. 55, pp. 953-985.
17. Noda, M.; Ikeda, T.; Kayano, T.; Suzuki, H.; Takeshima, H.; Kurasaki, M.; Takahashi, H.; Numa, S. *Nature*. 1986, Vol. 320, pp. 188-192.
18. Greenblatt, R.E.; Blatt, Y.; Montal, M. *FEBS Lett*. 1985, Vol. 193, pp. 125-134.
19. Guy, H.R.; Seetharamulu, P. *Proc. Natl. Acad. Sci.* 1986, Vol. 83, pp. 508-512.
20. Guy, H.R.; Conti, F. *Trends Neurosci*. 1990, Vol. 13, pp. 201-206.
21. Patton, D.E.; West, J.W.; Catterall, W.A.; Goldin, A.L. *Proc. Natl. Acad. Sci.* 1992, Vol. 89, pp. 10905-10909.
22. Hille, B. *Ionic Channels of Excitable Membranes*; Sinauer Assoc. Inc. Sunderland, MA, 1984, pp 76-98.
23. Hess, P. In *Annual Rev. Neurosci*; Cowan, W.M., Shooter, E.M., Stevens, C.F., Thompson, R.F., Eds.; Annual Rev. Inc: Palo Alto, CA, 1950, Vol. 13; pp 337-56.
24. Mori, Y.; Niidome, T.; Fugita, Y.; Mynieft, M.; Dirksen, R.T.; Beam, K.G.; Iwabe, N.; Miyota, T.; Furutama, D.; Furuichi, T.; Mikoshiba, K. In *Molecular Basis of Ion Channels and Receptors;* Higashida, H., Yoshioka, T., Mikoshiba, K., Eds., Annals of New York Acad. Sci.; New York Acad. Sci.; New York, NY, 1993, Vol. 707; pp 87-108.
25. Hille, B. *Ionic Channels of Excitable Membranes*; 2nd ed., Sinauer Assoc. Inc. Sunderland, MA, 1992, pp 607.
26. Kuffler, S.W.; Nicholls, J.G.; Martin, A.L. In *From Neuron to Brains, 2nd ed.,* Sinauer Assoc. Inc. Sunderland, MA, 1994, pp 293-320.
27. Macdonald, R.L.; Angelloti, T.P.; *Cell Physiol. Biochem.* 1993, Vol . 3, 352-373.
28. Barnard, E.A.; Sutherland, M.; Zaman, S.; Matsumoto, M.; Nayeem, N.; Green, T.; Parlison, M.G.; Batteson, A.N. In *Molecular Basis of Ion Channels and Receptors;* Higashida, H., Yoshioka, T., Mikoshiba, K., Eds., Annals of New York Acad. Sci.; New York Acad. Sci.; New York, NY, 1993, Vol. 707; pp 116-125.
29. Bloomquist, J.R. *Comp. Biochem. Physiol.* 1993, Vol. 106C, pp 301-314.
30. Barnard, E.A.; Seeburg, P.H.; In *Chloride Channels and Their Modulation by Neurotransmitters and Drugs*; Biggio, G., Costa, E., Eds.; Advances in Biochemical Psychopharmacology; Raven Press: New York, NY, 1988, Vol. 45, pp 1-18.
31. Schofield, P.R.; Darlison, M.G.; Fugita, N.; Burt, D.R.; Stephenson, F.A.; Rodriguez, H.; Rhee, L.M.; Ramachandran, J.; Reale, V.; Glencorse, T.A.; Seeburg, P.H.; Barnard, E.A. *Nature*. 1987, 328, pp 221-227.
32. Nistri, A.; Constanti, A. *Prog. In Neurobiol.* 1979, Vol. 13, pp. 117-235.
33. Usherwood, P.N.R. In *Neurotoxins and Their Pharmacological Implications*; Jenner, P., Ed.; Raven Press: New York, NY, 1987; pp 133-151.
34. Jonas, P. In *Molecular Basis of Ion Channels and Receptors;* Higashida, H., Yoshioka, T., Mikoshiba, K., Eds.; Annals of the New York Acad. Sci.; New York Acad. Sci., New York, NY, 1993, Vol. 707, pp 126-135.

35. Masum, M.; Nakajima, Y.; Ishii, T.; Akazawa, C.; Nakanashi, S. In *Molecular Basis of Ion Channels and Receptors;* Higashida, H., Yoshioka, T., Mikoshiba, K., Eds.; Annals of the New York Acad. Sci.; New York Acad. Sci.: New York, NY, 1993, Vol. 707, pp 153-164.

36. Usherwood, P.N.R. In *Neuropharmacology and Pesticide Action*; Ford, M.G., Lunt, G.G., Reay, R.C., Usherwood, P.N. R., Eds.; Ellis Hoarwood Series in Biomedicine; VCH Publishers; Deerfield Beach, FL, 1986, pp 137-152.

37. Usherwood, P.N.R.; Blagbrough, I.S. In *Insecticide Action*: From Molecule to Organism; Narahashi, T., Chambers, J.E., Eds.; Plenum Press; New York, NY, 1989, pp 13-31.

38. Aronson, A.I.; Beckman, W.; Dunn, P. *Microbiol. Rev.* 1986, Vol. 50(1), pp. 1-24.

39. Gill, S.S.; Cowles, E.A.; Pietrantonio, P.V. Annu. Rev. Entomol. 1992, Vol. 37, pp. 615-36.

40. English, L.; Slatin, S.L. *Insect Biochem. Molec. Biol.* 1992, Vol. 22, pp.1-7.

41. Honée, G.; Visser, B. *Entomol. Exp. Appl.* 1993, Vol. 69, pp. 145-155.

42. Li, J.; Carroll, J.; Ellar, D.J. *Nature.* 1991, Vol. 353, pp. 815-21.

RECEIVED January 23, 1995

VOLTAGE-GATED CHANNELS

Chapter 2

Sodium Channels and γ-Aminobutyric Acid Activated Channels as Target Sites of Insecticides

T. Narahashi, J. M. Frey[1], K. S. Ginsburg[2], K. Nagata, M. L. Roy[3], and
H. Tatebayashi

Department of Molecular Pharmacology and Biological Chemistry,
Northwestern University Medical School, Chicago, IL 60611

It is now well established that the voltage-activated sodium channels
and the GABA-activated chloride channels are the major target sites of
certain insecticides. Pyrethroids prolong the sodium current as a result
of the increase in open time of individual sodium channels thereby
causing hyperexcitation in both mammals and invertebrates.
Tetrodotoxin-sensitive (TTX-S) and tetrodotoxin-resistant (TTX-R)
sodium channels of rat dorsal root ganglion neurons exhibited
differential sensitivity to the pyrethroids, the former which is abundant
in the brain, being much less sensitive than the latter and invertebrate
sodium channels. The effects of pyrethroids on both types of the
mammalian sodium channels were more easily reversed after washing
than those on invertebrate sodium channels. These two factors,
together with the negative temperature dependence of pyrethroid actions
on the nervous system, the positive temperature dependence of
pyrethroid metabolism, and the difference in body size, are deemed
responsible for the selective toxicity to pyrethroids between mammals
and insects. Lindane and dieldrin suppressed the activity of the GABA-
activated chloride channels thereby causing hyperexcitation in animals.
However, the dieldrin action was more complex, including a reversible,
fast stimulating action and an irreversible, slow suppressive action.
These two actions required different subunit combinations of the GABA
receptor-channel complex. By contrast, δ-HCH enhanced the GABA-
induced current in keeping with its depressant action on animals.

[1]Current address: Department of Psychiatry, Johns Hopkins University Medical School,
5510 Nathan Shock Drive, Baltimore, MD 21224
[2]Current address: Department of Physiology, Loyola University Medical School,
2160 South First Avenue, Maywood, IL 60153
[3]Current address: Department of Neurology, Yale University School of Medicine,
New Haven, CT 06510

0097–6156/95/0591–0026$12.00/0
© 1995 American Chemical Society

Certain types of neuronal ion channels have been demonstrated to be the major target sites of insecticides. For example, pyrethroids and DDT modify the gating mechanisms of sodium channels resulting in their prolonged openings which in turn cause hyperactivity of the nervous system (*1-6*). The cyclodiene insecticides and lindane suppress the activity of the $GABA_A$ receptor-chloride channel complex thereby causing hyperactivity of animals (*7,8*). However, other types of ion channels, including potassium channels, calcium channels, acetylcholine-activated channels and glutamate-activated channels, have not been thoroughly documented to be the major targets of any insecticides (*2*). The present paper gives recent developments in our studies of modulation of sodium channels and GABA-activated channels by the pyrethroid insecticides and dieldrin and lindane insecticides, respectively.

Pyrethroid Modulation of Sodium Channels

The symptoms of pyrethroid poisoning in animals are characterized by various forms of hyperexcitation and are ascribed to repetitive discharges of the nervous system. Pyrethroids can be classified into type I and type II based on the absence and presence, respectively, of a cyano group at the α position (*1*). Although the symptoms in animals caused by intoxication with type I and type II pyrethroids are different somewhat, the basic mechanisms of action at the cellular and channel level are the same. In this paper, the experimental results with type I pyrethroids are described.

When tetramethrin or allethrin is applied at a low concentration to an isolated giant nerve fiber of the squid or crayfish, the action potential is not affected but the depolarizing after-potential is increased and prolonged, eventually reaching the threshold for initiation of repetitive action potentials. The next question is how the depolarizing after-potential is increased. This question can be answered by voltage clamp experiments which allow us to measure the sodium and potassium channel activities that underlie the generation of action potentials. At low concentrations of pyrethroids, the sodium current is greatly prolonged while the potassium current is not substantially changed. These ionic currents represent an algebraic sum of currents flowing through a large number of ion channels. Patch clamp experiments to record single sodium channel currents have indeed demonstrated that individual sodium channels are modified to remain open for long periods of time. Thus it has been clearly demonstrated that pyrethroids cause hyperactivity of the nervous system by modulating the gating kinetics of individual sodium channels (*1-3*).

Tetrodotoxin-sensitive and Tetrodotoxin-resistant Sodium Channels. Dorsal root ganglion (DRG) neurons of rats are endowed with two types of sodium channels, tetrodotoxin-sensitive (TTX-S) and tetrodotoxin-resistant (TTX-R) sodium channels (*9-12*). In addition to a large difference in TTX sensitivity which amounts to 10^5-fold, there are differences in voltage dependence and kinetics between them. TTX-R sodium channel currents are much slower in time course than TTX-S sodium channel currents. TTX-R channels are activated and inactivated at less negative membrane potentials than TTX-S channels. There also are pharmacological and toxicological differences. TTX-R channels are less sensitive than TTX-S channels to the blocking action of lidocaine, but are more sensitive than TTX-S channels to the blocking action of lead and cadmium (12).

Differential Actions of Pyrethroids on TTX-S and TTX-R Sodium Channels. TTX-S and TTX-R sodium channels responded differently to the action of allethrin and tetramethrin (13,14) (Fig. 1). The purified isomers, 1R, 3R, 3S (+)-*trans* allethrin and (+)-*trans* tetramethrin, were used. The data shown in Figures 1-5 were obtained at room temperature (21-23°C). In the presence of 1 μM tetramethrin, the peak current of TTX-S sodium channels was not much affected while the slow component of current during step depolarization was increased slightly (Fig. 1A). Upon step repolarization, there appeared a slowly rising and slowly decaying tail current in tetramethrin (Fig. 1A). In TTX-R sodium channels, the peak and slow currents were increased by tetramethrin, and step repolarization generated a large instantaneous tail current which decayed slowly (Fig. 1B).

The slowly rising and slowly decaying (hooked) tail current in tetramethrin-treated TTX-S sodium channels resembles that observed in frog myelinated nerve fibers (15). They interpreted the hooked tail current as being due to the activation (m) gate being stuck at the open position. When the membrane is suddenly repolarized, the inactivation (h) gate that has been almost unimpaired by pyrethroids and closed during step depolarization slowly opens, generating the rising phase of the hooked tail current. Meanwhile, the activation gate slowly closes at large negative potential causing the tail current to decrease. The slowing of opening and closing kinetics of both activation and inactivation gates was indeed demonstrated by single-channel experiments with neuroblastoma cells (16).

The behavior of TTX-R sodium channels in response to tetramethrin is similar to that of sodium channels of squid and crayfish giant axons which are highly sensitive to TTX (17,18). The sodium channel inactivation mechanism is partially inhibited and slowed, and the sodium channel activation kinetics are also slowed. Thus the peak sodium current is increased and the slow phase of sodium current during step depolarization is increased as a result of impairment of the sodium channel inactivation. Due to the large amplitude of slow current, a large instantaneous tail current appears upon repolarization and decays slowly, due to slowing of the sodium channel activation kinetics.

In the presence of tetramethrin, although both TTX-S and TTX-R sodium channels opened at potentials more negative than the respective control, tetramethrin was less efficacious in the former channel (14) (Fig. 2). This change, together with prolonged openings of single-channels, explains the membrane depolarization in the pyrethroid-exposed preparations. Fenvalerate, a type II pyrethroid, also causes a shift of activation voltage in the hyperpolarizing direction (19).

In contrast to the differential actions of tetramethrin and allethrin on TTX-S and TTX-R sodium channels as described above, the voltage dependence of the steady-state sodium channel inactivation was affected equally in both types of sodium channels (Fig. 3). It should be noted that the inactivation kinetics were slowed to a lesser extent in TTX-S channels than in TTX-R channels by the action of tetramethrin (Fig. 1). Thus despite the various differences in the tetramethrin on TTX-S and TTX-R sodium channels, there is a common denominator between the two types of channels with respect to the action of tetramethrin in shifting the steady-state sodium channel inactivation curve.

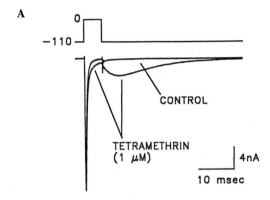

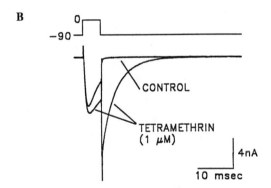

Fig. 1.   Effects of tetramethrin on TTX-S sodium current (A) and TTX-R sodium current (B) in rat dorsal root ganglion neurons.   A step depolarization to 0 mV was applied from a holding potential of -110 mV (A) or -90 mV (B) in control and in the presence of 1 μM tetramethrin (Reproduced with permission from ref. 14.   Copyright 1994 by the American Society for Pharmacology and Experimental Therapeutics).

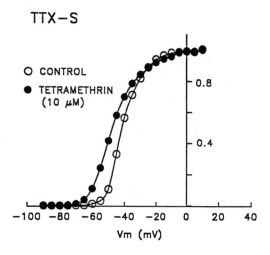

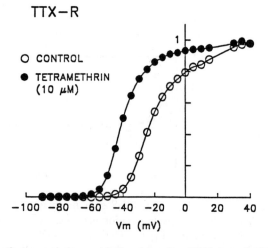

Fig. 2. Conductance-voltage relationships for TTX-S and TTX-R sodium channels of rat DRG neurons before and during application of 10 μM tetramethrin. Ordinate, normalized conductance; abscissa, membrane potential (Reproduced with permission from ref. 14. Copyright 1994 by the American Society for Pharmacology and Experimental Therapeutics).

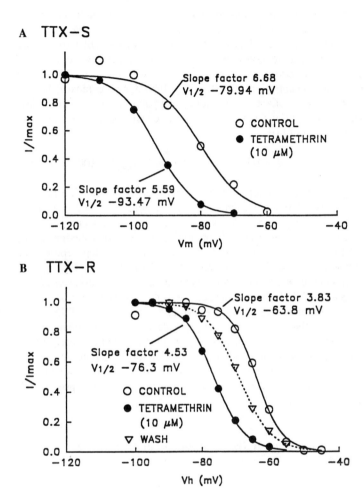

Fig. 3. Effects of 10 μM tetramethrin (TM) on the steady-state inactivation curve for TTX-S and TTX-R sodium channels of rat DRG neurons. The membrane potential was held at various levels for 20 sec and then sodium current was evoked by a step depolarization to 0 mV. A, TTX-S sodium current; the peak amplitude of sodium current during a step depolarization is plotted as a function of the conditioning voltage. O, control; ●, 10 μM TM. B, TTX-R sodium current; O, control; ●, 10 μM TM; ▽, washout. The curves were drawn according to the equation, $I/I_{max} = 1/\{1+\exp[(V_h - V_{\frac{1}{2}})/k]\}$, where V_h is the holding potential, $V_{\frac{1}{2}}$ is the potential at which sodium current is one-half maximum, and k is the slope factor (Reproduced with permission from ref. 14. Copyright 1994 by the American Society for Pharmacology and Experimental Therapeutics).

Difference in Pyrethroid Sensitivity of TTX-S and TTX-R Sodium Channels.
Tetramethrin and allethrin were more potent on TTX-R sodium channels than on
TTX-S sodium channels (13,14). Modification of the tail current kinetics was
noticeable at a minimum concentration of 0.1 μM of tetramethrin in TTX-S sodium
channels. By contrast, the minimum effect was observed in TTX-R sodium
channels at a concentration of 0.01 μM. In order to compare the tetramethrin
sensitivity in a more quantitative manner, the percentages of sodium channels that
are modified by various concentrations of tetramethrin were calculated by the
following equation (14):

$$M = [\{I_{tail}/(E_h\text{-}E_{Na})\}/\{I_{Na}/(E_t\text{-}E_{Na})\}] \times 100 \qquad (1)$$

where I_{tail} is the tail current amplitude obtained by extrapolation of the slowly
decaying phase of the tail current to the moment of membrane repolarization,
assuming a single exponential decay, E_h is the potential to which the membrane was
repolarized, E_{Na} is the equilibrium potential for sodium ions obtained as the reversal
potential for sodium current, and E_t is the potential of step depolarization. The
concentration-response data were fitted to the Hill equation (Fig. 4).

The maximum percentages of sodium channel modification were 15% and 85%
for TTX-S and TTX-R sodium channels, respectively, and the apparent dissociation
constants of tetramethrin were 2.9 μM and 0.44 μM for TTX-S and TTX-R
channels, respectively. The average percentages of sodium channel modification
at various concentrations of tetramethrin were calculated and are given in Table I
(14). Although the difference in the apparent dissociation constant between TTX-S
and TTX-R sodium channels is seven-fold, the differences in the concentrations to
modify 1% and 10% of sodium channels are estimated to be approximately 30- and
70-fold, respectively.

Table I. **Percentages of the Fraction of the TTX-S and TTX-R Sodium
Channels Modified by Various Concentrations of Tetramethrin**

	Modification of sodium channels (%)	
Tetramethrin (μM)	*TTX-S*	*TTX-R*
0.01	-	1.31 ± 0.28
0.03	0	5.15 ± 0.30
0.1	0.24 ± 0.10	15.35 ± 0.79
0.3	1.25 ± 0.13	35.48 ± 2.70
1	3.53 ± 0.66	57.82 ± 2.29
3	7.70 ± 1.20	74.85 ± 1.23
10	12.03 ± 1.89	81.20 ± 1.57

Mean ± S.E.M. (n=4) (Reproduced with permission from ref. 14. Copyright 1994
by the American Society for Pharmacology and Experimental Therapeutics).

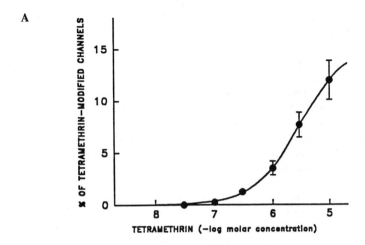

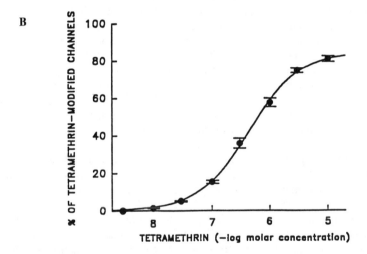

Fig. 4.   Concentration-dependent effect of tetramethrin (TM) on TTX-S and TTX-R sodium currents in rat DRG neurons. A, TTX-S currents were evoked by a 5-msec step depolarization to 0 mV from a holding potential of -110 mV under control condition and in the presence of 0.1, 0.3, 1 and 3 μM TM, and the dose-response relationship for induction of slow tail current is plotted. Each point indicates the mean $\pm$ S.E.M. (n=4).   Data were fitted by the Hill equation.   B, TTX-R currents were evoked by a 5-msec step depolarization to 0 mV from a holding potential of -90 mV under control condition and in the presence of 0.01, 0.03, 0.1, 0.3, and 1 μM TM, and the dose-response relationship for the induction of slow tail current is plotted.   Each point indicates the mean $\pm$ S.E.M. (n=4).   Data were fitted by the Hill equation (Reproduced with permission from ref. 14. Copyright 1994 by the American Society for Pharmacology and Experimental Therapeutics).

Toxicity Amplification and Comparison of Potency. A traditional method to evaluate whether an *in vitro* effect of a test compound is relevant to the symptoms in whole animal or human is to compare the *in vitro* EC_{50} with the concentration in the serum of animal or human exhibiting the symptoms. However, this would not be logical in the cases in which a threshold phenomenon is involved. We have previously found in squid giant axons that tetramethrin induces repetitive after-discharges at 0.03 μM which is much less than the high affinity component of the apparent dissociation constant of 0.15 μM and at which the percentage of modified sodium channels is estimated to be less than 1% (20). Thus a concentration much lower than the EC_{50} value modifies a very small percentage of sodium channels and increases the depolarizing after-potential to the threshold level to initiate repetitive discharges.

The calculations of the percentage of sodium channel modification by Lund and Narahashi (20) had several assumptions, including the single sodium channel conductance. The study of Tatebayashi and Narahashi (14) described in a preceding section has provided more accurate calculations of the percentage of sodium channel modification. If comparison is made of the apparent dissociation constants, the following potency order is obtained: squid axon sodium channels (0.15 μM, the high affinity component that is crucial for initiation of repetitive discharges) > rat TTX-R sodium channels (0.44 μM) > rat TTX-S sodium channels (2.9 μM). Thus the potency ratio for squid: TTX-R : TTX-S is 1 : 0.34 : 0.05. If comparison is made of the concentrations to modify approximately 1% of sodium channels, the following potency order is obtained: squid (~ 0.01 μM) $\approx$ TTX-R (0.01 μM) > TTX-S (0.3 μM). The potency ratio is: 1 : 1 : 0.03. It can be concluded that TTX-S sodium channels of rat DRG neurons are much less sensitive to tetramethrin than either TTX-R sodium channels of rat DRG neurons or squid axon sodium channels.

Two questions may be raised regarding behavior of mammalian neurons in response to pyrethroids. First, a question is asked how mammalian neurons in the brain respond to the actions of TTX and pyrethroids. Since DRG neurons are in the peripheral nervous system, the action of pyrethroids on them may or may not be relevant to the symptoms of pyrethroid poisoning in animals and humans. Our preliminary experiments with rat cerebellar Purkinje neurons clearly indicate that all of the sodium channels examined are TTX sensitive and relatively insensitive to tetramethrin (Song, J.-H., Narahashi, T., unpublished observation). Their tetramethrin sensitivity is in the same order of magnitude as that of the TTX-S sodium channel of rat DRG neurons. A second question is whether mammalian neurons generate repetitive after-discharges when intoxicated with pyrethroids as in the case of squid, crayfish or insect axons. This has proven to be the case in rat cerebellar Purkinje neurons exposed to low concentrations of tetramethrin (Song, J.-H., Narahashi, T., unpublished observation). Thus the conclusion given in the preceding paragraph may be generalized: the sodium channels of mammalian neurons are less sensitive to pyrethroids than those of invertebrate nerves.

Reversibility. Another significant difference between rat sodium channels and squid or crayfish sodium channels is the rate at which pyrethroid effects are reversed after washing. The recovery after washing was very slow in squid and crayfish axons with a time constant in the order of one hour (*18,20*). By contrast rat DRG neurons recovered much more quickly within 5-15 min after washing out tetramethrin (Fig. 5) (*14*).

Mechanisms of Selective Toxicity between Mammals and Invertebrates. Pyrethroids are much more potent on insects than on mammals. At least five factors are deemed important in contributing the selective toxicity. First and possibly the most important factor is temperature dependence of pyrethroid action. It is well established that pyrethroids as well as DDT which mimics type I pyrethroids in terms of the mechanism of sodium channel modulation (*21*) have negative temperature coefficients of action (*2*). The temperature dependence was very large and the potency increased 4- to 10-fold, as temperature was lowered by 10°C (*19,22*). Thus a 10°C difference in body temperature between mammals (37°C) and insects (27°C, ambient temperature) would cause very drastic differences in the potency of pyrethroids.

A second factor is the rate of metabolism. Since pyrethroids are metabolized by enzymes, the rate of metabolism is expected to be 2- to 3-fold faster in mammals than in insects. Differences in esterase levels in insects and mammals also are likely an important factor. A third factor is the sensitivity of sodium channels to pyrethroids. As has been discussed in the present chapter, there are large differences between mammals and invertebrates. Although data are not very quantitative, the sodium channels of squid, crayfish and insect nerves are highly sensitive to pyrethroids. In rat DRG neurons, although TTX-R sodium channels are almost equally sensitive to pyrethroids to invertebrate sodium channels, TTX-S sodium channels are much less sensitive with a pyrethroid potency of 3-5% of the former two groups. In rat brain, the majority of sodium channels is TTX sensitive, and our preliminary experiments have shown that pyrethroids are almost equipotent to rat DRG TTX-S channels (Song, J.-H., Narahashi, T., unpublished observation).

A fourth factor is reversibility of pyrethroid action. Although data are not quantitative, it is clear that rat DRG sodium channels recover much more quickly than invertebrate sodium channels after washing out pyrethroids. A fifth factor is the body size. Since mammals are much larger than insects, pyrethroids are likely to be detoxified more in mammals than in insects before reaching the target site. Thus when all of these factors are multiplied, the overall difference in pyrethroid potency between mammals and insects may amount to as much as one thousand-fold. These overall differences are in the same order of magnitude as those in LD_{50}.

Dieldrin and Lindane Modulation of GABA Receptor-channel Complex

Lindane and cyclodienes such as dieldrin were demonstrated some 40 years ago to stimulate the central nervous system by facilitating synaptic transmission (*23,24*). However, it was not until the early 1980s that the GABA system was identified as a target site by ^{36}Cl uptake and [^{35}S]TBPS binding experiments (*25,26*). We reported that lindane and cyclodiene insecticides block GABA-induced current in rat hippocampal neurons (*7*). Our recent patch clamp experiments have unveiled more complex actions of dieldrin and lindane on the GABA receptor-channel

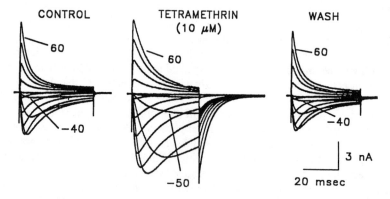

Fig. 5.   Effects of 10 μM tetramethrin (TM) on TTX-R sodium channel currents of rat DRG neurons.   Currents were evoked by 40-msec depolarizations to various levels from a holding potential of -90 mV.   Test potentials ranged from -90 to +60 mV in 5-mV increments and were delivered at a frequency of 0.1 Hz.   Currents under control condition, in the presence of 10 μM tetramethrin, and after washing with tetramethrin-free solution are shown (Reproduced with permission from ref. 14.   Copyright 1994 by the American Society for Pharmacology and Experimental Therapeutics).

complex (*27,28*). Furthermore, the roles of some of the GABA receptor subunits in these actions have been identified (*29*).

Dual Action of Dieldrin on GABA System. Dieldrin has been found to exert a dual action on the $GABA_A$ receptor-chloride channel complex of rat DRG neurons (*27*). When 1 μM dieldrin was co-applied with 10 μM GABA, the amplitude of GABA-induced current was greatly augmented, reaching over 200% of control (Fig. 6). However, repeated co-applications caused the current to decrease gradually, and after the 8th co-application it was reduced to a low steady state at about 50% of control (Fig. 6). An additional effect of dieldrin was the acceleration of current desensitization. This change was not merely due to the increase in current amplitude, as it was observed even after the current amplitude was reduced beyond the control level (Fig. 6A). The dose-response curves for dieldrin enhancement and suppression indicate EC_{50} values of 754 nM and 92 nM, respectively (Fig. 7). It was noted that the enhancing action of dieldrin was reversible after washing with dieldrin-free solutions while the inhibitory action was not. It appears that these two effects of dieldrin are exerted by two different mechanisms. The inhibitory action which is more potent than the stimulating action is directly responsible for the hyperactivity produced by dieldrin in the nervous system and whole animals. The role of the stimulating action in symptomatology remains to be seen, and it is expected that this action will suppress the animal behavior.

Binding Site of Dieldrin on GABA System. The $GABA_A$ receptor-chloride channel complex is endowed with several binding sites. These include the GABA site, the benzodiazepine site, the barbiturate site, the picrotoxin site, the Cu^{2+}/Zn^{2+} site and the La^{3+} site (*30,31*). Competition experiments were performed to determine the site of action of dieldrin (*27*). Dieldrin at 0.1 μM suppressed the GABA-induced current to 45% of control, and pentobarbital at 5 μM enhanced the current to 201% of control. When 0.1 μM dieldrin was added to 5 μM pentobarbital, the GABA-induced current was 102% of control, indicating 50% suppression of the pentobarbital-potentiated current by dieldrin. Similar results were obtained with chlordiazepoxide, a benzodiazepine, which also enhanced the current. Thus dieldrin does not bind to either the barbiturate or the benzodiazepine site.

Picrotoxin (PTX) at 0.3-10 μM antagonized the suppressive action of 0.1 μM dieldrin (Fig. 8A). However, the enhancing action of dieldrin at a higher concentration (1 μM) was not inhibited by 1 μM PTX (Fig. 8B). Dieldrin at 1 μM enhanced the GABA-induced current to 160% of control, and if coapplied with 1 μM PTX, it enhanced the current to 230% of the level in PTX alone. Thus PTX at 1 μM did not antagonize the enhancing action of 1 μM dieldrin. Only at a higher concentrtion of 10 μM, did PTX completely abolish the enhancing action of 1 μM dieldrin. These results indicate that dieldrin and PTX share a common binding site. This conclusion is in keeping with the results of experiments which showed that cyclodienes, picrotoxin and TBPS bind to the same site (*26,32-34*). It is possible that the binding site of these three groups of chemicals is closely associated with or allosterically linked to the chloride channel of the GABA receptor.

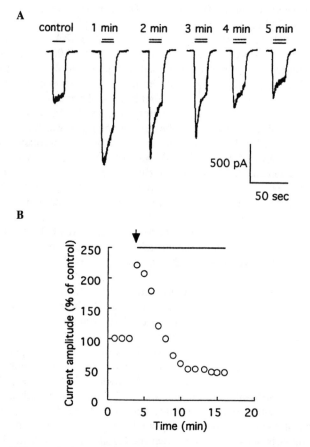

Fig. 6. Effects of dieldrin on GABA-induced chloride currents in a rat DRG neuron. A, current records in response to 20-sec application of 10 μM GABA (solid bar) and to co-application of 10 μM GABA and 1 μM dieldrin (dotted bar) at the time indicated after taking control record. The peak amplitude of current is greatly enhanced but gradually decreases during repeated co-applications. Desensitization of current is accelerated. B, time course of the changes in peak current amplitude before and during repeated co-applications (dotted line) (Reproduced with permission from ref. 27. Copyright 1994 by the American Society for Pharmacology and Experimental Therapeutics).

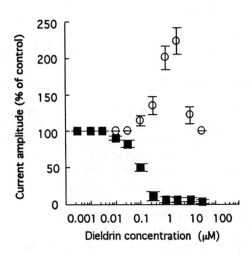

Fig. 7. Dose-response relationships for dieldrin enhancement (circles) and suppression (squares) of 10 μM GABA-induced chloride current in rat DRG neurons. Currents were induced by 5-sec applications of GABA or GABA plus dieldrin. Current enhancement was measured during the first co-application of GABA and dieldrin, and current suppression was measured when the current was decreased to a steady-state level following repeated co-applications. Enhancement: EC_{50} = 754 nM; Hill coefficient = 1.72; n=5. Suppression: EC_{50} = 92 nM; Hill coefficient = 1.34; n=5 (Reproduced with permission from ref. 27. Copyright 1994 by the American Society for Pharmacology and Experimental Therapeutics).

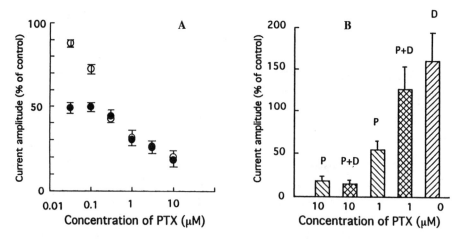

Fig. 8. The effect of 0.1 μM dieldrin on picrotoxin suppression of 10 μM GABA-induced chloride current in rat DRG neurons. A, dieldrin at 0.1 μM suppresses the current. The EC_{50} values of picrotoxin with (closed circles) and without (open circles) dieldrin are 2.98 and 0.35 μM, respectively. The dose-response relationship of picrotoxin block is shifted to higher concentrations in the presence of dieldrin indicating that dieldrin and PTX share a common binding site. n=4. B, the effect of 1 μM dieldrin on picrotoxin (PTX) suppression of 10 μM GABA-induced chloride current in rat DRG neurons. Dieldrin at 1 μM enhances the current. PTX suppression is not altered by the presence of dieldrin, indicating that dieldrin and PTX share a common binding site. n=4 (Reproduced with permission from ref. 27. Copyright 1994 by the American Society for Pharmacology and Experimental Therapeutics).

Role of GABA Receptor Subunits in Dieldrin Actions. The two separate actions of dieldrin described in the preceding section may be due to the ability of dieldrin to selectively bind to different GABA receptor subunits. The GABA receptor comprises at least sixteen subunits, including six αs, four βs, four γs, one δ and one ρ (35,36). In order to identify the subunit or subunits that distinguish the two separate actions of dieldrin, patch clamp experiments were performed by using human embryonic kidney cell line (HEK 293) expressing various combinations of subunits (29).

When co-applied with GABA to cells expressing the $\alpha1\beta2\gamma2S$ subunit combination, 3 μM dieldrin caused both enhancement and suppression of GABA-induced current as in the case of rat DRG neurons. Similar results were obtained with the $\alpha6\beta2\gamma2S$ combination. By contrast only the suppressive effect of dieldrin was observed in the $\alpha1\beta2$ combination. These results are illustrated in Fig. 9 as the dose-response curves of dieldrin actions. Thus it can be concluded that while any of the three combinations of subunits is sufficient for dieldrin to cause suppressive effect, the $\gamma2$ subunit is necessary for the enhancing effect.

Effects of HCH Isomers on GABA System. Lindane suppressed the GABA-induced current in rat hippocampal and DRG neurons (7,8,37). Further patch

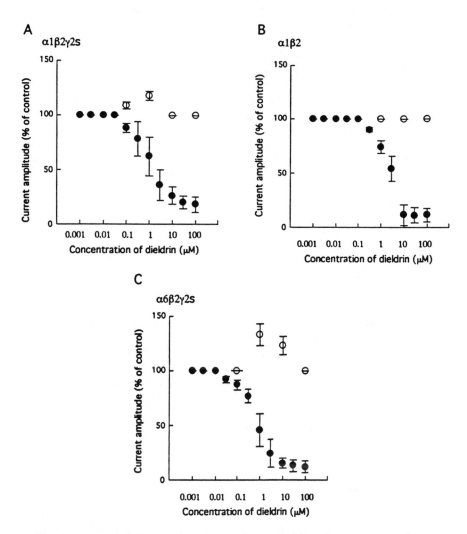

Fig. 9. Dose-response relationships for dieldrin effects on the GABA-induced chloride current in the $\alpha1\beta2\gamma2s$, $\alpha1\beta2$ and $\alpha6\beta2\gamma2s$ combinations of $GABA_A$ receptor expressed in HEK 293 cells. Current suppression (closed circles): A, $\alpha1\beta2\gamma2s$; EC_{50} = 2.1 μM; Hill coefficient = 0.6; n=5. B, $\alpha1\beta2$; EC_{50} = 2.8 μM; Hill coefficient = 1.1; n=5. C, $\alpha6\beta2\gamma2s$; EC_{50} = 1.0 μM; Hill coefficient = 0.8, n=5. Enhancement (open circles): Enhancement was observed in the $\alpha1\beta2\gamma2s$ and $\alpha6\beta2\gamma2s$ combinations, but not in the $\alpha1\beta2$ combination (Reproduced with permission from ref. 29. Copyright 1994 by Elsevier Science B.V.).

clamp experiments were performed to determine the effects of HCH isomers on GABA-induced currents and to identify the roles of the GABA receptor subunits in these actions (28).

Lindane (γ-HCH) exerted a suppressive effect on the GABA-induced current in rat DRG neurons and in HEK 293 cells expressing either the α1β2γ2S or the α1β2 combination. There also was a small and transient enhancing effect of lindane on the GABA-induced current. This suppressive effect can explain hyperactivity of lindane-intoxicated animals. By contrast, δ-HCH exerts only the enhancing effect in DRG neurons and in the α1β2γ2S and α1β2 combinations. The enhancing effect can account for depressive behavior of δ-HCH-intoxicated animals. Thus modulation of the GABA receptor-channel system by γ-HCH and δ-HCH is directly responsible for the symptoms of poisoning in animals.

Acknowledgments

We wish to thank Vicky James-Houff for her unfailing secretarial assistance. The author's work cited in this chapter was supported by NIH grant NS14143.

Literature Cited

1. Narahashi, T. *Neurotoxicology* **1985,** *6(2),* 3-22.
2. Narahashi, T. In *Insecticide Action: From Molecule to Organism;* Narahashi, T., Chambers, J. E.; Plenum Press: New York, 1989; pp. 55-84.
3. Narahashi, T. *Trends in Pharmacological Sciences* **1992,** *13,* 236-241.
4. Ruigt, G. S. F. In *Comprehensive Insect Physiology, Biochemistry and Pharmacology;* Kerkut, G. A., Gilbert, L. I.; Pergamon Press: Oxford, 1984, Vol. 12, Chapter 7; pp. 183-263.
5. Soderlund, D. M.; Bloomquist, J. R. *Annu. Rev. Entomol.* **1989,** *34,* 77-96.
6. Vijverberg, H. P. M.; van den Bercken, J. *Crit. Rev. Toxicol.* **1990,** *21(2),* 105-126.
7. Narahashi, T.; Frey, J. M. *Soc. Neurosci. Abstr.* **1989,** *15,* 1151.
8. Ogata, N.; Vogel, S. M.; Narahashi, T. *FASEB J.* **1988,** *2,* 2895-2900.
9. Elliott, A. A.; Elliott, J. R. *J. Physiol. (Lond.)* **1993,** *463,* 39-56.
10. Kostyuk, P. G.; Veselovsky, N. S.; Tsyndrenko, Y. *Neuroscience* **1981,** *6,* 2423-2430.
11. Ogata, N.; Tatebayashi, H. *J. Physiol. (Lond.)* **1993,** *466,* 9-37.
12. Roy, M.-L.; Narahashi, T. *J. Neuroscience* **1992,** *12,* 2104-2111.
13. Ginsburg, K. S.; Narahashi, T. *Brain Res.* **1993,** *627,* 239-248.
14. Tatebayashi, H.; Narahashi, T. *J. Pharmacol. Exp. Ther.* **1994,** *270,* 595-603.
15. Vijverberg, H. P. M.; van der Zalm, J. M.; van den Bercken, J. *Nature (Lond.)* **1982,** *295,* 601-603.
16. Chinn, K.; Narahashi, T. *J. Physiol. (Lond.)* **1986,** *380,* 191-207.
17. Lund, A. E.; Narahashi, T. *Neurotoxicology* **1981,** *2,* 213-229.
18. Lund, A. E.; Narahashi, T. *J. Pharmacol. Exp. Ther.* **1981,** *219,* 464-473.
19. Salgado, V. L.; Herman, M. D.; Narahashi, T. *NeuroToxicology* **1989,** *10,* 1-14.

20. Lund, A. E.; Narahashi, T. *Neurotoxicology* **1982**, *3*, 11-24.
21. Lund, A. E.; Narahashi, T. *Neuroscience* **1981**, *6*, 2253-2258.
22. Yamasaki, T.; Ishii (Narahashi), T. *Botyu-Kagaku (Scientific Insect Control)* **1954**, *19*, 39-46 (In Japanese). English translation (1957) Japanese Contributions to the Study of the Insecticide-Resistance Problem. Pub. by the Kyoto Univ. for the W.H.O., 155-162.
23. Yamasaki, T.; Ishii (Narahashi), T. *Botyu-Kagaku (Scientific Insect Control)* **1954**, *19*, 106-112 (In Japanese). English translation (1957) Japanese Contributions to the Study of the Insecticide-Resistance Problem. Pub. by the Kyoto Univ. for the W.H.O., 176-183.
24. Yamasaki, T.; Narahashi, T. *Botyu-Kagaku (Scientific Insect Control)* **1958**, *23*, 47-54.
25. Ghiasuddin, S. M.; Matsumura, F. *Comp. Biochem. Physiol.* **1982**, *73C*, 141-144.
26. Matsumura, F.; Ghiasuddin, S. M. *J. Environ. Health Sci.* **1983**, *B18*, 1-14.
27. Nagata, K.; Narahashi, T. *J. Pharmacol. Exp. Ther.* **1994**, *269*, 164-171.
28. Nagata, K.; Narahashi, T. *Eighth IUPAC Internat. Congr. Pesticide Chemistry* **1994**, Abstr. *1*, 240.
29. Nagata, K.; Hamilton, B. J.; Carter, D. B.; Narahashi, T. *Brain Res.* **1994**, *645*, 19-26.
30. Ma, J. Y.; Narahashi, T. *Brain Res.* **1993**, *607*, 222-232.
31. Olsen, R. W.; Bureau, M.; Ransom, R. W.; Deng, L.; Dilber, A.; Smith, G.; Krestchatisky, M.; Tobin, A. J. In *Neuroreceptors and Signal Transduction;* Kito, S., Segawa, T., Kuriyama, K., Tohyama, M., Olsen, R.W.; Plenum Press: New York, 1988; pp. 1-14.
32. Bloomquist, J. R.; Roush, R. T.; ffrench-Constant, R. H. *Arch. Insect Biochem. Physiol.* **1992**, *19*, 17-25.
33. Casida, J. E. *Arch. Insect Biochem. Physiol.* **1993**, *22*, 13-23.
34. Eldefrawi, A. T.; Eldefrawi, M. E. *FASEB J.* **1987**, *1*, 262-271.
35. Fritschy, J. M.; Benke, D.; Mertens, S.; Oertel, W. H.; Bachi, T.; Möhler, H. *Proc. Natl. Acad. Sci., USA* **1992**, *89*, 6726-6730.
36. Olsen, R. W.; Tobin, A. J. *FASEB J.* **1990**, *4*, 1469-1480.
37. Frey, J.; Dichter, M.; Narahashi, T. *The Toxicologist* **1989**, *9*, 149.

RECEIVED January 11, 1995

Chapter 3

Presynaptic Actions of Dihydropyrazoles

Russell A. Nicholson and Aiguo Zhang

Department of Biological Sciences, Simon Fraser University, Burnaby, British Columbia V5A 1S6, Canada

We have examined the actions of two dihydropyrazoles (RH-3421 and RH-5529) on changes induced by activating sodium channels (with veratridine) or calcium channels (with elevated K^+) in synaptosomal preparations isolated from mouse brain. Our studies on evoked release of γ-aminobutyric acid and L-glutamate showed the dihydropyrazoles not only to be effective inhibitors of veratridine-induced release but also provided evidence they can suppress the K^+-mediated (tetrodotoxin-insensitive) component of neurotransmitter efflux. At high concentrations the dihydropyrazoles were unable to influence the depolarization of nerve endings caused by exposure to elevated K^+. As expected, the dihydropyrazoles inhibited veratridine-stimulated increases in synaptosomal free $[Ca^{++}]$. K^+-induced rises in synaptosomal free $[Ca^{++}]$, which require external Ca^{++} and are unaffected by tetrodotoxin, are blocked by RH-3421 and RH-5529 in a dose-dependent fashion. K^+-stimulated $^{45}Ca^{++}$ uptake by synaptosomes is not influenced by tetrodotoxin, suppressed by Co^{++} and blocked by RH-3421 and RH-5529 both alone and in the presence of tetrodotoxin. RH-3421 and RH-5529 have negligible effects on the membrane potential, concentration of free Ca^{++} and basal $^{45}Ca^{++}$ accumulation in resting synaptosomes. It is proposed that in addition to their established inhibitory effects on sodium channels the dihydropyrazoles may interfere with the operation of calcium channels.

The insecticidal properties of the dihydropyrazoles have been known for approximately twenty years (1-3). As a class, these insecticides exhibit a fairly wide spectrum of activity and have been shown to be particularly effective against a number of coleopteran and lepidopteran pests (1,4). Despite this encouraging insecticidal profile and intensive research in the area, the dihydropyrazoles have not as yet achieved commercial significance as pesticides.

The majority of insecticides discovered so far are compounds which attack the nervous system, and the dihydropyrazoles are no exception. In cockroaches, the highly insecticidal analog RH-3421 (Figure 1 for structure) elicits a rather complex and protracted poisoning symptomology which is typical of a neurotoxicant (5). The

0097–6156/95/0591–0044$12.00/0

poisoning initially involves poor coordination and tremor. Symptoms then progressively develop into a condition which has the outward appearance of full paralysis; however, during this phase mechanical stimulation will initiate violent tremors and convulsions (5). In our laboratory we have shown (Nicholson, R. A., Simon Fraser University, Burnaby, B.C., unpublished observations) that RH-5529 (Figure 1 for structure) produces poisoning symptoms very similar to those described for RH-3421 and is also insecticidal when injected into house crickets at high doses.

In common with many other neurotoxic insecticides, the dihydropyrazoles appear to exert relatively specific effects at the level of the nervous system. In invertebrates, electrophysiological investigations with dihydropyrazoles have demonstrated that they cause a substantial reduction in the electrical activity of the central and peripheral nervous system which has been attributed to voltage-dependent block of neuronal sodium currents (5, 6). In mammals, the uptake of ^{22}Na into brain microvesicles that accompanies sodium channel activation is inhibited by RH-3421 (7), and this insecticide is also known to influence the binding of batrachotoxin A 20 α-benzoate to brain membranes by an allosteric mechanism (8). There is clearly compelling evidence that dihydropyrazoles act on sodium channels in the brain of invertebrates and mammals. Studies in our laboratory have strongly supported this mechanism of action. In addition, our studies have implicated presynaptic calcium channels as potential sites of action in the mammalian central nervous system (9, Zhang, A. and Nicholson, R. A. *Comp. Biochem. Physiol.*, in press).

This review describes observations that led us to consider that an alternative site of action for dihydropyrazoles might exist and then summarizes our lines of evidence for this proposal. Our experimental approach throughout this investigation has been to use synaptosomes (pinched-off nerve endings) isolated from mammalian brain. Synaptosomes display many of the properties attributed to nerve terminals found in functionally intact brain (10-12). The relative ease with which specific processes involved in the release of neurotransmitters can be pharmacologically activated in synaptosomal preparations makes them particularly suitable for studying the presynaptic actions of inhibitory compounds such as the dihydropyrazoles. Furthermore, synaptosomes are multifunctional entities and as such contain an abundance of possible neuronal sites of attack for toxic chemicals. This feature may increase the chances of identifying novel sites of action so offering advantage over less complex preparations.

Methods

Synaptosomes were isolated from mammalian brain according to published techniques (13, 14). Procedures for the release of neurotransmitters (γ-aminobutyric acid (GABA) and L-glutamate), determination of membrane potential, measurement of intrasynaptosomal free [Ca^{++}] and assay of $^{45}Ca^{++}$ uptake into nerve endings have been described in detail elsewhere (9, 15, 16, 17, 18). RH-3421 and RH-5529 were obtained from the Rohm & Haas Company, and other pharmacological agents were purchased from the Sigma Chemical Company. Rhodamine 6G and Fura-2 AM were supplied by the Kodak Company and Molecular Probes Incorporated respectively. $^{45}CaCl_2$ was obtained from Dupont Canada Incorporated.

Results and Discussion

Inhibition of GABA release. Our initial study focused on the effect of RH-3421 and RH-5529 on release of the neurotransmitter GABA from synaptosomes and was the first to implicate the sodium channel of mammalian brain as an important target for these compounds (*15*). The main evidence for this was that both analogs were highly effective inhibitors of the transmitter releasing effects of sodium channel activators such as veratridine (Figure 2). As expected, activation of GABA release by veratridine could be inhibited by tetrodotoxin (TTX) because it blocks Na^+ entry through sodium channels. Therefore, opening of voltage-sensitive calcium channels does not occur. In contrast, elevated K^+ directly depolarizes the synaptosomal membrane which opens voltage-sensitive calcium channels and triggers GABA release. At high concentrations TTX and RH-3421 are poor inhibitors of K+-evoked GABA release, but RH-5529 consistently gave significant inhibition (Table I). Thus when GABA-ergic nerve endings are assayed, RH-3421 emerged as a highly selective blocker of sodium channels whereas RH-5529 inhibited sodium channels and also blocked another neuronal process.

Inhibition of L-Glutamate release. The symptoms of poisoning observed after dietary administration of RH-3421 involve lethargy and pacificity. Such a syndrome would not be expected to result solely from a reduction in release of an inhibitory transmitter such as GABA. We therefore explored the effects of dihydropyrazoles on evoked release of the excitatory transmitter L-glutamate. As one might predict, veratridine-stimulated release was blocked by low concentrations of RH-3421 and RH-5529 (IC_{50}s = 0.5 μM and 4.7 μM respectively). However, micromolar concentrations of RH-5529 and, quite unexpectedly, RH-3421 were also effective at inhibiting K+-induced release of L-glutamate (Figure 3a & b). K+-evoked release of L-glutamate is unaffected by 3 μM TTX (Figure 3c). To summarize so far, our results on the effects of RH-3421 and RH-5529 on evoked release of GABA and L-glutamate demonstrated these compounds to be potent blockers of sodium channels but also strongly suggested they interfere with another presynaptic target which may facilitate evoked release of neurotransmitter such as calcium channels or a process 'downstream' from this.

Effects of dihydropyrazoles on synaptosomal membrane potential. We also considered that dihydropyrazoles may simply prevent elevated K^+ from depolarising the nerve ending. Although there are no precedents for an action of this type, it was felt important to check the possibility. Using the voltage-sensitive probe rhodamine 6G to measure synaptosomal membrane potential, we found that micromolar concentrations of RH-3421, RH-5529 and TTX were unable to influence elevated K+-induced depolarization (Figure 4). The study also showed that concentrations of dihydropyrazoles, at least up to 10 μM, do not affect the electrical stability of resting synaptosomes. The lack of any obvious inhibitory action on K+-induced depolarisation led us to examine the hypothesis that dihydropyrazoles were blocking depolarisation-coupled calcium accumulation in synaptosomes.

Studies on intrasynaptosomal free [Ca++]. The level of free [Ca++] in resting nerve terminals averaged 264 nM and was largely unaffected by high concentrations of dihydropyrazoles (Table II). The alkaloid neurotoxin veratridine (50 μM) typically produces an increase in free [Ca++] of approximately 350 nM and this response is fully sensitive to inhibition by 3 μM TTX (Figure 5). The figure also illustrates the inhibition of veratridine-induced rises in synaptosomal free [Ca++] with dihydropyrazoles. The IC_{50}s for RH-3421 and RH-5529 were estimated at 0.2 and 3 μM respectively. The results accord with the potency order against veratridine's responses observed in the transmitter release and membrane potential assays and are consistent with significant effects of these analogs on sodium channels.

Figures 6 and 7 show representative recordings of the effects of dihydropyrazoles on K+-induced changes to synaptosomal free [Ca++]. We found

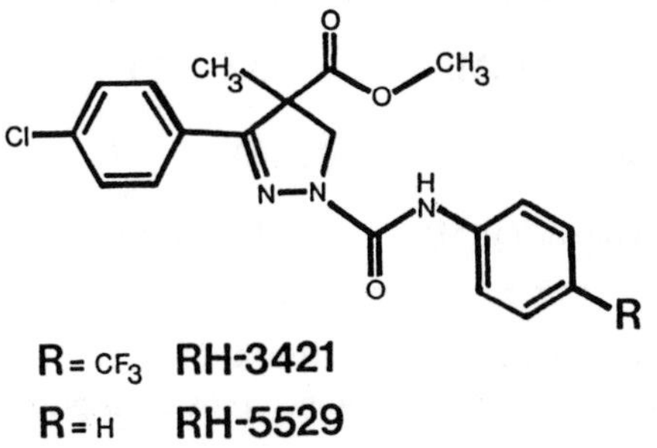

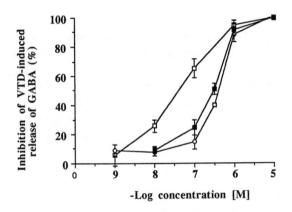

R = CF₃ **RH-3421**
R = H **RH-5529**

Figure 1.   Structures of the dihydropyrazoles RH-3421 and RH-5529.

Figure 2.   Concentration-dependent inhibitory effects of dihydropyrazoles (RH-3421(☐); RH-5529 (◯)) and tetrodotoxin (■) on veratridine (VTD)-induced release of ^{3}H-GABA from synaptosomes. Synaptosomes were exposed to neurotoxicants prior to challenge with VTD (10 μM). Values as mean ± standard error of three to five determinations.
(Reproduced with permission from ref. 15. Copyright 1990 Academic Press, Inc.).

Table I   The effect of dihydropyrazoles and tetrodotoxin on
release of ^{3}H-GABA induced by potassium ions[a]

Treatment[b]	% Inhibition of K$^+$-evoked release of ^{3}H-GABA[c]
RH-3421 (10 µM)	3.8 ± 8.4
RH-5529 (10 µM)	44.4 ± 6.7
Tetrodotoxin (1µM)	15.1 ± 3.8

[a] Depolarizing media contained 20 mM K$^+$
[b] Synaptosomes were exposed to neurotoxicants prior to challenge
with elevated K$^+$
[c] Values as mean ± standard error of 3 determinations

Adapted from ref. 15.

Table II   Inability of RH-3421 and RH-5529 to affect free
[Ca^{++}] in resting (non-depolarized) synaptosomes

Treatment	Synaptosomal free [Ca^{++}] (nM) [a]
Control	268.3 ± 18.0
RH-3421 (1 µM)	273.4 ± 25.7
RH-3421 (10 µM)	277.9 ± 9.1
Control	260.0 ± 29.2
RH-5529 (5 µM)	270.8 ± 28.2
RH-5529 (10 µM)	292.7 ± 15.6

[a]   Values as mean ± standard deviation of 4-8 experiments

Data on RH-5529 reproduced with permission from ref. 9.
Copyright 1993 Academic Press Inc. Data on RH-3421 adapted
from Zhang, A. and Nicholson, R. A. *Comp. Biochem. Physiol.*,
in press.

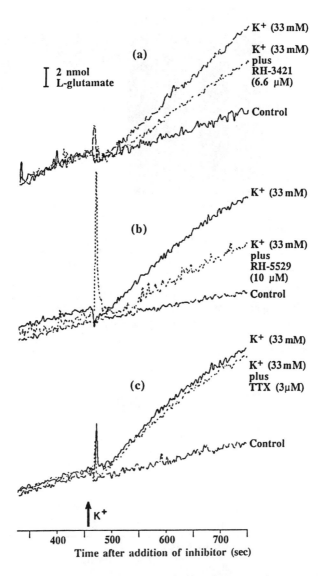

Figure 3.   Inhibition of K$^+$-evoked release of endogenous L-glutamate from synaptosomes by a) RH-3421 and b) RH-5529. K$^+$ (33 mM) added at arrow except in control assay. IC$_{50}$s for inhibition of K$^+$-evoked release of glutamate by RH-3421 and RH-5529 were estimated at 8.3 μM and 8.8 μM respectively. Figure 3c shows that tetrodotoxin is unable to influence K$^+$-induced release. Synaptosomes were exposed to neurotoxicants prior to challenge with K$^+$.

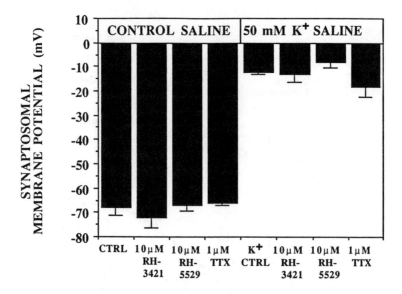

Figure 4.   Inability of RH-3421, RH-5529 and tetrodotoxin to influence the membrane potential of resting synaptosomes and the depolarization caused by elevated K$^+$. Values as mean $\pm$ standard error of three to five determinations.
(Adapted from ref. 17).

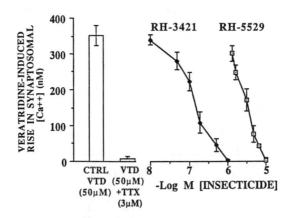

Figure 5.   Inhibition of the veratridine-stimulated increase in synaptosomal free [Ca^{++}] by RH-3421, RH-5529 and tetrodotoxin. Values as mean $\pm$ standard deviation of three to six determinations. Inhibitors added prior to veratridine
(50 μM).
(Histogram adapted from ref. 9.; concentration-response curve for RH-5529 reproduced with permission from ref. 9. Copyright 1993 Academic Press, Inc. Data on RH-3421 are from Zhang, A. and Nicholson, R. A. *Comp. Biochem. Physiol.*, in press).

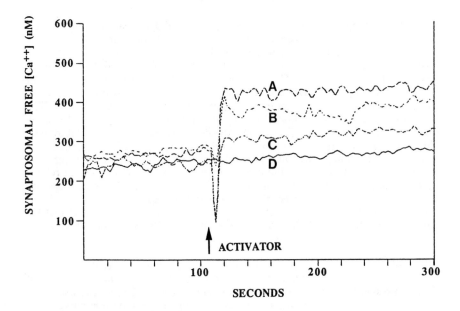

Figure 6.   Inhibitory effect of RH-3421 on the K^+-induced rise in synaptosomal free $[Ca^{++}]$. (A = 60 mM K^+; B = 60 mM K^+ plus 1 μM RH-3421; C = 60 mM K^+ plus 5 μM RH-3421; D = control).

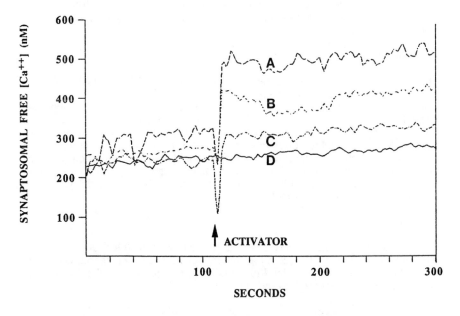

Figure 7.   Inhibitory effect of RH-5529 on the K^+-induced rise in synaptosomal free $[Ca^{++}]$. (A = 60 mM K^+; B = 60 mM K^+ plus 3.3 μM RH-5529; C = 60 mM K^+ plus 5 μM RH-5529; D = control).

significant inhibitory effects with RH-3421 and RH-5529, and the graph in Figure 8 defines the relationship between dihydropyrazole concentration and inhibition of the rise in [Ca^{++}]. The inhibitory potency of RH-3421 in the fura-2 assay with K^+ as the activator is five fold lower (IC_{50} = 1 μM) than when VTD is used. In contrast the inhibitory potencies of RH-5529 are very similar (IC_{50} ~ 3 μM) whether K^+ or veratridine are used. Furthermore, there appears to be no involvement of sodium channels in the rise in free calcium produced by elevated K^+ because 3 μM TTX, which totally supresses the veratridine response (Figure 5), fails to affect the rise in free [Ca^{++}] produced by K^+ (Figure 8). An additional feature of the inhibition by dihydropyrazoles of the rise in synaptosomal free [Ca^{++}] with K^+ is that it is also unaffected when a high concentration of TTX is present in the assay. Under these conditions, the inhibition by dihydropyrazoles of either sodium channel-related depolarization or low level Ca^{++} flux through sodium channels (*19*) can be regarded as extremely unlikely. The confirmation that the response to elevated K^+ is dependent on the presence of extrasynaptosomal calcium (*9*) suggests that the inhibitory effects of RH-3421 and RH-5529 can be explained by interference with the operation of calcium channels.

Effects of dihydropyrazoles on K^+-stimulated radiocalcium uptake. We carried out further studies on the effects of RH-3421 and RH-5529 on K^+-induced uptake of $^{45}Ca^{++}$ into synaptosomes. Threshold inhibition with each analogue was detected at low micromolar concentrations, and for each analogue, the inhibition curves were quantitatively similar (IC_{50}s: RH-3421 & RH-5529 ~ 11 μM. See Figure 9). Under our conditions of assay, ionic cobalt a known inhibitor of calcium channels, fully blocked K^+-evoked $^{45}Ca^{++}$ uptake. In agreement with previous findings using the fura-2 assay, suppression of sodium channel activity with TTX did not influence RH-3421 or RH-5529 in their ability to inhibit in the K^+-response. Concentrations of dihydropyrazoles which cause marked inhibition of depolarization-coupled $^{45}Ca^{++}$ uptake were unable to affect $^{45}Ca^{++}$ accumulation in resting synaptosomes (Table III).

Conclusions

The present results indicate that the dihydropyrazoles RH-5529 and RH-3421 are capable of interfering with the operation of calcium channels in presynaptic terminals isolated from mammalian brain. In situations where 'same assay' comparisons can be made, it is evident that RH-3421 affects sodium channel-dependent processes at lower concentrations than those required to affect calcium channels. Our results suggest RH-5529 is less discriminatory in this respect. Investigations with radioligands specific for the calcium channel should assist in finding out whether binding sites for dihydropyrazoles are present on this complex in the nerve terminal. Disruption of calcium channel function may contribute to the unusual 'delayed onset' neurotoxicity observed with RH-3421 in mammals.

Acknowledgements

Our studies described in this article were supported by grants (OGP 0042113, EQP 0092242 and EQP 0123003) from the Natural Sciences and Engineering Research Council of Canada. We thank the Rohm and Haas Company, Philadelphia, PA for providing the dihydropyrazoles used in this study.

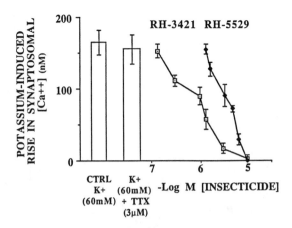

Figure 8. Concentration-dependent inhibition by RH-3421 and RH-5529 of the K^+-stimulated rise in synaptosomal free $[Ca^{++}]$ and failure of tetrodotoxin to influence this response. In presence of 3 μM tetrodotoxin RH-3421 and RH-5529 gave 101.5 ± 6.1 % and 112.3 ± 11.5 % inhibition of the rise in free $[Ca^{++}]$ elicited by elevated K^+. Synaptosomes were exposed to neurotoxicants prior to depolarizing challenge. Values as mean ± standard error of four to eight experiments.

(Adapted from ref. 9. Data on RH-3421 are from Zhang, A. and Nicholson, R. A. *Comp. Biochem. Physiol.*, in press).

Table III   Lack of effect of RH-3421 and RH-5529 on $^{45}Ca^{++}$ uptake in resting (non-depolarized) synaptosomes

Treatment	$^{45}Ca^{++}$ uptake into synaptosomes [a,b] (cpm / mg protein)
Control	$2,266 \pm 140$
RH-3421 (100 μM)	$2,472 \pm 275$
RH-5529 (100 μM)	$2,452 \pm 255$

[a] Values as mean ± standard error of three determinations
[b] K^+-stimulated uptake of $^{45}Ca^{++}$ by synaptosomes was $2,251 \pm 253$ cpm / mg protein above control value

Data on RH-3421 adapted from Zhang, A. and Nicholson, R. A. *Comp. Biochem. Physiol.*, in press.

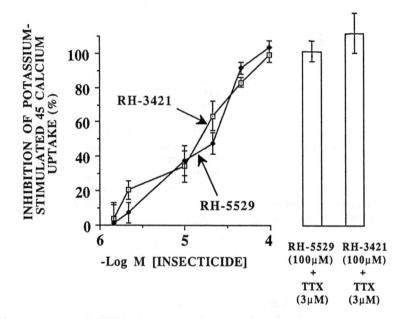

Figure 9.   Inhibition by RH-3421 and RH-5529 of K$^+$-stimulated uptake of
^{45} Ca^{++} into synaptosomes. The histogram shows that the dihydropyrazoles inhibitory action on K+-stimulated ^{45} Ca^{++} accumulation occurs when sodium channels are blocked by tetrodotoxin. Synaptosomes were incubated with neurotoxicants prior to exposure to elevated K$^+$ plus ^{45} Ca^{++}. In separate experiments Co^{++} (4 mM) gave complete inhibition of K$^+$-stimulated ^{45} Ca^{++} uptake.

(Data on RH-3421 are from Zhang, A. and Nicholson, R. A. *Comp. Biochem. Physiol.*, in press).

Literature cited

1. Mulder, R.; Wellinga, K.; van Dalen, J. J. *Naturwissenschaften* **1975**, *62*, 531-532.
2. Wellinga, K.; Grosscurt, A. C.; van Hes, R. *J. Agric. Fd. Chem.* **1977**, *25*, 987-992.
3. van Hes, R.; Wellinga, K.; Grosscurt, A. C. *J. Agric. Fd. Chem.* **1978**, *26*, 915-918.
4. Jacobson, R. M. In *Advances in the Chemistry of Insect Control* , St. Catherines College, Oxford, England, 1989.
5. Salgado, V. L. *Pestic. Sci.* **1990**, *28*, 389-411.
6. Salgado, V. L. *Mol. Pharmacol.* **1992**, *41*, 120-126.
7. Deecher, D.C.; Soderlund, D. M. *Pestic. Biochem. Physiol.* **1991**, *39*, 130-137.
8. Deecher, D.C.; Payne, G. T.; Soderlund, D. M. *Pestic. Biochem. Physiol.* **1991**, *41*, 265-273.
9. Zhang, A.; Nicholson, R. A. *Pestic. Biochem. Physiol.* **1993**, *45*, 242-247.
10. Dodd, P. R.; Hardy, J. A.; Oakley, A. E.; Edwardson, J. A.; Perry, E. K.; Delaunoy, J. P. *Brain Res.* **1981**, *226*, 107-118.
11. De Belleroche, J. S.; Bradford, H. F. In *Progress in Neurobiology*, Kerkut, G. A. and Phillis, J. W., Eds.; Pergamon: Oxford 1975, Vol. 1; 275-298.
12. Marchbanks, R. M.; Campbell, C. W. B. *J. Neurochem.* **1976**, *26*, 973-980.
13. Dunkley, P. R.; Jarvie, P. E.; Heath, J.W.; Kidd, G. J.; Rostas, J. A. P. *Brain Res.* **1986**, *372*, 115-129.
14. Hajos, F. *Brain Res.* **1975**, *93*, 485-489.
15. Nicholson, R. A.; Merletti, E. L. *Pestic. Biochem. Physiol.* **1990**, *37*, 30-40.
16. Nicholls, D. G.; Sihra, T. S.; Sanchez-Prieto, J. *J. Neurochem.* **1987**, *49*, 50-57.
17. Nicholson, R. A. *Pestic. Biochem. Physiol.* **1992**, *42*, 197-202.
18. Mendelson, W. B.; Scolnick, P.; Martin, J. V.; Luu, M. D.; Wagner, R.; Paul, S.M. Eur. *J. Pharmacol.* **1984**, *104*, 181-183.
19. Hille, B.; *Ionic Channels of Excitable Membranes*; Sinauer Associates Sunderland, Mass., 1992, 350-351.

RECEIVED January 14, 1995

Chapter 4

Insect Sodium Channel as the Target for Insect-Selective Neurotoxins from Scorpion Venom

E. Zlotkin[1], H. Moskowitz[1], R. Herrmann[1], M. Pelhate[2], and D. Gordon[1]

[1]Department of Cell and Animal Biology, Life Sciences Institute, Hebrew University, Jerusalem 91904, Israel
[2]Départment de Neurophysiologie, Faculté de Médecine, Université d'Angers, Angers F–49045, France

The combined employment of protein chemistry, electrophysiology and neurochemistry enabled the chemical and pharmacological characterization of two classes of neurotoxin polypeptides, the excitatory and the depressant, derived from the venom of Buthinae scorpions which selectively paralyze and kill insects. These insect selective neurotoxins:

1. Affect insect neuronal sodium conductance;
2. Serve as unique and exclusive probes of the insect voltage gated sodium channels;
3. Bind to these channels through multipoint attachment sites which include segments of external loops in domains I, III and IV of the insect sodium channel;
4. Distinguish among sodium channels of different groups of insects;
5. Are employed as pharmacological tools for the study of insect excitability and the design of future selective insecticides.

Insect Selective Neurotoxins Derived from Scorpion Venoms

Venom Neurotoxins. Venom is defined as a mixture of substances which are produced in specialized glandular tissues in the body of the venomous animal and injected, with the aid of a stinging-piercing apparatus, into the body of its prey in order to paralyze it. The majority of the venomous animals (such as snakes, spiders, scorpions, venomous snails, various coelenterates) are slow, and even static predators which feed on freshly killed prey of mobile and relatively vigorous animals. The locomotory inferiority of the venomous predator is largely compensated by the neurotoxic components of his venom, the neurotoxins, which are able to induce a rapid paralysis of the prey at a very low range of concentrations (10^{-9}-10^{-12} M).

NOTE: Abbreviations of scorpion toxin nomenclature are explained in the Figure 2 caption.

In this context a neurotoxin is defined as a substance which interferes with the function of excitable tissues due to a specific 'recognition' through high affinity binding to given sites in these tissues (*1*). A very common characteristic of venom neurotoxins is their polypeptide nature. This chemical characteristic has a double advantage for the survival of the venomous animal. First, polypeptides are the most readily available structures for adaptive modifications by genetic mutations and selection. Second, polypeptides, through their high diversity of covalent structures and the resulting tridimensional conformations, may reveal a highly diverse array of functional specificities.

When classified according to effect and site of action in the nervous system, venom neurotoxins are commonly divided into ion channel toxins which modify ion conductance, presynaptic toxins which affect neurotransmitter release, and postsynaptic toxins which block the neurotransmitter receptors (*2*). Each of these categories can be further classified according to more specific criteria. For example, the ion channel toxins are subdivide into sodium (such as the below-mentioned scorpion venom toxins), potassium and calcium channel toxins (*2,11*). Furthermore, some of the neurotoxins are able to distinguish between subtypes of ion channels such as the ω and μ venomous snail conotoxins which distinguish between the voltage-sensitive axonal and muscle calcium and sodium channels, respectively (*3*).

Where the vital prey-predator relationships serve as a target of preference for evolutionary selective pressure, additional specificities occur such as the animal group-selective toxins.

Animal Group Specificity of Scorpion Venom Neurotoxins. Animal group specificity refers to the phenomenon where an animal venom and/or its derived toxins reveal toxicity to a given group of organisms but are not effective to other groups of organisms. In cases where a venomous organism feeds on a given and limited group of animals, the animal group specificity is manifested already on the level of the whole venom. For example, the venomous marine Conidae snails are subdivided according to their feeding habits into fish, mollusc and worm eaters. A series of early bioassays (*4,5*) showed that the above feeding preference is associated with the specific toxicity of their venoms aimed at the respective groups of prey animals. On the other hand in the venom of animals, such as scorpions, spiders, marine nemertine worms and sea anemones, the animal group specificity is revealed in venom fractions and purified toxins (*1,6-8*).

In the past, when the study of Buthinae scorpion venoms was directed mainly by clinical and public health aspects, lethality to laboratory mice served as the exclusive method to monitor their fractionation and purification (*9,10*). This has resulted in the purification and chemical characterization of the so called α neurotoxins (Figure 2) responsible for strong toxicity to mammals and human envenomation (*9-11*). Zooecological considerations concerning the feeding and hunting behavior of scorpions in the field has motivated the choice of arthropods as test animals in monitoring the fractionation of scorpion venom. Such an approach coupled with molecular exclusion column chromatography and starch gel electrophoresis enabled the isolation of factors from buthid scorpion venoms which specifically affect insects, vertebrates and crustacea (12-14). The specific symptomatology revealed by blowfly larvae to the injection of various fractions of scorpion venom enabled the distinction among three categories of neurotoxins which affect insects: The first are the excitatory insect selective neurotoxins which induce in blowfly larvae an immediate fast and reversible paralysis (Figure 1A). The second are the depressant insect selective neurotoxins which cause a slow progressively developing flaccid paralysis (Figure 1B). The third group are the α toxins which include the extremely toxic neurotoxins to mammals (such as the AaH2, Figure 2), some of which possess a certain toxicity to insects (17, 18) and others (such as the LqhαIT, Figure 2) are strongly, but however not exclusively, toxic to insects (16). The LqhαIT induces in the blowfly larvae a syndrome defined as a delayed and sustained contraction paralysis (Figure 1C). It is noteworthy that the selective-exclusive neurotoxicity to insects of the excitatory and depressant toxins was established on three experimental levels: the intact animal, the neuromuscular preparation and binding assays to various neuronal and non-neuronal membranes (13).

Chemistry of Scorpion Toxins: Scorpion toxins were isolated and purified by the aid of various methods of low and high pressure column chromatography and characterized by electrophoresis, amino acid composition and sequence analyses (16,19). Figure 2 demonstrates the primary structures of representatives of the three groups of toxins derived from the venom of Buthinae scorpions and, for comparative purposes, one toxin (Ts7) which represents the so-called β-toxins derived from scorpions from Central and South America.

As shown (Figure 2), scorpion toxins are single chained polypeptides of about 65-70 amino acids, including 8 cysteins of similar allocation and are all involved in disulphide bridges (16,19) (Figure 2).

Effects on sodium conductance

Neuromuscular Effects. The above (Figure 1) and other (18) symptomatological observations suggested that scorpion toxins are neurotoxic and paralyze the insects by affecting their skeletal musculature. Studies on various neuromuscular preparations (17,20,27-29) revealed that the insects skeletal muscles are indirectly affected through a presynaptic action (17,28,29) (Figure 3A) on the peripheral branches of the motor neurons (17,30). This aspect is clearly exemplified in Figure 3A revealing a synchrony between the axonal (action potentials) and muscular (junction potentials) activities induced by the AaIT. In other words, the train of the junction potentials is caused by the repetitive firing of the motor axon (17) induced by the excitatory toxin.

It has recently been shown (20) that the depressant insect selective neurotoxins induce in blowfly larvae a short transient phase of contraction preceding the prolonged flaccid paralysis. A study on a prepupal housefly neuromuscular

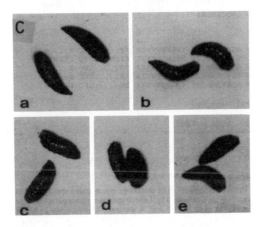

Figure 1. Responses of *Sarcophaga falculata* blowfly larva to the injection of various scorpion venoms and their derived insect toxins. **A**. A fast, immediate (seconds) and reversible contraction paralysis induced by excitatory insect selective neurotoxins such as AaIT and LqqIT1. **B**. Typical progressively developing (minutes) flaccid extended paralysis induced by depressant insect selective toxins such as BjIT2, LqqIT2 and LqhIT2. Bar corresponds to 3.8 mm. Taken from (*15*). **C**. Response at various time intervals to injection of a paralytic dose (PU_{50}=28ng) of a fraction including the LqhαIT toxin. (a) Before injection; (b) 1 min after injection - mobile and without detectable change in body forms; (c) 5 min - obvious contraction and immobility; (d) 8 min - fully contracted and paralyzed; (e) 20 min - still contracted but with partial recovery of mobility. Taken from (*16*).

AMINO ACID SEQUENCES

```
         1         2         3         4         5         6         7
         0         0         0         0         0         0         0
LqhIT2 ...DGYIKRR DGCKVACLIG NEG.CDKECK AYGG.SYGYC ...WTWGLAC WCEGLPDDET WK..SETNTC G
LqqIT2 ...DGYIRKR DGCKLSCLFG NEG.CNKECK SYGG.SYGYC ...WTWGLAC WCEGLPDEKT WK..SETNTC G
BjIT2  ...DGYIRKK DGCKVSCIIG NEG CRKECV AHGG.SFGYC ...WTWGLAC WCENLPDAVT WK..SSTNTC G
AaIT   .KKNGYAVDS SGKAPECLLS N..YCNNQCT KVHYADKGYC CLL.....SC YCFGLNDDKK VLEISDTRKS YCUTTIIN
LqqIT1 .KKNGYAVDS SGKAPECLLS N..YCYNECT KVHYADKGYC CLL.....SC YCNGLSDDKK VLEISDARKK YCDFVTIN
LqhaIT .VRDAYIAKN YNCVYEC.FR DA.YCNELCT KNGASS.GYC QWAGKYGNAC WCYALPDNVP IR...VPGKC R
Lqq4   GVRDAYIADD KNCVYTC.GS NS.YCNTECT KNGAES.GYC QWLGKYGNAC WCIKLPDKVP IR...IPGKC R
Lqq5   .LKDGYIVDD KNCTFFC.GR NA.YCNDECK KKGGES.GYC QWASPYGNAC WCYKLPDRVS IK...EKGRC N
AaH2   .VKDGYIVDD VNCTYFC.GR NA.YCNEECT KLKGES.GYC QWASPYGNAC YCYKLPDHVR TK...GPGRC H
Ts7    ..KEGYLMDH EGCKLSCFIR PSGYCGRECG .IKKGSSGYC AWP.....AC YCYGLPNMVK VWDRA.TNKC
```

PERCENT IDENTICAL RESIDUES

		LqhIT2	LqqIT2	BjIT2	AaIT	LqqIT1	LqhaIT	Lqq4	Lqq5	AaH2	Ts7
depressant	- LqhIT2	100	87	79	30	31	34	38	43	38	39
depressant	- LqqIT2		100	79	30	30	39	38	44	39	41
depressant	- BjIT2			100	25	28	38	38	43	39	41
excitatory	- AaIT				100	87	27	29	33	34	31
excitatory	- LqqIT1					100	25	29	33	34	33
α-insect	- LqhaIT						100	75	56	59	38
α-mammal	- Lqq4							100	83	83	38
α-mammal	- Lqq5								100	78	43
α-mammal	- AaH2									100	44
β-toxin	- Ts7										100

Figure 2. Comparison of scorpion toxin amino acid sequences. The depressant insect selective toxins affecting insects, namely LqhIT2, LqqIT2, BjIT2 (*20*) are compared to excitatory toxins (AaIT (*21*); LqqIT1 (*22*)). As shown the α toxin affecting insects (LqhαIT (*16*)) closely resembles the α toxins affecting mammals (Lqq4 (*23*); Lqq5 (*24*) and AaH2 (*25*)). Ts7 is a β-toxin isolated form a South American scorpion (*26*). Abbreviation of scorpion venom nomenclature: Aa - the North African scorpion *Androctonus australis Hector*; Bj - the Israeli black scorpion *Buthotus judaicus*, Lqh - the Israeli yellow scorpion *Leiurus quinquestriatus hebraeus*; Lqq - the African scorpion *Leiurus quinquestriatus quinquestriatus*; Ts - the Brazilian scorpion *Tityus serrulatus*.

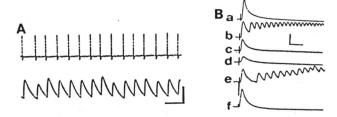

Figure 3. Neuromuscular effects of the excitatory (A) and depressant (B) insect selective neurotoxins. A. The excitatory toxin AaIT (125 nM) was applied to the locust hindleg *extensor tibiae* preparation. Simultaneous recording form the motor nerve (upper trace, showing a train of action potentials) and muscle (lower trace, showing a train of junction potentials produced by an activated muscle). Calibration: 50 msec; upper 0.15 mV, lower 10 mV. Taken from (*17*). B. The depressant insect toxin LqhIT2 (40 nM) causes a brief phase of repetitive motor neuron activity (b) followed by gradual suppression (b,c,d) of excitatory junction potentials (a) in prepupal *M. domestica*. Upon removal of toxin with a saline wash, repetitive activity reappears again only briefly (trace e) and the amplitude of the EJP partially recovers over a period of 30-60 min (trace f). Vertical calibration is 20 mV; horizontal calibration is 30 ms. Taken from (*20*).

preparation revealed that both phenomena, transient contraction and the prolonged flaccidity, follow from a presynaptic effect of the toxin. The effect of the depressant LqhIT2 toxin on a prepupal housefly neuromuscular preparation mimics the effects of the intact animal; i.e., a brief period of repetitive bursts of junction potentials is followed by suppression of their amplitude and finally by a block of neuromuscular transmission (Figure 3, *20*). Loose patch clamp recordings indicate that the repetitive activity has a presynaptic origin in the motor nerve and closely resembles the effect of the excitatory toxin AaIT. The final synaptic block is attributed to neuronal membrane depolarization, which results in an increase in spontaneous transmitter release; this effect is not induced by excitatory toxin (*20*).

To summarize, the neuromuscular studies have pointed out that the insect neuronal membrane is the target of both the excitatory and the depressant insect selective toxins. This conclusion, which was also supported by binding studies (*31,32*) and microscopical autoradiography (*30*), has directed our attention to an isolated insect axon as an experimental preparation (*37*) to study the mode of action of the scorpion toxins (*16,33,35,36*).

Studies on an Axonal Preparation. The isolated giant axon from the central nervous system of the cockroach *Periplaneta americana* in both current and voltage clamp experiments, using a double oil-gap, single fiber technique (*37*) was employed. Figure 4 presents a current clamp experiment which demonstrates the effects of the excitatory and depressant toxin on the action potentials of the cockroach giant axon. the excitatory toxin induces a small depolarization (Figure 4b) followed by either induced (Figure 4c) or spontaneous (Figure 4, upper trace) repetitive firing. The depressant toxin (BjIT2, Figure 4d-g) induces an obvious depolarization accompanied by a reduction of the action potential amplitude (Figure 4d,e) up to a complete blockage (Figure 4f). As shown (Figure 4g) the depolarization is prevented, in a reversible manner, by saxitoxin, indicating that it follows from an effect on sodium conductance. This depolarization, however, is not the cause or at least not the only cause for the action potential blockage (see below).

Figure 5 presents voltage clamp experiments which demonstrate the effects of the excitatory (AaIT) and depressant (LqhIT2) toxins on the sodium currents in the giant cockroach axon. The excitatory toxin does not effect the potassium current (Figure 5B). However, as revealed in Figure 5A, AaIT increases the sodium peak currents and slows their inactivation process (Figure 5Ab) in a voltage dependent manner (not shown, *33*). AaIT slowed the Na^+ current turnoff most effectively at low negative values of applied voltage and increased by 18% (S.D.±3, n=6) the peak sodium current (*33*). It was concluded (*33,35*) that the repetitive activity induced by AaIT results from the voltage dependent modulation of sodium inactivation coupled with an increase in sodium permeability. Figure 5C reveals that the depressant toxin blocks the sodium current in a dose and time dependent manner. Thus, the blockage of the action potential (Figure 4f) by the depressant toxin is a consequence of two effects of sodium conductance: Firstly, an increase in the resting sodium permeability,

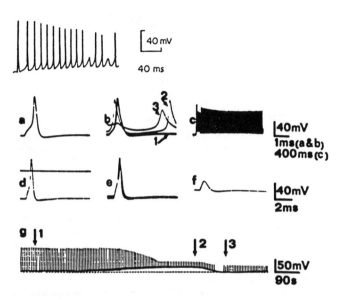

Figure 4. The effect of the excitatory (AaIT, BjIT1) and depressant (BjIT2) toxins on the action potentials recorded fro the isolated axon of the cockroach. Upper trace: Spontaneous repetitive firing induced by 1.3 μM AaIT. Taken from (*33*). (**a-c**) Action of BjIT1 (8.5 μM): (**a**) control action potentials; (**b**) after 8 min of superfusion with BjIT1 the three superimposed records revealing depolarization and repetitive activity. (**c**) Burst of repetitive activity of action potentials. (**d-g**) Action of BjIT2 (15 μM): (**d**) control action potential; (**e**) 2 min of superfusion with the toxin resulted in 5 mV depolarization accompanied by a progressive reduction in the amplitude of the action potential; (**f**) 2 min later an additional depolarization of 20 mV and a complete block of the evoked response were obtained; (**g**) continuous slow sweep recording indicating the gradual depolarization and blockage of the action potential induced by BjIT2 (applied at arrow 1). The depolarization was abolished by saxitoxin (applied at arrow 2) but has slowly reappeared after the removal of the saxitoxin by washing (applied at arrow 3).

(Reproduced with permission from reference 34. Copyright 1981 Elsevier.)

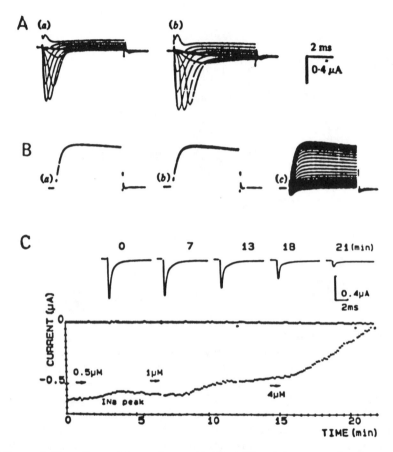

Figure 5. The effects of the excitatory (AaIT, A,B) and depressant (LqqIT2, C) insect selective toxins on sodium conductance in the cockroach axonal preparation under voltage-clamp conditions. **A**. AaIT (1.3 μM) increases the sodium peak current and slows its inactivation (b) when compared to control, before toxin (a) recording. Taken from (*33*). **B**. AaIT does not affect potassium currents. (a) before toxin; (b) 3.3 μM AaIT; (c) K current suppressed by 4-AP. Taken from (35). **C**. Depreessant toxin (0.5-4 μM) progressively suppresses sodium current as revealed by their direct recording (upper trace) and the plotting of peak current against time (bottom).

(Reproduced with permission from references 33 and 35. Copyright 1981 and 1982, respectively, Physiological Society.)

responsible for the depolarization (Figure 4e,f) and secondly a suppression of the activatable sodium permeability (Figure 5).

The effect of the α toxin affecting insects (LqhαIT) on the excitability of cockroach axon is demonstrated in Figure 6. The toxin induces the "classical" effect known for scorpion venoms and their derived toxins (*38,39*), namely the prolongation of the action potential, without affecting its amplitude (Figure 6A). As previously shown in various axonal preparations (*38,39*), also in the insect axon the above prolongation of the action potential can be entirely attributed to the slowing of the process of the turning off ("inactivation of the sodium current"). As shown (Figure 6B) the time constant of sodium inactivation increases from 2 msec in the control experiment up to 250 msec in the toxin treated preparation (Figure 6Bb).

It may thus be concluded that sodium conductance in an insect axon is affected by three different classes of scorpion venom toxins in three different manners, respectively.

The Vertebrate and Insect Voltage Gated Sodium Channels.

The voltage-sensitive sodium channels are integral membrane proteins responsible for the generation of action potentials in excitable cells. Sodium channels isolated in functional form contain a large α subunit with an apparent Mr of 260,000 (40,41). In the rat brain, α subunits are associated noncovalently with a β1 subunit (mr=36,000) and are disulfide linked to a β2 subunit with an Mr of 33,000, which can be removed upon reduction without loss of functional activity (*40,41*). The purified eel electroplax sodium channel is functionally active as a single α subunit (*40*).

Sodium channel polypeptides have been purified from various vertebrate excitable tissues and their distinct α subunit primary structures have been elucidated by cloning and sequence analysis of the cDNA (*41,42*). These sequence determinations reveal homology among the various vertebrate channels. In each case, the cDNA encode a large polypeptide of about 2,000 amino acids containing four conserved repeated domains. The amino acid sequence analysis reveals that each homologous domain contains six (*40*) or eight (*43*) transmembrane segments per domain, connected by internally as well as externally located segments of amino acid sequences (Figure 8). A high degree of conservation is present in a short internal segment linking homology domains III and IV, which is suggested to play an important role in sodium channel inactivation (*41,42*). Site-directed antibodies raised against a synthetic peptide (SP19) which corresponds to the above conserved segment (anti-SP19 antibodies) were shown to identify sodium channel α subunits in a wide range of vertebrate excitable tissues (*44*). The externally located amino acid segments are suggested to participate in formation of the binding sites of polypeptide neurotoxins (*45*).

As a critical element in excitability, sodium channels serve as specific targets for various neurotoxins included in animal allomonal systems (*39*). These toxins were shown to occupy at least four different receptor sites related to the unique functionality of the sodium channel and to serve as valuable tools for its identification and functional-structural characterization (*40*). Among them the α and β scorpion

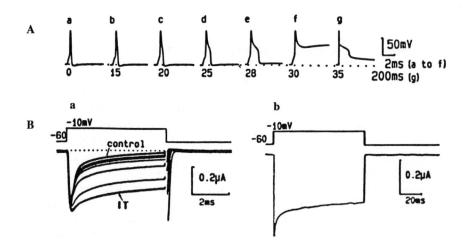

Figure 6. LqhαIT, the α toxin affecting insects, progressively prolongs the action potentials (**A**, 50 nM) and slows the sodium current inactivation (**B**, 1 μM). In A numbers under records indicate minutes following toxin application.

venom neurotoxins are well established markers of the voltage dependent sodium channels (*40,46*). The former slow sodium channel inactivation and bind to the receptor site 3 and the latter modify activation and bind to the receptor site 4 (*40,46*). It is noteworthy that the binding site of the α toxin on the rat brain sodium channel was recently localized to segments in domain I and IV by a combined use of sodium channel site-directed antibodies and photoaffinity labeling (*45*, and see below).

The close similarity among the vertebrate and insect sodium channels has been demonstrated by the following data:

1. The electrical activity of insect nerves and its ionic basis are in accordance with the stablished information of vertebrate neurophysiology (*47*).

2. The neuronal sodium conductance in insects is affected by the same blockers (such as tetrodotoxin and saxitoxin) and modifiers (such as the α scorpion and sea anemone toxins, veratridine and non-selective insecticides) as in the vertebrate systems (*48*).

3. In their gross chemical properties sodium channel polypeptides derived from the central nervous systems of insects resemble their vertebrate counterparts by (a) possessing the similar molecular weight (240-280 KDa), (b) being glycoproteins and (c) serving as a substrate of phosphorylation by cAMP-dependent protein kinase (*51,52* and Figure 7).

4. Sodium channels from locust, cockroach and fly were identified, through immuno-precipitation and radiophosphorylation (Figure 7) by certain site directed antibodies (Figure 8) which were raised against specific external as well as internal segments of the amino acid sequence of the rat sodium channel (Figure 8, *49,50*).

5. In *Drosophila* the para gene has been shown to encode a functional voltage dependent sodium channel (*53-56*) and the amino acid sequence within the four homologous domains (I-IV) of rat brain sodium channels is conserved in the deduced amino acid sequence of the insect sodium channel (39%, 66%, 58% and 62% identity, respectively (*55*)). These transmembrane segments of the conserved homologous domains were shown (by site-directed mutagenesis and expression studies) to be involved in the voltage dependent activation of sodium channels and cation transport (*57-59*).

With this background of the physiological, pharmacological, immunological and chemical similarities among the insect and vertebrate sodium channels, it appears that the only distinction among them is supplied by the fact that the former, in contrast to the latter, are affected by the excitatory and depressant insect toxins. Our next question is whether the insect sodium channel is the direct and primary target of the insect selective neurotoxins.

The Insect Sodium Channels as the Target of Insect Selective Neurotoxins

Indirect Evidence. Several lines of indirect evidence suggest that the insect selective toxins are directed to the insect neuronal sodium channels.

(1) The excitatory (AaIT) as well as the depressant (LqqIT2) insect toxins exclusively affect the sodium conductance of insect neuronal membranes, as

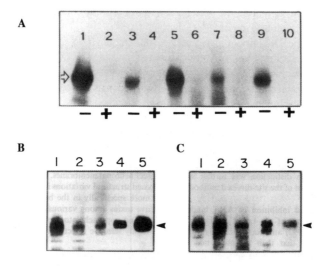

Figure 7. Identification of insect channel polypeptides by immunoprecipitation and phosphorylation using site directed antibodies prepared against specific segments of the rat brain sodium channel (Figure 8). Synaptosomal membranes from rat brain and the CNS of various insects were solubilized, immunoprecipitated, radiophosphorylated, resolved by SDS-PAGE and visualized by autoradiography. **A**. The anti-SP19 (1491-1508) antibody is able to recognize the sodium channel proteins from rat brain (1), locust (3), cockroach (5); fly heads (7) and moth larvae (9). (-) or (+) indicate the absence or presence, respectively, of 1 µM of the synthetic polypeptide to block the specific antibody, thus proving its specificity. Taken from (51). **B** and **C**: Sodium channel polypeptides of the cockroach (B) and fly head (C) were identified and immunoprecipitated by the various antibodies (Figure 8) 382-400, 1429-1449, 1686-1703, 1729-1748 and 1491-1508 (anti SP19) in lanes 1,2,3,4 and 5 respectively. Arrow points to the migration position of the sodium channel polypeptide with a Mr=260,000 Da. Taken from (*50*). (Reproduced with permission from references 51 and 50. Copyright 1990 Elsevier and 1994 Pergamon, respectively.)

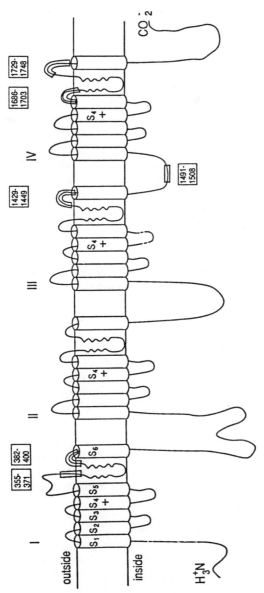

Figure 8. Two-dimensional model of the proposed transmembrane folding of a sodium channel α subunit depicting the sites of interaction of the site-directed antibodies (wide boxes). The framed numbers indicate the various site-directed antibodies used. Ab 1491-1508 corresponds to anti SP19 antibody and antibodies used. Ab 1491-1508 corresponds to anti SP19 antibody and represents part of the inactivation site. Taken from (*49*).

demonstrated by voltage clamp studies with an isolated single insect axon (Figure 4,5).

(2) It was shown that the insect neuronal membranes possess high affinity binding sites to STX (K_D=0.1-0.5 nM) (Figure 2B) and TTX (*60*), the well-known universal blockers of voltage sensitive sodium channels. The number of receptor sties for AaIT and STX was shown to be practically identical using the same insect synaptosomal membrane preparation (Figure 9, *60*).

(3) There is a similarity in binding properties between the insect selective toxin AaIT (*60*) (Figure 9) and the β scorpion toxins affecting mammals (Figure 2), the well-known markers of sodium channels (*62,63*). The binding of both groups of toxins is not dependent on membrane potential and is not affected by veratridine (in contrast to the α scorpion toxin binding (*61-63*).

(4) The excitatory insect toxin AaIT (*64,65*) as well as the depressant LqhIT2 (our unpublished results) were shown to be competitively displaced from their binding sites in insect neuronal membranes by the β scorpion toxin Ts7. The latter was shown to be toxic to mice as well as to insects and to compete with other β scorpion toxins on their mammalian sodium channel binding sites (*62-66*).

(5) Finally, it was shown that the AaIT toxin did not reveal a high affinity specific binding to rat brain synaptosomes (*61*).

Mutual displaceability Among the Excitatory and the Depressant Insect Selective Toxins. The direct evidence, namely the demonstration that the insect sodium channel is the binding receptor for the insect selective neurotoxin, was achieved by a study (*49*) which was aimed at clarifying the curious phenomenon of the mutual displaceability among the excitatory and depressant toxins (Figure 10).

The data presented in Figure 10 indicate that:

(1) An excitatory toxin (AaIT) is displaced with high affinity by both the excitatory and the depressant (BjIT2, LqIT2) toxins, with high affinity (Figure 10A,B).

(2) That the depressant toxin LqhIT2 possesses two binding sites in the locust neuronal membranes: a high affinity and low capacity site (K_{D1}=0.9±0.6 nM, B_{max1}=0.1±0.07 pmol/mg of protein) and a second low affinity and high capacity site (K_{D2}=185±13 nM, B_{max2}=10.0±0.6 pmol/mg of protein) (Figure 10C). The nature of the low affinity and high capacity LqhIT2 binding sites is presently unknown. The above binding constants, however, can not be related to the ion channels.

(3) The excitatory toxin, AaIT, displaces the radioiodinated depressant toxin only from its high affinity sites (Figure 10D,E).

We have employed sodium channel site-directed antibodies, shown to recognize the insect sodium channel (Figure 7), in order to clarify the phenomenon of the mutual displaceability among the insect selective neurotoxins. The specific question was whether the two categories of insect selective toxins, which differ in their induced symptomatology (Figure 1), effects on sodium conductance (Figures 3-5), and primary structure (Figure 2), compete for an _identical_ binding site, located on the insect sodium channel.

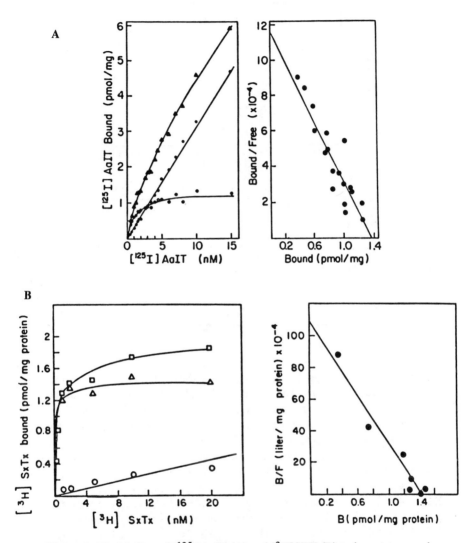

Figure 9. The binding of [^{125}I]AaIT (A) and [^{3}H]STX (B) to locust neuronal membrane preparation. Saturation equilibrium assays (left) and the respective Scatchard analysis (right) are presented, to determine the K_D and B_{max}.
A: [^{125}I]AaIT binding yielded a K_D=1.19 nM and B_{max}=1.37 pmol/mg of membrane protein. Taken from (*61*). B: [^{3}H]STX binding yielded a K_D=0.14 nM and B_{max}=1.43 pmol/mg protein.
(Reproduced with permission from reference 60. Copyright 1985 Elsevier.)

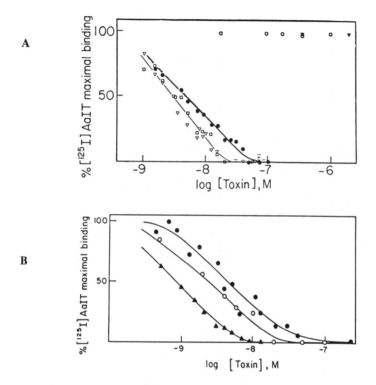

Figure 10. Mutual binding displaceability among the excitatory and depressant insect selective neurotoxins. **A,B**: Displaceability of the [^{125}I]AaIT (1.5 nM) by the excitatory and depressant insect toxins: **(A)** AaIT (□); BjIT1 (●); BjIT2 (▽); AaH2 (▼); *C. soffusus* β-mammal toxin 2 (○); **(B)** AaIT (○); LqqIT1 (▲); LqqIT2 (●). Taken from (*61,67*). **C**: Scatchard analysis of a saturation curve of the depressant toxin LqhIT2 binding to locust neuronal membranes. Scatchard plot analysis yielded the best fit (P<0.05) using a two binding site model. The K_D values, obtained for three separate experiments, are K_{D1}=0.9±0.6 nM, B_{max1}=0.1±0.07 pmol/mg of protein and K_{D2}=185±13 nM, B_{max2}=10.0±0.6 pmol/mg of protein . The total concentration of binding sties (R_T) used in the Hill plot was determined from Scatchard analysis nH=0.996. Taken from (*49*). **D,E**: Displacement of [^{125}I]LqhIT2 binding by LqhIT2 and AaIT from locust neuronal membranes. **(D)** AaIT, the excitatory toxin (○), displaces the depressant toxin [^{125}I] LqhIT2 (72 pM) only from its high affinity low capacity sites, in contrast to the LqhIT2 (●). The figure presents plots of [^{125}I]LqhIT2 bound as a function of log toxin concentration. **(E)** In presence of the unlabeled excitatory AaIT toxin (0.5 μM), the depressant [^{125}I]LqhIT2 toxin reveals only the second low affinity high capacity binding site with binding constants of K_D=194±36 nM and B_{max}=13.8±6.9 pmol/mg of protein. Taken from (*49*).

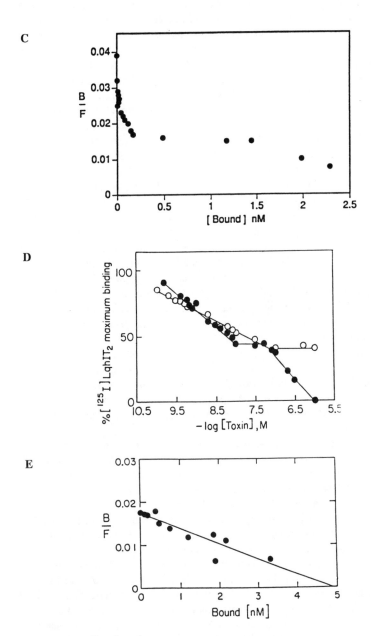

Figure 10. Continued.

Localization of Receptor Sites for Insect Selective Toxins on Sodium channels by Site- Directed Antibodies. To probe the binding sites of the depressant and excitatory toxins, we employed five site-directed antibodies that recognize different extracellular regions of the rat brain RII sodium channel α subunit corresponding to amino acid residues 355-371, 382-400, 1429-1449, 1686-1703 and 1729-1748 (42) (Figure 8). The previously described anti-SP19 antibody, corresponding tot he intracellular sequence 1491-1508 (44,68) was also employed. These sequences resemble the corresponding amino acid segments in the sodium channel polypeptide from *Drosophila para* locus (55) having 67, 79, 50, 56, 40 and 83% identity, respectively. Preincubation of the locust neuronal membrane with four out of the five externally directed antibodies (Ab) inhibited [^{125}I]LqhIT2 binding in a concentration-dependent manner. Maximum inhibition of about 50% was obtained with Ab 382-400 (55±17%), Ab 1429-1449 (56±18%), and Ab 1729-1748 (42±20%). Ab 355-371 inhibits [^{125}I]LqhIT2 binding by only 23±5% (data not shown, and Figure 12).

There was no correlation between the ability of the various site-directed antibodies to recognize the insect sodium channel polypeptide by the immunoprecipitation-radiophosphorylation method (Figure 7), and their ability to inhibit toxin binding to insect neuronal membranes. For example, Ab 382-400 significantly inhibited toxin binding (Figure 12) and recognized the sodium channel polypeptide by the immunoprecipitation-phosphorylation method (data not shown). Ab 1686-1703, on the other hand, did not reveal any significant inhibition of toxin binding to the locust neuronal membranes (Figure 12) but immunoprecipitated the sodium channel polypeptide similarly to Ab 1491-1508 (SP19) (data not shown). These results are consistent with specific inhibition of toxin binding by a subset of our antibodies that block the toxin receptor site. It is noteworthy that similar results concerning the immunoprecipitation and inhibition of LqhIT2 binding by site-directed antibodies were obtained also using the fly head neuronal preparation (69). The SP19 antibody was previously shown to specifically recognize sodium channel polypeptides in various insects (Figure 7).

The inhibitory effect of the site-directed antibodies on the binding of [^{125}I]LqhIT2 may result from a reduction in number of toxin binding sites or a decrease in the toxins' affinity. Figure 11 shows Scatchard analysis of saturation curves of [^{125}I]LqhIT2 binding to locust neuronal membranes with and without Ab 382-400. No significant change in Kd values was seen (Figure 11). The toxin was used in a low concentration range in order to occupy mainly the high-affinity sites of LqhIT2. Under these conditions, in the presence of a half-maximal concentration of Ab 382-400, about 25% reduction in the number of binding sites was obtained (Figure 11).

The competitive interaction observed between the two different insect toxins (Figure 10) could be interpreted as either competition at an identical binding site, steric interference between adjacent binding sites, or distant allosteric interactions. In the first possibility, the various site-directed antibodies would be expected to inhibit the binding of the two ^{125}I-labeled toxins to the same extent. Figure 12 presents antibody-mediated inhibition of [^{125}I]LqhIT2 and [^{125}I]AaIT binding to insect neuronal membranes. The membranes were preincubated in the presence of maximal

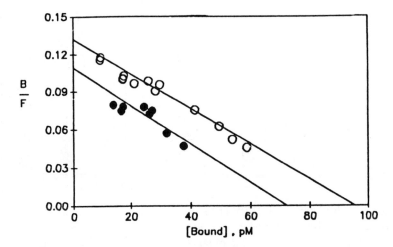

Figure 11. Scatchard plots of saturation curves of [125I]LqhIT2 binding to locust neuronal membranes in the presence or absence of site-directed antibody 382-400 indicating mainly a reduction in the toxin's binding capacity. The K_D and B_{max} values obtained in controls (○) were 0.72 nM and 0.21 pmol/mg protein respectively, and i the presence of Ab (●) 0.67 nM and 0.16 pmol/mg protein, respectively. two separate experiments yielded similar results. Taken from (*49*).

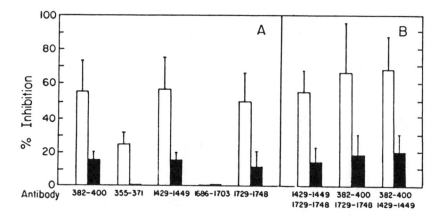

Figure 12. Inhibition of $[^{125}I]$LqhIT2 and $[^{125}I]$AaIT binding by the site-directed antibodies. Locust neuronal membranes (0.5 mg/ml) were preincubated with 20,30 and 40 μl of nonimmune antibodies (control, 100% binding) or 20 μl of the nonimmune antibodies together with 10 μl of each of the indicated site-directed antibodies (**A**). In the combined applications of two site-directed antibodies, each antibody was applied in 10 μl in the presence of 20 μl of nonimmune antibodies (**B**). Each bar represents the mean ± SD of three to five independent experiments, where triplicate determinations were made for each antibody. The concentrations of labeled toxins used were 224.7± 12.3 pM for $[^{125}I]$LqhIT2 (open bars) and 249.6±80.4 pM for $[^{125}I]$AaIT (solid bars). Taken from (*49*).

amounts of each Ab, and the binding of the two toxins was measured under similar conditions. Antibodies 382-400, 1429-1449 and 1729-1748, which cause over 50% inhibition of [^{125}I]LqhIT2 binding, were shown to inhibit [^{125}I]AaIT binding by less than 20%. Antibody 355-371, which inhibits the depressant toxin by 23%, did not reveal any significant effect on the binding of the excitatory ^{125}I-toxin, similarly to Ab 1686-1703 which did not affect the binding of either toxin (Figure 12A).

The data presented in Figure 12B demonstrates that the combined application of two Abs, each effective in toxin binding inhibition, was not additive. These results suggest that (a) each antibody was given in a saturable amount, and (b) the partial inhibition of toxin binding (Figure 12) indicates that all the antibodies used identify and bind to the same population of sodium channels. Otherwise, the combined presence of two effective antibodies would result in an additive effect (Figure 12B). It is noteworthy that the possibility of steric hindrance in the binding of the combined antibodies to a single sodium channel is not ruled out by our data. However, it would not alter our suggested interpretation of Figure 12B. The main conclusion, however, demonstrated in Figure 12 is the ability of the various site-directed Abs to quantitatively differentiate between the binding of the LqhIT2 and AaIT. This indicates that their binding sites are not identical although located on the same insect sodium channel. This conclusion is strongly supported by the below mentioned (see Figure 13) information revealing an "asymmetrical" binding interaction in reciprocal assays with the two toxins with different insect neuronal preparations.

To summarize, site-directed antibodies corresponding to conserved putative extracellular segments of sodium channels, coupled with binding studies of radiolabeled insect-selective scorpion neurotoxins, were employed to clarify the relationship between the toxin's receptor sites and the insect sodium channel. The binding of LqhIT2 was significantly inhibited in a dose-dependent manner by each of four site-directed antibodies. As exemplified with Ab 382-400 (Figure 11), the binding inhibition resulted from reduction in the number of binding sites. The antibody-mediated inhibition of [^{125}I]AaIT binding differs from that of LqhIT2: three out of the four antibodies which maximally inhibited LqhIT2 binding only partially affected AaIT binding. Two antibodies, one corresponding to extracellular and one to intracellular segments of the channel, did not affect the binding of either toxin. These data suggest that the receptors to the depressant and excitatory insect toxins (a) comprise an integral part of the insect sodium channel, (b) are formed by segments of external loops in domains I, III and IV of the sodium channel, and (c) are localized in close proximity but are not identical in spite of the competitive interaction between these toxins. It is noteworthy that the binding site of an alpha scorpion toxin affecting mammals corresponds to the extracellular loops between the transmembrane segments S5 and S6 in domains I and IV of the rat brain sodium channel (*45,76*).

The present results are in concert with the previously suggested concept of multipoint attachment of scorpion toxins. This concept was based on chemical modifications of selected amino acid residues, localized on various regions of the primary structure of the α scorpion toxins (*70*) as well as the insect toxins (*71*). Thus, it may be suggested that the different regions of the toxin essential for its attachment are complemented by several points of attachment in the receptor molecule, all simultaneously required to carry out the high-affinity binding reaction. This hypothesis, however, deserved further study.

The requirement of several segments of the sodium channel for toxin binding may reasonably explain the phenomenon of binding competition among chemically and pharmacologically distinct toxins such as the insect and β scorpion toxins. Thus, a partial overlap, even in one of the several attachment points of the various toxins to the receptor, may be sufficient to inhibit toxin binding. This notion is supported by the recent finding of a new scorpion toxin, AaHIT4 (72), which has a low homology with other scorpion toxins. AaHIT4, however, displaces both α- and β-toxins, which bind to distinct receptor sites (40,66), from rat brain sodium channels and also displaces the insect toxin AaIT from insect sodium channels.

Variability Among Insect Sodium Channels Revealed by Neurotoxins.

With this background of the mutual displaceability of the excitatory and depressant toxins in the locust neuronal membrane and their binding constants (Figure 10), neuronal preparations of several other insects were studied (50) and it has been shown:

(1) Similarly to locust neuronal membranes the membranes of cockroaches, flies and lepidopterous larvae possessed a single class of binding sites for the excitatory AaIT toxin and two classes of high and low affinities and low and high capacities respectively for the depressant LqhIT2 toxin (50).

(2) However, as shown in Figure 13, a mutual displaceability between the AaIT and LqhIT2 toxin occurred only in the cockroach preparation but not in the fly head and the *Spodoptera* larvae preparations.

The asymmetry observed in the mutual displacement of the excitatory and depressant toxins (Figure 13) can be attributed to the existence of several sodium channel subtypes in neuronal membranes of various insects. This consideration is supported by the notion that the mutual displacement between the depressant and excitatory toxins represents a steric interference between adjacent binding sites, as we suggested previously (49). These putative sodium channel subtypes may differ the spatial location or relative orientation of the binding sites to each of the two toxins. Accordingly, we presume the occurrence of at least three subtypes: the first, expressed in locust and in cockroach neuronal membranes, in which a mutual steric hindrance between the toxins occurs; the second, expressed in neuronal membranes of lepidopterous larvae, in which no displacement of the depressant toxin by the excitatory toxin takes place; the third subtype of channels, expressed in fly neuronal membranes, in which a decreased ability of the excitatory toxin to interfere with the binding of the depressant toxin is observed (Figure 13).

Our data suggest that different insect orders (or even species and developmental stages) express different and possibly variable sodium channels in their nervous system. Such differences and variability may be detected by the selective toxins as well as by the sodium channel site-directed antibodies. Elucidation of the structural

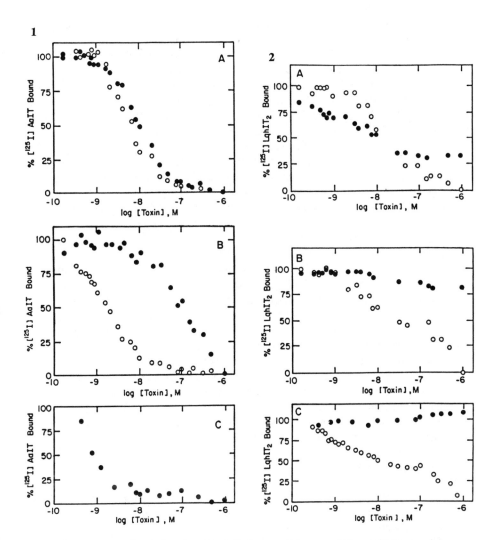

Figure 13. Reciprocal competitive displacement assays of the excitatory and depressant toxins. 1. Displacement of [^{125}I]AaIT binding by AaIT and LqhIT2. 2. Displacement of [^{125}I]LqhIT2 binding by LqhIT2 and AaIT. Neuronal membranes of *P. americana* (A) or *Sarcophaga falculata* (B) or *Spodoptera littoralis* (C) were incubated for 60 min in 22°C in the presence of [^{125}I]AaIT (1) or [^{125}I]LqhIT2 (2) and increasing concentrations of AaIT or LqhIT2. Non-specific binding, determined in the presence of 1 μM unlabeled toxin, was subtracted. In (1) ○ AaIT, ● LqhIT2; in (2) ○ LqhIT2, ● AaIT.
(Reproduced with permission from reference 50. Copyright 1994, Pergamon.)

basis of the unique pharmacology of insect sodium channels is an important area of future study.

The ability of the insect selective toxins to distinguish among sodium channels of various insects (*50*) is similar to the fact that α toxins affecting insects, such as the LqhαIT, are able to perform an interesting distinction between the vertebrate and the insect sodium channels (*74*). The α-toxins affecting mammals (*75*) constitute a family of structurally and functionally related polypeptide neurotoxins (Figure 2). These toxins (Figure 2) bind to receptor site 3 on the vertebrate sodium channel and interact synergically with the neurotoxin receptor site 2, that binds lipid soluble alkaloids (sodium channel activators) such as veratridine and batrachotoxin. The binding affinity of these site 3 toxins is reduced by depolarization in a degree similar to the voltage dependence of sodium channel activation (*40,77*).

With this background it is noteworthy that the interaction of LqhαIT with the locust neuronal membrane reveals the typical pharmacology of the α scorpion toxins by its: (a) electrophysiological effects on sodium conductance (Figure 6); (b) cooperativity with veratridine (Figure 14A); (c) displacement by the sea anemone ATX2 toxin (Figure 14B) and (d) absence of any effect on its binding by the various non-α sodium channel toxins (*74*) (Figure 14). LqhαIT, however, differs from the α scorpion toxins by possessing a voltage-independent binding site (Table I). The binding of ATX2 to the insect neuronal preparation is equally membrane potential-independent, in contrast to its binding to mammalian neuronal membranes (*74*). The latter may suggest that the receptor binding site of LqhαIT on insect sodium channels is structurally different from the homologous (or perhaps, analogous) receptor site 3 on vertebrate sodium channels. Clarification, on the molecular level, of LqhαIT receptor binding sites in the insect sodium channel may reveal the (a) unique properties of the insect sodium channel related to its inactivation; (b) structural features responsible for animal group specificity of scorpion toxins and (c) provision of a new target for future selective insecticides.

Concluding Remarks

1. The insect voltage gated sodium channels reveal obvious electrophysiological, pharmacological, immunological and chemical-structural similarities to their vertebrate counterparts.

2. The occurrence of the excitatory and depressant insect selective neurotoxin from Buthinae scorpion venoms supplies, so far, the strongest distinction between the vertebrate and insect sodium channels.

3. These insect selective toxins exclusively affect the insect sodium channel through their binding to their receptors which comprise an integral part of the insect sodium channel and are formed by segments of external loops in domains I, III and IV of the sodium channels, and thus form a multisite attachment to their receptors.

4. We propose that the competitive interaction between the depressant and excitatory insect toxins (Figure 10,13) is a consequence of a partial overlap in their points of attachment in the external segments of the insect sodium channel. Thus, the two groups of toxins may have closely localized, but not identical, binding sites. It may be further suggested that the insect selectivity of the insect

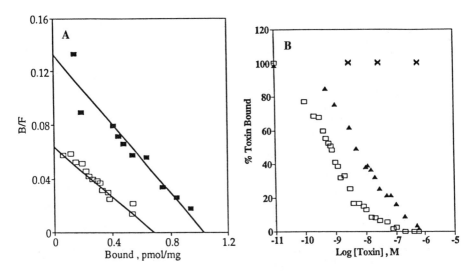

Figure 14. The pharmacology of binding of the insect toxic α scorpion toxin LqhαIT to locust neuronal membranes. **A.** Veratridine increases capacity and affinity of [^{125}I]LqhαIT to locust neuronal membranes. Scatchard analysis of specific binding yielded a single class of binding sites with the following constants: K_D=1.06±0.15 nM and B_{max}=0.7±0.19 pmol/mg protein in the absence of veratridine (□); K_D=0.88±0.17 nM and B_{max}=1.03±0.06 pmol/mg protein in the presence of veratridine (■). Taken from (74). B. Displacement of [^{125}I]LqhαIT binding by sodium channel neurotoxins. Locust neuronal membranes were incubated in the presence of [^{125}I]LqhαIT and increasing concentrations of each of the following toxins: LqhαIT (□); *Anemonia sulcata* toxin II (ATX2▲) and AaIT, LqhIT2, Ts7, and TTX toxins, which did not inhibit binding, are represented, for clarity, by a single symbol (X).

toxins may correspond to the recognition of minor modifications in the insect sodium channel when compared to its vertebrate counterpart, in one or more of the several points of attachment required for toxin binding. The difference may be expressed on the amino acid sequence and/or conformational levels. The structural basis of insect selectivity on the level of the toxins as well as their receptors deserves clarification on the molecular level (*73*).
5. The attachment sites of the insect selective toxins to the insect sodium channel should represent functionally critical sites which can be targeted by newly designed selective insecticides. The insect selective neurotoxins may serve as valuable tools for the study of the pharmacology of insect neuronal membranes. The understanding of the proposed structural variations in insect sodium channels, and more specifically in the binding sites of the insect selective toxins among various insects, may suggest new targets and approaches to the design and screening of new highly selective insecticides. The alpha toxins affecting insects, in spite of their lack of selectivity, may fulfill a similar role. These toxins reveal a difference, related to the sodium inactivation process, between the insect and the vertebrate sodium channels.

Table I. The binding of LqhαIT to insect neuronal membranes is not affected by depolarization

Conditions	LqhαIT bound (%)
Membrane vesicles (mvP2L)	100
Choline medium	111±4
Potassium medium	
Synaptosomes (P2L)	100
Choline medium	95±2
Potassium medium	100
Sodium medium	107±1
Sodium medium, 10 μg/ml gramicidin A	
Sodium medium, 10 μg/ml gramicidin A	161±6
veratridine 100 μM	

Locust synaptosomal membrane vesicles (mvP2L) loaded in 0.1 M potassium phosphate buffer or synaptosomes (P2L) were incubated with 0.18 nM [^{125}I]LqhαIT for 40 min at 22°C after 20-fold dilution in the following media: Choline medium (standard binding medium for mvP2L) or, for P2L, 120 mM choline chloride, 0.8 mM $MgCl_2$, 20 mM HEPES/Tris, pH 7.40, 0.1% BSA. In other media, the choline was replaced with potassium or sodium, respectively. Gramicidin A (10 μg/ml) alone or with veratridine (100 mM) were added to some incubation mixtures. Depolarization of the vesicles or synaptosomes is induced by potassium medium and by the sodium medium in the presence of the cation ionophor gramicidin. The results are reported as the percent of the binding measured in choline medium (100%) for each membrane preparation and represent mean ± S.E.M. of three separate experiments. Reproduced with permission from reference 74. Copyright 1993 Elsevier.)

6.   To summarize, the insect selective toxins are able to distinguish between the nervous system of an insect and a non-insect and to identify in the former a functionally critical target. As such, they can serve as research models for the design of insect selective insecticides. This approach may result in one or a combination of the following: their mimicry by synthetic non-peptide substances; the design of modified, metabolically stable neurotoxin polypeptides with oral toxicity, followed by cloning of their genes for the design of transgenic, insect-protected plants; the insertion and association of the insect-toxin genes into microorganisms (such as *Bacillus thuringiensis*) and/or viruses (such as baculoviruses) able to penetrate and inoculate within the insect body and produce the insect toxins there. The feasibility of these approaches was recently exemplified by significantly increasing the rate at which a baculovirus kills an insect pest due to the incorporation of toxin-producing genes from an insectivorous venomous mite (*78*) or scorpion (*79*).

Acknowledgments: The recent studies presented in this review were supported by grant 90-00186 from the US-Israel Binational Science Foundation, Grant IS-1982-91 from the US-Israel Binational Agricultural Research and Development Fund (BARD) and Joint French Israeli Research Grant No. 4403192.

Literature Cited

1. Zlotkin, E. *Endeavour,* 1987, *11,* 186-174.
2. *Neurotoxins in Neurochemistry;* Dolly, J. E., Ed.; Ellis Horwood Ltd.: Chichester, U.K., 1988
3. Gray, W. R.; Olivers, B. M.; Cruz, L. J. *Ann. Rev. Biochem.* 1988, *57,* 665-700.
4. Endean, R.; Rudkin, C. *Toxicon* 1963, *1,* 49-64.
5. Fainzilber, M. 1993. Ph.D. thesis, Hebrew University.
6. Frontali, N.; Grasso, A. *Arch. Biochem. Biophys.* 1964, *106,* 213-218.
7. Kem, W. R. *J. Biol. Chem.* 1976, *251,* 4184-4192.
8. Schweiz, H.; Vincent, J. P.; Barhanin, J.; Frelin, C.; Linden, G.; Hugues, M.; Lazdunski, M. *Biochem.* 1984, *20,* 5245-5252.
9. Rochat, C.; Rochat, H.; Miranda, F.; Lissitzky, S. *Biochem.* 1967, *6,* 578-588.
10. Miranda, F.; Kopeyan, C.; Rochat, C.; Rochat, H.; Lissitzky, S. *Eur. J. Biochem.* 1970, *16,* 514-523.
11. Zlotkin, E.; Miranda, F.; Rochat, H. In *Arthropod Venoms;* Bettini, S., Ed.; Springer: Berlin, Heidelberg, 1978, pp. 317-369.
12. Zlotkin, E. In *Comprehensive Insect Physiology, Biochemistry and Pharmacology;* Kerkut, G. A.; Gilbert, L. I., Eds.; Pergamon: Oxford, U.K., 1985, Vol. 10; pp. 499-546.
13. Zlotkin, E. In *Comparative Invertebrate Neurochemistry;* Lunt, G. G.; Olsen, R. W., Eds.; Croom Helm: London, U.K., 1988, pp. 256-324.
14. Zlotkin, E.; Miranda, F.; Lissitzky, S. *Toxicon* 1972, *10,* 211-216.
15. Zlotkin, E. *Insect Biochem.* 1983, *13,* 219-236.
16. Eitan, M.; Fowler, E.; Herrmann, R.; Duval, A.; Pelhate, M.; Zlotkin, E. *Biochem.* 1990, *29,* 5941-5947.
17. Walther, C.; Zlotkin, E.; Rathmayer, W. *J. Insect Physiol.* 1976, *22,* 1187-1194.
18. Zlotkin, E.; Rathmayer, W.; Lissitzky, S. In *Pesticide and Venom Neurotoxicity;* Shankland, D. L.; Hollingworth, R. M.; Smyth, T., Eds.; Plenum Press: New York, 1978, pp. 227-246.
19. Rochat, H.; Bernard, P.; Couraud, F.; *Adv. in Cytopharmac.* 1979, *3,* 325-333.
20. Zlotkin, E.; Eitan, M.; Bindokas, V. P.; Adams, M. E.; Moyer, M.; Burkhart, W.; Fowler, E. *Biochem.* 1991, *30,* 4814-4821.
21. Darbon, H.; Zlotkin, E.; Kopeyan, C.; Van-Rietschoten, J.; Rochat, H. *Int. J. Peptide Protein Res.,* 1982, *20,* 320-330.
22. Kopeyan, C.; Mansuelle, P.; Sampieri, F.; Brando, T.; Bahraoni, E. M.; Rochat, H.; Granier, C. *FEBS Lett.* 1990, *261,* 423-426.
23. Kopeyan, C.; Martinez, G.; Rochat, H. *FEBS Lett.* 1978, *89,* 54-58.
24. Kopeyan, C.; Martinez, G.; Rochat, H. *Toxicon* 1982, *20,* 71-75.
25. Rochat, H.; Rochat, C.; Sampieri, F.; Miranda, F.; Lissitzky, S. *Eur. J. Biochem.* 1972, *28,* 381-388.
26. Possani, L. D.; Martin, B. M.; Svendsen, I.; Rode, G. S.; Erickson, B. W. *Biochem. J.* 1985, *229,* 739-750.
27. Tintpulver, M.; Zerachia, T.; Zlotkin, E. *Toxicon* 1976, *14,* 370-377.

28. Ruhland, M.; Zlotkin, E.; Rathmayer, W. *Toxicon* 1977, *15*, 157-160.
29. Rathmayer, W.; Walther, C.; Zlotkin, E. *Comp. Biochem. Physiol.* 157-160.
30. Fishman, L.; Kagan, M. L.; Zlotkin, E. *J. Exp. Zool.* 1991, *257*, 10-23.
31. Teitelbaum, Z.; Lazarovici, P.; Zlotkin, E. *Insect Biochem.* 1979, *9*, 343-346.
32. Zlotkin, E.; Teitelbaum, Z.; Rochat, H.; Miranda, F. *Insect Biochem.* 1979, *9*, 347-354.
33. Pelhate, M.; Zlotkin, E. *J. Physiol. (Lond.)* 1981, *319*, 30-31.
34. Lester, D.; Lazarovici, P.; Pelhate, M.; Zlotkin, E. *Biochem. Biophys. Acta* 1981, *701*, 370-381.
35. Pelhate, M.; Zlotkin, E. *J. Exp. Biol.* 1982, *47*, 67-77.
36. Zlotkin, E.; Kadouri, D.; Gordon, D.; Pelhate, M.; Martin, M. F.; Rochat, H. *Arch. Biochem. Biophys.* 1985, *240*, 877-887.
37. Pichon, Y.; Boistel, J. *J. Exp. Biol.*, 1967, *47*, 343-355.
38. Koppenhofer, E.; Schmidt, H. *Experientia* 1968, *24*, 41-42.
39. Catterall, W. A. *Ann. Rev. Pharmacol. Toxicol.* 1980, *20*, 15-43.
40. Catterall, W. A. *Ann. Rev. Biochem.* 1986, *55*, 953-985.
41. Catterall, W. A. *Science* 1988, *242*, 50-61.
42. Gordon, D. *Curr. Opin. Cell Biol.* 1990, *2*, 695-707.
43. Guy, H. R.; Conti, F. *Trends Neurosci.* 1990, *13*, 201-206.
44. Gordon, D.; Merrick, D.; Wollner, D. A.; Catterall, W. A. *Biochemistry* 1988, *27*, 7032-7038.
45. Thomsen, W. J.; Catterall, W. A. *Proc. Natl. Acad. Sci. USA* 1989, *86*, 10161-10165.
46. Couraud, F.; Jover, E. In *Handbook of Natural Toxins;* Tu, A. T., Ed.; Marcel Dekker: New York, 1984, Vol. 2; pp. 659-678.
47. Pichon, Y. In *Insect Neurochemistry and Neurophysiology;* Borkovec, B.; Kelly T. J., Eds.; Plenum Press: New York and London, 1984, pp. 23-50.
48. Pelhate, M.; Sattelle, D. B. *J. Ins. Physiol.* 1982, *28*, 889-903.
49. Gordon, D.; Moskowitz, H.; Eitan, M.; Warner, C.; Catterall, W. A.; Zlotkin, E. *Biochem.* 1992, *31*, 7622-7628.
50. Moskowitz, H.; Herrmann, R.; Zlotkin, E.; Gordon, D. *Insect Biochem. Molec. Biol.* 1994, *24*, 13-19.
51. Gordon, D.; Moskowitz, H.; Zlotkin, H. *Biochim. Biophys. Acta*, 1990, *1026*, 80-86.
52. Moskowitz, H.; Zlotkin, E.; Gordon, D. *Neurosci. Lett.*, 1991, *124*, 148-152.
53. Ganetzky, B. *J. Neurogenet.* 1986, *3*, 19-31.
54. O'Dowd, D. K.; Germeraad, S. E.; Aldrich, R. W. *Neuron*, 1989, *2*, 1301-1311.
55. Loughney, K.; Kreber, R.; Ganetzky, B. *Cell* 1989, *58*, 1143-1154.
56. Amichot, M.; Castella, C.; Berge, J. B.; Pauron, D. *Insect Biochem. Molec. Biol.* 1993, *23*, 381-390.
57. Stuhmer, W. *Ann. Rev. Biophys. Chem.* 1991, *20*, 65-78.
58. Catterall, W. A. *Curr. Opin. Neurobiol.* 1991, *1*, 5-13.
59. Lopez-Barneo, J.; Hoshi, T.; Heinemann, S. H.; Aldrich, R. W. *Receptors and Channels* 1993, *1*, 61-71.

60. Gordon, D.; Zlotkin, E.; Catterall, W. A. *Biochim. Biophys. Acta* 1985, *821*, 130-136.
61. Gordon, D.; Jover, E.; Couraud, F.; Zlotkin, E. *Biochim. Biophys. Acta* 1984, *778*, 349-358.
62. Jover, E.; Couraud, F.; Rochat, H. *Biochem. Biophys. Res. Commun.* 1980, *95*, 1607
63. Bablito, J.; Jover, E.; Couraud, F. *J. Neurochem.* 1986, *46*, 1763
64. De Lima, M. E.; Martin, M. F.; Diniz, C. R.; Rochat, H. *Biochem. Biophys. Res. Commun.* 1986, *139*, 296-302.
65. De Lima, M. E.; Martin-Eauclaire, M. F.; Hue, B.; Loret, E.; Diniz, C. R.; Rochat, H. *Insect Biochem.* 1989, *19*, 413-422.
66. Cahalan, M. D. *J. Physiol. (Lond.)* 1975, *244*, 511-534.
67. Zlotkin, E.; Gordon, D. In *Neurochemical Techniques in Insect Research;* Breer, H.; Miller, T.A., Eds.; Springer: Berlin, 1985, pp. 243-295.
68. Vassilev, P.; Scheuer, T.; Catterall, W. A. *Science* 1988, *241*, 1658-1661.
69. Moskowitz, H.; Zlotkin, E.; Gordon, D. In *Recent Advances in Toxinology Research;* Gopalakrishnakone, P.; Tan, C.K., Eds.; National Univ. Singapore, 1992, Vol. 2; pp. 84-99.
70. Kharrat, R.; Darbon, H.; Rochat, H.; Granier, C. *Eur. J. Biochem.* 1989, *181*, 381-390.
71. Loret, E. P.; Mansuelle, P.; Rochat, H.; Granier, C. *Biochem.* 1990, *29*, 1492-1501.
72. Loret, E. P.; Martin-Eauclaire, M. F.; Mansuelle, P.; Sampieri, F.; Granier, C.; Rochat, H. *Biochem.* 1991, *30*, 633-640.
73. Zlotkin, E. *Phytoparasitica*, 1991, *19*, 177-182.
74. Gordon, D.; Zlotkin, E. *FEBS Lett.* 1993, *315*, 125-128.
75. Rochat, H.; Rochat, C.; Kupeyan, C.; Miranda, F.; Lissitzky, S. *FEBS Lett.* 1970, *10*, 349-351.
76. Tejedor, F. J.; Catterall, W. A. *Proc. Natl. Acad. Sci. USA* 1989, *85*, 8742-8746.
77. Strichartz, G.; Rando, T.; Wang, G. K. *Ann. Rev. Neurosci.* 1987, *10*, 237-267.
78. Tomalski, M. D.; Miller, L. K. *Nature (Lond.)* 1991, *352*, 82-85.
79. Stewart, L. M. D.; Hirst, M.; Ferber, M. L.; Merriweather, A. T.; Cayley, P. J.; Possee, R.D. *Nature (Lond.)*, 1991, *352*, 85-88.

RECEIVED February 8, 1995

Chapter 5

Restriction Fragment Length Polymorphism Analysis of a Sodium-Channel Gene Locus in Susceptible and Knockdown-Resistant House Flies, *Musca domestica*

C. A. Bell, M. S. Williamson, I. Denholm, and A. L. Devonshire

Institute of Arable Crops Research, Rothamsted Experimental Station, Harpenden, Hertfordshire AL5 2JQ, United Kingdom

Resistance to DDT and pyrethroid insecticides in the housefly *(Musca domestica)* often involves a common mechanism termed knockdown resistance (*kdr*). The voltage-sensitive sodium channel is generally regarded as the primary target for these insecticides, and has been implicated in nerve insensitivity to these compounds conferred by *kdr* alleles. Part of the sodium channel gene, designated *Msc*, previously cloned from the housefly, was used to identify restriction fragment length polymorphisms (RFLPs) at this locus in susceptible, *kdr* (resistant) and *super-kdr* (highly resistant) houseflies. These RFLPs showed tight linkage to resistance in controlled crosses, providing the first genetic evidence that *kdr*, and hence pyrethroid mode of action, is closely associated with the sodium channel. We report here that sodium channel RFLPs at the *Msc* locus show much diversity amongst susceptible insects, but are strongly conserved in pyrethroid resistant laboratory and field populations. This further implicates the sodium channel as the site of resistance and suggests a common origin for the *kdr* and *super-kdr* alleles.

Synthetic pyrethroid insecticides, developed from naturally-occurring pyrethrins, combine high insecticidal activity with low mammalian toxicity and lack of environmental persistence*(1)*. Their widespread use to control many agricultural and human health pests, combined with previous use of 1,1,1-trichloro-2,-2-bis(4-chlorophenyl)ethane (DDT), which is thought to share the same mode of action and mechanism of resistance, has led to the rapid development of resistance in many insect species *(2,3)*. Several lines of evidence point to the nervous system as the primary target of DDT and pyrethroid action. In insects, the symptoms of pyrethroid poisoning are indicative of an action on the nervous system: rapid loss of coordinated

0097–6156/95/0591–0086$12.00/0

movement, periods of convulsive activity and ultimate paralysis. These features were recorded in individual cockroaches using intracellular micro-electrodes *(4)*, and implicated directly a modification of transient sodium ion conductance mediated by voltage-sensitive sodium channels as the primary cause of the observed effects on nerve action potentials. Later work on whole cell membranes showed that the kinetics of sodium channel activation, as well as inactivation, are different in the presence of DDT *(5)*. Whilst other neuronal effects have been documented, effects on the sodium channel are considered to be the critical lesions mediating the action of these insecticides *(6)*.

DDT and Pyrethroid Resistance

Resistance to both the rapid paralytic ("knockdown") and lethal actions of DDT and pyrethrins was first reported in the housefly *(7)*, and shown to be controlled by a single gene (generally termed knockdown resistance or *kdr*) on autosome 3 *(8)*. Subsequent work *(9)* confirmed these findings, and identified other putative allelic variants at this locus including the more potent *super-kdr* factor *(10)*. Analogous mechanisms have been identified, though less well characterised, in several other pest species (reviewed in *11*).

The possible involvement of reduced neuronal sensitivity in *kdr* houseflies was suggested by the failure of synergists to increase the toxicity of DDT and pyrethroids in *kdr* resistant houseflies, and by the location of the *kdr* gene on autosome 3, not known to carry genes conferring metabolic mechanisms of resistance to insecticides *(9)*. Comparative studies of nerve preparations from susceptible and *kdr* type houseflies have shown that the latter survive longer periods of insecticide exposure or higher insecticide concentrations before nerve function is disrupted *(12, 13)*.

Two hypotheses have been advanced to explain the role of the sodium channel itself in the reduced sensitivity of the *kdr* phenotype. The first hypothesis proposes a reduced density of channels in the nerve membrane, as has been observed in the *nap^{ts}* (no action potential, temperature sensitive) mutation in *Drosophila melanogaster.* The presence of fewer channels was shown to be the cause of a low level of resistance to the knockdown effects of pyrethroids *(14, 15)*. This work prompted studies on the number of receptors binding to pyrethroids in *kdr* houseflies, but although one study reported a reduced sodium channel density associated with a *kdr* strain of housefly *(16)*, others showed little or no difference in density *(17-19)*.

The alternative hypothesis is that *kdr* resistance reflects alterations in the structure of the sodium channel protein, thereby reducing its affinity for DDT and pyrethroids. Until recently this hypothesis was only supported by indirect pharmacological evidence involving the effects of pyrethroids on the binding of known sodium channel ligands, demonstrating that DDT and neurotoxic pyrethroids enhance the binding of [^{3}H] batrachotoxin to rat and mouse brain sodium channels *(20, 21)*.

Cloning of a Sodium Channel Gene from the Housefly

A sodium channel gene was first isolated and sequenced from the electric eel *(22)*, and subsequently from rat brain *(23, 24)*, skeletal muscle *(25, 26)* and heart muscle *(27)*. Two putative sodium channel genes have now been identified in *Drosophila*, termed *DSC1 (28)* and *para (29, 30)*. By using homology to *para*, part of the equivalent housefly sodium channel gene encoding sequences towards the 3' (carboxy-terminal) end of the gene was cloned and designated *Msc* (Musca sodium channel) *(31)*. Restriction fragment length polymorphisms (RFLPs) at this locus were identified that differed between susceptible, *kdr* and *super-kdr* houseflies. Through a backcross strategy and discriminating dose bioassay, these sodium channel RFLPs were shown to segregate with resistance, providing clear genetic evidence that the sodium channel gene is tightly linked to knockdown resistance *(31)*. Using a similar approach, Knipple *et al (38)* have recently confirmed this linkage for the *kdr* strain *538ge*. Taken together, these results indicate that both the *kdr* and *super-kdr* traits are located within one map unit of the sodium channel gene locus. The aim of the work described in the present paper was to study the sodium channel RFLPs in a wider range of susceptible and resistant strains to assess their potential for diagnosing and monitoring *kdr* resistance in field populations.

Materials and Methods

Housefly Strains. The following strains were used and cultured using standard procedures. Cooper, a reference susceptible strain *(9)*, lacking visible markers and homozygous for the wild type allele at the *kdr* locus; 579 and 530, homozygous for the *kdr* and *super-kdr* resistance alleles respectively and derived from the multimarked strains (538ge and 3D respectively; see below) by repeated backcrossing to Cooper flies followed by re-selection for resistance *(32)*; 538ge, homozygous for the *kdr* resistance allele and the recessive markers *bwb* (brown body) and *ge* (green eye), all on autosome 3 *(32)*; 3D, homozygous for the *super-kdr* level of resistance and possessing the recessive visible markers *ac* (ali-curve; autosome 1), *ar* (aristapedia; autosome 2) and *ocra* (ocra-eye; autosome 5) *(32)*; A2, a wild-type strain collected in the Netherlands by F.J. Oppennorth and homozygous for a third type of knockdown resistance (*super-kdr*$_{A2}$); 5640sel, homozygous for *super-kdr* levels of resistance (derived from strain 3D) and the visible markers *ac*, *ar* and *ocra*, and containing other uncharacterised metabolic resistance factors which interact with *super-kdr* to confer virtual immunity to pyrethroids; 171sel, a pyrethroid-resistant strain originating from a pig farm on the outskirts of Harpenden, UK, and maintained under heavy selection with permethrin in the laboratory, homozygous for knockdown resistance but also possessing other resistance mechanisms; Royston, collected in March 1993 from a large poultry rearing unit in North East Hertfordshire, UK, and exposed to heavy selection pressure in the field with pyrethroid, organophosphate and carbamate insecticides; Holtwood, also collected in March 1993 from a small pig rearing unit in South Bedfordshire, UK, only sporadically exposed to insecticides and for which no resistance problems had been reported.

Bioassay Procedure. To assess resistance levels and attempt a diagnosis of any resistance alleles present in field populations, insects were bioassayed with DDT at two doses considered to be optimal for discriminating between susceptible and *kdr* homozygotes on the one hand, and *kdr* and *super-kdr* homozygotes on the other. These doses were 0.5 and 2.5 μg / fly, both applied in conjunction with 1 μg/fly PB (piperonyl butoxide) and 1 μg/fly FDMC (2,2-*bis*(4-chlorophenyl)-1,1,1-trifluoroethanol) to inhibit any other DDT resistance mechanisms present *(33)*. Bioassays were performed on 3 to 4 day old milk-fed female adults, with up to 200 insects tested at each dose. Flies were anaesthetized with diethyl ether, treated topically with 0.5 μl of an acetone solution containing the appropriate dose of insecticide and synergists. Mortality was scored after holding for 24 h at 20°C.

RFLP Analysis. Individual adult flies were homogenised in 300 μl extraction buffer (0.1M Tris-HCl pH 8.8, 0.1M EDTA, 1% SDS) using 1.5 ml microtubes with tight-fitting pestles (Kontes, supplied by Burkard Scientific UK) and incubated at 65°C for 30 min. 70 μl 5M potassium acetate was added, incubated on ice for 30 min and centrifuged at 12000 rpm for 20 min. The supernatant was extracted with phenol and DNA precipitated with ethanol. DNA samples (5 μg) were digested with *Eco*RI, separated on 0.8% agarose gels and transferred to nylon membrane using standard techniques *(34)*. The membranes were hybridised to the ^{32}P-labelled cDNA insert of pSCP2 *(31)* in a solution containing 50% (v/v) formamide, 10 x Denhardts, 4 x SSPE, 0.5% SDS and 200 μg/ml herring sperm DNA at 42°C for 16 h, and washed first in 2 x SSPE, 0.1% SDS at 60°C for 2 h then 0.5 x SSPE, 0.1% SDS at 60°C for 30 min. Membranes were exposed to X-ray film for 72 h.

Results and Discussion

RFLP Patterns of Laboratory Housefly Strains. RFLP patterns from standard 530 (*super-kdr*), 579 (*kdr*), and Cooper (susceptible) strains using the 'pSCP2' probe have been described previously *(31)*. The bands identified in these strains were 8.8 and 3.0 kb for Cooper, 6.0 and 3.4 kb for 579 and >20, 15.5, 6.5 and 5.7 kb for 530. These three RFLP patterns are compared with representative patterns obtained from five further resistant laboratory strains in Figure 1. All of the additional strains contained, in various combinations, at least one of the bands present in 579 and 530 flies. No new bands were detected in these strains, nor were either of the bands characteristic of the susceptible Cooper population. Based on the limited numbers of insects examined (20-40), three strains (3D, 171sel and 5640sel) appeared homozygous for a single RFLP variant yielding a consistent pair of RFLP fragments. Two others (538*ge* and A2) were, like 530, clearly polymorphic in this respect, possessing two allelic variants (each yielding a characteristic pair of RFLP fragments) present in homozygous or heterozygous condition.

Strain 538*ge* was polymorphic with two variants yielding fragment pairs

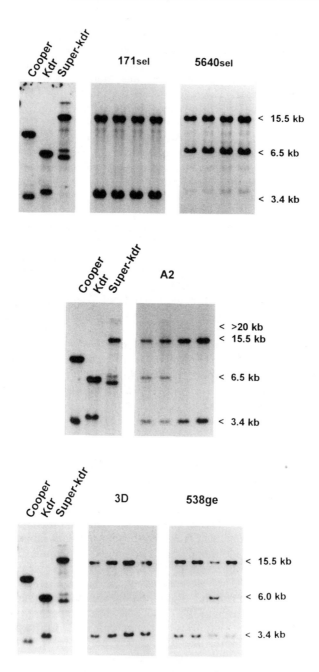

Figure 1. Examples of RFLP banding pattern in five resistant laboratory strains: 171sel, 5640sel, A2, 3D and 538*ge*. Standards strains shown on the left are, Cooper, 579 (*kdr*) and 530 (*super-kdr*).

of 15.5 with 3.4 kb, and 6.0 with 3.4 kb respectively. The 15.5 kb band was shared with 530, whereas the 6.0 and 3.4 kb bands were absent from 530 but occurred in all individuals of strain 579. A2 was polymorphic for two variants yielding fragments of >20 with 6.5 kb and 15.5 with 3.4 kb. The first of these variants was also present in 530, and the second appeared identical to the more common variant in 538*ge*. Strains 3D and 171sel were both homozygous for the 15.5 with 3.4 kb variant. 5640sel was homozygous for a variant yielding 15.5 with 6.5 kb bands, both of which were present but segregated independently in the 530 strain. In terms of individual bands, 15.5 kb was the most widely distributed and occurred in six of the resistant strains examined, the only exception being strain 579.

These results disclose a striking association between RFLP patterns, and the presence (though not necessarily the level) of knockdown resistance. Some similarities between laboratory strains were expected as a consequence of common descent from field populations in which *kdr* and *super-kdr* levels of resistance were first identified and characterised. Strain 579 was derived from 538*ge* by outcrossing to Cooper (to remove visible markers), followed by reselection with pyrethroids to recover the *kdr* resistance phenotype *(32)*, whereas strains 5640sel and 530 were derived from 3D using other outcrossing procedures (A.W. Farnham, personal communication). However, strains A2 and 171sel represented separate genetic lineages descended from field populations in Denmark and the UK respectively. The conservation of certain RFLP bands (those of 15.5 and 3.4kb in particular) in these strains also cannot be attributed purely to coincidence; it implies a more fundamental genetic relationship between resistance alleles in these populations.

Analysis of Field Populations. Once eggs had been collected from the Royston and Holtwood populations, insects from the original field collections were used for DNA analysis. RFLP patterns showed Holtwood to be highly variable at this locus; a representative sample is shown in Figure 2a. 14 different banding patterns were observed among the 16 Holtwood flies, 11 of which were putative heterozygote combinations. Many of the bands observed were shared by more than one individual, and, although it was difficult to ascertain whether any of these bands were the same as those previously observed, two individuals did seem to contain a band of 15.5 kb, similar to that found in 530 flies and other resistant laboratory strains.

Royston flies showed much greater homogeneity of RFLP patterns, (Figure 2b). Among the 16 flies tested, 6 different patterns were detected, of which three patterns accounting for 12 individuals were clearly related to each other. One pattern was an apparent heterozygote combination with bands of >20, 15.5, 6.5 and 3.4 kb, and the other two being the corresponding homozygote patterns, with the >20 and 6.5kb bands and the 15.5 and 3.4 kb bands segregating in combination. These banding patterns appeared identical to those observed in the polymorphic A2 strain (Figure 1). The other three banding patterns contained combinations of novel bands and ones already reported, but were not readily explained in terms of segregation of further allelic variants.

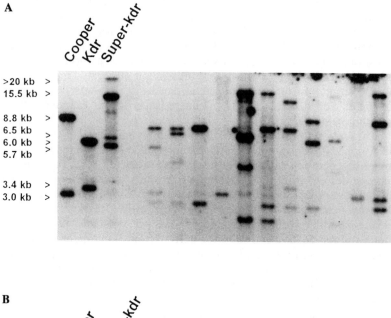

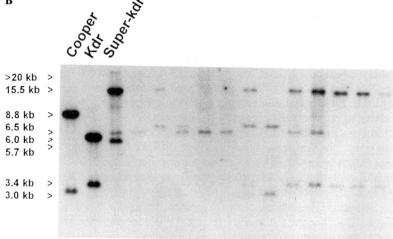

Figure 2. RFLP banding patterns of 12 representative individuals from each of the field collected housefly strains, A. Holtwood, B. Royston. Standards strains shown on the left are, Cooper, 579 (*kdr*) and 530 (*super-kdr*).

Discriminating dose bioassays with DDT + FDMC + PB disclosed marked differences in resistance between the Holtwood and Royston populations, Table I. The lower dose of 0.5 μg DDT/female (with both synergists applied a 1 μg/fly) was one calculated, from previous experience, to kill susceptible and heterozygous genotypes but to allow resistant homozygotes (whether *kdr* or *super-kdr* phenotypes) to survive treatment. As expected, all Cooper insects were killed by this dose, as were 100% of Holtwood flies. Mortality of 530 (*super-kdr*) and Royston insects was 0% and 0.4% respectively.

Table I. Mean percentage mortality of four housefly strains tested with DDT

	Dose of DDT	
Strain	0.5 μg/fly	2.5 μg/fly
Cooper	100	100
Holtwood	100	100
Royston	0.4	0
530	0	45 ± 9

The higher concentration of DDT (2.5 μg / fly) was one calculated to kill virtually all insects homozygous for *kdr* levels of resistance, but to allow at least 50% of *super-kdr* insects to survive exposure. The ca. 45% kill observed for 530 insects was in keeping with this expectation. Failure to kill any Royston flies at this dose was indicative of virtually all insects being homozygous for knockdown resistance at least equivalent to, and perhaps exceeding the *super-kdr* level of response. However, the response of Royston flies may have been modulated by the presence of the *pen* (reduced cuticular penetration) gene that alone confers very little resistance to DDT, but which may enhance the expression of any major resistance mechanism (including *kdr*) present (e.g. *35*). Any metabolic resistance should have been suppressed by the added synergists. The *pen* factor was not present in the 530 strain.

The Royston population was characteristic of others in the UK subjected to intensive treatment with pyrethroids, causing strong selection for *kdr* and leading in most cases to severe control difficulties (*36*). The flies exhibited little variation in RFLP banding pattern, and those bands present (especially 15.5 and 3.4 kb bands) resembled closely ones of resistant laboratory strains. The Holtwood population, in contrast, had rarely (if ever) been exposed to pyrethroids. No resistance homozygotes were detected in bioassays and insects exhibited much variation in RFLP banding patterns. Two of the sixteen flies tested showed a putative 15.5 kb band possibly indicative of heterozygotes possessing a single resistance allele. Since *kdr* resistance is effectively recessive in topical application tests, such insects would not have survived the discriminating dose bioassay.

The conservation of RFLP banding patterns within and between resistant strains, compared with extensive polymorphism within the susceptible Holtwood population, mirrors results of work on two overproduced esterases (A2 and B2), generally associated, and implicated in organophosphorus resistance in the mosquito, *Culex pipiens* (37). Resistance conferred by A2-B2 is now reported from four continents, and is a severe impediment to mosquito control in some countries. Mapping of the usually very highly polymorphic B esterase region for six resistant strains from Asia, Africa and North America using 13 restriction enzymes produced virtually identical banding patterns. This signified that A2-B2 overproduction had a single genetic origin, and has subsequently spread around the world through the passive transport of mosquitos on planes and boats (37). More extensive surveys of populations using further restriction enzymes and/or RFLP sites on the *Msc* gene could well implicate a similar phenomenon for knockdown resistance in houseflies. Results that are available support this hypothesis, and also imply that alleles conferring the different resistance patterns disclosed by bioassays (*kdr*, *super-kdr* and *super-kdr*$_{A2}$) arose as secondary modifications of the same original mutation. Sequencing work on susceptible and resistant strains is underway to explore these relationships in more detail.

A clear-cut association between RFLP patterns and knockdown resistance implicates *Msc* still further as the gene responsible for this mechanism in houseflies. However, the question of whether such RFLPs constitute reliable genetic markers for resistance is still uncertain at present, since none of the bands identified occurred consistently in all resistant populations. Studies on populations from a wider range of geographical localities, and on other restriction sites in the *Msc* gene (both now underway at Rothamsted) should clarify this issue further and may well identify markers even more closely conserved between resistant populations.

Acknowledgments

This work was supported by an AFRC studentship to C.A.B., and in part by the Ministry of Agriculture Fisheries and Food (MAFF). The Resistance Group at Rothamsted is a member of the European Network for Insect Genetics in Medicine and Agriculture (ENIGMA).

Literature Cited

1. Elliott, M.; Janes, N.F.; Potter, C. *Ann. Rev. Entomol.* **1978**, *23*, 443-469.
2. Sawicki, R.M. In *Progress in Pesticide Biochemistry and Toxicology*. Hutson D.H., Roberts T.R. Eds.; Wiley, New York, **1985**, vol 5, pp 143-192.
3. Georghiou, G.P. In *Managing Resistance to Agrochemicals*; Green, M.B.; LeBaron, H.M.; Moberg, W.K., Eds; American Chemical Society, Washington, **1990**, pp 18-41.
4. Narahashi, T. *J. Cell. Comp. Physiol.* **1962**, *59*, 61-65.

5. Lund, A.E.; Narahashi, T. *Neuroscience.* 1981, *6*, 2253-2258.
6. Narahashi, T. *TiPS.* 1992, *13*, 236-241.
7. Busvine, J.R. *Nature.* 1951, *168*, 193-195.
8. Milani, R. *Riv. Parasitol.* 1954, *15*, 513-542.
9. Farnham, A.W. *Pestic. Sci.* 1977, *8*, 631-636.
10. Sawicki, R.M. *Nature.* 1978, *275*, 443-444.
11. Soderlund, D.M.; Bloomquist, J.R. In *Pesticide Resistance in Arthropods.* Roush R.T., Tabashnik B.E. Eds; Chapman and Hall, New York, 1990, pp 58-96.
12. Miller, T.A.; Kennedy, J.M.; Collins, C. *Pestic. Biochem Physiol.* 1979, *12*, 224-230.
13. Gibson, A.J.; Osborne, M.P.; Ross, H.F.; Sawicki, R.M. *Pestic. Sci.* 1990, *30*, 379-396.
14. Kasbekar, D.P.; Hall, L.M. *Pestic. Biochem. Physiol.* 1988, *32*, 135-145
15. Bloomquist, J.R.; Soderlund, D.M.; Knipple, D.C. *Arch. Insect Biochem. Physiol.* 1989, *10*, 293-302.
16. Rossignol, D.P. *Pestic. Biochem. Physiol.* 1988, *32*, 146-152.
17. Grubs, R.E.; Adams, P.M.; Soderlund, D.M. *Pestic. Biochem. Physiol.* 1988, *32*, 217-223.
18. Sattelle, D.B.; Leech, C.A.; Lummis, C.R.; Harrison, B.J.; Robinson, H.P.C.; Moores, G.D.; Devonshire, A.L. In *Neurotox '88: Molecular Basis of Drug and Pesticide Action.* Lunt, G.G.; Ed.; Elsevier, Amsterdam, 1988 pp 563-582.
19. Pauron, D.; Barhanin, J.; Amichot, M.; Pralavorio, M.; Berge, J.B.; Lazdunski, M. *Biochemistry.* 1989, *28*, 1673-1677.
20. Brown, G.B.; Gaupp, J.E.; Olsen, R.W. *Mol. Pharmacol.* 1988, *34*, 54-59.
21. Payne, G.T.; Soderlund, D.M. *Pestic. Biochem. Physiol.* 1989, *33*, 276-282.
22. Noda, M.; Shimizu, S.; Tanabe, T.; Takai, T.; Kayano, T.; Ikeda, T.; Takahashi, H.; Nakayama, H.; Kanaoka, Y.; Minamino, N.; Kangawa, K.; Matsuo, H.; Raftery, M.A.; Hirose, T.; Inayamas, S.; Hayashida, H.; Miyata, T.; Numa, S. *Nature.* 1984, *312*, 121-127.
23. Noda, M.; Ikeda, T.; Kayano, T.; Suzuki, K.; Takeshima, H.; Kurasaki, M.; Takahashi, H.; Numa, S. *Nature.* 1986, *320*, 188-192.
24. Kayano, T.; Noda, M.; Flockerzi, V.; Takahashi, H.; Numa, S. *FEBS Lett.* 1988, *228*, 187-194.
25. Trimmer, J.S.; Cooperman, S.S.; Tomiko, S.A.; Zhou, J.; Crean, S.M.; Boyle, M.B.; Kallen, R.G.; Sheng, Z.; Barchi, R.L.; Sigworth, F.L.; Goodman, R.H.; Agnew, W.S.; Mandel, G. *Neuron.* 1989, *3*, 33-49.
26. Kallen, R.G.; Sheng, Z-H.; Yang, J.; Chen, L.; Rogart, R.B.; Barchi, R.I. *Neuron.* 1990, *4*, 233-242.
27. Rogart, R.B.; Cribbs, L.L.; Muglia, L.K.; Kephart, D.D.; Kaiser, M.W. *Proc. Natl. Acad. Sci. USA.* 1989, *86*, 8170-8174.
28. Salkoff, L.; Butler, A.; Wei, A.; Scavarda, N.; Giffen, K.; Ifune, C.; Goodman, R.; Mandel, G. *Science.* 1987, *237*, 744-749.
29. Loughney, K.; Kreber, R.; Ganetzky, B. *Cell 58*, 1143-1154.
30. Ramaswami, M.; Tanouye, M.A. *Proc. Natl. Acad. Sci. USA.* 1989, *86*, 2079-2082.

31. Williamson, M.S.; Denholm, I.; Bell, C.A.; Devonshire, A.L. *Mol. Gen. Genet.* **1993**, *240*, 17-22.

32. Farnham, A.W.; Murray, A.W.A.; Sawicki, R.M.; Denholm, I.; White, J.C. *Pestic. Sci.* **1987**, *19*, 209-220.

33. Farnham, A.W.; O'Dell, K.; Denholm, I.; Sawicki, R.M. *Bull. Ent. Res.* **1985**, *74*, 581-589.

34. Maniatis, T.; Fritsch, E.F.; Sambrook, J. *Molecular Cloning - A Laboratory Manual.* Cold Spring Harbour Laboratory Press, Cold Spring Harbour, New York, **1982**.

35. Sawicki, R.M. *Pestic. Sci.* **1970**, *1*, 84-87.

36. Denholm, I.; Sawicki, R.M.; Farnham, A.W. *Bull. Ent. Res.* **1985**, *75*, 143-158 .

37. Raymond, M.; Callaghan, A.; Fort, P.; Pasteur, N. *Nature.* **1991**, *350*, 151-153.

38. Knipple, D.C.; Doyle, K.E.; Marsella-Herrick, P.A.; Soderlund, D.M. *Proc. Natl. Aca. Sci. USA.* **1994**, *91*, 2483-2487.

RECEIVED October 18, 1994

Chapter 6

Actions of Insecticides on Sodium Channels

Multiple Target Sites and Site-Specific Resistance

David M. Soderlund and Douglas C. Knipple

Department of Entomology, New York State Agricultural Experiment Station, Cornell University, Geneva, NY 14456

Three classes of insecticides (pyrethroids/DDT analogues, N-alkylamides, and dihydropyrazoles) modify sodium channel function through effects by binding to three distinct target sites on the channel. The *kdr* resistance trait in the house fly, *Musca domestica*, confers reduced target site sensitivity to pyrethroids and DDT but does not diminish the efficacy of N-alkylamides or dihydropyrazoles, suggesting a domain-specific modification of the sodium channel as the mechanism of resistance. This paper reviews our recent research to test this hypothesis by cloning genomic DNA corresponding to a segment of the house fly homologue of the *para* sodium channel gene of *Drosophila melanogaster*, identifying restriction site polymorphisms within this stretch of DNA between the *kdr* and susceptible house fly strains, and employing these molecular markers in a genetic linkage analysis. The results of these studies show that the *kdr* trait is tightly linked to the *para* voltage-sensitive sodium channel gene. These findings are discussed in the context of continued exploitation of the sodium channel as an insecticide target site.

The opening of voltage-sensitive sodium channels mediates the transient sodium permeability of the cell membrane and is responsible for the rising phase of action potentials in most vertebrate and invertebrate nerves and in vertebrate skeletal and cardiac muscle (*1*). The crucial importance of sodium channels in the normal function of excitable cells is evident in the tremendous variety of naturally-occurring neurotoxins produced by plants and animals that disrupt sodium channel function, thereby contributing to the predation strategies or chemical defenses of these species. These neurotoxins have also been used as chemical probes of sodium channel function. The pharmacological profile of the sodium channel that has emerged from these studies is complex and includes evidence for at least five well-characterized neurotoxin binding sites plus circumstantial evidence for several other distinct sites (*2,3*).

The significance of the voltage-sensitive sodium channel as an insecticide target site has been widely recognized by virtue of the wealth of physiological evidence implicating sodium channels as the principal target site for DDT and pyrethroids (*4-6*). In addition, recent research has provided evidence for the action of other classes of experimental insecticides on sodium channels (*7,8*). Alterations in the pharmacological

0097–6156/95/0591–0097$12.00/0
© 1995 American Chemical Society

specificity of the sodium channel are also implicated as an important mechanism of insecticide resistance (6,9). This review summarizes evidence for the existence of multiple insecticide binding sites on the sodium channel and explores the specificity and significance of resistance mechanisms that involve reduced neuronal sensitivity to compounds acting at this target.

Multiple Insecticide Target Sites on the Sodium Channel

Neurotoxin Binding Sites on the Sodium Channel. Most of the pharmacological properties of sodium channels have been defined in the context of studies of the mode of action of naturally-occuring neurotoxins. Five principal neurotoxin recognition sites associated with sodium channels have been identified using a combination of functional and radioligand binding assays. These sites, each of which appears to be a physically distinct domain that is labelled specifically by a neurotoxin-derived radioligand, are designated Sites 1-5 in the widely-used classification scheme of Catterall (2) (Table I). The existence of additional neurotoxin recognition sites (Table I) has been inferred from studies of other classes of drugs and neurotoxins that are known to affect sodium channel function but fail to displace directly any of the radioligands that label Sites 1-5.

Insecticide Target Sites on the Sodium Channel. A wealth of physiological evidence documents the ability of DDT analogues and pyrethroids to modify sodium channel function, principally by prolonging the time course of sodium channel inactivation (4-6). Radiosodium flux and radioligand binding studies with mammalian brain preparations (10-14) have documented allosteric interactions between DDT or pyrethroids and several other classes of neurotoxins (Table I). These results imply that DDT analogues and pyrethroids bind to a domain of the sodium channel, designated Site 6 by Lombet et al. (13), that is distinct from other neurotoxin recognition sites. To date, attempts to label this site with a pyrethroid radioligand have been unsuccessful (13,15) due to the lipophilicity of potent pyrethroids, which contributes to extremely high levels of nonspecific binding. However, the search for new radioligands among the most potent pyrethroids (16) may yield more appropriate tools for such experiments.

More recently, two additional classes of experimental insecticides have been found to affect sodium channel function. Synthetic N-alkylamide insecticides (Figure 1), which are analogues of insecticidal natural products, produce excitatory effects on insect neurons in situ and in culture by prolonging sodium channel inactivation in a manner qualitatively similar to pyrethroids (7). In contrast, dihydropyrazole insecticides (Figure 1) suppress normal nerve activity by producing a voltage-dependent blockade of sodium currents (8).

Previous studies in our laboratories employed both radioligand binding and radiosodium uptake assays with mouse brain vesicle preparations to gain further insight into the binding sites for N-alkylamides and dihydropyrazoles on the sodium channel. The N-alkylamide BTG 502 acted as a partial agonist with respect to batrachotoxin in sodium uptake assays and as a competitive inhibitor of [^{3}H]batrachotoxinin A 20-α-benzoate (BTX-B) binding (17). These results identified Site 2 as the site of action of N-alkylamides (Table I). Sodium uptake experiments with the dihydropyrazole RH 3421 (18) confirmed the channel-blocking activity of this compound in mammalian brain preparations. RH 3421 was also a potent, noncompetitive inhibitor of BTX-B binding (19). These results, together with detailed biophysical studies (20), show that the action of dihydropyrazoles is qualitatively identical to that of local anesthetics, anticonvulsants, and antiarrhythmics. Dihydropyrazoles may therefore bind to the

Table I. Neurotoxin and Insecticide Binding Sites on the Voltage-Sensitive Sodium Channel

Site[a]	Active Neurotoxins[b]	Physiological effect	Allosteric coupling
1	Tetrodotoxin Saxitoxin	Inhibit ion transport	None
2	Veratridine Batrachotoxin * Aconitine N-ALKYLAMIDES	Cause persistent activation	Sites 3, 5, 6 and 9
3	α Scorpion toxins Sea anemone toxins	Prolong inactivation	Sites 2, 6 and 8
4	β Scorpion toxins	Enhance activation	Site 8
5	Brevetoxins Ciguatoxin	Enhance activation	Sites 2, 5, 6 and 8
6	* DDT AND ANALOGS * PYRETHROIDS	Prolong inactivation	Sites 2, 3, and 5
7	Gonioporatoxin	Prolong inactivation	None
8	Pumiliotoxin-B	Causes persistent activation	Sites 3, 4, and 5
9	Local anesthetics Anticonvulsants DIHYDROPYRAZOLES	Inhibit ion transport	Site 2

[a]Sites 1-5 after Catterall (2); Site 6 after Lombet *et al.* (13); Sites 7-9 numbered abritrarily to distinguish them from Sites 1-6.
[b]Insecticides shown in capital letters; compounds marked with asterisks exhibit reduced toxicity to *kdr* house flies.

same domain as these drugs (Table I); the lack of an appropriate radioligand for this site limits further efforts to clarify the specific site of action of dihydropyrazoles.

This brief summary of the pharmacology of sodium channels in the context of insecticide action highlights the diversity of neurotoxin binding domains that are known to exist on the sodium channel and summarizes evidence that three of these binding domains can serve as effective target sites for insecticides. Future efforts to exploit the sodium channel as an insecticide target site might productively focus on not only these three sites but also others that have not yet been identified as target sites for insecticide action.

Domain-Specific Resistance to Sodium Channel-Directed Insecticides

The long-term value of the sodium channel as a target site for new insecticides will be determined in part by whether the previous or current use of sodium channel-directed compounds has selected for target site-mediated resistance mechanisms in pest populations that will compromise the efficacy of future compounds acting at this target. House fly (*Musca domestica* L.) strains that are resistant to DDT and pyrethroids by virtue of reduced neuronal sensitivity represent well-characterized model systems for target site-mediated resistance to DDT and pyrethroids that may provide insight into the effect of prior selection for resistance on new sodium channel-directed agents.

kdr **in the House Fly.** Resistance to the rapid knockdown action and lethal effects of DDT and pyrethrins (*kdr*) was first documented in house flies in 1951 (*21*) and isolated genetically in 1954 (22). Subsequently, two similar resistance traits (*kdr-O* and *kdr-NPR*) were isolated genetically, mapped to chromosome 3 (*23,24*), and demonstrated by complementation analysis to be allelic to *kdr* (*25*). A single *kdr* strain (538ge, homozygous for *kdr* and the recessive visible markers *green eye* and *brown body*) constructed by Farnham in 1977 (*25*) has been widely disseminated and used in most of the studies undertaken to characterize the *kdr* trait over the past 15 years. The *kdr* trait confers resistance to both the rapid paralytic and lethal actions of all known pyrethroids, as well as the pyrethrins and DDT, but does not diminish the efficacy of other insecticide classes (*26*). Electrophysiological assays employing a variety of nerve preparations from larval and adult *kdr* insects (*6,27*) provide direct evidence for reduced neuronal sensitivity as the basis for the *kdr* trait.

super-kdr **in the House Fly.** A second resistance trait in the house fly (designated *super-kdr*) that confers much greater resistance to DDT and pyrethroids than that found in *kdr* strains has also been isolated genetically and mapped to chromosome 3 (*28,29*). Two distinct, independently-isolated *super-kdr* traits have been described, each having a unique pharmacological profile for relative resistance to a range of pyrethroids (*29*). Although *super-kdr* insects also exhibit reduced neuronal sensitivity to pyrethroids in physiological assays, the evidence that the *super-kdr* trait confers a greater level of resistance than *kdr* at the level of the nerve is ambiguous (*6*). The *kdr* and *super-kdr* traits are widely presumed to represent allelic variants at a single resistance locus on the basis of their similar spectra of resistance and their common localization to chromosome 3. However, genetic complementation studies that would demonstrate or refute allelism between *kdr* and *super-kdr* have not been reported.

Pharmacological Specificity of the *kdr* **and** *super-kdr* **Traits.** The *kdr* and *super-kdr* resistance traits exhibit a high degree of pharmacological specificity for the DDT/pyrethroid binding domain (Table I). Surveys of other classes of sodium channel-directed neurotoxins either in bioassays (*30*) or in physiological assays with nerve preparations *in vitro* (*31*) have identified aconitine as the only compound acting at a domain other than Site 6 whose potency is unambiguously reduced by the *kdr* mechanism. Resistance to aconitine appears to be compound-specific, because there is no firm evidence for resistance to other Site 2 compounds in either *kdr* or *super-kdr* insects (*6*). In particular, the *N*-alkylamide insecticides that act at Site 2 exhibit full or slightly enhanced insecticidal activity against both *kdr* and *super-kdr* insects (*32*). These resistance mechanisms in the house fly also do not confer resistance to compounds that produce voltage-dependent channel blockade such as the local anesthetic procaine (*30*) and the dihydropyrazole insecticide RH 3421 (D. M. Soderlund, unpublished data).

kdr-**Like Mechanisms in Other Species.** Resistance mechanisms similar to *kdr* in the house fly have been inferred in a number of agricultural pests and disease vectors on the basis of cross-resistance patterns and absence of synergism by esterase and oxidase inhibitors (*6,9,26*). Moreover, there is confirming electrophysiological evidence for reduced neuronal sensitivity to pyrethroids in at least five of these species: *Heliothis virescens, Spodoptera exigua, Culex quinquefasciatus, Anopheles stevensi,* and *Blattella germanica* (*6,9*). In cases where the pharmacological specificity of resistance has been examined there is evidence for specific resistance to pyrethroids and DDT analogues comparable to that found in *kdr* house flies (*6*). However, a few important differences in specificity have been noted. For example, the Ectiban-R (pyrethroid resistant) strain of *B. germanica* exhibits broader cross-resistance to Site 2 neurotoxins and low but reproducible levels of resistance to procaine and RH 3421 (*33*).

Molecular Analysis of the *kdr* Resistance Trait

The pharmacological specificity of the *kdr* trait strongly suggests that this type of resistance results from a mutation that confers reduced affinity for or responsiveness to pyrethroids and DDT analogues. To test the hypothesis that such a mutation occurs in the coding sequence of a sodium channel structural gene, we undertook the cloning and characterization of sodium channel genes from susceptible and *kdr* house flies and the identification of allele-specific molecular markers within these genes for use in genetic linkage analyses.

Isolation of House Fly Sodium Channel Gene Fragments. Two *Drosophila melanogaster* genes possessing structural homology to vertebrate sodium channels represented potential points of entry for the isolation of house fly sodium channel sequences. One of these, *DSC1*, was isolated independently by several groups by low stringency hybridization of *Electrophorus electricus* sodium channel cDNA probes (*34-37*). The other, which is encoded by the *para* locus, is isolated by molecular genetic methods (*38*). The coding regions of these two genes are quite divergent, in that their inferred amino acid sequences for the four conserved homology domains exhibit only ~50% amino acid sequence identity. Thus, in these regions the two *D. melanogaster* sequences are as similar to the corresponding regions of vertebrate sodium channel genes as to each other. Genetic and physiological criteria implicate the *para*+ gene product as the physiologically predominant sodium channel of the *D. melanogaster* nervous system (*39*). In contrast, *DSC1* maps to a cytogenetic locus that has not been identified by either genetic or physiological criteria as being involved in sodium channel function (*34*).

In view of the evidence for the *para*+ gene product as a physiologically important sodium channel in *D. melanogaster*, we developed a polymerase chain reaction (PCR)-dependent homology probing strategy that was specifically biased toward the isolation of *para*-homologous sequences from the house fly. We identified target amino acid sequences for oligonucleotide design that were completely conserved between the *para* locus of *D. melanogaster* and vertebrate brain sodium channels but were not conserved in the *DSC1* sequence *(40)*. PCR amplification on genomic DNA isolated from susceptible (NAIDM strain) flies yielded a single 128 base pair product with complete predicted amino acid sequence homology and a high degree of nucleotide sequence identity to the corresponding region of the *para* locus of *D. melanogaster* *(40)*. Further survey experiments employing DNA isolated from several other arthropod species resulted in the isolation of similar *para*-homologous fragments *(41)*, thereby confirming the general applicability of this approach.

In order to identify RFLPs that could be used as allele-specific markers of the *para* gene in the NAIDM and 538ge strains of the house fly, we isolated a 4.4 kb segment of genomic DNA of the NAIDM strain using our conspecific PCR-derived house fly *para* probe under high stringency hybridization conditions *(42)*. DNA sequence analysis of this clone identified an exon (Figure 2) that exhibited greater than 90% amino acid sequence identity to the homologous region of the *para* gene product of *D. melanogaster* *(38)* and lesser degrees of identity to the corresponding regions of rat brain sodium channel I *(43)* and the *DSC1* locus of *D. melanogaster* *(34)*. The high degree of predicted amino acid sequence identity between this house fly exon and the *para* locus of *D. melanogaster*, which extends beyond highly conserved domains into regions of low overall sequence conservation in this gene family, unambiguously identifies the house fly genomic DNA segment as the homologue in this species of *para* rather than *DSC1*.

PCR amplification of this entire exon and the adjacent upstream intronic region from susceptible (NAIDM strain) and resistant (538ge strain) genomic DNA templates and sequencing of the amplification products revealed numerous DNA sequence polymorphisms, most of which were localized in the upstream intronic region *(42)*. Three nucleotide substitutions in this region resulted in *Xba*I site polymorphisms that we used as markers for the *para*^{NAIDM} and *para*^{538ge} alleles (Figure 3). As an expedient alternative to detecting these *Xba*I site polymorphisms as RFLPs by Southern blot analysis, we developed a high-throughput genotyping procedure based on the electrophoretic analysis of the *Xba*I-digested PCR products of individual flies. As predicted by DNA sequence analysis, primer pairs A/B1 and A/B2 (see Figure 3) yielded PCR amplification products of NAIDM and 538ge DNA that showed subtle but reproducible mobility differences. Digestion with *Xba*I of the A/B2 amplification products yielded restriction fragments of approximately 85, 150 and 390 bp for NAIDM and approximately 70 and 540 bp for 538ge, which provided much more robust diagnostic gel patterns for the purposes of genetic analysis.

Linkage Analysis. The use of these PCR/RFLP markers in linkage analysis required the development of an unambiguous and highly reliable scoring system to identify individual flies among the backcross progeny that carried the resistant phenotype. Preliminary studies of the relative sensitivity of *kdr/+* and *kdr/kdr* flies to the lethal effects of DDT an deltamethrin failed to identify a dose that would discriminate adequately between these classes due to the overlap of the dose-response curves (P. M. Adams and D. M. Soderlund, unpublished data). To resolve this problem, we treated individual flies topically with 10 μg DDT, an extreme overdose for both strains *(28)*, and scored paralysis ("knockdown") after 2 h *(42)*. The only flies remaining unparalyzed 2 h after this treatment were *kdr* homozygotes. However, the parazyzed class contained not only all of the heterozygotes but also 5-15% of the treated *kdr* homozygotes and therefore was uninformative with respect to linkage.

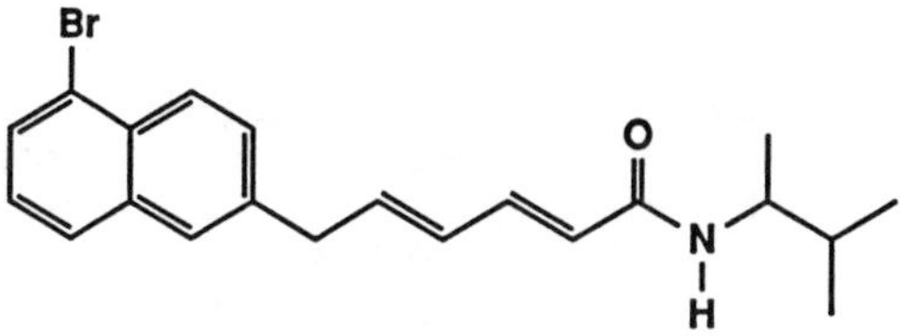

BTG 502

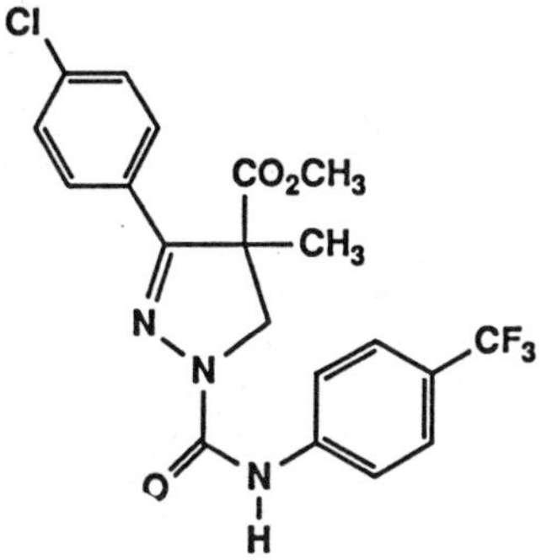

RH 3421

Figure 1.   Structures of the *N*-alkylamide insecticide BTG 502 and the dihydropyrazole insecticide RH 3421.

```
              *  *        *  ***     ** ** *  *     *   * * *    * **    *
NAIDM   QSGEDYVC LQGFGPNPNY DYTSFDSFGW AFLSAFRLMT QDFWEDLYQH    48
para    QCDDDYVC LQGFGPNPNY GYTSFDSFGW AFLSAFRLMT QDFWEDLYQL   396
ratI    QCPEGYMC VKA.GRNPNY GYTSFDTFSW AFLSLFRLMT QDFWENLYQL   390
DSC1    HCPFEYVC L.CVGENPNH GYTNFDNFMW SMLTTFQLIT LDYWENVYNM   359
                                                 ■■■■■■■■■■■■■■■■

 *     *        **     * * **** ** **   **   * *
VLQAAGPWHM LFFIVIIFLG SFYLVNLILA IVAMSYDEFQ KKAEEEEAAE EEAIR   103
VLRAAGPWHM LFFIVIIFLG SFYLVNLILA IVAMSYDELQ RKAEEEEAAE EEAIR   451
TLRAAGKTYM IFFVLVIFLG SFYLINLILA VVAMAYEE.Q NQATLEEAEQ KEAEF   445
VLATCGPMSV SFFTVVVFFG SFYLINLMLA VVALSYEEEA EITNEVSPSN LY...   411
```

Figure 2.   Inferred amino acid sequence of an exon from the *para*-homologous sodium channel gene of the house fly (NAIDM) *(42)* compared with the sequences of the corresponding regions of the *D. melanogaster para* locus *(38)*, rat brain sodium channel I (rat I) *(43)*, and the *D. melanogaster DSC1* locus *(34)*. Amino acids conserved in all four sequences are identified with asterisks. A putative pore-forming domain *(3)* is underlined with a dashed line and the conserved IS6 transmembrane domain is underlined with a solid line.

Because crossing over occurs during meiosis in female house flies, the results of the backcross between F1 females and *kdr/kdr* males reveal the genetic distance between *kdr* and the *para* gene. Of the 465 flies from this cross subjected to PCR/RFLP analysis, all 228 flies that were scored as DDT-resistant in the paralysis bioassay also displayed the *para*^{538ge} allele (*42*). Statistical analysis of the fit of the data from the informative (unparalyzed) class to the Poisson distribution showed that a map distance between *kdr* and *para* of as much as 1.15 map units could still permit the recovery of no recombinants within 95% confidence limits for a sample of this size (Table II). This result demonstrates the tight genetic linkage between *para* and *kdr* and is consistent with the hypothesis that the *kdr* mutation lies within the *para* sodium channel gene.

Concurrent with these studies, Williamson *et al.* (*44*) identified other RFLPs in the *para* locus of the house fly and employed these markers to demonstrate tight linkage of a *super-kdr* resistance trait and the *para* gene. Analysis of their data (Table II) shows that *super-kdr* is linked to *para* at a degree of genetic resolution identical to that found in our analysis of the *kdr* trait. Williamson *et al.* (*44*) also attempted to map the *kdr* trait of the 538ge strain relative to the *para* locus, but their results were ambiguous because the mortality bioassay they employed to score resistance, though effective in the linkage analysis of *super-kdr* and *para*, failed to distinguish reliably between the *kdr/kdr* and *kdr/+* genotypes obtained from the F1 female parent backcross. This result is not surprising in view of lower overall levels of resistance afforded by the *kdr* trait, which result in relatively small differences in the sensitivity of F1 and *kdr* flies to the lethal effects of DDT and pyrethroids (see above).

Taken together, the results of these two studies provide important new insight into the molecular basis of reduced nerve sensitivity to DDT and pyrethroids in the house fly. Clearly, both the *kdr* and *super-kdr* traits arise from mutations at or near the *para* locus, a voltage-sensitive sodium channel gene. These findings therefore implicate an alteration in the target macromolecule as the primary mechanism of reduced nerve sensitivity resistance. Moreover, the tight linkage of both the *kdr* and *super-kdr* traits to the *para* gene provides the first genetic evidence consistent with the widely-presumed allelism of these two traits.

Structural and Functional Analysis of the *para*^{NAIDM} and *para*^{538ge} Alleles. Further insight into the mechanism of *kdr* resistance requires a detailed structural and functional analysis of the gene products of the *para* locus of susceptible and resistant strains. Our strategy for such studies is based on the use of first-strand cDNA as the template for PCR amplification of moderately-sized segments of coding sequence, which are then employed as templates in direct automated DNA sequencing. To date, we have amplified five overlapping PCR fragments that span the entire coding region of the *para* coding sequence (Figure 4). The sequencing of the *para*^{NAIDM} and *para*^{538ge} alleles from these templates is in progress.

Because the *para* alleles of our *kdr* and susceptible strains were derived from different populations, any resistance-associated mutations are likely to be present in a background of other mutations that are neutral with respect to insecticide sensitivity. Data available so far from our sequence analysis of *para*^{NAIDM} and *para*^{538ge} cDNA fragments (D. M. Soderlund and D. C. Knipple, unpublished data) confirm the existence of multiple amino acid substitutions beween these alleles. These considerations suggest that structural analysis alone will not be able to identify resistance-causing mutations. Therefore, the demonstration of resistance at the level of the sodium channel and the identification of specific mutations that confer this resistance will require the pharmacological characterization of channels from resistant

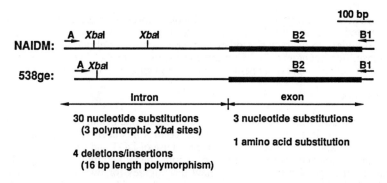

Figure 3.   Diagram of PCR products obtained from NAIDM and 538ge genomic DNA templates (*42*) showing sites of annealing for oligonucleotide primers (A, B1, B2) and summarizing structural polymorphisms identified in the exonic (see Figure 2) and adjacent upstream intronic regions of the amplification products.

Table II.   Analysis of Genetic Linkage Between   *kdr,   super-kdr* and *para.*

Comparison	*n*	Number of recombinants	Maximum distance, (map units)[a]
para/kdr [b]	228	0	1.15
para/super-kdr [c]	210	0	1.30

[a]Based on fit to Poisson distribution ($\chi^2, P \leq 0.05$).
[b]Data and analysis from Knipple *et al.* (*42*).
[c]Our analysis of data from Williamson *et al.* (*44*).

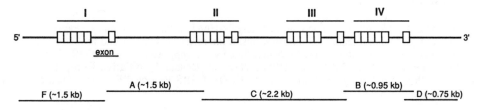

Figure 4.   Top: diagram of the structural organization of a sodium channel coding sequence showing repeated homology units I-IV and transmembrane domains (squares). Bottom: Five overlapping partial cDNAs (A-E), obtained by PCR amplification on first strand cDNA templates, that span the entire house fly *para* coding sequence. The location and relative size of the exon identified from sequencing genomic DNA (see Figures 2 and 3) is also shown.

and susceptible insects in a suitable heterologous expression assay. It remains to be demonstrated whether the *Xenopus* oocyte expression system, which has been employed widely to characterize other vertebrate and invertebrate ion channels and neurotransmitter receptors (45), will prove to be suitable for the characterization of the effects of insecticides and other neurotoxins on expressed insect sodium channels.

Conclusions

The voltage-sensitive sodium channel remains one of only a small number of insecticide target sites whose value for insect control has been validated in practice. Despite heavy exploitation due to the widespread use of DDT and synthetic pyrethroids for crop protection and vector control, the sodium channel continues to be a valuable and important target for the design of new insecticides for two reasons. First, the diversity of binding domains on the sodium channel provides a variety of structurally distinct "target sites" other than the DDT/pyrethroid site that can be attacked to achieve profound disruption of nerve function. The multiplicity of useful target domains is illustrated by the discovery of the *N*-alkylamides and dihydropyrazoles and the elucidation of their modes of action on the sodium channel. Second, target site-mediated resistance to pyrethroids (exemplfied by the *kdr* mechanism of the house fly) does not necessarily preclude the continued exploitation of other sodium channel domains that remain unaffected by mutations that confer resistance to pyrethroids.

Despite recent progress, the molecular basis of reduced neuronal sensitivity to DDT and pyrethroids remains to be determined, and the extent to which the house fly can serve as a mechanistic model for this type of resistance in other pest species is unknown. Although genetic linkage analyses place both the *kdr* and *super-kdr* traits of the house fly in close proximity to the *para* sodium channel gene, these results alone do not prove that reduced nerve sensitivity in this species arises from a mutation in the coding sequence of this gene. However, the tools to address this issue are now in hand, and we expect that the the mutations that cause the *kdr* and *super-kdr* traits and their impact on the pharmacology of pyrethroid action will be identified in the near future. These results, together with our present capabilities to isolate and characterize *para*-homologous genes from any insect species of interest, will permit us to determine whether *kdr*-like resistance in other species exhibits the same molecular basis and pharmacological specificity as in the house fly. The outcome of such studies will determine whether the prior selection for *kdr*-like resistance in pest populations will constitute a limiting factor in the continued development and use of insecticides that act at the sodium channel.

Acknowledgments

We thank P. Adams, K. Doyle, and P. Marsella-Herrick, K. Nelson, and L. Payne for their conceptual and technical contributions to the research summarized in this paper. Studies of the molecular basis of *kdr* resistance were supported by grants (89-37263-4425 and 92-37302-7792) from USDA National Research Initiative Competitive Grants Program and by funds allocated to USDA CSRS Regional Research Projects NE-115 and NE-180.

Literature Cited

(1)   Hille, B. *Ionic Channels of Excitable Membranes;* second ed.; Sinauer: Sunderland, MA, 1992, pp 607.
(2)   Catterall, W. A. *Science* 1988, *242*, 50-61.
(3)   Catterall, W. A. *Physiol. Rev.* 1992, *72*, S15-S48.

(4)   Soderlund, D. M.; Bloomquist, J. R. *Annu. Rev. Entomol.* **1989**, *34*, 77-96.
(5)   Narahashi, T. *Trends Pharmacol. Sci.* **1992**, *13*, 236-241.
(6)   Bloomquist, J. R. In *Reviews in Pesticide Toxicology*; Roe, M. and Kuhr, R. J., Ed.; Toxicology Communications: Raleigh, NC, 1993; Vol. 2; pp 181-226.
(7)   Lees, G.; Burt, P. E. *Pestic. Sci.* **1988**, *16*, 189-191.
(8)   Salgado, V. L. *Pestic. Sci.* **1990**, *28*, 389-411.
(9)   Soderlund, D. M.; Bloomquist, J. R. In *Pesticide Resistance in Arthropods*; Roush, R. T. and Tabashnik, B. E., Ed.; Chapman and Hall: New York, NY, 1990; pp 58-96.
(10)   Ghiasuddin, S. M.; Soderlund *Pestic. Biochem. Physiol.* **1985**, *24*, 200-206.
(11)   Bloomquist, J. R.; Soderlund, D. M. *Mol. Pharmacol.* **1988**, *33*, 543-550.
(12)   Brown, G. B.; E., G. J.; Olsen, R. W. *Mol. Pharmacol.* **1988**, *34*, 54-59.
(13)   Lombet, A.; Mourre, C.; Lazdunski, M. *Brain Res.* **1988**, *459*, 44-53.
(14)   Payne, G. T.; Soderlund, D. M. *Pestic. Biochem. Physiol.* **1989**, *33*, 276-282.
(15)   Soderlund, D. M.; Ghiasuddin, S. M.; Helmuth, D. W. *Life Sci.* **1983**, *33*, 261-267.
(16)   Latli, B.; Greenfield, L. J.; Casida, J. E. *J. Labelled Comp. Radiopharm.* **1993**, *33*, 613-625.
(17)   Ottea, J. A.; Payne, G. T.; Soderlund, D. M. *Mol. Pharmacol.* **1989**, *36*, 280-284.
(18)   Deecher, D. C.; Soderlund, D. M. *Pestic. Biochem. Physiol.* **1991**, *39*, 130-137.
(19)   Deecher, D. C.; Payne, G. T.; Soderlund, D. M. *Pestic. Biochem. Physiol.* **1991**, *41*, 265-273.
(20)   Salgado, V. L. *Mol. Pharmacol.* **1992**, *41*, 120-126.
(21)   Busvine, J. R. *Nature* **1951**, *168*, 193-195.
(22)   Milani, R. *Riv. Parassitol.* **1954**, *15*, 513-542.
(23)   Hoyer, R. F.; Plapp, F. W. J. *J. Econ. Entomol.* **1966**, *59*, 495-501.
(24)   Farnham, A. W. *Pestic. Sci.* **1973**, *4*, 513-520.
(25)   Farnham, A. W. *Pestic. Sci.* **1977**, *8*, 631-636.
(26)   Oppenoorth, F. J. In *Comprehensive Insect Physiology Biochemistry and Pharmacology*; Kerkut, G. A. and Gilbert, L. I., Ed.; Pergamon Press: Oxford, 1985; Vol. 12; pp 731-773.
(27)   Bloomquist, J. R. In *Neurotox '88: The Molecular Basis of Drug and Pesticide Action*; Lunt, G. G., Ed.; Elsevier: Amsterdam, 1988; pp 543-551.
(28)   Sawicki, R. M. *Nature* **1978**, *275*, 443-444.
(29)   Farnham, A. W.; Murray, A. W. A.; Sawicki, R. M.; Denholm, I.; White, J. C. *Pestic. Sci.* **1987**, *19*, 209-220.
(30)   Bloomquist, J. R.; Miller, T. A. *NeuroToxicology* **1986**, *7*, 217-224.
(31)   Salgado, V. L.; Irving, S. N.; Miller, T. A. *Pestic. Biochem. Physiol.* **1983**, *20*, 100-114.
(32)   Elliott, M.; Farnham, A. W.; Janes, N. F.; Johnson, D. M.; Khambay, B. P. S.; Sawicki, R. M. In *Combating Resistance to Xenobiotics: Biological and Chemical Approaches*; Ford, M. G., Holloman, D. W., Khambay, B. P. S. and Sawicki, R. M., Ed.; Ellis Horwood: Chichester, 1987; pp 306-313.
(33)   Dong, K.; Scott, J. G. *Pestic. Biochem. Physiol.* **1991**, *41*, 159-169.
(34)   Salkoff, L.; Butler, A.; Wei, A.; Scavarda, N.; Giffen, N.; Ifune, K.; Goodman, R.; Mandel, G. *Science* **1987**, *237*, 744-749.
(35)   Ramaswami, M.; Tanouye, M. A. *Proc. Natl. Acad. Sci. USA* **1989**, *86*, 2079-2082.
(36)   Okamoto, H.; Sakai, K.; Goto, S.; Takasu-Ishakawa, E.; Hotta, Y. *Proc. Japan Acad. Ser. B* **1989**, *63*, 284-288.

(37) Soderlund, D. M.; Bloomquist, J. R.; Wong, F.; Payne, L. L.; Knipple, D. C.
      Pestic. Sci. **1989**, *26*, 359-374.
(38) Loughney, K.; Kreber, R.; Ganetzky, B. *Cell* **1989**, *58*, 1143-1154.
(39) Stern, M.; Kreber, R.; Ganetzky, B. *Genetics* **1990**, *124*, 133-143.
(40) Knipple, D. C.; Payne, L. L.; Soderlund, D. M. *Arch. Insect Biochem.
      Physiol.* **1991**, *16*, 45-53.
(41) Doyle, K. E.; Knipple, D. C. *Insect Biochem.* **1991**, *21*, 689-696.
(42) Knipple, D. C.; Doyle, K. E.; Marsella-Herrick, P. A.; Soderlund, D. M.
      Proc. Natl. Acad. Sci. USA **1994**, *91*, 2483-2487.
(43) Noda, M.; Ikeda, T.; Kayano, T.; Suzuki, H.; Takeshima, H.; Kurasaki, M.;
      Takahashi, H.; Numa, S. *Nature* **1986**, *320*, 188-192.
(44) Williamson, M. S.; Denholm, I.; Bell, C. A.; Devonshire, A. L. *Mol. Gen.
      Genet.* **1993**, *240*, 17-22.
(45) Lester, H. A. *Science* **1988**, *241*, 1057-1063.

RECEIVED July 19, 1994

Chapter 7

Molecular Biological Investigations of the Property of Sodium Channels of *Blattella germanica*

Fumio Matsumura, Akira Mizutani, Koichiro Kaku, Masahiro Miyazaki, Shuichiro Inagaki, and Debra Y. Dunlap

Departments of Environmental Toxicology and Entomology, Center for Environmental Health Sciences, University of California, Davis, CA 95616–8588

Polymerase chain reaction (PCR) studies were carried out to compare the difference in the DNA and/or amino acid sequence of the sodium channels between the susceptible and the DDT-resistant, *kdr* strain of *Blattella germanica*. Using various PCR primers, 1179-base fragments of a *Drosophila* DSC type sodium channel DNAs, representing the section corresponding roughly to S6c through S6d domains, were sequenced. At least 4 different variants of the SC type sodium channel DNAs were found to exist in each strain. Of these, only one strand found in the resistant strain showed a T to C mutation which induced a change in amino acid sequence of phenylalanine (TTC) to serine (TCC) at amino acid number 1595. This area coincides with the proposed intracellular Ca^{2+} binding region of the sodium channel. Based on this information antibodies have been constructed to specifically interact with two of the putative calcium binding areas of sodium channels. Model experiments using rat brain synaptosomal membrane and [^{3}H]-batrachotoxin binding tests showed that one antibody directed at the C terminal calcium binding site, corresponding to the site of the above mutation, has the property to modify the effect of deltamethrin. Another experiment using a liposome preparation from the synaptosomal membrane and $^{22}Na^+$ transport has also shown the same tendency. These results support the view that this calcium binding region at the C-terminus could be an important site for the action of pyrethroids.

The *kdr* (knock-down resistance) factor is one of the major genes responsible for DDT-and pyrethroid-resistance in several insect species (*1*). This mechanism was first found in the Orlando housefly strain (*2*). This type of resistance in the housefly is related to the *kdr*-gene on chromosome 3, causing resistance to DDT, pyrethrins (*3*) and pyrethroids (*4*). The *kdr*-gene directly or indirectly affects the nervous system which makes the site of action less vulnerable toward these insecticides. Tsukamoto *et al.* (*5*) found that the *kdr*-factor was responsible for reduced sensitivity of the nervous system toward DDT; the observation was later confirmed by Miller et al. (*6*). The *kdr*-type DDT-resistance in the housefly has been

0097–6156/95/0591–0109$12.00/0

shown to cause some degree of cross-resistance to pyrethroids (7). A DDT-resistant strain of *Boophilus microplus* (8), and a DDT-resistant mosquito strain (9) have also shown cross-resistance to pyrethroids. In addition to the housefly, *kdr*-type resistance has been found in *Culex pipiens quinquefaiatus* (10), *Spodoptera littoralis* (11) *Blattella germanica* (12, 13) and the predatory mite *Amblyseius fallacis* (14).

The mechanism of *kdr*-resistance has been studied by many workers, but so far conflicting theories have been offered to explain the mechanism of this type of resistance in insects. Salgado et al. (15) have suggested that *kdr*-type resistance is due to a modified sodium channel. Binding studies with [^{14}C]DDT and [^{14}C]permethrin to the housefly central nervous system have led Chang and Plapp (16) to a conclusion that pyrethroids affect a putative "receptor"and the *kdr*-resistant houseflies have fewer receptors than the susceptible ones. Ghiasuddin et al. (17) found that the DDT-resistant strain of German cockroach has an altered Ca-ATPase with a lower affinity to Ca^{2+} than that from the susceptible strain. It has been reported (18, 19) that *kdr*-resistant German cockroaches and houseflies exhibit cross-resistance to calcium mimics or agents affecting calcium regulatory mechanism, and that the degree of stimulation of the Ca^{2+}-ATPase by exogenously added calcium was much higher in the two susceptible strains than in the DDT-resistant strains. The tendency for calcium insensitivity was more pronounced in the S*kdr* than in the *kdr* strain of housefly. Furthermore, Matsumura (20, 21) demonstrated that DDT and pyrethroids affect protein phosphorylation processes, particularly those involved in calcium regulatory processes.

Thus Charalambous and Matsumura (22) initiated the study on calcium sensitive protein phosphorylation systems in these insects in order to understand the meaning of these cross-resistance phenomena (i.e., *kdr*-resistant insects being more tolerant against agents known to increase intracellular [Ca^{2+}] or mimic the action of calcium). For this purpose, the effects of calcium, calmodulin and 1R-deltamethrin on protein phosphorylation processes were investigated using lysed synaptosomal preparations isolated from the nervous system of the susceptible and *kdr*-resistant German cockroaches and houseflies. By comparing interstrain differences in phosphorylation activities of isolated, lysed synaptic membranes, they concluded that the stimulatory effect of calcium alone on protein phosphorylation was the same in the preparations from the susceptible and the *kdr*-resistant strains of the German cockroach. However, calmodulin, added in the presence of Ca^{2+}, significantly increased the level of phosphorylation of the two putative subunits of calcium-calmodulin-dependent protein kinase (CCPK, or CaM kinase) from the susceptible strains, but its effect on the same enzyme in the *kdr*-resistant strains of the German cockroach and housefly was much less. 1R-deltamethrin at 10^{-8} M inhibited both the total protein phosphorylation and the phosphorylation on the two subunits of CaM Kinase in the susceptible and *kdr*-resistant strains of the German cockroach. Depolarization induced in intact synaptosomes by veratridine or "high K$^+$" in the presence of 1R-deltamethrin at 10^{-10} M and 10^{-6} M had the effect of significantly increasing the total level of endogenous protein phosphorylation in the susceptible strain, but the increase was not significant in the *kdr*-resistant strain of the German cockroach. An identical experimental approach on *kdr* and the susceptible strain of the housefly produced essentially the same results, indicating that the CaM Kinase enzyme from the *kdr*-type resistant insects behave in a different manner from the susceptible counterparts in terms of their responsiveness to calmodulin.

Since it has been well established that such calcium- and calmodulin-induced rise in CaM Kinase activity and resulting protein phosphorylation on key synaptic proteins are the driving forces to initiate and maintain synaptic transmitter releasing activities, and since type II pyrethroids have been shown to amplify these activities (23, 24), it is logical to expect that the reduction in the maximum level of activation of CaM Kinase in the resistant insects gives them an effective protection mechanism to counter the action of these neuroactive insecticides. Furthermore, we have recently obtained the experimental evidence that the mRNA levels for CaM Kinase in the *kdr*-resistant strains (VT and VP) are much lower than that of CSMA, indicating that the strain difference is due to the degree of the gene expression of CaM Kinase. Otherwise we could not find any strain differences in the quantity or quality of calmodulin and calcium channels so far identified from cDNA library of the resistant and susceptible *B. germanica* (Inagaki and Matsumura, unpublished data). On the other hand, *kdr*-resistant insects do not usually exhibit cross-resistance to organophosphates or cyclodienes, despite the fact that both groups of agents are known to stimulate synaptic events. Therefore, the above protective measures do not appear to be universally effective to all synaptic stimulators. Nevertheless, these insects exhibit cross-resistance against veratridine, grayanotoxin II and A23187 (*18*) and, therefore, the resistance mechanism found here works effectively against agents affecting Na channels and a Ca^{2+} ionophore.

An important question we must raise is whether this biochemical change itself is directly or indirectly related to the reduction in sensitivity of sodium channel to DDT and pyrethroids in the *kdr* insects. While such a relationship is a possibility, our current opinion is that this CaM Kinase-involved resistance mechanism should be dealt with as an independent resistance factor. Earlier Osborne and Pepper (*25*) have found in S*kdr* houseflies that the factor, which makes this strain more resistant to pyrethroids than the corresponding *kdr*-houseflies, is not related to the sodium channel. Also, in the above study (*22*) the inter-strain difference in the sensitivity of the synaptosome deltamethrin could be observed in the preparation depolarized by high K^+ treatment, which bypasses the sodium channel operation, indicating that this CaM Kinase-involved resistance mechanism could operate independently from the sodium channel. On the other hand, we have no evidence to totally rule out the possibility that the change in CaM Kinase operations is directly or indirectly related to the function of the sodium channel.

The main question we must now ask, therefore, is whether there is an additional mutational change in *kdr* strains involving the sodium channel that is independent from the CaM Kinase type resistance mechanism. Fortunately, the structures of the voltage-gated sodium channels have been determined by many workers in several species (*26-28*). When mRNA for one of the sodium channel α-subunits transcribed from a full-length cDNA was injected to *Xenopus* oocytes a full voltage-activated sodium channel current could be produced, indicating that the full-length α-subunit proteins alone are capable of constituting the normal sodium channel functions.

In planning this research project, we have chosen a polymerase chain reaction approach to obtain the sequence information on the cockroach sodium channels, instead of regular cloning approaches, since the former is quicker, and also offers the possibility of finding various forms of sodium channels including those mutated. Also, we have chosen the homology domain c and d (S III and S IV in the terminology of Noda et al. (*26*)) as the main study region, since this region contains two of the possible target sites for DDT and pyrethroids; i.e. S-19 (1266-1295) and the internal calcium binding region (*28*).

MATERIALS AND METHODS

Chemicals and Biochemicals. [³H]batrachotoxin A 20α-benzoate (56.8 Ci/mmol) ([³H]IBTX) and ²²NaCl (5.1 Ci/mmol) were obtained from Amersham. Aconitine, grayano-toxin-I, TPA (12-0-decanoyl phorbol 13-acetate), scorpion venom (*Androctonus australis*), tetrodotoxin and veratridine were purchased from Sigma Chemical Co. N-succinimidyl-3-

(2-pyridyldithio)propionate (SPDP) and m-maleimidobenzo-p-sulfo-sucinimide ester (sulfoMBS) were purchased from Pierce. Keyhole limpet hemocyanin (*Megathrua crenuluta* lyoph.) was purchased from Calbiochem. Two specific sodium channel protein antibodies, RNa32C7 and LNa2F9, were kindly provided by Dr. Y. Ishikawa of the Department of Agricultural Biology, University of Tokyo, Japan. The former was obtained by using purified rat brain sodium channel and the latter using lobster nerve sodium channel proteins. 1R- and 1S-deltamethrin were generous gifts from Roussel Uclauf.

Preparation of Synaptosomes and Synaptosomal Membranes. Crude synaptosomes were prepared according to the method described by Robinson and Dunkley (*29*). Young Sprague-Dawley rats (males) were used for this purpose. Brains were removed, chilled on ice, weighed and homogenized gently in 10 volumes of 0.32 M sucrose containing 1 mM EDTA, 1 mM iodoacetamide, 0.1 mM phenylmethylsulfonyl fluoride, and 1 μg/ml aprotinin with 10 up and down strokes using a glass-glass homogenizer. The homogenate was centrifuged at 1500 x g for 10 min at 4°C with a refrigerated Sorvall centrifuge, and the supernatant was saved. The pellet was resuspended in 20 volumes (60 ml) of the homogenate solution and recentrifuged. The combined supernatants were centrifuged at 15,000 x g for 30 min. For [^{3}H]BTX binding tests the pellet (P$_2$) was resuspended in the same volume of Krebs-Finger solution containing 25 mM NaHCO$_3$, 5 mM KCl, 1 mM CaCl$_2$, 1.5 mM MgSO$_4$, 124 mM NaCl and 10 mM glucose, and centrifuged at 15,000 x g for 10 min. The Krebs-Ringer solution was pre-bubbled with 95% O$_2$-5% CO$_2$. The resultant pellet (P$_2$) was resuspended in the same solution using a 10 ml glass-glass homogenizer and protein concentration was adjusted to 3 mg/ml. This preparation was used as intact synaptosomes or as a source of synaptosomal membranes. To prepare the latter, the P2 fraction was resuspended in 5 ml of Krebs-Ringer solution and homogenized using a 10 ml glass-glass homogenizer. The suspension was diluted 10-fold with 95 ml of a dilution buffer containing 0.5 mM EGTA, 2 mM EDTA, 50 mM β-mercaptoethanol and 1 μg/ml aprotinin in 20 mM Hepes buffer (adjusted to pH 7.4 with tris base), and homogenized vigorously for 1 min. The homogenate was stirred for 1 hr at 4°C and centrifuged for 1 hr at 100,000 x g using a Beckman L-65 ultracentrifuge. The resultant pellet was resuspended in 20 mM Hepes buffer (pH 7.4) and protein concentration was adjusted to 2 mg/ml. The suspension was divided into 3 ml portions and stored at -80°C until use. All the procedures were performed at 4°C or on ice and all the solutions were prechilled before use. Protein concentration was determined by the method of Peterson (*30*) using bovine serum albumin as a standard.

For ^{22}Na$^+$ uptake tests, the above P$_2$ fraction was resuspended in the same volume (60 ml) of 25 mM Hepes-Tris buffer (pH 7.4) containing 130 mM choline chloride, 5.4 mM KCl, 0.8 mM MgSO$_4$, 5.5 mM glucose and 1 mg/ml bovine serum albumin, and centrifuged for 10 min at 15,000 x g using a Sorvall centrifuge. The final pellet was resuspended in the above Hepes-Tris buffer using a glass-glass homogenizer and protein concentration was adjusted to 10 mg/ml. This preparation was used as the source for rat brain synaptosomes.

[^{3}H]BTX Binding Assay for Synaptosomal Membranes. Measurement of [^{3}H]BTX binding to synaptosomal membranes was performed as follows: synaptosomal membranes (1 mg of protein) were suspended in 20 mM tris-HCl (pH 7.4) and preincubated with various effectors of appropriate concentrations (as shown in each table) for 10 min at 24°C. Except for scorpion venom, all effectors were dissolved in ethanol, and a 1 μl aliquot was added to each test tube containing the membrane suspension. Control tubes received the same amount of ethanol alone. All tubes were incubated for 10 min at 24°C. Scorpion venom was dissolved in 20 mM Hepes-Tris (pH 7.4) and 1 μl of the solution was added to the membrane suspension. Assay runs without scorpion venom received the same amount of the same Hepes-Tris solution. Following preincubation, 30 nCi of [^{3}H]BTX was added to each tube and incubated for 20 min at 24°C. A small glass tube (10x75 mm) was used for each test and the total assay volume was 1 ml. After incubation, the suspension was centrifuged at 2,000 x g for 3 min using a microcentrifuge (Micro-centrifuge Model 235C, Fischer Scientific) and the pellet was washed twice with 1 ml of ice-cold 20 mM Hepes-Tris (pH 7.4) using the same centrifugal procedures. The final pellet was suspended in 100 μl of distilled water using a vortex mixer, and the entire content was transferred into liquid scintillation fluid (3.5 ml), to assess [^{3}H]dpm. The level of nonspecific binding of [^{3}H]BTX was determined in the presence of 300 μM veratridine and routinely subtracted from the total binding values. In the experiment with antibodies, 1 μl of antisera containing 0.3 μg of protein was added first to the suspension containing 1 mg of protein and preincubated prior to addition of effectors for 5 min at 24°C which was followed by the above method. Assay runs in the absence of antibody received the same amount of serum from nonimmunized control rabbit containing 0.02% sodium azide.

Binding Assays for Intact Rat Brain Synaptosomes. [^{3}H]BTX binding to intact synaptosomes was measured as follows: 50 μg protein suspended in Krebs-Ringer solution (pre-bubbled with 95% O_2-59% CO_2 for 3 min) were preincubated with effectors for 15 min at 36°C. Deltamethrin (0.1 μM) with 1 μl of ethanol was added to the suspension 5 or 10 min after the start of the preincubation. Assay runs without these materials received the same amount of ethanol or distilled water. The total volume of preincubation was made to 150 μl with pre-bubbled Krebs-Ringer solution. After preincubation, 150 μl of Krebs-Ringer solution (for undepolarized tests) or the same amount of high potassium solution (150 mM KCl, 25 mM NaHCO$_3$, 1.5 mM MgSO$_4$, 1 mM CaCl$_2$ and 10 mM glucose for depolarized tests) each containing 30 nCi/ml [^{3}H]BTX was added to initiate the binding reaction which lasted for 15 min at 36°C. The reaction was stopped by addition of 3 ml of ice-cold Krebs-Ringer solution followed by immediate filtration through a glass filter (Whatman GF/F 0.7 μm), and washing two times with 3 ml of ice-cold Krebs-Ringer solution. The filters were suspended in liquid scintillation fluid and [^{3}H] dpm was determined after overnight extraction.

Measurement of ^{22}Na Uptake by Rat Brain Synaptosomes. Rate of uptake of ^{22}Na into synaptosomes was studied according to the method of Hartshorene and Catterall (*31*) and Matsumura and Clark (*33*) with minor modification. A 50 μl aliquot of synaptosomal suspension containing 500 μg of protein was placed in a small glass tube (10 x 75 mm). All effectors were added to the suspension at appropriate concentrations, and preincubated for 10 min at 36°C under Na-free conditions. In the experiment with antibodies, they were added first to the synaptosomal suspension and preincubated for 5 min at 36°C before the addition of other effectors (i.e., total 15 min preincubation). The antibodies were suspended in a solution containing 20 mM Hepes-Tris buffer (pH 7.4), 0.02% sodium azide and 28 mM NaCl. A 0.5 μl portion of the antibody suspension, containing 0.3 μg protein, was added to 50 μl of the synaptosomal suspension. Tetrodotoxin was dissolved in 10 mM acetic acid. Scorpion venom was dissolved in 10 mM Hepes-Tris buffer (pH 7.4). In all cases an 0.5 μl aliquot of each stock solution was added to the synaptosomal suspension. Other compounds were dissolved in ethanol, and 0.2 μl of each stock solution was added to the synaptosomal suspension. In all cases total reaction volume was 51.4 μl. After preincubation, Na$^+$ influx was initiated by the addition of 150 μl of ^{22}Na uptake solution consisting of 5.4 mM KCl, 0.8 mM $MgSO_4$, 5.5 mM glucose, 1.33 mM NaCl, 5 mM ouabain, 0.2 mM EGTA, 128 mM choline chloride, 30 mM Hepes-Tris (pH 7.4), 1 mg/ml bovine serum albumin and 250 nCi/ml of $^{22}NaCl$. Incubation was continued for 30 sec, and terminated by the addition of 3 ml of ice-cold "wash solution," containing 163 mM choline chloride, 0.8 mM $MgSO_4$, 1.8 mM $CaCl_2$, 5 mM Hepes-Tris buffer (pH 7.4), and 1 mg/ml bovine serum albumin, which was followed by a rapid filtration through a glass filter (Whatman GF/F) under vacuum to collect synaptosomes, and washed twice with 3 ml of ice-cold "wash solution." The filters were air dried and suspended in liquid scintillation fluid. ^{22}Na influx was measured by determining ^{22}Na collected on the filters. In the experiment with high concentration of potassium ions, a flux buffer containing 100 mM KCl, 0.8 mM $MgSO_4$, 5.5 mM glucose, 33 mM choline chloride, 1.3 mM NaCl, 5 mM ouabain, 0.2 mM EGTA, 30 mM Hepes-Tris buffer (pH 7.4), 1 mg/ml bovine serum albumin and 250 nCi/ml of $^{22}NaCl$ was used to initiated ^{22}Na influx instead of the usual uptake solution, followed by the above procedures.

Preparation of Liposomes. To prepare protein samples for reconstitution of sodium channel into liposomes, the above P_2 pellet was resuspended in 10 ml of Krebs-Ringer solution and diluted 10-fold with 5 mM Tris-HCl buffer (pH 8.2) containing 1 mM EDTA, 1 mM iodoacetoamide, 0.1 mM phenylmethylsulfonyl fluoride and 1 µg/ml aprotinin, homogenized using a glass-Teflon homogenizer, and incubated for 15 min on ice. The suspension was homogenized again with four up and down strokes and centrifuged at 27,000 x g for 40 min. The resultant pellet was homogenized in 3 ml of a solution containing 67.5 mM Na_2SO_4, 150 mM sucrose, 25 mM Hepes-Tris buffer (pH 7.4) and 1% Triton X-100 for 5 min on ice, and then centrifuged at 120,000 x g for 50 min using a Beckman L-65 ultracentrifuge. After centrifugation, the supernatant was dialyzed in a cold room against 25 mM Hepes-Tris buffer (pH 7.4) containing 67.5 mM Na_2SO_4 and 150 mM sucrose for 4 hrs, using dialysis tubing (Spectra/Pov, M.W. cutoff 6000-8000). After dialysis, the protein concentration was adjusted to 10 mg/ml and divided into 0.5 ml portions. They were frozen by immersion in dry ice-acetone for 1 min and stored at -80°C until use. All the procedures were performed at 4°C or on ice and all the solutions were prechilled before use.

The method of incorporation of sodium channel proteins into liposomal preparation was adopted from the original method of Kasahara and Hinkle (*33*). Crude soybean phospholipids (asolectin, from Concentrated Associates, Woodside, NY) were used without further purification. They were suspended at 30 mg/ml in a solution containing 67.5 mM Na_2SO_2, 0.5 mM $MgSO_4$, 150 mM sucrose, and 25 mM Hepes-Tris buffer (pH 7.4). A 100 µl aliquot of the suspension was transferred into a small glass tube (10 x 75 mM), the air was replaced with nitrogen (10 s), the test tube capped with Parafilm, and sonicated for 10 min at room temperature in a bath type sonicator (Branson 1200).

Measurement of ^{22}Na Remaining Inside Liposomal Vesicles. A frozen protein sample (500 µl containing 5 mg protein kept at -80°C) was thawed, rehomogenized and divided into 40 µl aliquots. Each aliquot in a small glass tube was filled with N_2, capped with a piece of parafilm and kept on ice until use. All stocks of effectors and antibodies were prepared as above and were added to each tube with 0.2 µl of appropriate medium for each agent. Control tubes received the corresponding medium only. The protein sample treated with effectors was capped with Parafilm again and preincubated for 10 min on ice. After preincubation, 10 µl of phospholipid suspension (asolectin) containing 1.1 µCi/ml $^{22}NaCl$ was added to the sample. The mixture was immediately frozen by immersing in dry ice-acetone for 1 min, thawed by vortexing in cold water, and sonicated for 30 s using a bath-type sonicator (Branson 1200) at $0°C$. To initiate the $^{22}Na^+$ reaction, the liposomal vesicle suspension was diluted 15-fold with 1260 µl of 25 mM Hepes-Tris buffer (pH 7.4) containing 105 mM tris-SO_4 and 150 mM sucrose. The diluted sample was incubated for 30 s at $36°C$, transferred to a Dowex AG50W-X8 (tris form) column (2 x 0.5 cm) pretreated with 100 µl of 10% bovine serum albumin, and immediately eluted with 1 ml of 0.38 M sucrose containing 1 mg/ml bovine serum albumin. The eluates were collected and assayed for their ^{22}Na radioactivity. Each batch of liposomal preparation consisted of 9 tubes which were used up for one efflux test. Such a test was normally repeated three times each with fresh liposomal preparation to avoid deterioration of liposomes. Frozen protein samples were thawed only once.

Generation of Antibodies to Synthesized Peptides. Two peptides were synthesized by the UC Davis Protein Structure Laboratory. The amino acid sequence of the first peptide (antigen I) was identical to the segment between S5a and S6a (AGRNPNYGYTSFD-TFSWAFL) in the rat sodium channel (26). The amino acid sequence of antigen II (FYEVWEKFDPDATQFIEFEKLSDFAAAL) was identical to the segment between S6d and the end of sequence in the rat sodium channel.

These synthesized peptides were conjugated to a carrier protein, keyhole limpet hemocyanin (KLH), via sulfoMBS as a coupling agent as follows: 3 mg of the lyophilized peptides were dissolved in 0.4 ml of 100 mM Hepes buffer (pH 7.5). To the solution 40 M SPDP was added to make 7.25 mM and the system was kept at room temperature overnight. β-mercaptoethanol (1.5 M in final conc.) was added to the SPDP-linked peptide, and the system was kept at room temperature for an additional 2 hr. The sample was transferred to a 5 cm Sephadex G10 column, and eluted with 50 mM sodium phosphate buffer (pH 6.0). The eluate was immediately added to 2 ml of the carrier protein–sulfoMBS mixture, adjusted to pH 7.0–7.5 with 1 N NaOH, and incubated at room temperature with gentle shaking for 1 hr and an additional 15 hr in a cold room. After conjugation, the sample was dialyzed in a cold room against 100 mM Hepes buffer (pH 7.4) containing 100 mM NaCl overnight, using dialysis tubing (Spectra/Pov MW cutoff 6000–8000).

The carrier protein-sulfoMBS mixture was prepared as follows: 40 mg of KLH was dissolved with 2.5 ml of 10 mM sodium phosphate buffer (pH 7.2) and 7.5 mg sulfoMBS was added to the solution. The mixture was incubated at room temperature with gentle shaking. After 30 min the mixture was purified using a Sephadex G-10 column. A white New Zealand rabbit (2 kg male) was immunized at weekly intervals with 2 ml (1 mg) of peptide conjugate mixed with an equivalent volume of Freund's complete adjuvant (Sigma Chemical Co.) for the first immunization, and Freund's incomplete adjuvant for all subsequent injections. After 40 days, 20 ml of blood was taken from the ear vein. The blood was kept first at room temperature for 2 hr, and then in a cold room overnight. To obtain the serum, the blood sample was centrifuged at 30,000 x g for 30 min at $4°C$.

The IgG class antibodies were purified using a protein A column as follows: the serum was passed through a protein A–sepharose (CL-4B) column (0.5 x 2 cm), and the column was washed with 10 column volumes of 10 mM tris (pH 8.0). The antibodies were eluted with 100 mM glycine-HCl (pH 3.0) buffer, and 0.5 ml aliquots of eluate were collected in 1.5 ml Eppendorf tubes already containing 50 µl of 1 M tris (pH 8.0). Each tube was mixed gently to bring the pH back to neutral. The immunoglobulin-containing fractions were identified by absorbance at 280 nm. The main fractions were combined, sodium azide (to make 0.02% final conc.) was added, divided into 1 ml aliquots, and stored at $-80°C$. Antigen-antibody interactions were detected by Oucherlony test (double-diffusion precipitation) which indicated a strong and specific reaction to the original antigen and a weak reaction to the carrier protein KLH by each antibody preparation.

RESULTS

Studies on the DNA Sequences of the Sodium Channels. Fig. 1a illustrates the primary sequences and strategies we used to obtain PCR generated DNA fragments. This approach was successful in yielding *Drosophila melanogaster* DSC1 (*27*) type DNA fragments. The deduced sequences of both DNAs and amino acids are shown in Fig. 1b. The first amino acid (alanine) corresponds to #1234 of DSC1. Four similar, but distinct, variants of 1179-base pair fragments were recognized (Fig. 1c). They were designated as GCSC 1A, B, C and D. In Fig. 2, we have compared the sequence of *B. germanica* sodium channels to those of *D. melanogaster*. It is apparent that GCSC of the former is indeed very similar to DSC 1 of the latter species, the major difference region being first in the connecting region between S5d and S6d (#1526-1531 of DSC 1) and the tail part (#1620-23). Other differences are sporadic. In terms of differences of critical areas, the frontal part of S19, #1265-1268 showed some difference: RRSI for DSC1 and KKKY for GCSC. Also, the calcium binding region had one amino acid difference (#1597 H for DSC1 and R for GCSC).

For interstrain comparison, we have studied the DNA sequences of the same region of DSC type sodium channels of a permethrin resistant, *kdr*-resistant *B. germanica* strain. The results showed that their sequences were identical in all cases except in one DNA base where T to C conversion (DNA #1857) has caused a shift in amino acid phenylalanine (#1595) to serine. This was found only in GCSCD strand, not in any other GCSC strands.

Thus the only mutation which results in the amino acid shift in the GCSC type sodium channel sequence of the *kdr* German cockroach was found to be the one which occurred at DNA #1857 (conversion of T to C and thereby F to S) in the calcium binding region of the GCSCD strand. Based on this information, we have obtained a specific antibody aimed at this region of the sodium channel and assessed its effects on the action of deltamethrin using two biochemical criteria, binding of ^{3}H-batrachotoxin A 20α-benzoate and ^{22}Na$^+$ uptake through the sodium channel.

The actual functions or the localization of the SC type sodium channels are not known whereas those of the para-type sodium channels are acknowledged to be the main voltage sensitive one operating the depolarization coupled sodium ion permeability changes on the axolemmas. Therefore, we have extended our PCR studies on the latter type in *B. germanica* as well. Using various combinations of PCR primers we were successful in sequencing the entire region between IS3 through IVS6 and 206 bases down stream from the end of IVS6 domain (altogether 5068 bases, Miyazaki and Matsumura, unpublished data). As a result of a side by side comparison of DNA sequence between susceptible (CSMA) and *kdr* resistant (VT strain) we could locate one base mutation leading to a shift in amino acid composition. The location of this shift was in the center of SII6 region where a G to C mutation created a change of L to F. The actual sequence of *B germanica* SII6 region of para SC in CSMA strain is WSCIPFFLATVVIGN**L**VVLNLFLALLLS, where the central L (indicated by a bold letter) was F in the equivalent *kdr* DNA strand. Note that the above susceptible strain's amino acid sequence of this membrane spanning region is identical to that of *Drosophila* para-type sodium channel with one exception of the first amino acid which is V in *Drosophila*. Among all known sodium channel sequences, this region must be regarded as one of the most conserved areas particularly in the central to the immediately downstream region where all of the 17 sequenced sodium channels possess the

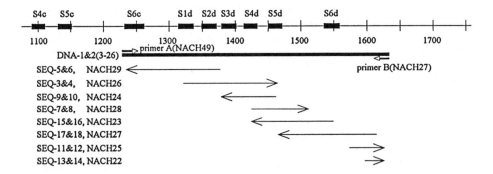

Fig. 1. The primer combinations used for isolation and identification of the DNA and amino acid sequence of the DSC type sodium channel DNAs from the German cockroach cDNA library through polymerase chain reactions.

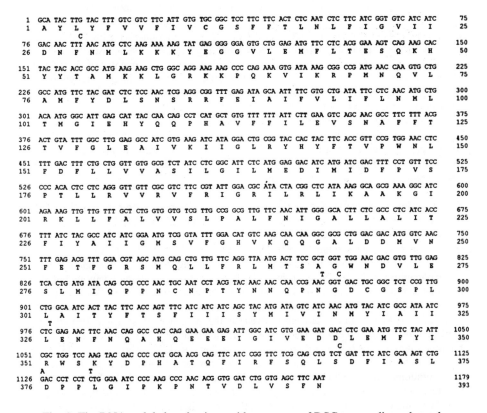

Fig. 2. The DNA and deduced animo acid sequences of DSC type sodium channels (GCSC) of the German cockroach.

sequence of "NLV" with only one exception of Drosophila DSC which, interestingly show a "NFM" sequence. Certainly, finding a mutation does not automatically indicate that this is a functionally important change conferring *kdr* insects the insensitivity to DDT and pyrethroids. Much more work lie ahead to determine whether it is related to resistance or not. Three pieces of circumstantial evidence supporting some relationship between this mutation and *kdr* resistance are: (a) this is a very conserved region where every functionally proven neural and muscle sodium channel so far identified has the susceptible configuration, (b) this VT strain has been genetically purified by 8 generations of backcrossing to CSMA strain making the former genetically very close to the latter (the fact that throughout the entire sequence of over 5000 bases, only one mutation leading to the amino acid change attests to this interpretation), and (c) this is also one of the most hydrophobic regions of the entire α subunit of the sodium channel and, therefore, lipophilic compounds such as DDT and pyrethroids could be attracted to. Thus, it appears to be worthwhile to pursue this line of investigation. The details of this particular study on para sodium channel will be published elsewhere.

While it is difficult to assess the functional meaning of any mutational change without the availability of full-length cloned genes, from a pharmacological point of view, it is possible to utilize a number of specific agents known to attack different sites within the molecule of the main body of the sodium channel (α-subunit) to assess the functional differences among different channels (e.g., *34, 35*). Of particular interest is the finding that these pesticides stimulate [^{3}H]batrachotoxin binding (*36-38*). The cause for such a stimulatory action of these insecticides has been implicated to be the opening of the sodium channel, exposing the batrachotoxin binding site which is otherwise largely hidden in the closed conformation (*39*). Such changes in the level of [^{3}H]batrachotoxin binding could be used as the measurement of the degree of openness of the sodium channel.

Another approach we have taken is to assess the part of ^{22}Na$^+$ uptake that is affected by tetrodotoxin, deltamethrin and veratridine using an experimental methodology similar to that used by several workers (e.g., *32, 34, 40*). This approach allows one to assess directly the gating operation of the sodium channel providing that specific actions of tetrodotoxin and veratridine are ascertained.

Construction of Specific Antibodies and [^{3}H]Batrachotoxin Binding Studies. The effects of active (1R) and nonactive (1S) isomer of deltamethrin on [^{3}H]BTX binding to synaptosomal membranes were compared to those of neuroactive agents already known to affect its binding to the sodium channel. The results (Table I) showed that scorpion venom, as expected (*22*), is a strong promotor of [^{3}H]BTX binding to synaptosomal membranes. 1R-deltamethrin at the 10 µM level showed activity as strong as scorpion venom, but 1S-deltamethrin showed no sign of interactions at the same concentration level. When scorpion venom and 1R-deltamethrin were added together, there was a very significant rise in the level of [^{3}H]BTX binding. However, even in the presence of scorpion venom 1S-deltamethrin showed no effect. These data clearly indicate that the main site of [^{3}H]BTX binding studied here is the sodium channel and that 1R-deltamethrin affect [^{3}H]BTX binding to this protein. The effects of deltamethrin on [^{3}H]BTX binding at various concentrations were tested next (Fig. 3). The threshold concentration of 1R-deltamethrin to cause stimulation was determined to be 0.1 µM. 1S-deltamethrin did not stimulate [^{3}H]BTX binding at all concentrations tested. Rather it slightly inhibited [^{3}H]BTX binding at higher concentrations in the presence of scorpion venom.

Table I. Effect of Sodium Channel Neurotoxins on [³H]BTX Binding
in Rat Brain Synaptosomal Membranes

Treatment	[³H]BTX bound (fmol•mg⁻¹) Mean ± S.D.	No. of Experiments
Scorpion venom	1.9 ± 0.2*	3
1R-deltamethrin	2.0 ±0.2*	3
1S-deltamethrin	0.8 ± 0.3	3
1R-deltamethrin + scorpion venom	9.0 ± 0.6*	3
1S-deltamethrin + scorpion venom	0.9 ± 0.1	3
None (control)	1.0 ± 0.1	3

Synaptosomal membranes were preincubated with neurotoxins for 10 min. After pre-incubation, 30 nCi of [³H]BTX was added to the sample and incubated for 20 min at 24°C. Veratridine (300 µM) was used to measure the nonspecific binding. Concentrations are: scorpion venom, 0.5 µg/ml; 1R- and 1S-deltamethrin, 10 µM. Most scorpion venom preparations are known to contain toxins that bind also to other channels than sodium channels.
*Significant difference from control (none) at $P ≤ 0.01$.

In the next series of experiments, two antibodies known to bind with specific parts of the sodium channel protein were tested to gain an insight to the action site of BTX and deltamethrin. Both antibodies are aimed at possible Ca^{2+} binding sites. Antibody I (FM-AbI) is aimed at an external site and antibody II (FM-AbII) is designed to bind at an internal site. The results shown in Table II indicated that FM-AbII is effective in reducing [³H]BTX binding to synaptosomal membranes: i.e., the total binding level decreased to about 80% of the value of deltamethrin-treated membranes. Upon heat denaturation such a property of FM-AbII was lost, indicating that blocking the stimulatory action of deltamethrin is not due to the presence of heat-stable contaminants. FM-AbII had no effect on the [³H]BTX binding to membranes pretreated by deltamethrin.

Synaptosomal membranes were preincubated with antibodies for 5 min at 24°C prior to the addition of deltamethrin, followed by the standard [³H]BTX binding assay. Veratridine (300 µM) was used to measure the nonspecific binding. At Ouchterlony double diffusion test, significant precipitations were formed between antisera and their immunogens. Antisera reacted slightly with keyhole limpet hemocyanin. No precipitation was observed between sera from the rabbit before immunization and immunogens. Concentrations are: antibodies, 0.3 µg/ml; heated antibodies, boiled for 3 min; deltamethrin, 10 µM; scorpion venom, 0.5 µg/ml.

²²Na⁺ Uptake By Rat Brain Synaptosomes. Deltamethrin stimulated the ²²Na⁺ uptake triggered by veratridine in synaptosomes from rat brain. In contrast, the treatment with 0.1 µM deltamethrin alone produced no stimulatory effect on the uptake (Table III). The stimulatory effects of deltamethrin and veratridine were suppressed by tetrodotoxin.

Under a high external potassium condition, ²²Na uptake in synaptosomes was not stimulated with deltamethrin (data not shown). Aconitine was antagonistic against veratridine in terms of stimulation of veratridine-induced ²²Na uptake in synaptosomes. Grayanotoxin I and DDT, however, worked synergistically with veratridine, though none was as effective as scorpion venom or deltamethrin (Table IV).

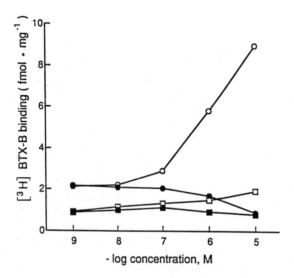

Fig. 3. Effect of 1R- and 1S-deltamethrin on the specific binding of [³H]BTX to rat brain synaptosomal membranes. binding of 30 nCi [³H]BTX was measured at 24°C in the presence of increasing concentration of 1R- and 1S-deltamethrin, and either with or without scorpion venom (0.5 μg/ml). Nonspecific binding was determined in parallel assay tubes containing 300 μM veratridine and has been subtracted from the data. Data variance was within 10% in all treatments. □ = 1R-deltamethrin; ■ = 1S-deltamethrin; O = 1R-deltamethrin plus scorpion venom; ● = 1S-deltamethrin plus scorpion venom.

Table II. Effect of Antibodies on [^{3}H]BTX Binding Enhanced by Scorpion Venom in Synaptosomal Membranes

Treatment	[^{3}H]BTX bound (fmol•mg^{-1}) Mean ± S.D.	No. of Experiments
Deltamethrin + antibody-I (FM-AbI)	6.0 ± 0.1	6
Deltamethrin + antibody-II (FM-AbII)	5.2 ± 0.3*	6
Deltamethrin + heated antibody-II (FM-AbI)	6.3 ± 0.3	6
Deltamethrin + heated antibody-II (FM-AbII)	6.3 ± 0.4	6
Deltamethrin	6.3 ± 0.3	6
None (scorpion venom alone)	1.9 ± 0.2**	6

*Significantly different from treatment 1 or 5 at P ≤ .001 according to Student's t-test.
** Significantly different from all other treatments at P ≤ .001.

Table III. Effect of Neurotoxins on Uptake of ^{22}Na in Rat Brain Synaptosomes [a]

Treatment[b]	^{22}Na uptake (nmol•min^{-1}•mg^{-1})	
	Mean ± S.D.	No. of exp.
Veratridine, deltamethrin plus tetrodotoxin	16.2 ± 0.8	6
Veratridine plus deltamethrin	22.9 ± 2.8*	6
Veratridine plus tetrodotoxin	13.9 ± 1.0	6
Veratridine	16.1 ± 1.8	6
Deltamethrin	14.5 ± 1.2	6
Tetrodotoxin	14.5 ± 1.2	6
None	16.4 ± 2.4	6

[a]Synaptosomes were preincubated with neurotoxins for 10 min. at 36°C.
[b]Concentrations are: veratridine 100 μM, deltamethrin 0.1 μM, tetrodotoxin 1 μM.
*Significantly different from veratridine alone or veratridine + deltamethrin + tetrodotoxin at P ≤ 0.01.

Table IV. Effect of Aconitine, Deltamethrin, DDT and Grayanotoxin I on [22]Na Uptake Affected by Veratridine in Rat Brain Synaptosomes [a]

Treatment[b]	[22]Na uptake (nmol•min[-1]•mg[-1])[c]	
	Mean ± S.D.	No. of exp.
Veratridine plus scorpion venom	10.9 ± 1.6*	6
Veratridine plus aconitine	1.0 ± 0.1*	6
Veratridine plus grayanotoxin-I	4.1 ± 0.4*	6
Veratridine plus deltamethrin	5.2 ± 0.2*	6
Veratridine plus DDT	3.2 ± 0.2*	6
Veratridine	2.9 ± 0.2*	6

[a] Synaptosomes were preincubated with effectors for 10 min at 36°C.
[b] Concentrations are: veratridine 30 μM; scorpion venom 0.5 μg/ml; aconitine 100 μM; grayanotoxin-I 100 μM; deltamethrin 3 μM; DDT 100 μM.
[c] [22]Na uptake in the presence of tetrodotoxin (1 μM) was subtracted from the total [22]Na uptake to obtain tetrodotoxin-sensitive [22]Na uptake.
*Significantly different from veratridine alone at $P \leq 0.01$.

Table V. Effect of Antibodies on [22]Na Uptake Enhanced by Veratridine and Deltamethrin in Rat Brain Synaptosomes [a]

Treatment[b]	[22]Na uptake (nmol•min[-1]•mg[-1])[c]	
	Mean ± S.D.	No. of exp.
Veratridine, deltamethrin plus FM-AbI[d]	4.6 ± 0.5	6
Veratridine, deltamethrin plus RNa32C7	11.4 ± 1.2*	6
Veratridine, deltamethrin plus PLa2F9	11.9 ± 1.4*	6
Veratridine plus FM-AbI	3.7 ± 0.5	6
Veratridine plus RNa32C7	5.8 ± 0.1	6
Veratridine plus LNa2F9	7.5 ± 0.6	6
Veratridine plus deltamethrin	4.0 ± 0.04	6
Veratridine	2.9 ± 0.2	6

[a] Synaptosomes were preincubated with effectors for 5 min at 36°C, and then incubated with neurotoxins for 10 min at 36°C.
[b] Concentrations are: veratridine 30 μM; deltamethrin 3 μM; antibodies (FM-AbI, RNa32C7, LNa2F9) 0.3 μg.
[c] [22]Na uptake in the presence of tetrodotoxin (1 μM) was subtracted from the total [22]Na uptake to obtain tetrodotoxin-sensitive [22]Na uptake.
[d] At Ouchterlony double diffusion tests for FM-AbI and -AbII (see Table 7), remarkable precipitins were formed between antisera and their immunogens. Antisera reacted slightly with keyhole limpet hemocyanin. No precipitation was observed between sera from the rabbit before immunization and immunogens.
*Significantly different from the corresponding antibody + veratridine control (without deltamethrin) at $P \leq 0.01$.

Table V shows the effect of antibodies on the synaptosomal $^{22}Na^+$ uptake stimulated with deltamethrin and veratridine. FM-AbI is directed to the extracellular segment between S5a and S6a in rat sodium channel. It had no effect on the ^{22}Na uptake stimulated with deltamethrin. However, PNa32C7 and LNa2F9 antibodies enhanced deltamethrin-induced increase in the ^{22}Na uptake in the presence of veratridine. Interestingly, all these antibodies themselves had the properties to stimulate ^{22}Na uptake, particularly in the presence of deltamethrin and veratridine (Table V).

In the experiment with liposomal vesicles containing sodium channel, 24% of the internal $^{22}Na^+$ preloaded was lost in the vesicles when they were treated with veratridine. Also, 26.4% of the internal $^{22}Na^+$ was lost in the vesicles treated with deltamethrin in the presence of veratridine. However, only 11.4% of the internal $^{22}Na^+$ was lost in the vesicles treated with tetrodotoxin, indicating that there is a small but definitely significant component of $^{22}Na^+$ efflux that is regulated by tetrodotoxin-sensitive sodium channels (Table VI).

FM-AbI had no effect on the ^{22}Na efflux in the liposomal vesicles treated with deltamethrin in the presence of veratridine (Table VII). However, RNa32C7 and LNa2F9 inhibited slightly the ^{22}Na efflux. Furthermore, only 15% of the internal ^{22}Na was lost in the vesicles treated with FM-AbII in the presence of deltamethrin and veratridine. Thus, FM-AbII significantly inhibited the ^{22}Na efflux, reducing the efflux to a level similar to that observed in the presence of tetrodotoxin.

DISCUSSION

Previously, Jacques et al. (*41*), Lazdunski et al. (*42*), Lombet et al. (*37*) and Brown et al. (*36*) have shown that pyrethroids and DDT have the property to increase binding of [^{3}H]BTX to the sodium channel from several biological sources. Also shown by Bloomquist et al. (*38*) is the stimulatory effect of BTX on DDT-induced increase in ^{22}Na uptake. Thus, the interactive properties of BTX and this type of insecticidal chemicals have been well established. Our current findings have confirmed such an interactive phenomenon between [^{3}H]BTX and 1R-deltamethrin in the synaptosomal membranes from rat brain synaptosomes, and further demonstrated that the noninsecticidal stereoisomer, 1S-deltamethrin, does not cause any stimulatory action on [^{3}H]BTX binding, indicating the strict stereospecificity of interaction of the type II pyrethroids with sodium channels. That the main target site of BTX in synaptosomal membranes is the sodium channel was also clearly established by showing the inhibitory action of veratridine, which is known to specifically bind to the same binding site as BTX, and by demonstrating the stimulatory effects of sodium channel specific scorpion venom (Table I).

As for the $^{22}Na^+$ uptake study we have confirmed the generally acknowledged view that type II pyrethroids are capable of keeping the sodium channel open for longer time periods than expected from the normal course of depolarization-induced channel opening and closing (*43-45*). Also confirmed is the "use-dependent" action of deltamethrin; i.e., the action of type II pyrethroids is observed only when the sodium channel is opened through depolarization. We have, furthermore, established a "liposomal" methodology for studies on antibody action on sodium channels by improving the earlier model developed by Matsumura and Clark (*32*). The uptake of $^{22}Na^+$ could be specifically inhibited by tetrodotoxin and stimulated by agents known to keep the sodium channel open. Moreover, specific antibodies directed against particular segments of the sodium channel proteins

Table VI. Effect of Neurotoxins on ^{22}Na Efflux in Liposomal Vesicles Containing Reconstituted Sodium Channel [a]

Treatment[b]	% ^{22}Na remaining inside vesicles[c]	
	Mean ± S.D.	No. of exp.
Veratridine plus deltamethrin	73.7 ± 0.4*	6
Veratridine	76.0 ± 4.0	3
Tetrodotoxin	88.6 ± 1.1	6

[a] Solubilized protein samples were preincubated with neurotoxins for 10 min at 0°C before reconstitution into liposome.

[b] Concentrations are: veratridine 50 μM, deltamethrin 1.5 μM, tetrodotoxin 2 μM.

[c] Percent of ^{22}Na remaining inside liposomal vesicles was calculated using the following equation:

$$\frac{100 \; ([^{22}Na]_{30 \; s} - [^{22}Na]_{10 \; min})}{[^{22}Na]_{timezero} - [^{22}Na]_{10 \; min}}$$

*Significantly different from tetrodotoxin at P ≤ 0.01.

Table VII. Effect of Antibodies on ^{22}Na Efflux Stimulated with Neurotoxins in Liposomal Vesicles Containing Reconstituted Sodium Channel [a]

Treatment[b]	% ^{22}Na remaining inside vesicles[c]	
	Mean ± S.D.	No. of exp.
Veratridine, deltamethrin plus FM-AbI	73.6 ± 4.5	3
Veratridine, deltamethrin plus FM-AbII	85.0 ± 2.5*	3
Veratridine, deltamethrin plus RNa32C7	79.2 ± 1.6	3
Veratridine, deltamethrin plus LNa2F9	79.0 ± 1.5	3
Veratridine plus deltamethrin	73.7 ± 0.4	3
Tetrodotoxin	88.6 ± 1.1*	3

[a] Solubilized protein samples were preincubated with effectors for 10 min at 0°C before reconstitution into liposomes.

[b] Concentrations are: veratridine 50 μM, deltamethrin 1.5 μM, antibodies (FM-AbI and -AbII, RNa32C7, LNa2F9) 0.3 μg, tetrodotoxin 2 μM.

[c] Percent of ^{22}Na remaining inside liposomal vesicles was calculated using the same equation as shown in Table VI.

*Significantly different from veratridine + deltamethrin at P ≤ 0.01.

clearly affected the operation of the tetrodotoxin-sensitive portion of $^{22}Na^+$ efflux. Thus, we have shown that the portion of $^{22}Na^+$ efflux that was affected by these agents is likely through the specific sodium channels within the reconstituted liposomes.

The main objectives of this project have been to test the feasibility of establishing a methodology to study the effect of specific antibodies directed against known segments of the sodium channels.

The effects of various antibodies on $^{22}Na^+$ are expected to be relatively specific and straight forward, since antibodies are known to directly bind with certain parts of the sodium channel proteins and thereby cause either conformational changes or physically obstruct normal operations of the channel. One problem associated with the use of antibodies for their effects on channel operations is their accessibility to the specific parts of the channel proteins. For instance, FM-AbII is directed at an intracellular site which is not accessible in intact synaptosomal preparations. These antibodies are bulky and hydrophilic, and therefore, are not expected to penetrate through the plasma membrane or through the sodium channel. The sites of binding of RNa32C7 and LNa2F9 are not known, though that of FM-AbI is definitely extracellular. In the intact synaptosomal experiment (Table V) RNa32C7 and LNa2F9 synergized the effect of deltamethrin and veratridine to increase ^{22}Na uptake by the synaptosomes, indicating that the action sites of these antibodies may include some extra-cellular sites of the sodium channel.

FM-AbI had only modest stimulatory effects in this regard and furthermore such effects were statistically insignificant, indicating that either this external site is not functionally important or it did not actually bind with the site in intact synaptosomes.

The "liposomal" experiments (Tables VI and VII) are designed to circumvent the problem of accessibility by allowing these antibodies to first interact with lysed synaptosomal membranes and then incorporate them into liposomes. Under our experimental conditions the "inside out" preparation (i.e., ^{22}Na efflux study from $^{22}Na^+$ preloaded liposomes) worked better than "outside out" preparations. By this approach FM-AbII was found to be very potent in preventing veratridine-deltamethrin triggered $^{22}Na^+$ efflux. The level of its inhibition was almost equal to that achieved by 2 μM of tetrodotoxin. FM-AbI showed no effect, confirming the result obtained with intact synaptosomes.

The same question on the accessibility of antibodies to the [^3H]BTX binding site also must be considered. The synaptosomal membrane preparations are expected to offer the accessibility to both inside and outside surfaces of the plasma membrane. Furthermore, both batrachotoxin and veratridine are known to keep the channel in an open position. The calcium binding region, at any rate, is expected to be protruding from the inner surface of the plasma membrane (*26, 28*). Thus it should be accessible to the FM-AbII at least.

Judging by the Tufty-Kretsinger EF-hand test (*46*), this site could accommodate Ca^+ as a binding site as originally suggested by Babitch (*28*), though no actual observation has been made. This region is also a highly conserved one among various biological systems (*26-29*), indicating the essential nature of the presence of this region in normal functioning of the sodium channel. The fact that the stimulatory action of 1R-deltamethrin was blocked by this antibody indicates that the blockage and possibly impairment of the function of this portion of the sodium channel could affect in some way the effectiveness of this insecticide's action to keep the sodium channel open. This conclusion was derived from the observations that this group of pyrethroids is known to prolong the voltage depolarization-triggered increase in Na^+ current (Noda et al., 1986) and that the reason why they increase [^3H]BTX binding is the same, i.e., opening the sodium channel and, thereby, exposing the BTX binding site within the sodium channel. On the other hand, we have not proven or even implicated that the primary binding site of 1R-deltamethrin is this region of the sodium channel that is specifically blocked by FM-AbII. There are other possibilities such as allosteric interactions or rigidification of the gating machinery of the sodium channel, etc.

Nevertheless, we have demonstrated in this study that the immunological approach to test a specifically targeted area of the sodium channel is a powerful technique with great future potential for understanding the action mechanisms of those chemicals interacting with important target proteins such as the sodium channel.

Finally it must be emphasized that the above immunological test results show only the importance of this region in the operation of the rat brain sodium channel and do not show that this Ca^{2+} binding region is involved in *kdr* resistance. To be sure the above FM-AbII preparation is directed against the rat brain sodium channel and not against the corresponding channels of the German cockroach. Besides, the roles of SC type sodium channels in insects are not known, and instead the consensus among scientists is that para-type sodium channels are the ones carrying out the function of voltage dependent sodium gating on the axonic membrane of insect nervous system. Thus in the absence of my information regarding the nature of mutational changes within the structural component of para-type radium channels in *kdr* insects, it does not appear to be prudent to implicate this Ca^{2+} binding region for the *kdr* resistance phenomenon. In conclusion, we could show that the Ca^{2+} binding region of the rat brain sodium channel appears to be important and that there is a mutational change in an SC type sodium channel in the *kdr* resistance German cockroach. The relationship between these two events as well as that to the second resistance factor, the decreased CaM Kinase (CAM Kinase) liter in the resistant strains must be clarified in the future through vigorous work and critical evaluation.

Acknowledgments

Supported by California Agricultural Experiment Station, College of Agricultural and Environmental Sciences, University of California, and by research grants ES01963 and ES05707 from the National Institute of Environmental Health Sciences, Research Triangle Park, North Carolina.

Literature Cited

1. Georghiou, G.P.; Saito, T. (Eds). *Pest Resistance to Pesticides*; Plenum Press: New York, NY, 1983, 809 pp.
2. Milani, R.; Travaglino, A. *Riv. Parasitol.* 1957, *18*, 199.
3. Farnham, M.W. *Pestic. Sci.* 1973, *4*, 513-520.
4. Sawicki, R.M. *Pestic. Sci.* 1973, *4*, 501-512.
5. Tsukamoto, M.; Narahashi, T.; Yamasaki, T. *Botyukagaku*, 1965, *30*, 128-132.
6. Miller, T.A.; Salgado, V.L.; Irving, S.N. In *Pest Resistance to Pesticides*; Georghiou, G.P; Saito, T., Eds.; Plenum Press: New York, NY, 1983, pp. 353-366.
7. DeVries, D.H.; Georghiou, G.P. *Pestic. Biochem. Physiol.* 1981, *15*, 234-241.
8. Nolan, J.; Roulston, W.H.; Wharton, R.H. *Pestic. Sci.* 1977, *8*, 484-486.
9. Prasittisuk, C.; Busvine, J.R. *Pestic. Sci.* 1977, *8*, 527-533.
10. Priester, T.M.; Georghious, G.P. *J. Econ. Ent.* 1978, *71*, 197-200.
11. Gammon, D.W. *Pestic. Biochem. Physiol.* 1979, *2*, 53-62.
12. Matsumura, F. *Proc. 2nd Pestic. Chem. Congr.* 1971, *2*, 95-116.
13. Scott, J.G.; Matsumura, F. *Pestic. Biochem. Physiol.* 1983, *19*, 141-150.
14. Scott, J.G.; Croft. B.A.; Wagner, S.A. *J. Econ. Ent.* 1983, *76*, 6-10.

15. Salgado, V.L.; Irving, S.N.; Miller, T.A. *Pestic. Biochem. Physiol.* 1983, *20*, 100-114.
16. Chang, C.P.; Plapp, F.W. Jr. *Pestic. Biochem. Physiol.* 1983, *20*, 86-91.
17. Ghiasuddin, S.M.; Kadous, A.A.; Matsumura, F. *Comp. Biochem. Physiol.* 1981, *68C*, 15-20.
18. Rashatwar, S.; Matsumura, F. *Comp. Biochem. Physiol.* 1985, *81C*, 97-103.
19. Rashatwar, S.; Ishikawa, Y.; Matsumura, F. *Comp. Biochem. Physiol.* 1987, *88C*, 165-170.
20. Matsumura, F. In *Membrane Receptors and Enzymes as Targets of Insecticide Action*; Clark, J.M.; Matsumura, F., Eds.; Plenum Press: New York, NY, 1986; pp. 173-178.
21. Matsumura, F. *Comp. Biochem. Physiol.* 1988, *89C*, 179-183.
22. Charalambous, P.; Matsumura, F. *Insect Biochem. Molec. Biol.* 1992, *22*, 721-734.
23. Enan, E.; Matsumura, F. *Pestic. Biochem. Physiol.* 1991, *39*, 182-195.
24. Clark, J.M.; Brooks, M.W. *Biochem. Pharmacol.* 1989, *38*, 2233-2245.
25. Osborne, M.P.; Pepper, D.R. Agrochemical Division, Amer. Chem. Soc. Fourth Chemical Congress of North America, New York, NY, 1991, Abstract No. 150.
26. Noda, M.; Ikeda, T.; Katano, T.; Suzuki, H.; Takashima, H.; Kurasaki, M.; Takahashi, H.; Numa, S. *Nature* 1986, *320*, 188-192.
27. Salkoff, L.; Butler, A.; Wei, A.; Scavarda, N.; Giffen, K.; Ifune, C.; Goodman, R.; Mandel, G. *Science* 1987, *237*, 744-749.
28. Babitch, J. *Nature* 1990, *346*, 321-322.
29. Robinson, P.J.; Dunkley, P.R. *J. Neurochem.* 1983, *41*, 909-918.
30. Peterson, G.L. *Anal. Biochem.* 1977, *83*, 346-356.
31. Hartschorne, R.P.; Catteral, W.A. *Proc. Nat. Acad. Sci. USA* 1981, *78*, 4620-4624.
32. Matsumura, F.; Clark, J.M. *Neurotoxicology* 1985, *6*, 271-288.
33. Kasahara, M.; Hinkle, P.C. In *Biochemistry of Membrane Transport*; Semenza, G.; Caratoli, E., Ed.; Springer-Verlag, New York, NY, 1977; pp. 346-350.
34. Soderlund, D.M.; Bloomquist, J.R.; Ghiasuddin, S.M.; Stuart, A.M. In *Sites of Action for Neurotoxic Pesticides*; Hollingworth, R.M.; Green, M.B., Ed.; Am. Chem. Soc., Washington, DC, 1987; pp. 251-261.
35. Soderlund, D.M.; Bloomquist, J.R.; Payne, G.T.; Ottea, J.A. In *Insecticide Action: From Molecule to Organism*; Narahashi, T.; Chambers, J.E., Ed.; Plenum Press, New York, NY, 1989; pp. 85-98.
36. Brown, G.B.; Gaupp, J.E.; Olson, R.W. *Molec. Pharmacol.* 1988, *34*, 54-59.
37. Lombet, A.; Mourre, C.; Lazdunski, M. *Brain Res.* 1988, *459*, 44-53.
38. Bloomquist, J.R.; Soderlund, D.M. *Molec. Pharmacol.* 1988, *33*, 543-550.
39. Catterall, W.A.; Morrow, C.S.; Daly, J.W.; Brown, G.B. *J. Biol. Chem.* 1981, *256*, 8922-8927.
40. Villegas, R.; Villegas, G.M.; Barnola, F.V.; Racker, E. *Biochem. Biophys. Res. Commun.* 1977, *79*, 210-217.
41. Jacques, Y.; Romey, G.; Cavey, M.T.; Kartalovski, B.; Lazdunski, M. *Biochim. Biophys. Acta* 1980, *600*, 882-897.
42. Lazdunski, M.; Lombet, A.; Mourre, C. In *Neurotox '88: Molecular Basis of Drug and Pesticide Action*; Lunt, G.G., Ed.; Elsevier Science Publishers BV (Biomedical Division), Amsterdam, 1988; pp. 289-300.
43. Narahashi, T. *Neurotoxicology* 1985; *6*, 3-22.
44. Narahashi, T. In *Insecticide Action: From Molecule to Organism*; Narahashi, T.; Chambers, J.E., Eds.; Plenum, Press, New York, NY, 1989; pp. 55-58.
45. Vijverberg, H.P.M.; DeWeille, J.R. *Neurotoxicology* 1985, *6*, 23-34.
46. Tufty, R.M.; Kretsinger, R.H. *Science* 1975, *187*, 167-169.

RECEIVED December 5, 1994

Chapter 8

Site-Insensitive Mechanisms in Knockdown-Resistant Strains of House Fly Larva, *Musca domestica*

M. P. Osborne, D. R. Pepper, and P. J. D. Hein

Department of Physiology, University of Birmingham Medical School,
Birmingham B15 2TT, United Kingdom

Electrophysiological studies were carried out on the effects of
pyrethroids on the segmental nerves and presynaptic terminals of
neuromuscular junctions of the body wall muscles from larvae of
susceptible (Cooper) and resistant strains (*kdr*; *super-kdr*) of the
housefly (*Musca domestica* L.). We were unable to find any differences
in sensitivity between *kdr* and *super-kdr* axons to deltamethrin which
would account for the higher resistance shown by *super-kdr* over *kdr*
flies. Assay of presynaptic terminals, which is based on the ability of
pyrethroids to induce massive increases in miniature excitatory
postsynaptic potentials, showed that *super-kdr* junctions were no less
sensitive to the type I pyrethroid, fenfluthrin, than those of *kdr*. By
contrast *super-kdr* junctions were markedly insensitive to the type II
pyrethroid, deltamethrin, compared with *kdr* junctions. We propose
that *kdr* resistance encompasses at least two site-insensitive
mechanisms. The first is associated with the voltage-dependent Na^+
channel, and is similarly effective in *super-kdr* and *kdr* strains against
both type I and type II pyrethroids. The second is associated with
pyrethroid stimulation of Ca^{2+}-activated phosphorylation of key
proteins involved in release of neurotransmitters. Stimulation of
phosphorylation is suppressed in *kdr* strains, but such suppression is
particularly enhanced in *super-kdr* against type II compounds.
Ultrastructural studies of pyrethroid-poisoned neuromuscular junctions
indicate that this resistance mechanism is more likely to be associated
with processes involved with docking and fusion of vesicles with the
presynaptic membrane than with those associated with breaking links
between vesicles and the presynaptic cytoskeleton.

It has long been held that the major target site for DDT and the pyrethroid
insecticides is the voltage-dependent Na^+ channel responsible for the propagation of

0097–6156/95/0591–0128$12.25/0
© 1995 American Chemical Society

action potentials in nerve cells (*1*). This view has been challenged by several groups of workers (*2-7*) who have suggested that additional sites, namely, voltage-dependent Ca^{2+} channels, GABA and ACh receptors and various enzymes, ATP-ase, *Ecto* ATP-ase and Ca^{2+}-dependent calmodulin-activated presynaptic phosphorylases/phosphatases are also important target sites for these pesticides. Recent biochemical evidence (*8-10*) has strongly implicated presynaptic calcium-dependent phosphorylation reactions as target sites. The picture of pyrethroid toxicity is further complicated by the recognition of two classes of pyrethroid (*11-13*). Type I compounds, which lack the cyano group on the α-carbon atom of the 3-phenoxy benzyl alcoholic moiety, are in general less toxic than the type II compounds which posses this group. Whilst clear-cut differences in action between type I and II compounds upon neuronal preparations have been described, these differences are not always overt and are by no means consistent even within the same preparation (*14*). Type II compounds, unlike type I, have potent effects upon Ca^{2+}-activated phosphorylation of certain key proteins that are involved in release of neurotransmitters from presynaptic terminals (*8-10*). Moreover, much of this activity, although dependent upon depolarisation, is independent of Na^+ channel activation, since this activity persists in the presence of the Na^+ channel blocking agent, tetrodotoxin (TTX) (*7,15,16*).

Electrophysiological studies on the knockdown resistance factor, *kdr*, in houseflies and in other knockdown resistant species of insects (see *6* for references) have reinforced the key role of the Na^+ channel in pyrethroid toxicity. Several groups of investigators have confirmed that the Na^+ channel in resistant strains is relatively insensitive to DDT and pyrethroids. This insensitivity factor ranged between 2 and 10,000 fold depending on the species investigated, the type of assay employed and on which particular pyrethroid or DDT analogue was used (*6*). However, recent studies on the more resistant *super-kdr* strains of housefly have been unable to correlate this enhanced resistance with a further increase in Na^+-channel insensitivity to pyrethroids (*17-19*). Na^+ channels from *super-kdr* strains are, if anything, slightly *more* sensitive to pyrethroids than the less resistant *kdr* strains (*17-19*).

The *super-kdr* allele is much more effective against type II than type I pyrethroids. Furthermore, type II pyrethroids are more effective than type I in disrupting presynaptic phosphorylation reactions. Consequently we have undertaken a series of electrophysiological and electron microscopic investigations, using the neuromuscular junctions of housefly larvae as a model synapse, to clarify both the mechanism of action of pyrethroids and the *super-kdr* resistance factor at presynaptic terminals. Some of the neurophysiological results presented here have recently been published elsewhere (*19*).

Experimental Procedures

Insects. All investigations were carried out on 3rd (last) instar larvae of the housefly, *Musca domestica* L. The following genetically defined strains were used. A susceptible wild phenotype, Cooper strain; a *kdr* strain, 538ge (phenotypic markers *bwb:ge*), resistance factor *kdr*~Latina~; a *super-kdr* strain (wild phenotype), resistance factor *super-kdr*~3D~.

Pyrethroids. Two compounds were used, deltamethrin ((S)-α-cyano-3-phenoxy-benzyl (1R)-cis-3-(2,2-dibromovinyl)-2,2-dimethylcyclopropanecarboxylate (99%)) and fenfluthrin (pentafluorobenzyl (1R)-$trans$-3-(2,2-dichlorovinyl)-2,2-dimethylcyclo-propanecarboxylate). They were chosen because the two resistant strains showed no differential resistance to fenfluthrin (RF for *super-kdr* relative to *kdr* = 1.07), whereas, *super-kdr* flies were much more resistant (RF for *super-kdr* relative to *kdr* = 18.7) than those of the *kdr* strain to deltamethrin (A. W. Farnham, pers. commn.). Stock solutions of these compounds were made up in acetone. Application of compounds to experimental preparations was made via a final dilution step by adding a small volume of acetone/pesticide solution (0.5 ml litre^{-1}) to insect saline (sodium chloride, 172; potassium chloride, 13.3; calcium chloride, 1.00; magnesium chloride, 6; sodium hydrogen carbonate, 6; sodium dihydrogen phosphate, 0.7; sucrose 35 mM; pH 7.1)(*20*). Control preparations were challenge with 0.5 ml litre^{-1} acetone in saline.

Electrophysiological Experiments. Two series of experiments were undertaken. The first utilised pieces of isolated segmental nerves and was designed to study the effects of deltamethrin upon voltage sensitive Na$^+$ channels since these nerves contain neither nerve cell bodies nor synaptic contacts (*21*). The second series was carried out on the neuromuscular junctions of body wall muscles (*22*) and was concerned with investigating the effects of deltamethrin and fenfluthrin on the release process of neurotransmitters from presynaptic terminals by studying their effects upon miniature excitatory postsynaptic potentials (mEPSPs) (*23,24*). Since a full account of these experimental procedures has been published recently (*19*), no further details are given here.

Electron Microscopic Investigations. Larvae were dissected in plastic baths under saline to reveal the nervous system and body wall musculature. The saline in the bath was replaced with fresh saline to which was added a deltamethrin/acetone solution to give final bath concentrations of either 5 x 10^{-9}, 5 x 10^{-7} or 5 x 10^{-6} M deltamethrin. Controls were exposed to acetone only (0.5 ml litre^{-1}) dissolved in saline. All larvae were incubated in these solutions for 1 h following which they were fixed for 1-2 h in 2.5% glutaraldehyde in phosphate buffer (0.05M, pH 7.2-7.4) with sucrose added to raise the molarity of the fixative to 0.34 M. Post-fixation for a further 1 h was achieved in 1% osmium tetroxide in the same buffer (pH 7.2-7.4; 0.34 M). Larvae were placed in 70% methanol where individual ventral longitudinal muscles were dissected free. Dehydration of these muscles was completed in methanol and they were transferred via propylene oxide to the embedding medium, a modification of Epon 812 (Agar 100 resin). Polymerisation was carried out overnight at 60° C. Transverse sections of the muscle cells were cut on a Reichert Ultracut E ultramicrotome and stained in uranyl acetate and lead citrate before viewing in a JEOL 120CXII electron microscope.

Results

Isolated Segmental Nerves. Action potentials from about 2 to 6 axons were regularly obtained from these nerves. Application of acetone only to these nerves produced no obvious changes. There was a slow progressive decline in the frequency of action potentials over the duration of the experiments but nerve block did not occur. Deltamethrin was deemed to have a positive effect if it increased the rate of firing by 100% compared to that of the controls (see *19*). The smallest concentration of deltamethrin which produced a 100% increase in firing rate for Cooper strain was 5×10^{-13} M (Figure 1, Table I). The equivalent concentration for both the *kdr* and *super-kdr* strains was ten-fold lower at 5×10^{-12} M (Figures 2, 3, Table I). At these concentrations the increase in frequency following application of deltamethrin began almost immediately in *super-kdr*, rising to the 100% level within about 20 min (Figure 3). With *kdr* the rise in frequency to the 100% line did not occur until some 40 min after addition of deltamethrin and even then was only maintained transiently (Figure 2).

The maximum firing rate induced by deltamethrin differed between *kdr* and the other two strains (Figure 4). For *kdr* this was about 700 % above initial firing rates, but was much lower at 412% and 474% respectively for Cooper and *super-kdr*. This maximum percentage increase was obtained at a concentration of 5×10^{-7} M for both Cooper and *super-kdr*. The maximum percentage increase with *kdr* was obtained at a lower concentration of 5×10^{-8} M deltamethrin. With Cooper and *kdr* larvae the maximum increase in firing rate was obtained almost immediately upon application of deltamethrin, following which there was a rapid decline to levels well below the firing rate prior to application of the pyrethroid. However, with *super kdr* larvae the rise in the firing rate to the peak level was slower and more erratic and the fall in firing rate following the peak percentage increase did not fall below the pre-application firing rate as in the other two strains.

Neuromuscular Junctions.   Resting membrane potentials for these muscles lay between 60-70 mV. Spontaneously evoked mEPSPs of varying amplitude were evident in all three strains (Figure 5) and these continued without change in frequency or amplitude for several hours. Addition of suprathreshold concentrations of both pyrethroids usually produced within 5-10 min a massive, sustained increase in the frequency of the mEPSPs (several hundred fold) accompanied by increases in amplitude (Figure 6A and B) which lasted until the experiments were terminated, about 50 min later. There was, with one exception, no gradual and progressive increase in this pyrethroid-induced mEPSPs activity, the onset of this increase was always very sudden and sharply defined. The exception proved to be that of deltamethrin upon junctions of *super-kdr* larvae. In this case the increase was not so massive and was of a sporadic nature even when saturated solutions of deltamethrin as high as 5×10^{-6} M were employed. Only isolated bursts of high amplitude mEPSPs were obtained (Figure 7A). Fenfluthrin on the other hand was able to induce massive and prolonged discharges of mEPSPs in all strains including *super-kdr* (Figure 7B). The minimum concentrations required to induce mEPSPs activity with deltamethrin were 1.5×10^{-10}, 5×10^{-9} and 1.5×10^{-6} M respectively for Cooper, *kdr* and *super-kdr* larvae. With fenfluthrin the equivalent values were Cooper, 5×10^{-7} M and for both *kdr* and *super-kdr*, 5×10^{-6} M (Table I).

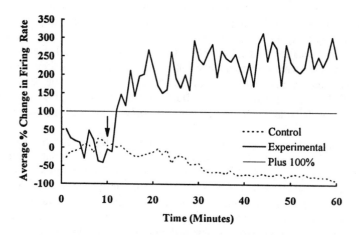

Figure 1. Graph to show effect of 10^{-13} M deltamethrin on Cooper larva isolated nerve. Arrow denotes point of application of deltamethrin. Data for this figure and figures 2,3 and 5-7 taken from ref *19*.

Table I. Comparison of Electrophysiological Responses and Toxicity Data

Compound	Strain	Nerve Threshold Concentration (M)	mEPSPs Threshold Concentration (M)	Neurophysiological Resistance Factor*	Toxicological Resistance Factor**
Deltamethrin	Cooper	$< 5\times10^{-13}$	$\cong 1.5\times10^{-10}$	-	-
	Kdr	$\cong 5\times10^{-12}$	$\cong 5\times10^{-9}$	Nerve ≥ 10 mEPSPs ≅ 33	$\cong 30$
	Super-kdr	$\cong 5\times10^{-12}$	$\cong 5\times10^{-6}$	Nerve ≥ 10 mEPSPs ≅ 10^4	$\cong 560$
Fenfluthrin	Cooper	-	$\cong 5\times10^{-7}$	-	-
	Kdr	-	$\cong 5\times10^{-6}$	mEPSPs ≅ 10	$\cong 14$
	Super-kdr	-	$\cong 5\times10^{-6}$	mEPSPs ≅ 10	$\cong 15$

*Neurophysiological factors are relative to Cooper threshold values (data from *20*).
**Toxicological factors are relative to Cooper LD_{50} values (data kindly furnished by A. W. Farnham, Rothamsted).

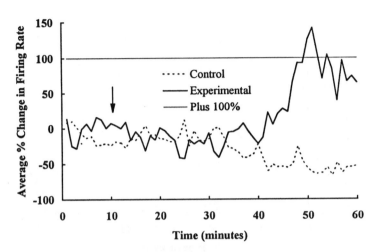

Figure 2. Effect of 10^{-12} M deltamethrin on isolated nerve from *kd*r larva. Arrow as before.

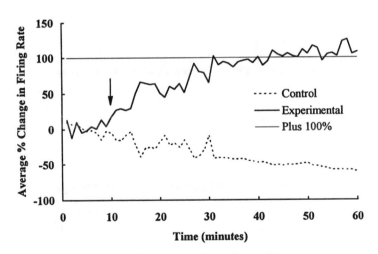

Figure 3. Effect of 10^{-12} M deltamethrin on isolated nerve from *super-kdr* larva. Arrow as before.

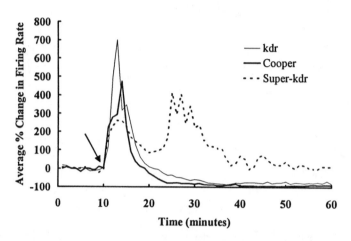

Figure 4. Isolated nerve. Maximum increases in rates of firing shown for all three strains. Arrow as before.

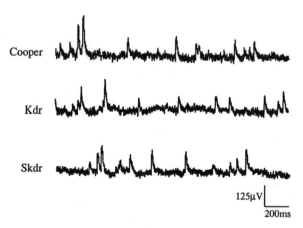

Figure 5. Control mEPSPs recorded from all three strains.

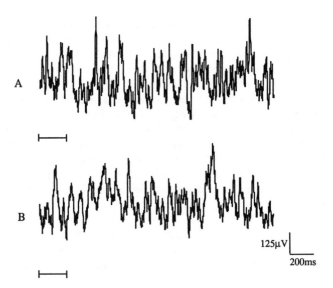

Figure 6. Massive trains of mEPSPs induced by pyrethroids. A, Cooper (5 x 10^{-9} M deltamethrin). B, Cooper (5 x 10^{-6} M fenfluthrin). Bars denote levels of resting membrane potentials.

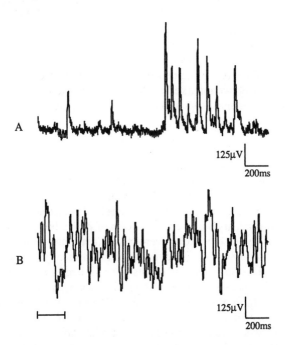

Figure 7. Recordings from *super-kdr* larvae. A, low frequency train of high amplitude mEPSPs induced by 5 x 10^{-6} M deltamethrin. B, massive increase in mEPSPs induced by 5 x 10^{-6} M fenfluthrin. Bar denotes level of resting membrane potential.

Electron Microscopy. Transverse sections of neuromuscular junctions showed typical features which have been described in detail for fly larvae (25,26). The presynaptic terminals contain mitochondria and classic electron lucent synaptic vesicles. These vesicles are associated with dense bodies (active zones) that are adpressed to the inner surface of the axolemma (Figure 8). In all three strains, concentrations of deltamethrin that were below those needed to produce massive increases in mEPSPs did not noticeably affect the fine structure of presynaptic terminals. However, with Cooper and *kdr* larvae, doses sufficient to elevate mEPSPs frequencies, reduced considerably, the numbers of synaptic vesicles in nerve terminals .

In Cooper junctions, 5 x 10⁻⁹ M deltamethrin produced almost total depletion of vesicles in nerve terminals (Figure 9). Vacuolation of mitochondria was also apparent together with accumulations of granular material in the axoplasm. Similar features were not seen with *kdr* nerve terminals until the concentration of deltamethrin was raised to 5 x 10-7 M.

A concentration of 10⁻⁹ M, deltamethrin induced an interesting effect upon *super-kdr* nerve terminals (This dose lies well below that which elevates mEPSPs frequencies in this strain). Although the numbers of synaptic vesicles were not depleted, many were distended or swollen and gathered in clumps adjacent to the presynaptic membrane in the neighbourhood of the active zones (Figure 10). The remainder of the axoplasm was occupied by granular material identical to that seen in Cooper nerves also treated with 5 x 10⁻⁹ M deltamethrin (Figure 9). Such clumping and swelling of vesicles was not seen in nerve terminals from Cooper or *kdr* larvae regardless of the concentration of deltamethrin used. With deltamethrin at 5 x 10⁻⁸ M there was still no significant loss of vesicles from nerve terminals of *super-kdr* larvae. Clumps of vesicles, similar to those seen with 5 x 10⁻⁹ M deltamethrin, were still present, but swelling of vesicles was by no means as prevalent or pronounced. Only at the very high concentration of 5 x 10⁻⁶ M deltamethrin, i.e. a saturated solution, did disruption of nerve terminals occur with marked depletion of vesicles. Even then, some nerve-terminal profiles still contained appreciable numbers of vesicles.

Discussion

Knockdown Resistance and Axonal Transmission. There is little doubt that *kdr* and *super-kdr* resistance is associated with insensitivity of the voltage dependent Na⁺ channel to the pyrethroids and to DDT. The dilemma appears to be that of explaining the enhanced resistance shown by *super-kdr* strains by this mechanism (6).

In an earlier preliminary study on the effects of permethrin upon larval nerves it was indicated that *super-kdr* axons were about ten times less sensitive to this compound than those of *kdr* (27). The results of this present study with deltamethrin do not agree with this. Although our results upon isolated nerves clearly show that both resistant strains posses a nerve insensitivity factor, being some 10 fold less sensitive to deltamethrin than the susceptible Cooper strain, we were unable to separate the two resistant strains by this type of assay (Table I). However, our results consolidate our previous studies upon nerves of adult houseflies in which we used several pyrethroids as well as DDT. As before, (17,18) we found that the more

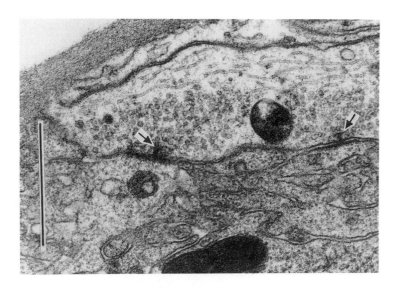

Figure 8. Electron micrograph (EM) of Cooper control nerve terminal. Terminal contains many synaptic vesicles clustered about active zones (arrows). Scale bar = 1 μM.

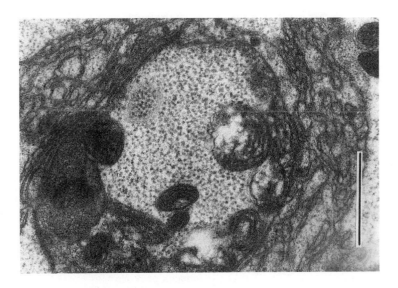

Figure 9. EM of Cooper nerve terminal exposed to 5 x 10^{-9} M deltamethrin. Note absence of synaptic vesicles, axoplasm packed with granular material and vacuolated mitochondria. Scale bar = 1 μM.

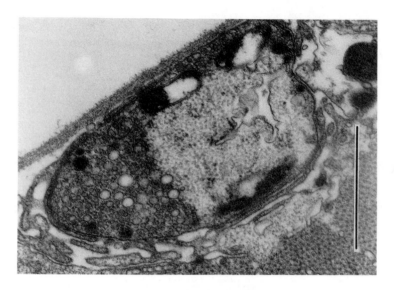

Figure 10. EM of *Super-kdr* nerve terminal exposed to 5 x 10^{-9} M deltamethrin. Note clumping of large numbers of swollen vesicles adjacent to presynaptic membrane, numerous granules in axon and vacuolated mitochondria. Scale bar = 1 μM.

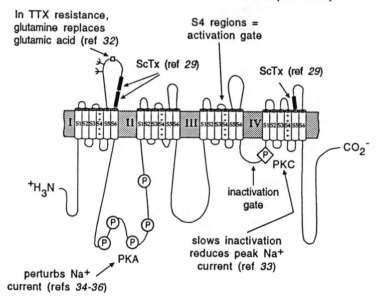

Figure 11. Transmembrane diagram of α-subunit of vertebrate Na^+ channel based on DNA sequencing studies. For explanation see text.

resistant *super-kdr* strain has Na$^+$ channels which appear to be marginally *more* sensitive to pyrethroids than those of *kdr* (Compare figures 2 and 3). It was apparent in the isolated larval nerve preparation that the maximum percentage increase in firing rate of *kdr* axons following deltamethrin treatment was much higher than in the other two strains. This result substantiates our earlier work on adult houseflies (*6*) where we found that the percentage increase in the rate of firing of action potentials induced by the pyrethroid, kadethrin (RU15525), in the neck connectives of *super-kdr* flies was twice that obtained in Cooper and *kdr* strains. This suggests that the *kdr* factor produces alterations in the Na$^+$ channels that are not simply concerned with insensitivity to pyrethroids. It might also induce changes associated with gating kinetics or the relative refractory period of the Na$^+$ channel.

Genetic Sequencing of Na$^+$-Channel Structure and Site of Pyrethroid Action. The structure of the α-subunit of the Na$^+$ channel has been sequenced for both vertebrates (rat brain) and insects (*Drosophila* brain) (Figures 11,12) (*28-31*). Both consist of four homologous repeating membrane-spanning domains (I-IV) comprising six α-helical hydrophobic segments, S1-S6, connected by hydrophilic extracellular and intracellular links (Figures 11,12). Several sites of functional significance have been identified on the vertebrate Na$^+$ channel. The voltage sensor and activation gate is considered to be associated with the positively-charged S4 segments. The inactivation gate has been localised to the intracellular link connecting the S6 and S1 segments of domains III and IV, and a scorpion toxin binding site has been identified on the extracellular links between the S5 and S6 segments on both domains I and IV. A genetic mutation consisting of a single amino acid substitution, replacing glutamic acid (residue no. 387) with glutamine, on the extracellular link between segments S5 and S6 of repeat domain I confers resistance towards TTX and saxitoxin in the rat (*32*). Furthermore this change also reduced the inward current flow through the Na$^+$ channel. This evidence suggests that the S5-S6 loop lies near to the channel mouth.

The identification and functional significance of phosphorylation sites on Na$^+$ channels have recently received attention. In rat brain (Figure 11) four sites have been located on the intracellular link between domains I and II. These sites are phosphorylated by protein kinase A (PKA). One other site, located on the inactivation gate which comprises the intracellular link between domains III and IV, is phosphorylated by protein kinase C (PKC). Phosphorylation of the inactivation gate phosphorylation site by PKC is accompanied by slowing of inactivation and a reduction of the peak Na$^+$ current (*33*). Evidence concerning the physiological effects of PKA-activated phosphorylation is contradictory (*34,35*) although there is no disagreement that PKA-phosphorylation does perturb Na$^+$ channels. A more complex pattern of Na$^+$-channel phosphorylation has recently emerged (*36*). It appears that phosphorylation of the PKA sites and its subsequent effects upon the Na$^+$ channel is dependent upon prior phosphorylation of the PKC site. Thus the action of PKA phosphorylation in reducing peak Na$^+$ currents is dependent on phosphorylation by PKC of the site on the inactivation gate.

In insects, Gordon *et al* (*37*) have reported that the α-subunit of sodium channels in four orders, Orthoptera, Dictyoptera, Lepidoptera and Diptera, can be phosphorylated by PKA. Phosphorylation sites in *Drosophila* Na$^+$ channels (Figure

12) occur one on each of the intracellular links between domains I and II and III and IV, and two on the inactivation gate (*31*). It is at present not known if phosphorylation of these sites has any functional significance. In crustacean (lobster) axons, Ca^{2+}-calmodulin-dependent phosphorylation of the α-subunit of the Na^+ channels has been reported. Furthermore, deltamethrin at a concentration as low as $10^{-11}M$ inhibited this phosphorylation (*38*). This raises the question whether some of the effects of pyrethroid poisoning could result from perturbation of phosphorylation of Na^+-channels.

In this context the effects of an anticalmodulin agent, W-7, as well as those of a pyrethroid (meta-methylbenzyl pyrethrate) and a methoxychlor DDT analogue were recently studied on action potentials in crayfish giant axons (*39*). W-7 and the pesticides all prolonged the fall phase of the action potential. W-7 also suppressed both the inward and outward membrane currents. However, whilst both pesticides prolonged the opening time of the Na^+ channels, W-7 did not. It apparently delayed the fall phase of the action potential by suppressing K^+ channel currents. These results suggested that the target sites of the pesticides and W-7 are not the same. Nevertheless, more work needs to be done with respect to pyrethroid action and phosphorylation of Na^+ channels.

In housefly Na^+ channels (Figure 12), *kdr* and *super-kdr* resistance has recently been mapped to a locus on gene *Msc* that lies between segment S2 of domain IV and the intracellular carboxy terminus (*40*). In pyrethroid-resistant strains of the tobacco budworm, *Heliothis*, changes in structure of gene *hscp* map to a region of the Na^+ channel situated between S5 of domain III and S4 of domain IV (*41*). Both of these sequences include an S4 component of the activation gate; the sequence in *Heliothis* also includes the inactivation gate. A study of *kdr*-like resistance in *Drosophila* found only a single substitution, in which asparagine replaced aspartic acid, in the *DSC1* gene that was localised to the extracellular loop between S5 and S6 segments of domain III (*42*) (Figure 12). There is a similarity between the pyrethroid resistant mutation in *Drosophila* and that of the TTX-resistant mutation in the rat Na^+ channel (*32*). Both occur on S5 to S6 linkers, although from different domains, and both reduce the net negative charge on these extracellular loops. These similarities may eventually turn out to be significant. However, the facts that TTX and pyrethroids have such very different physicochemical properties, have separate binding sites and dissimilar actions on Na^+ channels, make it extremely difficult to correlate the recently described structural changes in the α-subunit with knockdown resistance.

Knockdown Resistance and Synaptic Transmission. It has previously been shown that neuromuscular junctions of dipteran larvae with *kdr* or *kdr*-like resistance factors are less sensitive to pyrethroids and to DDT than junctions of susceptible strains (*23,24,43,44*). The strength of the mEPSP assay was that it gave good correlation both with respect to the relative toxicity of pyrethroids and for *kdr* resistance factors. At that time the actions of pyrethroids upon presynaptic terminals in both susceptible and resistant strains were ascribed to effects upon Na^+ channels (*23,24*). The more resistant *super-kdr* strains were not investigated using the mEPSP assay. However, the results reported here with this assay clearly show that it is also able to distinguish between *kdr* and *super-kdr* resistance factors. Toxicological data from adult flies

(Table I) indicate that *super-kdr* is no more effective than *kdr* against the type I pyrethroid, fenfluthrin, both being about 14-15 times more resistant to this compound than Cooper. This is in good agreement with values obtained from the mEPSP assay, both strains being about 10 times less sensitive to pyrethroids (Table I). On the other hand *super-kdr* junctions were extraordinarily insensitive to the type II compound deltamethrin. They were relatively unperturbed (c.f. figures 6,7) even by saturated solutions of deltamethrin being some 300 times less sensitive than *kdr* neuromuscular junctions. *Kdr* junctions themselves were about 30 times less sensitive than those of Cooper. Thus *super-kdr* junctions were, in this assay, about 10,000 times less sensitive to deltamethrin than those of Cooper! Toxicological estimates show that *super-kdr* flies are only about 20 times more resistant than *kdr*, a figure somewhat less than our electrophysiologically determined figure.

The mEPSP assay is thus a far superior indicator of the efficacy of *kdr* and *super-kdr* resistance factors than assays based on effects of pyrethroids on axonal transmission. What is encouraging is that *super-kdr* is particularly effective against type II compounds and relatively ineffective or non-effective against some type I pyrethroids. It is precisely these differences that are highlighted by the mEPSP assay.

It is apparent, therefore, that some mechanism which exhibits insensitivity to pyrethroids, particularly type II compounds, is located within the presynaptic terminal. Such a mechanism could result from modification of presynaptic Na^+ channels since these may be different from those associated with axonal conduction, or it may be related to modification of voltage-gated Ca^{2+} calcium channels since the latter are also targeted by pyrethroids (*45,46*). Neither of these possibilities can be completely excluded, but recent biochemical evidence, reported below, suggests that other factors, particularly those associated with phosphorylation/dephosphorylation of key presynaptic proteins involved with release of neurotransmitters, are probably involved.

Effects of Pyrethroids on Phosphorylation at Presynaptic Terminals. Although membrane-associated Na^+-K^+-ATPase and Ca^{2+}-ATPase and mitochondrial Mg^{2+}- and CA^{2+}-ATPase are inhibited by DDT and pyrethroids it is doubtful if this inhibition is important in their toxicity since it only occurs at relatively high doses, ≥ 1 µM, well above that required to disrupt axonal conduction (*6*). A more likely target site at presynaptic terminals is perturbation of the phosphorylases, PKC, calcium-calmodulin stimulated protein kinase II (CaM-Kinase II) and the phosphatase, calcineurin, especially by type II pyrethroids. Type I compounds and DDT are much less effective (*8-10,17,47-49*). Significantly, some of these effects occurred at concentrations of deltamethrin as low as 10^{-13} M in both squid and mammalian synaptosomes (*9,50*), that is at similar levels to those that perturb voltage dependent Na^+ channels. A further point is that these phosphorylation reactions are dependent upon depolarisation of the presynaptic terminal. However, most of this phosphorylation activity was still retained when Na^+ channels were blocked with TTX and depolarisation was triggered by high K^+ or excess of Ca^{2+} (*7,8,17*). Moreover, deltamethrin-induced phosphorylation was not completely abolished even in the presence of both Na^+ and Ca^{2+} channel blockers or when the extracellular $[Ca^{2+}]$ was zero (*17*). Some of these effects are still likely to be triggered by a rise in intracellular $[Ca^{2+}]$ since deltamethrin is able to enhance release of this cation indirectly from intracellular stores by stimulating the PKC/phosphoinositide pathway (*10*) and the production of inositol phosphates from phosphoinositides. Inositol triphosphates in turn liberate Ca^{2+} from intracellular sites (*49*). Clearly then, disruption of enzymes associated with presynaptic phosphorylation reactions appears to be an important target site for type II pyrethroids that is quite distinct from their effects on Na^+ channels.

The major proteins that are phosphorylated are CaM-Kinase II, synapsin I and B50 (Growth associated protein [GAP]-43 or neuromodulin). Dephosphosynapsin I anchors vesicles within the presynaptic terminal by binding to them to actin and to fodrin (see reviews *51,52*). The vesicles are thus locked within the cytoskeletal framework and are prevented from interacting with the active zones at the release sites on the presynaptic membrane (Figure 13). The combined effect of raising intraterminal $[Ca^{2+}]$ and phosphorylation of synapsin I via stimulation of CaM-Kinase II with deltamethrin would be to reduce binding of vesicles to the cytoskeleton, enabling them to approach the presynaptic membrane to facilitate transmitter release. The vesicle-presynaptic membrane fusion process involved with exocytosis is also in part regulated by CaM-Kinase II. For instance, synaptophysin and synaptophorin, two integral proteins of the synaptic vesicle membrane which can be phosphorylated by this kinase, are implicated with the formation of the exocytotic fusion pore (see *53*).

Stimulation of PKC by deltamethrin could also enhance release of neurotransmitter via phosphorylation of synaptotagmin which is an integral protein of the synaptic vesicle membrane. Synaptotagmin is considered to be a docking-fusion protein, facilitating exocytosis perhaps by interacting with another docking protein, syntaxin, which is associated with the presynaptic membrane (*53,54*)(Figure 14). PKC-stimulated phosphorylation of other presynaptic membrane associated-proteins such as myristoylated, alanine-rich C-kinase substrates (MARCKS) and neuromodulin could also be involved (*52*)(Figure 14). Phosphorylation of MARKS reduces its affinity for binding actin which may facilitate rearrangement of actin-based cytoskeleton in the immediate vicinity of the active zones, thereby modulating the access and docking of vesicles with the presynaptic membrane. Phosphorylation of neuromodulin is likely to release calmodulin and thus facilitate other phosphorylation reactions involved in release of neurotransmitters. All or some of these phosphorylation reactions instigated by deltamethrin might explain the appearance of massive and prolonged bursts of mEPSPs at insect neuromuscular junctions (*23,24,55,56*)(Figures 7, 8). In the longer term they would also account for subsequent depletion of vesicles from neuromuscular junctions seen by electron microscopy (*57*)(Figure 9).

More recent work with rat brain synaptosomes has shown that the pattern and time course of deltamethrin-induced-protein phosphorylation are complicated (*7*). Depolarisation of synaptosomes pretreated with deltamethrin for 5-20 min caused marked elevation in phosphorylation of synaptic proteins. Prolonged incubation with deltamethrin over 30-40 min significantly reduced depolarisation-induced protein phosphorylation. Consequently depending on the length of treatment, the effects of deltamethrin on protein phosphorylation could be stimulatory or inhibitory (*7*). Moreover, pyrethroids do not only perturb protein kinases. Calcineurin, the most significant presynaptic phosphatase, is specifically inhibited by type II but not by type I pyrethroids (*8*). Such an effect would preserve proteins in their phosphorylated state.

INSECT Na+ CHANNEL

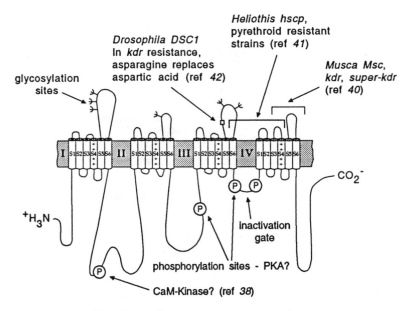

Figure 12. Transmembrane diagram of α-subunit of insect Na+ channel. Regions where changes in structure associated with resistance have been identified are indicated. See text.

PRESYNAPTIC TERMINAL

Ca2+-Calmodulin-Dependent Protein Kinase II
Phosphorylated Proteins

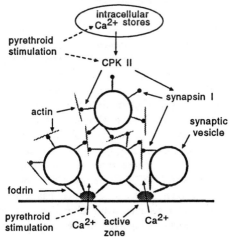

Figure 13. Protein molecules of nerve terminal that are phosphorylated by pyrethroid stimulation of CaM-Kinase II. For explanation see text.

PRESYNAPTIC TERMINAL

Protein Kinase C
Phosphorylated Proteins

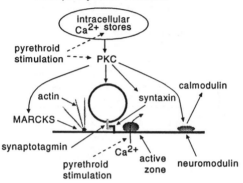

Figure 14. Protein molecules of nerve terminal that are phosphorylated by pyrethroid stimulation of PKC. For explanation see text.

Thus early in pyrethroid toxicity, when protein phosphorylation would be elevated and phosphatase activity reduced by deltamethrin, neurotransmitter release would be enhanced. Later, when inhibition of protein phosphorylation occurred, release of neurotransmitter would be curtailed. The latter has been correlated (7) with the phenomenon of late block (23,24). However, late block may well be caused by depletion of neurotransmitter. This is supported by direct observation of pyrethroid-treated presynaptic terminals; many of which are depleted or devoid of synaptic vesicles (Figure 9)(55,57).

Phosphorylation and Knockdown Resistance. Ghiasuddin et al (58) reported that a DDT-resistant strain of cockroach had a Ca-ATPase (subsequently shown to be a protein kinase-phosphatase) with a reduced affinity for Ca^{2+} compared with the susceptible strain. This strain was later shown to have a *kdr*-like factor with cross resistance to pyrethroids (13). Studies on the stimulatory effects of Ca^{2+} upon Na^+ Ca^{2+}-protein kinase-phosphatases in synaptosomes of housefly brain (59) have shown that these stimulatory effects were less in both *kdr* and *super-kdr* resistant strains. Moreover insensitivity to Ca^{2+} was even more marked in *super-kdr* than in *kdr* flies. Charalambous and Matsumura (60) have shown in both housefly and German cockroach that calmodulin was a better activator of synaptosomal CaM-Kinase II in susceptible than in *kdr*-type resistant strains, indicating that *kdr* resistance is associated with an alteration in CaM-Kinase. Indeed these resistant insects also exhibit cross-resistance to agents such as A23187, which are Ca^{2+} ionophores that increase intracellular [Ca^{2+}]. Clearly then sufficient evidence has emerged which strongly implicates *kdr* and *super-kdr* resistance with alterations in sensitivity of Ca^{2+}-stimulated phosphorylation reactions involved in the release of neurotransmitters at presynaptic terminals.

Nerve Terminal Structure and Knockdown Resistance. Biochemical and electrophysiological evidence indicate that *kdr* resistance and particularly that of *super-kdr* involves suppression of pyrethroid-stimulation of phosphorylation process associated with release of neurotransmitters from the presynaptic terminals. At least two major effects have been considered (Figures 13, 14). One concerns the release of synaptic vesicles from the cytoskeletal skeleton allowing them to approach the presynaptic membrane. The other is involved with docking and attachment of the vesicles to the presynaptic membrane as a prelude to exocytosis. Clearly *kdr* or *super-kdr* resistance mechanisms could affect either or both of these processes. Electron micrographs of nerve terminals treated with subthreshold doses of deltamethrin show that the synaptic vesicles are not prevented from approaching the presynaptic release sites. On the contrary they appear to be aggregated in large numbers against the presynaptic membrane (Figure 10). This suggests at least with *super-kdr* that the resistance mechanism interferes with docking of vesicles or their fusion with the presynaptic membrane rather than interfering with their attachment to the cytoskeleton. A similar phenomenon has been reported at the 'early block' phase in cypermethrin-poisoned neuromuscular junctions of non-resistant housefly larvae (*20*). Clusters of vesicles were found in association with 'active zones' on the presynaptic membrane. However, in this case block was not attributed to interference with vesicle/membrane interactions but to block of conduction in the presynaptic terminal (*20*).

Electrophysiological recordings from *super-kdr* junctions show that when mEPSPs are evoked by deltamethrin, they are of relatively large amplitude (Figure 7A). This correlates well with the enlarged synaptic vesicles which are present in nerve terminals of *super-kdr* larvae poisoned with deltamethrin (Figure 10).

Conclusions

There is no doubt that *kdr* and *super-kdr* resistance factors are linked with insensitivity of the voltage dependent Na^+ channel to DDT and to both type I and type II pyrethroids. However, there is strong evidence that *kdr* and *super-kdr* resistance embraces a second mechanism that is located at the presynaptic terminal. This second mechanism is associated with pyrethroid-stimulation/inhibition of Ca^{2+}-dependent phosphorylase/phosphatase reactions involved with release of neurotransmitters. It entails insensitivity of these phosphorylation reactions to pyrethroids and is especially effective in *super-kdr* resistance against type II pyrethroids. It has yet to be determined if the resistance mechanism operates at the level of synaptic vesicle/cytoskeleton interactions or synaptic vesicle/membrane-docking-fusion interactions, or, perhaps both. The relative efficacies of *kdr*/*super-kdr* resistance factors would therefore depend on the combined effectiveness and interplay of these two mechanisms against DDT and both types of pyrethroid.

Mutants of the *kdr* type could well provide useful tools which would be helpful in unravelling fundamental processes associated with release of neurotransmitters from presynaptic terminals. Recent findings clearly highlight the importance of Ca^{2+}-activated phosphorylases/phosphatases as important target sites for pyrethroids in addition to that of the Na^+ channel. Whether perturbation of phosphorylation of the Na^+ channel is significant in pyrethroid toxicity remains to be resolved.

The genetic evidence that *kdr* resistance maps to structural modifications of the Na^+ channel is encouraging. Further studies along these lines are necessary to fully identify and understand these modifications in relation to pyrethroid toxicity. Perhaps an even greater challenge is to understand how two seemingly separate mechanisms, insensitivity of the Na^+ channel- and Ca^{2+}-activated phosphorylation-of proteins, are reconciled at the genetic level by the *kdr/super-kdr* alleles.

Acknowledgements

We thank the AFRC for a link grant with Rothamsted Experimental Station in aid of this work. We are also indebted to Rothamsted for supplying the pesticides and susceptible and resistant strains of flies and to Staff members of this station, particularly Drs. A. Devonshire, A. Farnham and I. Denholm for very helpful collaboration and discussions throughout the course of this work. Also greatly appreciated is the computing and statistical advice given by H.F. Ross of the Physiology Department, University of Birmingham.

Literature Cited

1. Vijverberg, H. P. M.; Vandenbercken, J. *Critical Revs. Toxicol.* **1990**, *21*, 105-126.
2. Clark, J. M.; Brooks, M. W. *Biochem. Pharmacol.* **1988**, *38*, 2234-2245.
3. Clark, J. M.; Brooks, M. W. *Environ. Toxicol. Chem.* **1989**, *8*, 361-372.
4. Sattelle, D. B.; Yamamoto, D. *Adv. Insect Physiol.* **1988**, *20*, 147-213.
5. Soderland, D. M.; Bloomquist, J. R.; Wong, F.; Knipple, D. C.; Payne, L. *Pestic. Sci.* **1988**, *26*, 359-374.
6. Osborne M. P.; Pepper, D. R. In *Molecular Mechanisms of Insecticide Resistance*; Mullin, C. A., Scott, J. G., Eds.; ACS Symp. Ser. No. 505; American Chemical Society: Washington, DC, 1992, pp 71-89.
7. Kanemoto, Y.; Enan, E. E.; Matsumura, F.; Miyazawa, M. *Pestic. Sci.* **1992**, *34*, 281-290.
8. Enan, E.; Matsumura, F. *Biochem. Pharmacol.* **1992**, *43*, 1777-1784.
9. Enan, E.; Matsumura, F. *Pestic. Sci.* **1993**, *37*, 21-30.
10. Enan, E.; Matsumura, F. *Biochem. Pharmacol.* **1993**, *45*, 703-710.
11. Narahashi, T. In *Pyrethroid Insecticides: Chemistry and Action*; Mathieu, J., Ed.; Roussel-UCLAF: Romainville, 1980, pp 15-17.
12. Gammon, D. W.; Brown, M. A.; Casida, J. E. *Pestic. Biochem. Physiol.* **1981**, *15*, 181-191.
13. Scott, J. G.; Matsumura, F. *Pestic. Biochem. Physiol.* **1983**, *19*, 141-150.
14. Ruight, G. S. F. In *Insect Physiology, Biochemistry and Pharmacology*; Kerkut, G., Gilbert, L., Eds.; Pergamon Press: Elmsford, NY, 1985, Vol. 12; pp 194-251.
15. Clark, J. M.; Matsumura, F. *Pestic. Sci.* **1991**, *31*, 73-90.
16. Enan, E. E.; Matsumura, F. *Pestic. Biochem. Physiol.* **1991**, *39*, 182-195.
17. Gibson, A J.; Osborne, M. P.; Ross, H. F.; Sawicki, R. M. *Pestic. Sci.* **1988**, *23*, 283-292.

18. Gibson, A J.; Osborne, M. P.; Ross, H. F.; Sawicki, R. M. *Pestic. Sci.* 1990, *30*, 379-396.
19. Pepper, D. R.; Osborne, M. P. *Pestic. Sci.* 1993, *39*, 279-286.
20. Seabrook, G. R.; Duce, I. R.; Irving, S. N. *Pestic. Sci.* 1988, *24*, 179-180.
21. Osborne, M. P.; Smallcombe, A. In *Pesticide Chemistry-Human Welfare and the Environment*; Miyamoto, J., Ed.; Pergamon Press: Oxford, 1983, pp 103-107.
22. Osborne, M. P.; Hart, R. J. *Pestic. Sci.* 1979, *10*, 407-413.
23. Salgado, V. L.; Irving, S. N.; Miller, T. A. *Pestic. Biochem. Physiol.* 1983, *20*, 100-114.
24. Salgado, V. L.; Irving, S. N.; Miller, T. A. *Pestic. Biochem. Physiol.* 1983, *20*, 169-182.
25. Osborne, M. P. *J Insect Physiol.* 1967, *13*, 827-833.
26. Osborne, M. P. In *Insect Muscle*; Usherwood, P. N. R., Ed.; Academic Press: London, 1975, pp 151-205.
27. Nicholson, R. A,; Hart, R. J.; Osborne, M. P. In *Insect Neurobiology and Pesticide Action (Neurotox '79)*; Soc. Chem. Ind.: London, 1980, pp 465-471.
28. Catterall, W. A.; Scheuer, T.; Thomsen, W.; Rossie, S. *Ann. N. Y. Acad. Sci.* 1991, *625*, 174-180.
29. Catterall, W. A. *Science* 1988, *242*, 50-61.
30. Nakayama, H.; Shikano, H.; Kanaoka, Y. *Biochim. Biophys. Acta* 1992, *1175*, 67-72.
31. Loughney, K.; Kreber, R.; Ganetzky, B. *Cell* 1989, *58*, 1143-1154.
32. Noda, M.; Suzuki, H.; Numa, S.; Stühmer, W. *FEBS Lett.* 1989, *259*, 213-216.
33. West, J. W.; Numann, R.; Murphy, B. J.; Scheuer, T.; Catterall, W. A. *Biophys. J.* 1992, *62*, 31-33.
34. Rossie, S.; Catterall, W. A. *J. Biol. Chem.* 1989, *264*, 14220-14224.
35. Smith, R. D.; Goldin, A. L. *Amer. J. Physiol.* 1992, *263*, 660-666.
36. Li, M.; West, J. W.; Numann, R.; Murphy, B. J.; Scheuer, T.; Catterall, W. A. *Science* 1993, *261*, 1439-1442.
37. Gordon, D.; Moskowitz, H.; Zlotkin, E. *Arch. Insect Biochem. Physiol.* 1993, *22*, 41-53.
38. Miyazawa, M.; Matsumura, F. *Pestic. Biochem. Physiol.* 1990, *36*, 147-155.
39. Matsuda, K.; Okimoto, H.; Hamada, M.; Nishimura, K.; Fujita, T. *Comp. Biochem. Physiol.* 1993, *104C*, 181-186.
40. Williamson, M. S.; Denholm, I.; Bell, C. A.; Devonshire, A. L. *Mol. Gen. Genet.* 1993, *240*, 17-22.
41. Taylor, M. F.; Heckel, D. G.; Brown, T. M.; Kreitman, M. E.; Black, B. *Insect Biochem. Molec. Biol.* 1993, *23*, 763-775.
42. Amichot, M.; Castella C.; Cuany, A.; Berge, J. B.; Pauron, D. *Pestic. Biochem. Physiol.* 1992, *44*, 183-190.
43. Omer, S. M.; Georghiou, G. P.; Irving, S. N. *Mosquito News* 1980, *40*, 200-209.
44. Nicholson, R. A.; Miller, T. A. *Pestic. Sci.* 1985, *16*, 561-570.
45. Orchard, I.; Osborne, M. P. *Pestic. Biochem. Physiol.* 1974, *10*, 197-202.
46. Narahashi, T. In *Site of Action for Neurotoxic Pesticides*; Hollingsworth, R. M., Green, M. B., Eds.; ACS Symp. Ser. No. 356, American Chemical Society: Washington, DC, 1987, pp 226-251.

47. Sahib, I. K. A.; Desaiah, D.; Rao, K. S. P. *J. Appl. Toxicol.* **1987**, *7*, 75-80.
48. Gusovsky, F.; Secunda, S. I.; Daly, J. W. *Brain Res.* **1989**, *492*, 72-78.
49. Daly, J. W.; McNeal, E. T.; Gusovsky, F. *Biochim. Biophys. Acta* **1987**, *930*, 470-474.
50. Matsumura, F.; Clark, J. M.; Matsumura, F. M. *Comp. Biochem. Physiol.* **1989**, *94C*, 381-390.
51. Trimble, W. S.; Linial, M.; Scheller, R. H. *Annu. Rev. Neurosci.* **1991**, *14*, 93-122.
52. Talvinder, S. S.; Nichols, R. A. *Neurochem. Res.* **1993**, *18*, 47-58.
53. Benfenati, F.; Valtorta, F. *Neurochem. Int.* **1993**, *23*, 27-34.
54. Popoli, M. *Neuroscience* **1993**, *54*, 323-328.
55. Seabrook, G. R.; Duce, I. R.; Irving, S. N. *Pestic. Sci.* **1988**, *24*, 179-180.
56. Seabrook, G. R.; Duce, I. R.; Irving, S. N. *Pflügers Arch.* **1989**, *414*, 44-51.
57. Schouest, L. P.; Salgado, V. L.; Miller, T. A. *Pestic. Biochem. Physiol.* **1986**, *25*, 381-386.
58. Ghiasuddin, S. M.; Kadous, A. A.; Matsumura, F. *Comp. Biochem Physiol.* **1981**, *68C*, 15-20.
59. Rashatwar, S.; Ishikawa, Y.; Matsumura, F. *Comp. Biochem Physiol.* **1987**, *88C*, 165-170.
60. Charalambous, P.; Matsumura, F. *Insect Biochem. Molec. Biol.* **1992**, *22*, 721-734.

RECEIVED October 25, 1994

Chapter 9

Possible Role for Guanosine 5'-Triphosphate Binding Proteins in Pyrethroid Activity

Daniel P. Rossignol

Eisai Research Institute, 4 Corporate Drive, Andover, MA 01810–2441

Pyrethroid insecticides exert a variety of toxic actions, making elucidation of the mechanism(s) of pyrethroid activity difficult. Described below is an approach taken to analyze pyrethroid interactions with target proteins. A tritiated photoreactive aryl-azide analog of fenvalerate- decyanoazidofenvalerate (DeCAF) has been used to study photoaffinity labeling of neuronal membrane preparations as well as crude and purified preparations of GTP-binding proteins. Under optimal labeling conditions, photolabeling predominantly occurred at a 36 kDa membrane protein. In preparations of rat brain membranes, binding was stimulated by the addition of the sodium channel-specific blockers tetrodotoxin and saxitoxin, although photolabelling of voltage-dependent sodium channels was undetectable. Further studies revealed that DeCAF labeling of a 36 kDa protein was also apparent in bovine retina, rod outer segments (ROS) and purified β subunits of G protein (transducin-β) from ROS. These results suggest that pyrethroids may interfere with the ability of G-proteins to transduce intracellular signals from cell surface receptors to target enzymes. In light of the ever-expanding roles for β as well as α subunits of G-proteins in signal transduction, it is possible that pyrethroids could express their broad array of toxic effects by modifying G-proteins.

Biochemical analysis of the mechanism of activity of pyrethroids has proven to be extremely difficult. Extreme hydrophobicity of pyrethroids may be one of their physical characteristics responsible for their potency, however, it is also a characteristic that renders their activity difficult to analyze by traditional biochemical methods such as equilibrium binding analysis. While a main target of their action is probably the voltage-dependent sodium channel, pyrethroids have also been shown to modify the activity of a broad range of other targets. These include ion channels and other membrane proteins including potassium channels, calcium channels, calmodulin, peripheral benzodiazepine receptors, ATPases, nicotinic acetylcholine receptors, a Na^+/Ca^+ exchanger, and receptors for γ-aminobutyric acid (GABA), GABA-activated Cl⁻ channels (these targets are reviewed in ref. 1,2), as well as

*Studies reported herein were performed at E.I. DuPont de Nemours, Wilmington, DE

0097–6156/95/0591–0149$12.00/0
© 1995 American Chemical Society

phosphoinositides (*3*) and protein kinases such as protein kinase C (*4*). In the case of voltage-dependent sodium channels, evidence for direct interaction with the purified protein has been recently published (*5*). However, it is not clear how pyrethroids interact with other potential targets. Is it possible that pyrethroids alter a central control (cascade) system that next modifies these enzymes? Interestingly, certain pyrethroid targets are part of the large number of enzymes that are either directly or indirectly modulated by G-proteins (*6-8*). This report reviews experiments characterizing photolabeling by DeCAF (structure in inset, Figure 2) with neuronal membranes (rat brain) as well as partially purified and purified G-proteins from retina and bovine brain. Under optimal conditions established for detection of specific and saturable photolabeling, a 36 kDa protein was the major target of photoderivatization (*9,10*) [^{3}H]DeCAF bound to a 36 kDa protein in rat brain and liver membranes. Binding to brain membranes was stimulated by addition of the sodium channel specific toxins saxitoxin and tetrodotoxin, and altered by non-hydrolyzable analogs of GTP indicating that binding was to a protein that was linked to both voltage-sensitive sodium channels and G-proteins. The possibility that the 36 kDa β subunit of G-proteins (Gβ) was investigated using the β subunit of transducin (Tβ). The emerging central role for Gβγ and their modification by pyrethroids implicates pyrethroid toxicity in key diverse cellular regulatory processes.

Materials and Methods

Materials. Scorpion toxin was purified from venom of *Leuirus quinquestriatus* var. Sudan as described (*11*). Batrachotoxin and pumiliotoxin-B were generous gifts of John W. Daly (NIH). *Ptychodiscus brevis* toxin-B was a gift from K. C. Nicolau (Scripps Institute, La Jolla Calif.). Tetracaine, *Leuirus quinquestriatus* venom, GDPβS and GppNHp were from Sigma. GTPγS (tetralithium salt) was from Boehringer Mannheim Biochemicals. Pertussis toxin and cholera toxin were from List Biological Laboratories. [^{3}H]saxitoxin was from New England Nuclear. Molecular weight standards for SDS-PAGE were purchased from Biorad. Pyrethroids (>95% pure) were purchased from Chem Serve (West Chester, Pa.), while crystalline S,S-fenvalerate was from Dupont Agricultural Products Dept. [^{3}H]m-phenoxybenzylalcohol was made by catalytic displacement of bromine from the corresponding brominated alcohol using ^{3}H gas (Custom Tritiation Service of New England Nuclear). Synthesis of DeCAF and [^{3}H]DeCAF as well as analysis of [^{3}H]DeCAF binding to rat brain membranes, bovine retina, rod outer segments, and purified retinal G-proteins was described (*9,10*).

Preparation of Tissues. Rat brain synaptoneurosomes were prepared by minor modifications of the method of McNeal, et al. (*9,12*). Tetrodotoxin (TTX) and batrachotoxin-sensitive uptake of ^{22}Na by these synaptoneurosomes has been measured in this laboratory as well as others (*3,13*), indicating that the preparation contains intact vesicles. Other preparations, and conditions for ADP-ribosylation and quantitation of release of T$_\alpha$ by ADP-ribosylation has been previously described (*9,10*). Antibodies (U-49 and S-217) to β subunit of GTP binding protein were the generous gifts of Suzanne Mumby (Univ. of Texas, Dallas). Immunoprecipitation of [^{3}H]DeCAF-labeled Tβ was performed as described (*10*). Purified βγ subunit complex of G-proteins from bovine brain was the generous gift of Patrick Casey (Univ. of Texas, Dallas).

Binding of [^{3}H]Saxitoxin ([^{3}H]STX) and [^{3}H]Batrachotoxinin A benzoate ([^{3}H]BTX) to particulate fractions was measured by filtration assays (*12,14,*). Uptake of ^{22}Na into N18 cells and synaptoneurosomes was performed as described (*11*), and protein was assayed by the method of Peterson, et al. (*15*).

Results and Discussion

Activity of DeCAF. DeCAF was toxic to houseflies (LD$_{50}$~0.5 μg/g) and induced frequency-dependent depolarization and repetitive firing in crayfish nerve cord (9). The original intent of these studies was to elucidate the interaction of DeCAF with voltage-dependent sodium channels. As pyrethroids have been found to interact with voltage-dependent sodium channels in rat brain, rat brain membranes were chosen as the experimental system to study photolabeling. To determine if sodium channel activity of DeCAF could be detected in rat brain membranes, DeCAF binding was tested for effects on [^{3}H]BTX binding (1,16). In this assay, type II pyrethroids stimulate [^{3}H]BTX binding, while type I compounds inhibit. As shown in Figure 1, DeCAF demonstrates inhibitory activity similar to that of allethrin (a type I pyrethroid), and opposite to that of fenvalerate and deltamethrin (type II, stimulatory activity). This demonstration of type I activity by DeCAF compared to the type II activity of its "parent" fenvalerate may be coincident with the loss of the cyano group. For this reason, it is interesting that both type I and type II pyrethroids inhibited [^{3}H]DeCAF photolabeling (see below).

Characterization of [^{3}H]DeCAF Labeling. The protocol established for photoaffinity labeling of proteins by this hydrophobic probe utilized low temperature photolysis (liquid nitrogen) in the presence of a free radical scavenger (DTT) followed by extraction of the labeled protein with 90% acetone to remove lipid-associated [^{3}H]DeCAF. The extracted protein was analyzed by SDS-PAGE and radioactivity was quantitated by slicing the gel and determining radioactivity in each slice.

When these optimized conditions were used to measure labeling of rat brain synaptoneurosomes, [^{3}H]DeCAF covalently labeled a 36 kDa protein (Figure 2). Labeling was dependent on photolysis, and was abolished by heat-pre treatment of synaptoneurosomes (100°C, 5 min.) prior to incubation with probe. In addition, labeling was specific, reproducible and dose-dependent for DeCAF (0.1 to 1 μM) and for protein (9).

Unlabeled DeCAF (100μM), fenvalerate (50 μM) and the most active stereo isomer of fenvalerate (S,S-fenvalerate; 50 μM) completely inhibited labeling by [^{3}H]DeCAF. Deltamethrin, resmethrin, and permethrin blocked only 32 to 58% of total DeCAF labeling. The four isomers of tetramethrin were 52 to 34% effective at up to 100 μM, with inactive and active isomers being roughly equal in ability to inhibit labeling. Lack of differences in inhibition of labeling indicate that the inactive isomers can inhibit binding of their active counterparts. These results are supported by electrophysiological studies demonstrating that all isomers of tetramethrin bind to the pyrethroid binding site(s), with the inactive isomers inhibiting active isomers (17).

Several conclusions can be derived from these results. First, competition for DeCAF labeling was most effective with pyrethroids having structures closely related to DeCAF. Second, saturability and specificity (ability to be inhibited by the unlabeled molecule) indicated that labeling was not due to artifactual non-specific hydrophobic interaction. Third, specific labeling did not require an aryl-azido moiety because fenvalerate inhibited labeling. Finally, complete inhibition of labeling by the purified active stereo isomer of fenvalerate (S,S-fenvalerate) indicated that labeling was not specific for a type I pyrethroid "binding" site, or that no stereospecific labeling occurred that is unique to the "inactive" isomer of DeCAF.

DeCAF Labeling is Associated with Sodium Channels but DeCAF does not Label Voltage-Dependent Sodium Channels. Other sodium channel toxins, batrachotoxin, veratridine, α-scorpion toxin, pumiliotoxin-B, brevitoxin-B, and DDT had no effect on labeling. This independence of binding for alkaloid toxins

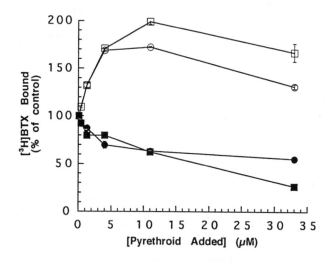

Figure 1. Effect of Pyrethroids on the Binding of [3H]BTX to Rat Brain
Membranes. Binding of [3H]BTX (20 nM) was analyzed as described (12) in
the presence of scorpion toxin plus the indicated amount of deltamethrin (O) S,S
fenvalerate (□), allethrin (■) or DeCAF (●). In these experiments, binding
ranged from 20 to 80 fmol/mg protein and results were calculated as percent of
specific binding (total binding minus nonspecific binding) measured in the
absence of pyrethroid. Nonspecific binding was determined in the presence of
300 μM veratridine.

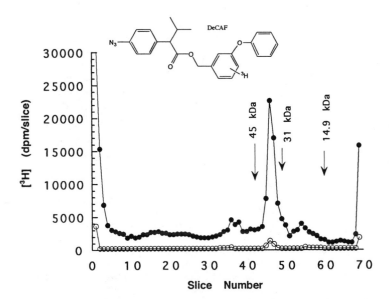

Figure 2. Labeling of a 36 kDa protein from rat brain. Rat brain membranes (2 mg protein) was equilibrated with [^{3}H]DeCAF (0.9 µM) in the absence (●) or presence (O) of 1 mM unlabeled DeCAF, photolyzed, separated by SDS-PAGE and analyzed by measuring bound radioactivity from slices (2 mm) of 4-17% acrylamide gel. Total binding was 578 fmoles/ mg while non-specific binding was 30 fmoles/ mg. Inset: structure of DeCAF. (Reproduced with permission from ref. 9. Copyright 1991).

and pyrethroids is consistent with pyrethroid enhancement of batrachotoxin binding (here and references *3,16*) and batrachotoxin or veratridine-induced Na influx into neuroblastoma cells (*18,19*). Even though DDT demonstrated activity similar to type II pyrethroids in the [³H]BTX binding assay described above (under our labeling conditions, binding was increased 190% by 33 μM DDT; see also ref. *20*), DDT was unable to inhibit DeCAF labeling, indicating its activity is exerted through a site separate from that of pyrethroids.

Preincubation of synaptoneurosomes with tetrodotoxin (TTX) dose-dependently stimulated labeling of the 36 kDa protein(s) (Fig. 3), while labeling of other proteins and non-specific labeling were unaffected. Saxitoxin (10 μM) was also effective at stimulating binding (*9*). TTX has remarkable specificity for voltage-dependent sodium channels, thus stimulation of labeling by TTX is a strong indication that DeCAF labels a protein affected by conformational changes in voltage-dependent sodium channels. Both STX and TTX have been shown to inhibit binding of batrachotoxin to sodium channels in synaptoneurosomes (*16,21*), presumably by inducing a conformational change (*22*). This latter observation has been confirmed in this laboratory. Under conditions used for photolabeling, 100 nM TTX or 20 nM STX inhibited [³H]BTX binding by 30% indicating that under our reaction conditions, TTX and STX can allosterically modify voltage-dependent sodium channels. TTX/STX-modified photolabeling of voltage-dependent sodium channels by DeCAF would be a convenient explanation for DeCAF activity. However, four other results strongly argue that the 36 kDa DeCAF-labeled protein is not a known sodium channel subunit. Voltage-dependent sodium channels are heterotrimers consisting of an α subunit (260 kDa), a β1 subunit (36 kDa), and a β2 subunit (33 kDa) that is disulfide linked to the α subunit (*23*). [³H]DeCAF labeling of the 36 kDa protein(s) and [³H]STX binding to voltage-dependent sodium channels were compared during various purification steps utilized for voltage-dependent sodium channels. DeCAF labeling of membranes separated on sucrose density gradients indicated that while some labeling was found at the same density as membranes containing [³H]STX binding, most DeCAF labeling activity was found at a higher density fraction. Secondly, omission of reducing agent did not reduce the amount of labeling found at 36 kDa on SDS-PAGE indicating that the DeCAF-labelled protein was not the 33 kDa β2-subunit which is disulfide-linked to the 260 kDa- α subunit of the sodium channel. Third, binding to β1 subunit was ruled out because [³H]DeCAF-labeled protein did not bind to wheat-germ agglutinin under conditions where >90% of solubilized nerve membrane sodium channels were adsorbed (*24*). Finally, [³H]DeCAF labeled a 36 kDa protein in rat liver membranes at 60 % the efficiency of brain membranes, whereas saxitoxin binding was <0.5%.

Thus, while DeCAF labeling of rat brain membranes is sensitive to TTX/STX, DeCAF does not directly label (or directly modify?) voltage-dependent sodium channels. Without direct interaction, it must be postulated that DeCAF labels a protein that is associated with voltage-dependent sodium channels, and is conformationally altered by binding of TTX or STX.

Effect of G-Protein Modifiers on DeCAF Binding. Sodium channels and muscarinic acetylcholine receptors may interact through G-proteins (*25*). The possibility that pyrethroids work at this G-protein interaction provides an attractive explanation for the broad activity demonstrated by pyrethroids. To study possible G-protein relationships to the DeCAF labeling described above, analogs of GTP were tested for effects on [³H]DeCAF labeling. Addition of the non-hydrolyzable triphosphate analogs, GTPγS and GppNHp, stimulated labeling as much as two-fold (*9*). This stimulation was found to be optimal at GTPγS concentrations greater than 1 mM with DeCAF concentrations greater than 0.5 mM. Activation was specific for triphosphate; 100 μM GDP, GDPβS or ATP had no effect.

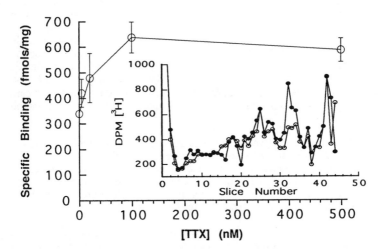

Figure 3. Stimulation of DeCAF binding by tetrodotoxin. DeCAF labeling was performed at the indicated concentration of TTX and labeling of the 36 kDa binding protein(s) was assayed as described (9,10). Values are means ± S.E.M. for triplicate samples. Non-specific binding was determined in parallel incubations containing 100 μM unlabeled DeCAF and was subtracted from all values this non-specific binding was unaffected by TTX. Inset: membranes were incubated for 30 min. at 22°C in the absence (o) or presence (●) of 2.5 μM TTX and analyzed for bound radioactivity by SDS-PAGE. (Reproduced with permission from ref. 9. Copyright 1991).

These observations indicate that DeCAF may bind to either GTP-binding proteins or to a protein that is modified by GTP-binding proteins. The possibility that pyrethroids bind to G-proteins is intriguing. In view of their diverse regulatory roles, demonstrating alteration of G-proteins could help explain their broad diversity of toxic activity.

Does GTP Alter Sodium Channel Activity? Efforts have been directed towards determining if sodium channel activity is regulated by addition of GTP or GTPγS. When influx of ^{22}Na was measured in the presence of batrachotoxin plus scorpion toxin or veratridine plus scorpion toxin, a TTX-sensitive component of flux (~1.2 nMoles/10 sec./ mg protein) was observed. Homogenization of membranes in the presence of GTPγS stimulated influx approximately 2-fold. Similar increases in influx were not seen with GDPβS, or ATP. Unfortunately, while stimulation was reproducible, it was variable in degree between experiments, and was not seen with pyrethroid as agonistic agent (stimulation of sodium influx with any agonist other than batrachotoxin or veratridine was weak). Finally, interpretation of these results is difficult in light of recent results demonstrating the existence of multiple potential targets for stimulation by GTP such as phosphorylation of voltage-dependent sodium channels by cyclic AMP-dependent protein kinase which reduces peak Na$^+$ current and protein kinase C which reduces peak current and rate of channel inactivation (26,27).

Does DeCAF Interact Directly with G-Proteins? G-proteins are signal transducing messengers that relay signals from activated cell-surface (transmembrane) receptors to intended cellular target or "effector" molecules. G-protein-linked receptors are characterized as having seven transmembrane-spanning segments. While the number and diversity of these receptors is large (far greater than 100), and still being defined, they include receptors for light (transducin), odorants, and neurotransmitters.

The wide variety of G-proteins linked to these receptors include stimulatory G-protein (G$_S$) which activates adenylyl cyclase, inhibitory G-protein (G$_i$)- which inhibits adenylyl cyclase, Gq which activates phospholipase-Cβ (PLCβ), and other G-proteins (G$_0$ etc.) with activities that are still being defined.

As a class, these receptor-linked G-proteins are heterotrimeric containing a single α subunit and a βγ subunit complex. The G$_\alpha$ subunit (39-50 kDa), when active, binds and slowly hydrolyzes GTP. In addition, the G$_\beta$ subunit (35-36 kDa) is tightly associated with a third subunit, Gγ (7-10 kDa), which is normally isoprenylated. The G$_{\alpha\beta\gamma}$ trimer is associated with non-activated transmembrane receptors, and to the membrane (via the βγ subunits) in an inactive GDP-bound form. When a signal is recognized by the transmembrane receptor, G-protein is activated, and the α subunit replaces its bound GDP with GTP (Figure 4). At this point, G$_\alpha$– GTP, recognizes and activates its target effector molecule (some are described above). Activation of effector molecules by G$_\alpha$ occurs as long as it has not hydrolyzed its bound GTP to GDP. Once GTP is hydrolyzed, G$_\alpha$ reassociates with G$_{\beta\gamma}$ and its cognate transmembrane receptor.

G-proteins have been purified from a variety of sources including bovine brain and retina. In the latter case, retinal G-protein (transducin or T$_{\alpha\beta\gamma}$) is well characterized and (except for lack of isoprenylation of the γ subunit) are highly homologous to their counterparts in other tissues. In this system, T$_{\beta\gamma}$ as well as T$_\alpha$ dissociates from ROS membranes after GTP is added in the presence of light. Soluble T$_{\beta\gamma}$ can then be readily purified (28,29). These attributes make tranducin an attractive model system for G-protein study and to assess the specificity of DeCAF labeling.

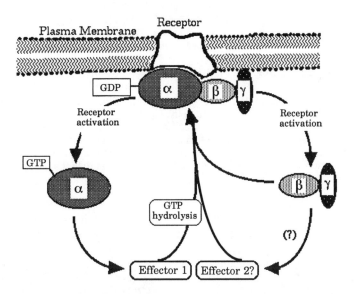

Figure 4. Role of G-proteins in signal transduction. Inactive G-protein heterotrimer (G$_{\alpha\beta\gamma}$) binds GDP when associated with inactive receptor. Upon activation of the receptor by ligand binding, G$_\alpha$ replaces GDP with GTP which it slowly hydrolyzes while activating effector molecules. After GTP is hydrolyzed, G$_\alpha$ reassociates with G$_{\beta\gamma}$ and receptor to complete the cycle. Besides assisting in coupling of G$_\alpha$ to receptor, new roles for G$_{\beta\gamma}$ in direct activation or inhibition of effector molecules have been elucidated and are described in figure 5.

Binding of [^{3}H]DeCAF to Retina, Retinal Rod Outer Segments (ROS), Transducin-β (T$_\beta$) and β-Subunit of GTP-Binding Proteins (G$_\beta$) from Bovine Brain. [^{3}H]DeCAF demonstrated specific binding to a 36 kDa protein(s) in homogenized retina and ROS. As would be expected if T$_\beta$ was labeled, hypotonic lysis and treatment of ROS with GTP nearly quantitatively released the 36 kDa DeCAF-labeled protein from the membrane fraction, and after purification of T$_{\beta\gamma}$ by DEAE, a 195-fold enrichment of labeling was observed. Side-by-side SDS-PAGE analysis of [^{3}H]DeCAF labeled protein and T$_\alpha$ subunit labeled by cholera toxin-catalyzed [^{32}P]ADP-ribosylation suggested that [^{3}H]DeCAF labeling was distinct from the α-subunit.

As in rat brain membranes, labeling was dependent on the concentration of [^{3}H]DeCAF and was inhibited by DeCAF and fenvalerate. Finally, labeled 36 kDa-protein could be immunoprecipitated by the anti- T$_\beta$ antibodies S217 (44% immunoprecipitation) and U49 antibodies (89% immunoprecipitation).

Binding of [^{3}H]DeCAF to Purified G$_\beta$ (G$_{\beta\gamma}$ complex) from Bovine Brain. Like T$_\beta$ subunit from retina, purified bovine brain G$_{\beta\gamma}$ could be photoaffinity labeled with [^{3}H]DeCAF. Binding was specific and saturable at greater than 0.8 μM DeCAF and corresponded to the single silver-stained protein band on SDS-polyacrylamide gels. Maximum specific activity of labeling of G$_\beta$ by DeCAF was approximately 23 pmoles/mg protein (slightly less than maximal labeling of holotransducin at ~100 pmoles/mg). In comparison, labeling of bovine serum albumin or cytochrome C under similar conditions was less than 0.37 pmoles/mg or 1.6 % of labeling seen with G$_\beta$.

What are Potential Consequences of Pyrethroid Modification of G$_{\beta\gamma}$? Purified holotransducin (T$_{\alpha\beta\gamma}$) was used to study possible effects on the ability of DeCAF to alter T$_\alpha$ function. DeCAF was found to have no effect on equilibrium binding, association rate, or rate of dissociation of [^{35}S]GTPγS from T$_\alpha$. DeCAF also had no effect on cholera toxin or pertussis toxin catalyzed ADP-ribosylation of purified T$_\alpha$. This latter result justified the use of cholera toxin and [^{32}P]NAD to quantitate release of T$_\alpha$ from hypotonically lysed ROS membranes, and conditions were established where ADP-ribosylation correlated to levels of T$_\alpha$ (9). Using this method, release of T$_\alpha$ was found to be two-fold stimulated by DeCAF [2.03±0.2-fold (average ± S.E.M.; n=13)]. DeCAF also stimulated the release of T$_\alpha$ from the membrane when GTPγS was present at limiting concentration (3.3 μM). Under these conditions, GTPγS-stimulated release of T$_\alpha$ by DeCAF varied from 1.2 to 3.8-fold with an average of 2.24±0.31 (n=10), and was dose-dependent for DeCAF and maximal at about 0.5 μM. These results suggest that DeCAF may modify the ability of T$_{\beta\gamma}$ to couple with T$_\alpha$, thus stimulating its release from the membrane.

As described above, G$_{\beta\gamma}$ subunits assist in mediating the coupling of G$_\alpha$ with membrane receptors. It was originally believed that G$_{\beta\gamma}$ had no other function. More recently however, it has been shown that G$_{\beta\gamma}$ has signal transducing properties of its own (Figure 5). The first observation of effector activation by purified G$_{\beta\gamma}$ suggested that it could activate potassium channels (I$_{K.ACH}$) in atrial myocytes (30). More recently, as reviewed by Iniguez-Lluhi, et al. (31) and Clapham and Neer (32), G$_{\beta\gamma}$ can also modify other effectors both directly and indirectly by:
• Synergizing with G$_\alpha$ to activate adenylyl cyclase [type II adenylyl cyclase (brain) or type IV (widely distributed)].
• Activating effectors independently of G$_\alpha$. Phospholipase Cβ can be activated by G$_{\beta\gamma}$ alone (33). In turn, the action of PLCβ on phosphoinositides generates inositol phosphates (e.g. IP3) and diacylglycerol.

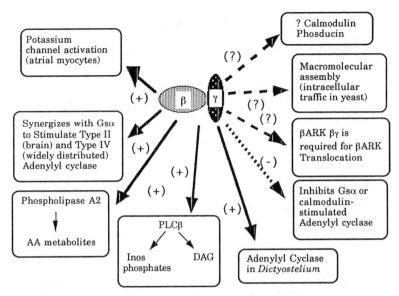

Figure 5. Diverse roles for direct activity of G$_{\beta\gamma}$. As described in recent reviews (*31,32*), G$_{\beta\gamma}$ has been found to activate and inhibit a variety of effector molecules that have broad implications in cellular homeostasis.

• Antagonizing the activation of type I adenylyl cyclase that is activated by Gs$_\alpha$ and calmodulin.
• Other activities of G$_{\beta\gamma}$ are less-well defined, but G$_{\beta\gamma}$ has been shown to interact with β-adrenergic receptor kinase (βARK), phosducin, calmodulin, and molecules that direct intracellular traffic in yeast.

What are the consequences of modifying these effectors? It is unclear that the interaction of pyrethroids with G$_{\beta\gamma}$ would result in stimulation or inhibition of G$_{\beta\gamma}$. In most cases, G$_\gamma$ is modified by isoprenylation, this anchors G$_{\beta\gamma}$ to the membrane. Recently, it has been shown that this anchoring is necessary for coupling of an olfactory G-protein receptor kinase to its substrate (*34*). The hydrophobicity of pyrethroids may enable them to act at the interface of G$_{\beta\gamma}$ and the membrane and inhibit the ability of G$_{\beta\gamma}$ to carry out its signal-transducing functions. Because the protein-protein interactions of G$_{\beta\gamma}$ and effector can result in activation, inhibition or even "desensitization" (*34*), it is difficult to speculate on the ultimate downstream effects of modification of G$_{\beta\gamma}$. However, Figure 5 outlines some of the known direct and further "downstream" effects of G$_{\beta\gamma}$. This allows us consider some of the normal consequences of G$_{\beta\gamma}$ activities, and speculate on the consequences of their modification.
• Adenylyl cyclase. To date, six forms of cyclase have been cloned. All are stimulated by the α-subunit of stimulatory G-protein (Gs$_\alpha$), but are differently modified by G$_{\beta\gamma}$. Calmodulin-and Gs$_\alpha$-stimulated type I cyclase are inhibited by G$_{\beta\gamma}$. In contrast, type II and type IV cyclase are conditionally stimulated by G$_{\beta\gamma}$, i.e. G$_{\beta\gamma}$ must be at high concentration, and only when G$_\alpha$ is also present (*31*). Activation of cyclase and resultant increases in cyclic AMP stimulates cAMP-

dependent protein kinase, which has a wide variety of substrates. Interestingly, cAMP-dependent protein kinase has been shown to phosphorylate rat brain sodium channels. This phosphorylation results in decreasing peak current amplitude and delay of inactivation (26,27).
• Phopholipase $C\beta$. Stimulation of PLC would increase levels of diacylglycerol and inositol phosphates (e.g. IP_3). Diacylglycerol activates protein kinase C. As reviewed in (4) many neuronal effects are modulated by protein kinase C. Deltamethrin has been shown to stimulate the protein kinase C pathway (3). In addition, IP_3 stimulates release of Ca^{++} from internal storage sites. This release of Ca^{++} can activate calmodulin-dependent protein kinases as well as muscle contraction.
• Ion currents. Representative examples of G-proteins (α subunits) from almost every defined class of G-proteins (G_s, G_i, G_o, transducin), have effects on ion currents through direct interaction of their G_α subunits with channels. Other second messenger-derived effects (cAMP, cGMP, AA, $IP3/Ca^{++}$), and/or diacylglycerol (35) are indirect and still debatable. Alteration of the ability of $G_{\beta\gamma}$ to couple G_α subunits to their targeted ion channels would inhibit G_α activity and subsequent control of ion flux.

Through direct and indirect mechanisms, G-proteins alter many enzymes that are pyrethroid "targets", including ion channels and protein kinases. Binding of pyrethroids to G_β may directly alter stimulatory or inhibitory functions of $G_{\beta\gamma}$ or alter coupling of G_α to membrane receptors. In this way, pyrethroid poisoning of G-proteins would explain the apparent diversity of pyrethroid activity. Thus, neurotoxicity of pyrethroids could be manifested as the result of alteration of one or more of the many systems that use G-proteins for signal transduction.

Acknowledgments

I am grateful to Dr. John Bisaha for the synthesis and characterization of the pyrethroid precursors and m-phenoxybromobenzyl alcohol. Dr. Albert Lund, Doug Hines and Steve Irons performed electrophysiological and toxicity experiments. Excellent technical assistantship of Paul Pipenberg, Chris Lynch, and Lisa Chapaitis is gratefully acknowledged. I would like to thank Dr. Gabriel Berstein, Dr. Lynn Hawkins, Dr. Fabian Gusovsky, Jeff Rose and Colleen Ramsden for assistance with and review of this manuscript.

Literature Cited

1   Soderland, D.M.; Bloomquist, J.R. Ann. Rev. Entomol. 1989, 34, 77-96.
2   Narahashi, T. Trends in Parm. Sci. 1992, 13, 236-241.
3   Gusovsky F.; Secunda S.; Daly J. W. Brain Research 1989, 492, 72-78.
4   Enan, E.; Matsamura, F. Bioch. Pharmacol. 1993, 45, 703-710.
5   Trainer, V.L.; Moreau, E.; Guedin, D.; Baden, D. G.; Catterall, W.A. J. Biol. Chem. 1993, 23, 17114-17120.
6   Doherty, J.D.; Nishimura, K.; Kurihara, N.; Fujita , T. Pestic. Biochem. Physiol.. 1987, 29, 187-196.
7   Abbassy, M.A.; Eldefrawi, M.E.; Eldefrawi A.T. Pestic. Biochem. Physiol 1983, 19, 299-306.
8   Ramadan, A.A.; Bakay,N.M.; Marei, A-S.; Eldefrawi, M.; Eldefrawi, A.T.; Eldefrawi, M.E. Pestic. Biochem. Physiol. 1988, 32, 106-113.

9   Rossignol, D.P. *Pest. Bioch. Physiol.* **1991,** *41,* 103-120.
10   Rossignol, D.P. *Pest. Bioch. Physiol.* **1991,** *41,* 121-131.
11   Gusovsky, F.; Rossignol, D.P.; McNeal, E.T.; Daly, J.W. *Proc. Nat. Acad Sci. USA* **1988,** *85,* 1272-1276.
12   McNeal, E.T.; Lewandowski, G.A.; Daly J.W.; Creveling, C.R. *Med. Chem.* **1985,** *28,* 381-388.
13   Gusovsky, F.; Hollingsworth, E.B.; Daly, J.W. *Proc. Nat. Acad Sci. USA.* **1986,** *83,* 3003-3007.
14   Rossignol, D.P. *Pestic. Biochem. Physiol.* **1988,** *32,* 146-152.
15   Peterson, G.L. *Analytical Bioch.* **1977,** *83,* 346-356.
16   Brown, G.B.; Gaupp, J.E.; Olsen, R. *Molec. Pharm.* **1988,** *34,* 54-59.
17   Narahashi, T.; Frey, J.M.; Gindburg, K.S.; Roy, M.L. *Tox. Let.* **1993,** 429-436.
18   Jacques, Y.; Romey, G.; Cavey, M.T.; Kartalovski, B.; Lazdunkski, M. *Bioch. Biophys. Acta* **1980,** *600,* 882-897.
19   Lombet, A.; Mourre, C.; Lazdunski, M. *Brain Research.* **1988,** *459,* 44-53.
20   Bloomquist ,J.R.; Soderlund, D.M. *Molec. Pharm.* **1988,** *33,* 543-550.
21   Garritsen, A.; Ijzerman, A.P.; Sondijn, W.*Eur. J. Pharmacol.* **1988,** *145,* 261-266.
22   Tejedor, F.J.; McHugh, E.; Catterall, W.A. *Biochemistry* **1988,** *27,* 2389-2397.
23   Hartshorne, R.P.; Catterall, W.A. *J. Biol. Chem.* **1984,** *259,* 1667-1675.
24   Messner, D.J.; Catterall, W. A.*J. Biol. Chem.* **1985,** *260,* 10597-10604.
25   Cohen-Armon, M.; Garty, H.; Sokolovsky, M. *Biochemistry.* **1988,** *27,* 368-374.
26   Numann, R.; Catterall, W.A.; Scheuer, T. *Science* **1991,** *254,* 115-118.
27   Murphy, B.J.; Rossie, R.; De Jongh, K.S.; Catterall, W.A. *J. Biol. Chem.* **1993,** *268,* 27355-27362.
28   Papermaster, D.S. *Methods in Enzymol.* **1982,** *81,* 48-53.
29   Baehr, W.; Morita, E.A.; Swanson, R.J.; Applebury, M.L. *J. Biol. Chem.* **1982,** *257,* 6452-6460.
30   Logothetis, D.E.; Kurachi, Y.; Galper, J.; Neer, E.; Clapham, D.E. *Nature* **1987,** *325,* 321-326.
31   Iniguez-Lluhi, J.; Kleuss, C.; Gilman, A.G. *Trends in Cell Biol.* **1993,** *3,* 230-236.
32   Clapham, DE.; Neer, E.J. *Nature* **1993,** *365,* 403-406.
33   Blank, J.L.; Brattain, K.A.; Exton, J.H. *J. Biol. Chem.* **1993,** *267,* 23069-23075.
34   Boekhoff, I.; Inglese, J.; Schleicher, S.; Kock, W.J.; Lefkowicz, R.L.; Breer, H. *J. Biol. Chem.* **1994,** *269,* 37-40.
35   Sternweiss, P.C.; Pang, I-H. *Trends in Neurobiol.* **1990,** *13,* 122-126.

RECEIVED September 27, 1994

Chapter 10

Calcium Channel as a New Potential Target for Insecticides

L. M. Hall[1], D. Ren[1], G. Feng[1], D. F. Eberl[1], M. Dubald[2], M. Yang[1], F. Hannan[1], C. T. Kousky[1], and W. Zheng[1]

[1]Department of Biochemical Pharmacology, State University of New York at Buffalo, Buffalo, NY 14260–1200
[2]Rhone Poulenc AG, P.O. Box 12014, 2 TW Alexander Drive, Research Triangle Park, NC 27709

Pharmacological, electrophysiological, and ligand-binding studies have indicated the presence of diverse voltage-sensitive calcium channels in insects which appear to be pharmacologically distinct from those of vertebrates. To define the molecular structure of these channels, we have used a polymerase chain reaction (PCR) cloning strategy to identify and sequence a cDNA encoding an α_1 subunit of *Drosophila* calcium channels. Quantitative Northern blotting studies have shown that this subunit is most highly expressed in legs and heads. Genetic analysis has demonstrated that a premature stop codon in the α_1 subunit causes an embryonic lethal phenotype demonstrating that function of this subunit (which is the target for organic calcium channel blockers) is required for survival of the organism. The structural differences between this insect calcium channel and those from vertebrates may prove useful for the design of new insect-specific calcium channel blockers.

Voltage-sensitive calcium channels play two important roles in all excitable cells. First, they play a key role in control of cell excitability by either contributing to the shape of regenerative action potentials or by generating the action potential in cells which lack sodium channels. The second major role of calcium channels is to serve as the link which transduces a depolarization into the nonelectrical processes that are controlled by excitation. Some of these processes include: synaptic transmission, endocrine secretion, and muscle contraction. (See review by Hille(1).) Calcium channels are ubiquitous and have been reported in excitable cells in species ranging from *Paramecium* to humans.

We are interested in calcium channels as potential targets for the development of new insecticides. In determining their suitability as targets, we have used molecular genetic and pharmacological approaches to consider 4 questions:

1. What is the structure of insect calcium channels?

2. Are insect calcium channels structurally and pharmacologically different enough from mammalian calcium channels so that insect selective agents can be developed?

3. Are insect calcium channels accessible to insecticides?

4. What are the physiological consequences of blocking calcium channels in insects?

0097–6156/95/0591–0162$12.00/0

What is the Structure of Insect Calcium Channels?

Voltage-dependent calcium channels in mammalian skeletal muscle are comprised of 5 subunits (α_1, α_2, β, γ, and δ). As reviewed by Hofmann and coworkers(2), the α_1 subunit itself forms the calcium channel conducting pore and also contains the binding sites for all known calcium channel blockers. The smaller β subunit enhances calcium current expression when coexpressed with the α_1 subunit in a variety of heterologous expression systems. In some expression systems the β subunit alters channel kinetics. The α_2/δ subunits are encoded by a single gene. The two subunits are formed by posttranslational proteolytic processing. Coexpression of α_1, α_2/δ, and β cDNAs together is required for maximum calcium channel expression in some, but not all, expression systems. The addition of the γ subunit to coexpression experiments generally has little affect on overall channel expression levels, but does shift the steady state inactivation of I_{Ba} by 40mV to negative membrane potentials *(3)*. It is, however, not yet clear whether in vertebrates the γ subunit is present in tissues other than skeletal muscle *(2)*.

Although molecular analysis of voltage-dependent calcium channels is just beginning in insects, gene cloning studies from our laboratory have provided evidence that calcium channels in the fruitfly *Drosophila melanogaster* have a similar subunit structure to the mammalian channels. We have sequenced complete cDNA clones encoding α_1 and β subunits (D. Ren, G. Feng, W. Zheng, D.F. Eberl, F. Hannan, M. Dubald and L.M. Hall, unpublished results). In addition, using reduced stringency PCR and probing Southern blots with a mammalian α_2 cDNA, we have obtained preliminary evidence for the existence of an α_2 subunit in *Drosophila*. Because of the key role that α_1 subunits play in the function of calcium channels and their utility as targets for pharmacologically active agents in mammalian systems, we have focused our initial efforts on the characterization of this component.

Figure 1 illustrates the successful strategy used to clone the first α_1 subunit from *Drosophila*. Initially, we designed a set of 9 PCR primer pairs from the most highly conserved regions of the known vertebrate calcium channel α_1 subunits focusing on those conserved regions which would also give the least codon degeneracy. These primer pairs were used to amplify from a genomic DNA template so that no assumptions were made concerning the time in development or the tissue of expression. PCR products of approximately the same size or larger than that predicted from the vertebrate cDNA sequences were sequenced directly to determine whether any encoded a peptide with amino acid sequence similarity to the vertebrate subunit. Codon preference analysis was used to identify introns. Three of the most promising PCR products were sequenced and one of these showed a high degree of similarity to the vertebrate α_1 subunit. The *Drosophila* genomic sequence did contain two small but easily identifiable introns of 59 and 60 base pairs. The positions of the successful primers are given as black boxes numbered 6 and 7 in Figure 1A.

Once the PCR amplification product was shown to encode a portion of an α_1 subunit, this product was used to screen an adult head cDNA library to obtain the full coding sequence. Our previous calcium channel ligand binding studies *(4)* plus Northern analysis had suggested that this channel was expressed in significant amounts in adult heads which are enriched in neuronal tissue. In *Drosophila* as in other organisms, the α_1 subunit mRNA is very long (>8kb). To obtain the full open reading frame, it was necessary to isolate three overlapping cDNA clones (N1, W8A, and SH22C) as shown in Figure 1B. Although there was excellent match in most of the overlapping regions of these clones, there is a region of 149 nucleotides in the 3' end of clone W8A (indicated by the downward pointing arrow) which

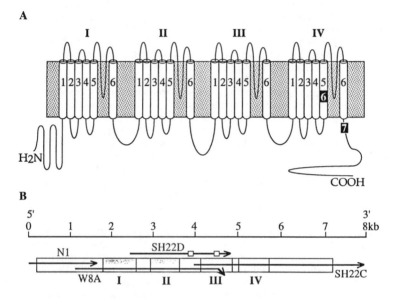

Figure 1. Strategy for cloning and relationship of cDNA clones encoding a full length calcium channel α_1 subunit from *Drosophila*. (A) Cartoon showing the general structure of all α_1 subunits of voltage-sensitive calcium channels cloned to date including the α_1 subunit from *Drosophila melanogaster*. The black boxes labeled 6 and 7 designate the positions of primers used for successful amplification of the *Drosophila* α_1 subunit. (B) cDNA clones used to obtain the full open reading frame for the *Drosophila* α_1 subunit. The small open boxes within clone SH22D indicate regions of known alternative splicing.

shows no overlap with the sequence in SH22C. Sequence analysis of genomic clones has indicated alternatively spliced exons in this region. Additional regions of alternative splicing are indicated as open boxes in clone SH22D. These alternatively spliced regions provide a means to generate functional diversity of calcium channel subunits from a single gene.

Complete sequencing of the α_1 subunit revealed that the *Drosophila* subunit is similar in overall plan to the mammalian homologues. This is revealed dramatically in the hydropathy plots shown in Figure 2. The *Drosophila* subunit contains 4 repeat regions labeled in Figures 1 and 2 as I, II, III, and IV. Within each repeat there are 6 transmembrane domains labeled 1-6 (and referred to in the literature as S1-S6) which show structural similarity when compared among the 4 repeat regions. For example, the S4 regions all show positively charged amino acid side chains alternating every 3 to 4 amino acids. These positively charged side chains have been proposed to all lie on one side of an α helix and to constitute at least part of the voltage sensing mechanism of these channels.

Another structural feature that is conserved between this *Drosophila* α_1 subunit and those from other species is a hydrophobic loop (generally referred to as SS1 and SS2) which is located between transmembrane segments S5 and S6 in each repeat segment. This region has been modeled as dipping part way through the membrane as shown in Figure 1A. This region contributes to the ion selectivity filter since changing single amino acid residues in these regions can shift channel selectivity from sodium to calcium or vice versa *(5, 6)*. In this crucial region the *Drosophila* subunit contains the glutamic acid residues diagnostic of calcium channels. This, along with the high sequence similarity to vertebrate calcium channels (ranging from 63.4% to 78.3% similarity), firmly establishes this clone as encoding an insect calcium channel α_1 subunit.

Are there structural and pharmacological differences between insect and vertebrate calcium channels which would allow the development of insect specific channel blockers?

Although the general structural plan between insect and vertebrate calcium channels is conserved, there are many regional differences. One striking structural difference is revealed in the comparison of the hydropathy plots in Figure 2. It is apparent that the amino and carboxy terminal cytoplasmic domains are much longer in the *Drosophila* subunit than in the illustrated mammalian skeletal muscle homologue. When these sequences are compared at the amino acid level, there is no similarity at all in the insect sequence after about 225 amino acids upstream of the beginning of IS1 and after 160 amino acids downstream of the carboxyl region IVS6. It is not yet clear whether these regions are functionally significant in the insect calcium channel. If they are, they might be targeted by membrane permeant agents to generate insect specific toxins. In mammals the very effective phenylalkylamine class of calcium channel blockers are thought to penetrate membranes and interact with the cytoplasmic domains of the α_1 subunit *(7)*. Therefore, there is a precedent for designing highly effective, membrane permeant calcium channel blockers.

A more complete and systematic analysis of the regional differences between the rabbit skeletal muscle subunit and the *Drosophila* subunit is given in Figure 3 where the cytoplasmic, transmembrane and extracellular domains have been compared sequentially across the molecule and the percent difference in amino acid sequence has been plotted. It should be noted that the first (highly nonconserved) 493 amino acids in the *Drosophila* have not been included in this analysis. This figure shows that in addition to the nonconserved, cytoplasmic amino and carboxy termini, there are numerous extracellular domains such as those between IS1 & IS2, IS3 & IS4, IVS1 & IVS2, and IVS3 & IVS4 which are only weakly conserved and therefore are possible targets for insect-specific agents which act extracellularly.

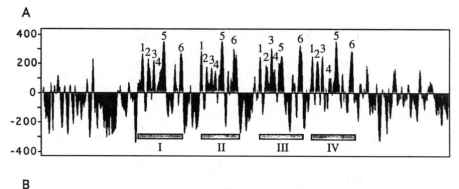

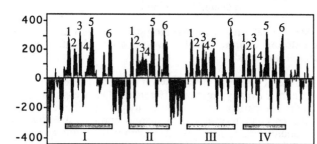

Figure 2. Comparison of hydropathy plots for α_1 subunits from rat brain type D *(17)* (panel B) with the *Drosophila* subunit (panel A). Hydropathy plots were determined using the method of Kyte and Doolittle *(18)* with the GeneWorks software. Up is hydrophobic and down (negative numbers) is hydrophilic.

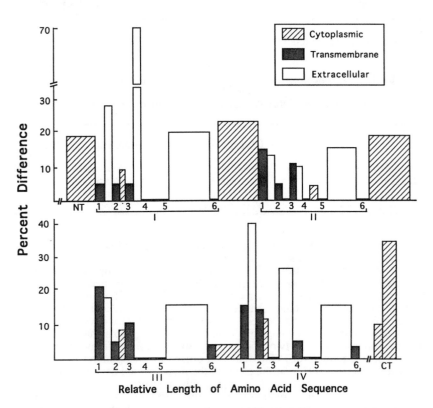

Figure 3. Comparison of regional sequence differences in calcium channel α_1 subunits from *Drosophila* versus rabbit skeletal muscle. An amino acid pair was scored as different only if it represented a nonconservative substitution. The width of each bar is proportional to the length of the amino acid sequence except in the region labeled CT where the scale is 1/10 that of the rest of the chart. The 4 repeat domains are designated I, II, III, and IV. Within each repeat the proposed transmembrane segments (S1-S6) are designated as 1-6. NT = amino terminal segment (beginning with methionine 494). CT = Carboxy terminus. Note that parts of the "extracellular" loop between S5 and S6 form the membrane embedded SS1/SS2 loop shown diagrammatically in Figure 1A.

The dihydropyridine class of calcium channel blockers are thought to act against the mammalian calcium channel α_1 subunit by blocking from the outside of the channel *(8-10)*.

From the point of view of amino acid sequence there appear to be numerous areas of differences. These gene cloning and sequencing studies provide a means to define the molecular basis for previously observed pharmacological similarities and differences between insect and mammalian calcium channels *(4, 11-13)*. Some highlights of these differences are summarized in Table I below.

**Table I. Pharmacology of Calcium Channels Predominant in *Drosophila*
Heads versus
Mammalian Skeletal Muscle L-type Calcium Channels *(4, 13)***

	Drosophila	Mammalian L-type
Phenylalkylamines	Sensitive	Sensitive
Dihydropyridines	Insensitive	Sensitive
Benzothiazepines	Insensitive	Sensitive
Tetrandrine	Very sensitive	Sensitive

As summarized in Table I, at least some subtypes of both *Drosophila* and mammalian calcium channels are sensitive to the phenylalkylamine class of calcium channel blockers. Photoaffinity labeling coupled with immunoprecipitation studies have identified a peptide including the cytoplasmic domain immediately adjacent to IVS6 and extending into the transmembrane domain of IVS6 which is involved in binding this class of calcium channel blocker. Phenylalkylamines act from the cytoplasmic side of the channel *(9)*. Our sequencing studies have shown that the *Drosophila* α_1 subunit is completely conserved in the cytoplasmic portion of this peptide and shows only two conservative amino acid differences within the transmembrane domain of IVS6. Thus, in order to develop insect-specific agents targeted against this region of the channel, the ligand would have to have a domain of action that extended outside of this highly conserved area.

In contrast, the predominant calcium channel activity expressed in *Drosophila* heads is insensitive to the dihydropyridines which are very effective at blocking mammalian L-type calcium channels. Since the cloned calcium channel we describe here has not yet been expressed, we do not know its exact pharmacological specificity. We do know that it is highly expressed in *Drosophila* heads and therefore is likely to encode the phenylalkylamine-sensitive, dihydropyridine-insensitive channel found in heads *(4)*. Consistent with this idea is our finding that this *Drosophila* subunit shows many nonconservative amino acid substitutions in the regions thought to encode the dihydropyridine binding site. This binding site is thought to reside in part in the regions beginning in the extracellular domain between IIIS5 & IIIS6 and between IVS5 & IVS6 and extending into the adjacent S6 transmembrane domain *(8-10)* Thus, this drug binding site is thought to extend into the channel from the outside.

Binding sites for the other two classes of calcium channel drugs (benzothiazepines and tetrandrine) shown in Table I have not yet been defined in any species. Chimeras between the insect and mammalian subunits could be used in expression studies to identify the position of these sites in the future. Of particular interest is the very high sensitivity of the insect calcium channel ligand binding to inhibition by tetrandrine, an active component found in the Chinese herb used to treat cardiac arrythmias *(14)*. In nature tetrandrine might contribute to the defense mechanism of the plant by affecting pest insect calcium channels much the same way that the pyrethrins from chrysanthemums act on insect sodium channels to protect the plant from invaders. Thus, tetrandrine is potentially a useful lead compound for the development of effective insecticides targeted against insect calcium channels.

Are insect calcium channels accessible to the actions of insecticides ?

We have taken two approaches to this question. One is to use quantitative Northern blotting and *in situ* hybridization to ask where this calcium channel α_1 subunit is expressed. The second approach is to ask whether *Drosophila* is susceptible to orally administered phenylalkylamines.

As summarized in Table II, quantitative Northern blotting using mRNA prepared from adult *Drosophila* body parts shows that this α_1 subunit is most highly expressed in legs. (The level of calcium channel subunit mRNA has been standardized against ribosomal protein-49 [RP49] message *(15)*.) The second highest level of expression is in heads which are enriched for neuronal tissue. This table also illustrates that there was more apparent heterogeneity in calcium channel message size classes in heads than in other body parts. The enrichment in heads was determined by combining the two different size classes. Enrichment of message expression in heads relative to bodies is generally indicative of a nervous system specific expression. Thus, this calcium channel is likely to show a distribution pattern similar to the *para* sodium channel which is a target for pyrethroid action *(16)*. This sodium channel is expressed throughout the nervous system.

The high level of expression in legs suggests that this channel may be susceptible to the action of insecticides which could penetrate through the legs as an insect walks across a surface. Neuronal specific expression of this subunit has been confirmed by whole mount *in situ* hybridization to embryos which shows widespread expression throughout the central nervous system.

Table II. Quantitative Expression of Calcium Channel mRNA

Tissue	Calcium Channel α_1 mRNA size class	α_1 subunit/RP49
Body	9.6 kb	0.21
Head	10.2 kb	0.18
Head	9.5 kb	0.16
Leg	9.5 kb	1.23

To further address the issue as to whether these calcium channels are susceptible to action of blocking agents, we raised flies from egg to adult on media containing various amounts of the phenylalkylamine verapamil. This drug was designed for use in treating cardiac arrythmias in humans and so would not be expected *a priori* to be extremely effective against insect calcium channels. Nevertheless, as shown in Figure 4 we found a dose dependent killing effect with 100% lethality for wild-type flies at the highest verapamil concentration tested. Interestingly, we also observed a dramatic, drug-dependent delay in time required to mature from egg to adult. There was no obvious sexual dimorphism in the effects of verapamil on either the viability or the developmental delay.

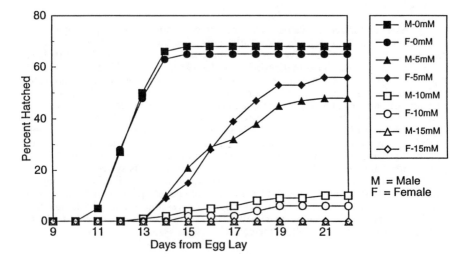

Figure 4. Effects of the calcium channel blocker verapamil on *Drosophila* survival and development. Wild-type Canton-S adults were allowed to lay eggs on standard *Drosophila* medium *(19)*. Shortly after laying, groups of 50 eggs were transferred to fresh shell vials containing 7 ml of Formula 4-24 Instant *Drosophila* Medium (Carolina Biological Supply Co.) prepared with an equal volume of 10% ethanol containing the indicated concentration of verapamil. Vials were incubated in total darkness at 25°C until adults began eclosing. The number and sex of the flies hatching each day was recorded. Two hundred eggs (~50% male, ~50% female) were treated at each concentration.

What are the consequences of functionally inactivating this specific α_1 subunit in insects?

Although the above verapamil feeding experiment suggests that blockade of calcium channels will lead to lethality, the interpretation of this type of drug feeding experiment is complicated by the fact that we cannot be certain that the cloned α_1 subunit is the target causing lethality. Since it requires high concentrations of drug administered in the food and since we do not know what the actual concentration is at the target site, there is a concern that the lethality might be due to effects at a site other than the expected target. Interpretation is further complicated by the fact that in mammals, there are multiple genes encoding different α_1 subunits *(2)*. There is preliminary evidence for at least one additional gene in *Drosophila* encoding a different α_1 subunit (L. Smith and J.C. Hall, personal communication). From a simple feeding experiment we cannot tell whether the lethality is due to an effect on calcium channels. Even if the killing is through a calcium channel mechanism, we cannot distinguish whether the lethality is due to effects on this channel or on another channel or due to effects on multiple channels. Indeed, with the multiplicity of neuronal calcium channel subtypes in mammals, it is an important question to determine whether they functionally overlap.

We have used a genetic approach to address the question of the consequences of inactivating this specific α_1 subunit. First, we mapped the gene to a position on the left arm of the third chromosome using *in situ* hybridization of the cloned cDNA to salivary gland chromosomes. This initial localization was followed by mapping of the cDNA with respect to chromosome deletions to obtain a more precise location. This physical mapping of the locus was correlated with mutant analysis to determine which of the exising mutations in the region showed the same deletion mapping pattern as the cDNA. This approach allowed us to identify one candidate complementation group which causes embryonic lethality. DNA sequencing studies showed that one allele at this locus causes a premature stop codon in the cytoplasmic loop between IVS4 and IVS5. (See Figure 1.) This point mutation would result in the formation of a truncated protein lacking carboxy portion of homology region IV and the long carboxy terminal cytoplasmic tail.

Since the genetic mutation specifically affects this particular α_1 subunit subtype, we can conclude that agents which inactivate this subunit have the potential to kill the insect as early as the late embryonic stage. For pest insects, if an egg-permeable compound can be developed, it would have the potential to kill the insect before the destructive larval stages emerge. These experiments establish the utility of this subunit as a potential insecticide target. The availability of a complete cDNA encoding this subunit provides a basis for developing a heterologous expression system which can be used for the initial, rapid screening of test compounds for action on this target.

Calcium Channels as Targets for New Insecticides

Our gene cloning and sequencing studies have shown that although there is enough sequence similarity to allow cloning of insect calcium channel subunits across species using appropriately designed PCR primers and reduced stringency conditions for amplifications, there are substantial regional differences between the insect and mammalian channels. These regions could be targeted to develop species-selective agents. The expression of these channels in peripheral structures such as legs offers the hope that these channels may be susceptible to compounds as the insect walks across a surface. The existing calcium channel blockers as well as natural products such as tetrandrine may provide useful lead compounds for the development of new insecticidal compounds which target this calcium channel subunit. Screening for such new compounds will be facilitated by the expression of this cloned gene in cells which can be readily mass produced for large scale screening.

Acknowledgments

We thank Dr. Maninder Chopra for comments and help in the preparation of this manuscript. This work was supported by grants from NIH (HL39369) and the New York State Affiliate of the American Heart Association (92329GS) to LMH who is a Jacob Javits Scholar in Neuroscience. Salary support for GF was from NIH grant NS16204 and from a Pharmaceutical Manufacturers Association Predoctoral Fellowship. A part of this study was conducted under the BIOAVENIR program financed by RHONE-POULENC with the contribution of the Ministère de l'Enseignement Supérieur et de la Recherche. DFE was supported by a Postdoctoral Fellowship from the Natural Sciences and Engineering Research Council of Canada. CTK was supported by a Howard Hughes Predoctoral Fellowship and WZ received salary support from NIH grants HL16003 and GM42859 to D.J. Triggle.

Literature Cited

(1)   Hille, B. *Neuron* **1992**, *9*, 187-195.
(2)   Hofmann, F.; Biel, M.; Flockerzi, V. *Annu. Rev. Neurosci.* **1994**, *17*, 399-418.
(3)   Singer, D.; Biel, M.; Lotan, I.; Flockerzi, V.; Hofmann, F.; Dascal, N. *Science* **1991**, *253*, 1553-7.
(4)   Greenberg, R. M.; Striessnig, J.; Koza, A.; Devay, P.; Glossmann, H.; Hall, L. M. *Insect Biochem* **1989**, *19*, 309-322.
(5)   Heinemann, S. H.; Terlau, H.; Stuhmer, W.; Imoto, K.; Numa, S. *Nature* **1992**, *356*, 441-443.
(6)   Tang, S.; Mikala, G.; Babinski, A.; Yatani, A.; Varadi, G.; Schwartz, A. *J. Biol. Chem.* **1993**, *268*, 13026-13029.
(7)   Striessnig, J.; Glossmann, H.; Catterall, W. A. *Proc Natl Acad Sci U S A* **1990**, *87*, 9108-9112.
(8)   Nakayama, H.; Taki, M.; Striessnig, J.; Glossmann, H.; Catterall, W. A.; Kanaoka, Y. *Proc Natl Acad Sci U S A* **1991**, *88*, 9203-9207.
(9)   Striessnig, J.; Murphy, B. J.; Catterall, W. A. *Proc Natl Acad Sci U S A* **1991**, *88*, 10769-73.
(10)   Catterall, W. A.; Striessnig, J. *Trends Pharmacol Sci* **1992**, *13*, 256-62.
(11)   Pauron, D.; Qar, J.; Barhanin, J.; Fournier, D.; Cuany, A.; Pralavorio, M.; Berge, J. B.; Lazdunski, M. *Biochemistry* **1987**, *26*, 6311-6315.
(12)   Pelzer, S.; Barhanin, J.; Pauron, D.; Trautwein, W.; Lazdunski, M.; Pelzer, D. *EMBO J* **1989**, *8*, 2365-2371.
(13)   Glossmann, H.; Zech, C.; Striessnig, J.; Staudinger, R.; Hall, L.; Greenberg, R.; Armah, B. I. *Br J Pharmacol* **1991**, *102*, 446-452.
(14)   King, V. F.; Garcia, M. L.; Himmel, D.; Reuben, J. P.; Lam, Y.-K. T.; Pan, J.-X.; Han, G.-Q.; Kaczorowski, G. J. *J Biol Chem* **1988**, *263*, 2238-2244.
(15)   O'Connell, P.O.; Rosbash, M. *Nucleic Acids Res.* **1984**, *12*, 5495-5513.
(16)   Hall, L. M.; Kasbekar, D. P. In *Book* ; T. Narahashi and J. E. Chambers, Eds.; Plenum Press: New York, NY, 1989.
(17)   Hui, A.; Ellinor, P. T.; Krizanova, O.; Wang, J. J.; Diebold, R. J.; Schwartz, A. *Neuron* **1991**, *7*, 35-44.
(18)   Kyte, J.; Doolittle, R. F. *J. Mol. Biol.* **1982**, *157*, 105-132.
(19)   Lewis, E. B. *Drosophila Inform. Serv.* **1960**, *34*, 117-118.

RECEIVED October 19, 1994

Chapter 11

Action of DDT and Pyrethroids on Calcium Channels in *Paramecium tetraurelia*

J. Marshall Clark[1], Sean J. Edman[1], Scott R. Nagy[1], Alfredo Canhoto[1], Frank Hecht[2], and Judith Van Houten[2]

[1]Department of Entomology, University of Massachusetts, Amherst, MA 01003
[2]Department of Zoology, University of Vermont, Burlington, VT 05405

DDT and neurotoxic pyrethroids are potent agonists of the voltage-sensitive Ca^{2+} channel associated with the ciliary membrane of *Paramecium tetraurelia*. As such, these insecticides resulted in increased avoidance behaviors and mortality. Deciliated *Paramecium* and defective Ca^{2+} channel *pawn* mutants are unaffected by these insecticides whereas hyperactive *Dancer* mutants are hypersensitive. This channel is more sensitive to the phenethylamine-type and the dihydropyridine-type Ca^{2+} channel blockers but unaffected by *omega*-conotoxin GIVA. Electrophysiologically, 10^{-9} M deltamethrin rapidly produces membrane destabilization, repetitive Ca^{2+} action potentials and membrane depolarization. All the above effects are produced in the absence of any involvement of a voltage-sensitive Na^{+} channel.

The symptoms of pyrethroid poisoning are characterized by hyperexcitation, convulsions, seizures, and finally paralysis. The biophysical mechanisms responsible for these symptoms have been elucidated using a variety of experimental protocols, in a wide array of organisms, with essentially the same result. It is now well accepted that pyrethroid insecticides, in common with DDT, have a major action at voltage-sensitive Na^{+} channels associated with excitable cells, most notably the nerve cell (1). This interaction causes a modification in the kinetics of both the inactivation and activation gating processes of the channel. The resulting increased inward sodium ion flux produces nerve cell depolarization and overall hyperexcitability in the nervous system. Although this membrane depolarization aspect is considered to be a primary action of DDT and pyrethroids, the exact molecular entity that is modified in this interaction has not been established (2). Additionally, it is unclear whether or not this is the sole mechanism of action for pyrethroids and DDT in all excitable tissues and in all organisms.

0097–6156/95/0591–0173$12.00/0

Besides the voltage-sensitive Na^+ channel, there are a number of other cellular events that also are perturbed by pyrethroids. These include: potassium channels (3); calcium channels (4); nicotinic acetylcholine receptor (5); GABA-chloride ionophore complex (6); ATP-hydrolyzing processes (e.g., adenosine triphosphatases), (7); protein kinases and phosphatases, (8); inositol triphosphates (9); GTP-binding proteins (l0); and neurotransmitter release (ll).

This last aspect, neurotransmitter release, is a particularly robust toxicologic endpoint to assess the various actions of DDT and the pyrethroids. The isolated presynaptic nerve terminal preparation (i.e., synaptosomes) used in this assay possesses all the above mechanisms for study. Selection of specific assay conditions and inhibitors provides a means of assessing the relative importance of each in the mode of action of these ion channel-directed insecticides.

Previous studies have established that Type II pyrethroids are extremely potent enhancers of neurotransmitter release (12, 13) and that this effect is particularly evident during depolarization (14, 15). The deltamethrin-dependent release of neurotransmitter during depolarization was highly correlated with the uptake of external Ca^{2+} by the synaptosome with a stoichiometry of approximately 1 to 5 (16). This release was also shown to be only partially inhibited by tetrodotoxin (i.e., approximately 50%) but was completely abolished by a phenethylamine-type Ca^{2+} channel antagonist, D595 (17). Since tetrodotoxin is a specific Na^+ channel blocker, deltamethrin may be enhancing neurotransmitter release by increasing the uptake of Ca^{2+} via other voltage-sensitive channels (e.g. calcium) or exchange mechanisms in addition to its action at the voltage-sensitive Na^+ channel.

Because of these results, we hypothesized that deltamethrin and other Type II pyrethroids may also function as phenethylamine-type Ca^{2+} channel agonists resulting in a potent enhancement of neurotransmitter release (17) and producing a positive inotropic action at the myocardium (18). To evaluate this hypothesis, we have chosen first to study the effects of DDT and pyrethroids on the ciliate protozoan, *Paramecium tetraurelia.*

Voltage-sensitive Ca^{2+} Channels and Avoidance Reaction Behaviors

A huge volume of research exists on the ciliary movement and locomotion in *Paramecium* (19). In normally forward swimming *Paramecium*, the power stroke of the rigid cilia beats in an anterior to posterior manner that propels the organism forward. Because the somatic cilia beat in a wave-like, metachronic fashion, the forward swimming *Paramecium* rotates on its long axis in the form of spirals. The biophysical and biochemical mechanism of ciliary movement is best described by the sliding microtubule model (20). The cilia, which are comprised of a 9 + 2 arrangement of paired microtubules called the axoneme, are covered with a ciliary membrane and are attached at their proximal end to the *Paramecium* by a cylindrical basal body or kinetosome. Associated with the tubulin proteins are contractile proteins called dyneins that possess Mg^{2+}-dependent ATPase activity. Although not completely understood, the hydrolysis of ATP to ADP by the dyneins results in the mechanochemical energy transduction that allows the ciliary microtubules to slide by each other resulting in the wave-like undulations of the cilia and swimming.

Given favorable environmental stimuli, *Paramecium* swims in the forward direction for a large proportion of its time. However, unfavorable environmental stimuli (e.g. water temperature changes, chemical gradients, physical obstacles, etc.) result in a backward swimming behavior that is called the *avoidance reaction* (21). *Paramecium* swims backward by reversing its ciliary action usually for a short period of time (e.g. 1-2 sec). It will then spin around through an angle of less than 180° by looping its anterior end about in a large circle and swim away in the forward mode. This *avoidance reaction* thus consists of three distinct behaviors; backward swim, spin, and looping. Under natural conditions, the time to complete these behaviors is brief but can be greatly extended for experimental examination by exposing *Paramecium* to high K^+ solutions. This K^+-induced signal produces membrane depolarization and is very similar to that used to release neurotransmitter from isolated presynaptic nerve terminals (12).

Similar to the Ca^{2+}-dependent release of neurotransmitter, the transitory reversal of cilia that gives rise to the *avoidance reaction* is explicitly dependent upon a Ca^{2+} action potential via voltage-sensitive Ca^{2+} channels associated with the ciliary membrane (22-24). Upon receiving an electrical, chemical, or mechanical stimulation, the ciliary Ca^{2+} channels open giving rise to a depolarizing Ca^{2+} action potential and a transient increase in intracellular Ca^{2+} concentration. This increase in internal Ca^{2+} ions is absolutely necessary for the reorientation of the axonemal movements and results in the reversal of the ciliary power stroke. The activation and inactivation processes that give rise to the Ca^{2+} action potentials in *Paramecium* are similar to those of the voltage-sensitive Na^{2+} channel of other organisms and *Paramecium* membranes have the same basic properties as neurons, sensory cells, and other excitable cells (18). Thus, by correlating the mortality response of *Paramecium* to selected insecticides and other toxins to their action on the avoidance behaviors elicited by Paramecium, one can easily and rapidly assess which compounds modify voltage-sensitive Ca^{2+} channels and which ones do not.

Paramecium Strains

Types and Rearing Conditions. Wild-type and *Dancer* and *Pawn* mutant strains of *Paramecium tetraurelia* were supplied by Dr. J. Van Houten (University of Vermont at Burlington). Each strain was reared under identical conditions (24). A 5 mL aliquot of improved wheat culture medium was added to an autoclaved culture tube. This medium was then inoculated with bacteria (*Enterobacter aerogenes*), and allowed to incubate at room temperature for 24-48 h. After the incubation, the medium was then inoculated with 500 µL of *Paramecium*. The paramecia were cultured at 28°C. After 14 days, the paramecia were reinoculated into another 5 mL aliquot of bacteria-rich medium.

Standard Mortality Bioassay. Using a 12 well Corning cell-well plate (22 mm diam., Cat. No. 25815), 12 watch glasses (26 mm diam.) were placed onto each of the 12 wells of the plate. If the treatment chemical in the assay was to be an insecticide, the wells were first soaked for five min in a 0.5% solution of carbowax (PEG 20,000, Fisher Scientific) and allowed to air dry overnight. A 500 μL aliquot of buffer solution was added to 11 of the watch glasses. Either a resting buffer (1 mM Ca(OH)$_2$, 1 mM Tris base, 1 mM citric acid, 15 mM NaCl, 5 mM KCl) or a depolarizing buffer (1 mM Ca(OH)$_2$, 1 mM Tris base, 1 mM citric acid, 20 mM KCl) was used. In the remaining watch glass, 250 μL of resting buffer was combined with 250 μL of culture media containing the desired strain of *Paramecium tetraurelia*. Under a dissecting microscope, five paramecia were removed from the last well using a plastic 1 cc tuberculin syringe which had been pulled to a fine tip under a low heat flame and placed into each of the 11 watch glasses. The first 6 wells were injected with a solution of the treatment compound using a Hamilton 10 μL glass syringe. The solution in each watch glass was gently stirred during the injection and a timer was started after the injection was complete. A *Paramecium* was considered dead when it permanently ceased swimming and the time was recorded. The last 5 wells were injected in an identical fashion as the first 6 except that the solvent contained no treatment chemical (i.e., solvent controls).

Deciliation Mortality Bioassay. Three changes were made to the procedures of the standard mortality bioassay in order to examine the effects of deciliation on treatment mortality. One, an additional 12 well plate was used so that the number of wells in the treatment group and the control group could be doubled (i.e., from 6 and 5 to 12 and 10, respectively). Two, half of the watch glasses in each group contained 5 deciliated paramecia while the other half each contained the usual 5 normal (ciliated) paramecia. Paramecia were deciliated by placing them in a solution of 10% ethanol in resting buffer for 5 min before transferring them into the watch glasses (24). Finally, since lack of movement was normally the criterion used to indicate mortality, enough time was allowed to elapse (2 h) for sufficient reciliation to occur before mortality recording began.

Behavioral Bioassay. Three watch glasses were placed onto a Corning cell-well plate. In the first watch glass, 250 μL of resting buffer was combined with 250 μL of culture media containing *Paramecium tetraurelia*. A 500 μL aliquot of resting buffer was placed into the second watch glass and a 500 μL aliquot of depolarizing buffer was placed into the third. A single *Paramecium* was removed from the first watch glass and placed into the second for 5 min. The third watch glass was injected with treatment compound (or just solvent in the case of the controls) to give the desired concentration. The paramecium was then transferred from the second watch glass to the third and a timer was started. The time the organism spent backward swimming, spinning, and looping was recorded until normal (forward) swimming behavior resumed or the paramecium died. The third watch glass was then replaced and another 500 μL aliquot of depolarizing buffer was added. This entire process was repeated until the desired number of repetitions had been reached.

Electrophysiological Assay. A small drop of resting buffer containing a single *Paramecium* was transferred into the recording chamber fixed on an inverted microscope (Olympus CK, Tokyo). The buffer was carefully reduced until the cell was immobilized on the bottom of the chamber. A single microelectrode was inserted into the immobilized cell and the chamber flooded with resting buffer. The microelectrode was made from borosilicate glass filament (outer diameter-1.0 mm; inner diameter-0.5 mm, World Precision Instruments, Sarasota, FL) and was filled with 0.5 M KCl. The bath was grounded via a 0.5 M KCl / 2% agar bath electrode. The membrane potential recorded with the microelectrode was amplified 10 times with a dual micro-probe amplifier (model 750, World Precision Instr., Sarasota, FL) , and recorded with a chart recorder (model 220, Gould) at 125 mm/min and 2 mV/division. At the indicated time, deltamethrin was added to the chamber with an automatic pipet.

Pyrethroids and DDT act as Ca^{2+} Channel Agonists on *P. tetraurelia*

Deltamethrin is toxic to *P. tetraurelia* in a Dose- and Use- dependent Manner. The results of mortality bioassays conducted in both resting and high K^+ depolarizing buffers are illustrated in Figure 1. Lethal time to 50% mortality values ($LT_{50}s$) are determined by logit mortality versus log dose transformations (POLO PC, Berkeley, CA).

Deltamethrin is toxic to *P. tetraurelia* in a dose-dependent manner under both experimental conditions but is particularly so when depolarized by high K^+ buffer (i.e., 20 mM). This aspect is reminiscent to the use-dependent action of pyrethroids on voltage-sensitive Na^+ channels (2). Under depolarizing conditions, concentrations of deltamethrin as low as 10^{-10} M are significantly more toxic than the concurrent ethanol-treated controls (likelihood ratio test, $P < 0.05$). This makes *P. tetraurelia* one of the most sensitive organisms that are intoxicated by pyrethroids.

Neurotoxic Na^+ Channel Ligands are not toxic to *P. tetraurelia*. Mortality assessment of *Paramecium* by deltamethrin is also not a trivial process. Because *Paramecium* has no voltage-sensitive Na^+ channels (19), the most notable site of action for pyrethroids is not available for modification in this organism. As expected, neurotoxic ligands that are selective to voltage-sensitive Na^+ channels, such as tetrodotoxin, aconitine, and veratridine, produced no overt toxicity in *Paramecium* (Table I) compared to ethanol-treated controls.

Neurotoxic Pyrethroids and DDT are selectively toxic to *P. tetraurelia*. As judged by the likelihood ratio test (POLO PC, P < 0.05), 1*R* deltamethrin (10^{-7} M) significantly reduced the LT_{50} value compared to a concurrent ethanol-treated control value under depolarizing conditions (Table II). A similar significant reduction is apparent in the presence of resmethrin and DDT at the same concentration. Neither the non-toxic 1S isomer of deltamethrin nor the less toxic DDE metabolite of DDT are significantly different in their toxicity to *P. tetraurelia*

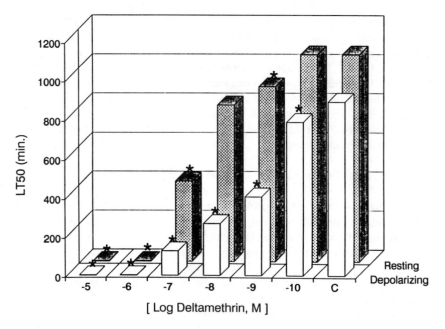

[Log Deltamethrin, M]

Figure 1.   Dose- and use-dependent mortality of   *P. tetraurelia* to deltamethrin. Asterisks indicate significant differences between treated and concurrent control groups (likelihood ratio test, P < 0.05).

Table I. Comparative mortality responses of *Paramecium tetraurelia* in bioassays with Na$^+$ channel ligands in depolarizing buffer

Compound [10^{-7}M]	N	Slope ± SE	LT$_{50}$ (95% CL)	LT$_{97}$(95% CL)	Likelihood Ratio Test[a]
Control	26	4.4 + 0.7	546 (492,606)	1761 (1302,3157)	--
Tetrodotoxin	32	6.3 + 0.9	47 (412,519)	1641 (1182,2946)	accepted
Aconitine	30	4.8 + 0.7	1061 (921, 1270)	5573 (3620,11875)	accepted
Veratridine	30	6.7 + 1.0	527 (479, 578)	1728 (1312,2825)	accepted

[a] Likelihood that the slope and intercepts are the same as in the control (p < 0.05).

than their concurrent ethanol-treated controls. These results indicate that *Paramecium* elicits the same stereospecific response that has been reported for mammals (25), other vertebrates (26) and invertebrates (27, 28) to these compounds.

Toxicity of Insecticides with no action on Voltage-Sensitive Ion Channels. To ascertain whether the toxic actions of DDT and pyrethroids are selective in *Paramecium* or due to a more generalized response, insecticides known not to interact with voltage-sensitive ion channels were examined (Table III). Neither abamectin nor dieldrin resulted in toxicity statistically different from the ethanol-treated controls. Both insecticides interact with ligand-gated chloride channels (e.g. GABA and glutamate, 29, 30, respectively). Similar ligand-gated channels or currents have not been reported in *Paramecium* (19). Diflubenzuron, a chitin synthesis inhibitor, likewise did not produce any enhancement in toxicity. Being a protozoan, *Paramecium* has no chitin synthesizing ability.

Two insecticides that have no action at voltage-sensitive ion channels, however, are quite toxic to *Paramecium*. Paraoxon, an acetylcholinesterase inhibitor, is the most toxic insecticide tested (Table III). Chlordimeform, a formamidine insecticide that most probably functions as an octopamine mimic (31), is also very toxic to *Paramecium* (Table III).

Toxicity of Organic Ca^{2+} Channel Ligands. Nifedipine, a 1,4-dihydropyridine-type Ca^{2+} channel antagonist, produced significant toxicity in *Paramecium* as judged by its LT_{50} value and a likelihood ratio test (Table IV). D595, a phenethylamine-type Ca^{2+} channel blocker, produced similar results (Table IV). Both compounds have potent antagonistic effects on L-type and T-type calcium channels (32) and D595 had been found previously to be a potent inhibitor of deltamethrin-dependent neurotransmitter release (17). Interestingly, the polypeptide toxin from the piscivarous marine snail, *Conus geographus*, ω-conotoxin GVIA, is not toxic to *Paramecium* (Table IV). In studies with cultured chick DRG cells and in cultured rat sympathetic neurons, ω-conotoxin GVIA blocked both N- and L-type Ca^{2+} currents but not T-type (33).

In view of its inhibition of deltamethrin-dependent neurotransmitter release (17) and its apparent interaction with the ciliary Ca^{2+} channels in *Paramecium* (Table IV), an experiment was conducted to determine whether pre-treatment with D595 would protect *Paramecium* from the toxic action of deltamethrin. As indicated in Table V, 10^{-7} M deltamethrin produced a substantial decrease in the LT_{50} value for treated *Paramecium* compared to ethanol controls. However, a 5 min pretreatment with 10^{-7} M D595, prior to exposure to deltamethrin, greatly attenuated the toxicity of deltamethrin and resulted in a dose-response that was not significantly different form the ethanol control (Table V). Because D595 is a specific Ca^{2+} channel antagonist, its protective action in the presence of deltamethrin may be due to it occupying a similar site on the ciliary Ca^{2+} channel of *Paramecium* as deltamethrin.

Table II. Comparative mortality responses of *P. tetraurelia* in insecticide bioassays with depolarizing buffer

Compound [10^{-7}M]	N	Slope ± SE	LT$_{50}$ (95% CL)	LT$_{97}$(95% CL)	Likelihood Ratio Test[a]
Control	34	4.9 ± 0.7	889 (692,1073)	4601 (2952,12210)	--
1R-Deltamethrin	113	2.8 ± 0.2	125 (105, 144)	2131 (1331,4402)	REJECTED
1S-Deltamethrin	32	2.3 ± 0.5	263 (202,359)	7561 (2589,113220)	accepted
Resmethrin	32	2.7 ± 0.5	148 (104,191)	3015 (1391,16429)	REJECTED
DDT	30	10.2 ± 3.0	329 (308,351)	534 (466,703)	REJECTED
DDE	30	7.0 ± 1.3	246 (221,274)	767 (575,1305)	accepted

[a]Likelihood that the slopes and intercepts are the same as in the control (p < 0.05).

Table III. Comparative mortality responses of *P. tetraurelia* in insecticide bioassays with depolarizing buffer

Compound [10^{-7}M]	N	Slope ± SE	LT$_{50}$ (95% CL)	LT$_{97}$(95% CL)	Likelihood Ratio Test[a]
Control	34	4.9 ± 0.7	889 (692,1073)	4601 (2952,12210)	--
Dieldrin	31	5.9 ± 0.6	252 (230,277)	979 (756,1452)	accepted
Abamectin	26	5.2 ± 1.0	280 (245,330)	1310 (823,3627)	accepted
Diflubenzuron	29	8.2 ± 2.0	379 (329,523)	1004 (655,3432)	accepted
Paraoxon	33	2.1 ± 0.9	21.4 (0.5,44.8)	455 (235,13461)	REJECTED
Chlordimeform	30	5.5 ± 1.4	30.1 (18.6,38.2)	129 (84,455)	REJECTED

[a]Likelihood that the slopes and intercepts are the same as in the control (p < 0.05).

Table IV. Comparative mortality responses of *P. tetraurelia* in bioassays with Ca^{2+} channel ligands in depolarizing buffer

Compound [10^{-7}M]	N	Slope ± SE	LT$_{50}$ (95% CL)	LT$_{97}$(95% CL)	Likelihood Ratio Test[a]
Control	25	7.0 ± 1.2	1025 (914,1195)	3223 (1249,5582)	--
Nifedipine	30	3.4 ± 1.0	315 (196,402)	3259 (1740,16813)	REJECTED
D595	30	5.4 ± 0.8	803 (724,909)	3535 (2519,6176)	REJECTED
ω-conotoxin	30	7.8 ± 1.2	1152 (1045,1287)	3206 (2515,4760)	accepted

[a]Likelihood that the slopes and intercepts are the same as in the control (p < 0.05).

Correlation of Pyrethroid-dependent Mortality and Effects on Avoidance Reaction Behaviors that are controlled by Ciliary Voltage-sensitive Ca^{2+} Channels. In *Paramecium*, the avoidance reaction (i.e., backward swimming, spinning, and looping behaviors) and swimming velocity are controlled by the bioelectrical processes of the plasma membrane. The avoidance reaction is a transitory reversal of cilia that corresponds explicitly with a Ca^{2+} action potential via voltage-sensitive Ca^{2+} channels and increase in intracellular Ca^{2+} (19). Similar to its mortality response, only the neurotoxic 1R isomer of deltamethrin significantly increased all aspects of the avoidance behaviors in *P. tetraurelia* (Figure 2). The 1S isomer was not significantly different from ethanol-treated control behaviors (Figure 2). Conversely, the phenethylamine-type Ca^{2+} channel blocker, D595, significantly reduced backward swim and spin behaviors, as expected, but had no significant effect on the looping process (Figure 2).

We can use this behavioral assay also to correlate the lack of mortality of insecticides that do not act on voltage-sensitive ion channels with the lack of effect on the avoidance reaction. Abamectin, which resulted in no increased mortality, was likewise found to have no effect on the avoidance reaction in *Paramecium* (Figure 3). These findings are consistent with the widely-held belief that abamectin acts on ligand-gated Cl^- channels and not on voltage-sensitive Ca^{2+} channels.

Chlordimeform, which was toxic to *Paramecium*, did not significantly alter any of the behaviors of the avoidance reaction (Figure 3). Thus, chlordimeform does not elicit its toxicity via any modification of ciliary voltage-sensitive Ca^{2+} channels. This finding is intriguing, nonetheless, because previous reports of octopaminergic responses have not been found for *P. tetraurelia.*.

Paraoxon, which was particularly toxic to *Paramecium*, altered the backward swim and spin behaviors (Figure 3) of the avoidance response. When compared to the effects elicited in the presence of deltamethrin, paraoxon treatment resulted in a decreased backward swim behavior, a greatly increased spin behavior, but looping behavior was unaffected. These findings indicate that paraoxon may be altering voltage-sensitive Ca^{2+} channels in the cilia but in a way quite distinct from that caused by deltamethrin. However, cholinesterase activity has been demonstrated in *Paramecium* (34) and organophosphate and carbamate insecticides have been previously reported as toxic (35). Thus, it is possible that paraoxon is acting as a cholinesterase inhibitor and its effects on avoidance behaviors are due to an indirect modification of ciliary Ca^{2+} channels.

Because the behaviors involved in the avoidance reaction are explicitly due to modifications of ciliary Ca^{2+} channels, the above results are consistent with the hypothesis that deltamethrin may be acting as a phenethylamine-type Ca^{2+} channel agonist in *Paramecium*. However, ligand binding data is necessary to clearly establish this aspect.

Comparative Mortality of Behavioral Ca^{2+} Channel Mutants of *Paramecium* to Deltamethrin. Two genic mutants with altered systems of excitation are included in this comparison (36, 37). The *pawn* mutant has a mutation that has inactivated its ciliary voltage-sensitive Ca^{2+} channels. It shows no avoidance response and can

Table V. Mortality responses of *P. tetraurelia* in insecticide bioassays with and without D595 in depolarizing buffer

Compound [10^{-7}M]	N	Slope ± SE	LT$_{50}$ (95% CL)	LT$_{97}$(95% CL)	Likelihood Ratio Test[a]
Control	26	4.6 ± 0.7	417 (355,537)	2112 (1232,6846)	--
1R-Deltamethrin	30	4.9 ± 1.0	280 (239,334)	1575 (1005,3629)	REJECTED
1R-Deltamethrin + D595	32	5.1 ± 0.8	358 (278,553)	1700 (867,22171)	accepted

[a] Likelihood that the slopes and intercepts are the same as in the control (p < 0.05).

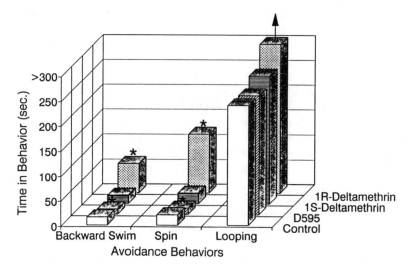

Figure 2. Effect of 1R- and 1S-deltamethrin and D595 on avoidance reaction behaviors. Asterisks and arrows indicate significant differences between treated and concurrent control groups (likelihood ratio test, P< 0.05).

only move forward like the chess piece it is named after. The *dancer* mutant has a mutation that has made its avoidance response hyperexcitable. Single Ca^{2+} action potentials now produce extended periods of backward swimming behavior. As illustrated in Figure 4 (top panel), the *dancer* mutant is hypersensitive to the toxic action of 10^{-7} M deltamethrin compared to wild type *P. tetraurelia* in resting buffer. Comparatively, the *pawn* mutant responds no differently than the ethanol-treated control. Upon depolarization (Figure 4, bottom panel), wild type *P. tetraurelia* becomes equally sensitive to deltamethrin as the *dancer* mutant but the *pawn* mutant is still recalcitrant to the action of this potent pyrethroid insecticide. In this case, a functional Ca^{2+} channel must be available for deltamethrin to elicit a toxic response to *Paramecium*. The functional removal of this channel in the *pawn* mutant eliminates the toxic action of deltamethrin. The availability of a Ca^{2+} channel that remains in the open state more often (e.g. *dancer* mutant) is consistent with the use-dependency of the action of deltamethrin and with the increased sensitivity of this mutant to deltamethrin.

Deltamethrin produces no Enhanced Toxicity on Deciliated *P. tetraurelia.* Ethanol pre-treatment (10%) provides a rapid, efficient, and non-lethal means of removing the somatic ciliature of *P. tetraurelia* (16). Subsequent exposure of deciliated *P. tetraurelia* to 10^{-7} M deltamethrin resulted in no increased toxicity under either resting or depolarized conditions (Figure 5). Statistically, neither of the deltamethrin-treated lines are significantly different from their concurrent deciliated but ethanol-treated control lines (likelihood ratio test, $P < 0.05$).

The chemical removal of the somatic cilia effectively removes the associated voltage-sensitive Ca^{2+} channels. In the physical absence of these channels, deltamethrin no longer elicits a toxicological response in deciliated *Paramecium*.

Deltamethrin causes Membrane Destabilization, Repetitive Ca^{2+} Action Potentials and Membrane Depolarizations. The effect of deltamethrin (10^{-9} M) on the soma resting potential of *P. tetraurelia* using intracellular recording is illustrated in Figure 6. Immediately upon application, Ca^{2+} action potentials (AP) are evident and the plasma membrane begins to depolarize from its untreated resting potential (-31 mV). At approximately 2 min post-application, a repetitive Ca^{2+} action potential (RAP) is seen that is followed by an extremely intense RAP about one min later. Approximately 4 min post-application, the membrane potential is at -4mV and the organism dies. These electrophysiological symptoms of poisoning (i.e. membrane destabilization, increased spontaneous action potentials, repetitive discharges, and membrane depolarization) are similar to those used to describe the action of pyrethroids on voltage-sensitive Na^+ channels (38). Nevertheless, these symptoms occur in the complete absence of voltage-sensitive Na^+ channels.

These and the above results are consistent with the hypothesis that Type II pyrethroids, such as deltamethrin, act as potent voltage-sensitive Ca^{2+} channel agonists in *Paramecium*.

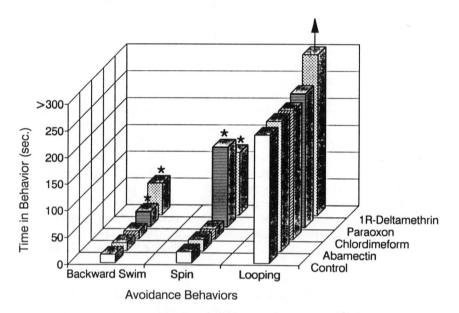

Figure 3.   Effects of insecticides that do not interact with voltage-sensitive ion channels on advoidance reaction behaviors. Asterisks and arrows indicate significant differences between treated and concurrent control groups (likelihood ratio test, P < 0.05).

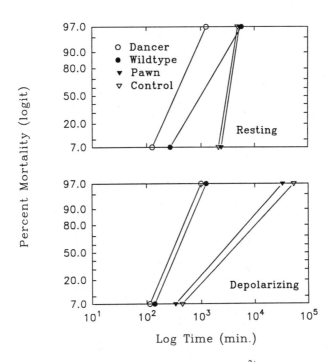

Figure 4. Comparative mortality of behavioral Ca^{2+} channel mutants of *Paramecium* to 10^{-7}M deltamethrin (Reproduced with permission from ref. 49. Copyright 1993 Pesticide Science Society of Japan).

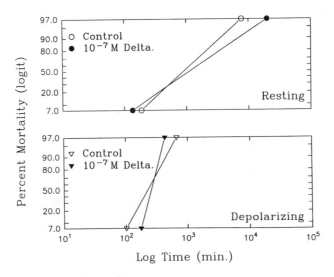

Figure 5. Effect of 10^{-7}M deltamethrin on deciliated *P. tetraurelia* (Reproduced with permission from ref. 49. Copyright 1993 Pesticide Science Society of Japan).

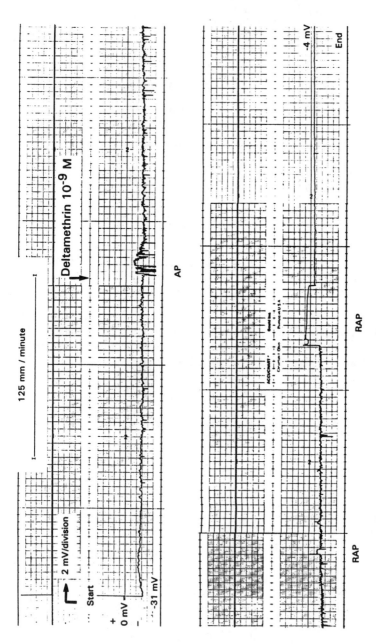

Figure 6. Intracellular recording response of the plasma membrane of *P. tetraurelia* to 10^{-7}M deltamethrin (Reproduced with permission from ref. 49. Copyright 1993 Pesticide Science Society of Japan).

Conclusions and Future Research

Results are presented that substantiate the role of ciliary voltage-sensitive Ca^{2+} channels as a site of action of pyrethroids that act as Ca^{2+} channel agonists in *Paramecium*. Most germane is the fact that both mortality and avoidance behavior changes are produced in the complete absence of any involvement of voltage-sensitive Na^+ channels. Insecticides such as abamectin, chlordimeform, dieldrin, diflubenzuron, paraoxon and non-toxic analogues such as DDE and 1S-deltamethrin do not interact with this Ca^{2+} channel in a similar fashion. 1R-deltamethrin (10^{-9} M) causes rapid membrane destabilization, increased spontaneous action potentials, repetitive discharges, and membrane depolarization. These findings reflect a high level of similarity in the pharmacology, toxicology, and electrophysiology of tetrodotoxin-sensitive, voltage-gated Na^+ channels and the voltage-gated Ca^{2+} channels in the cilia of *Paramecium*.

The fact that Type II pyrethroids may interact with both Na^+ and Ca^{2+} voltage-sensitive channels should not be that contradictory to the existing sodium channel theory of DDT and pyrethroid action in that there appears to be a great deal of structural homology between them. As Curtis and Catterall (39, 40) point out, both the isolated sodium channel ionophore (41) and the dihydropyridine calcium antagonist receptor complex of the voltage-sensitive Ca^{2+} channel (42) are large membrane glycoproteins of 200-300 kD that consist of one large subunit and two smaller subunits. Indeed, Tanabe *et al.* (43) have reported close structural and primary sequence similarities of the dihydropyridine receptor of the voltage-sensitive Ca^{2+} channel to the voltage-sensitive Na^+ channel in support of this contention. Such overall structural similarities indicate similar requirements for rapid movement of ions across membranes via voltage-gated ionophores and may indicate similar binding regions for DDT and pyrethroids. This apparent commonality in action has already been extended to the phenethylamine class of Ca^{2+} channel blockers (e.g., D595) which also have an inhibitory action at other voltage-gated channels including the Na^+ channel, although at much higher dosages (44-47).

A final point that is worthy of some discussion is illustrated in Figure 7. As indicated, *Paramecium* plays a central role in detritus-based food webs and serves as a link in plantonic food chains in aquatic ecosystems (19). Ciliates as a group, and *Paramecium* in particular, feed on an enormous array of bacteria that grow in turn by feeding on decomposed organic matter. *Paramecium* then serve as a food source themselves and are fed upon by other protozoans and a variety of aquatic metazoan organisms. Although this relationship has been recognized for some time, little quantitative research has been reported on these ecological relationships (19).

Given the well-recognized aquatic toxicity associated with the pyrethroids (48), the present finding that pyrethroids are extremely potent toxins to *Paramecium* may be axiomatic. A possible, and probable, ramification of this finding may be that we have greatly underestimated the overall ecological impacts

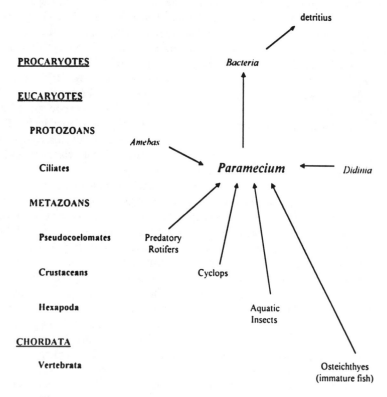

Figure 7. Central role of *Paramecium* in aquatic ecosystems.

from pyrethroids in aquatic areas by focusing the majority of our investigations on macroinvertebrates (e.g., insects) and vertebrates (e.g., fish). A toxicological evaluation of the impact of pyrethroids on protozoans and other microinvertebrates, the role of voltage-sensitive Ca^{2+} channels in these interactions, and the effects of this toxicity to other trophic levels in the aquatic ecosystem, would seem pertinent.

Acknowledgments

We thank Dr. Susan Barry, Dept. of Biology, Mount Holyoke College, for her assistance in our behavioral experiments. This work was supported by MAES (USDA NE180), UMASS, Amherst, MA.

Literature Cited

1. Narahashi, T. *J. Cell. Comp. Physiol.* **1962,** 59, 61.
2. Narahashi, T. *Trends in Pharmacol. Sci.* **1992,** 13, 236.
3. Narahashi, T. In *Advances in Insect Physiology;* Beament, J.W.L.; Treherne, J.E.; Wiggleworth, V.B., Eds.; Academic Press: New York, NY, 1971, pp 1-93.
4. Orchard, I; Osborne, M.P. *Pestic. Biochem. Physiol.* **1979,** 10, 197.
5. Abbassy, M.A.; Eldefrawi, M.E.; Eldefrawi, A.T. *Life Sci.* **1982,** 31, 1547.
6. Lawrence, L.J.; Casida, J.E. *Science.* **1983,** 221, 1399.
7. Clark, J.M.; Matsumura, F. *Pestic. Biochem. Physiol.* **1982,** 18, 180.
8. Matsumura, F.; Clark, J.M.; Matsumura, F.M. *Comp. Biochem. Physiol.* **1989,** 94C, 381.
9. Gusovsky, G.; Secunda, S.I.; Daly, J.W. *Brain Res.* **1989,** 492, 72.
10. Rossignol, D.P. *Pestic. Biochem. Physiol.* **1991,** 41, 121.
11. Aldridge, W.N.; Clothier, B.; Forshaw, P.; Johnson, M.K.; Parker, V.H.; Price, R.J.; Skilleter, D.N.; Verschoyle, R.D.; Stevens, C. *Biochem. Pharmacol.* **1978,** 27, 1703.
12. Nicholson, R.A.; Wilson, R.G.; Potter, C.; Black, M.H. In *Pesticide Chemistry, Human Welfare and the Environment;* Miyamoto, J; Kearney, P.C., Eds.; Pergamon Press: Oxford, UK, 1983, pp. 75-90.
13. Salgado, V.L.; Irving, S.N.; Miller, T.A. *Pestic. Biochem. Physiol.* **1983,** 20, 169.
14. Schouest, L.P., Jr.; Salgado, V.L.; Miller, T.A. *Pestic. Biochem. Physiol.* **1986,** 25, 381.
15. Brooks, M.W.; Clark, J.M. *Pestic. Biochem. Physiol.* **1987,** 28, 127.
16. Clark, J.M., Brooks, M.W. *Env. Toxicol. and Chem.* **1989,** 8, 361.
17. Clark, J.M.; Brooks, M.W. *Biochem. Pharmacol.* **1989,** 38, 2233.
18. Berlin, J.R.; Akera, T.; Brody, T.M; Matsumura, F. *Eur. J. Pharmacol.* **1984,** 98, 313.
19. Wichterman, R. In *The Biology of Paramecium;* Plenum Press, New York, NY, 1986, pp 223-229.
20. Satir, P. In *Behaviour of Micro-organisms: Proceedings of the 10th International Congress*; Perex-Miravete, A., Ed.; Plenum Press, New York, NY, 1973, pp 214-228.
21. Jennings, H.S. *J. Comp. Neurol.* **1905,** 15, 138.
22. Eckert, R. *Science.* **1972,** 176, 473.
23. Naitoh, Y.; Kaneko, H. *Science.* **1972,** 176, 523.
24. Kung, C.; Chang, S.Y.; Satow, Y.; van Houten, J.; Hansma, H. *Science.* **1975,** 188, 898.
25. Gray, A.J; Connors, T.A.; Hoellinger, H.; Hoang-Nam, N. *Pestic. Biochem. Physiol.* **1980,** 13, 281.
26. Vivjerberg, H.P.M.; Oortgiesen, M. In *Stereoselectivity of Pesticides, Biological and Chemical Problems*; Ariëns, E.J.; van Rensen, J.J.S.; Welling, W., Eds.; Elsevier Science Publishers B.V., Amsterdam, 1988, pp 151-182.
27. Soderlund, D.M.; Bloomquist, J.R. *Annu Rev. Entomol.* **1989,** 34, 77.

28. Clark, J.M.; Marion, J.R. In *Insecticidal Action: From Molecule to Organism;* Narahashi, T.; Chambers, J.E., Eds.; Plenum Press, New York, NY, 1989, pp., 139-168.
29. Pong, S.S.; Wang, C.C. *J. Neurochem.* **1982,** 38, 375.
30. Arena, J.P.; Liu, K.K.; Paress, P.S.; Schaeffer, J.M.; Cully, D.F. *Molecular Brain Res.* **1992,** 15, 339.
31. Hollingworth, R.M.; Murdock, L.L. *Science.* **1980,** 208, 74.
32. Triggle, D.J. In *Calcium Regulation by Calcium Antagonists*; Rahwan, R.G.; Witiak, D.T., Eds.; ACS Symposium Series 201, Washington, D.C., 1982, pp 17-37.
33. Fox, A.P.; Hirning, L.D.; Kongsamus, S.; McCleskey, E.W.; Miller, R.J.; Olivera, B.M.; Perney, T.M.; Thayer, J.A.; Tsien, R.W. In *Neurotoxins and their Pharmacological Implications*; Jenner, P., Ed.; Raven Press, New York, NY, 1987, pp 115-131.
34. Andrivon, C. *J. Med.* **1975,** 123, 87A.
35. Lejczak, B. *Pol. Arch. Hydrobiol.* **1977,** 24, 583.
36. Kung, C. *Z. Verlag. Physiol.* **1971,** 71, 142.
37. Kung, C. *Genetics.* **1971,** 69, 29.
38. Narahashi, T. *Pharmacol. Sci.* **1992,** 13, 236.
39. Curtis, B.M.; Catterall, W.A. *Biochemistry.* **1984,** 23, 2113.
40. Catterall, W.A. *Science.* **1984,** 223, 653.
41. Tamkun, M.M.; Talvenheine, J.A.; Catterall, W.A. *J. Biol. Chem.* **1984,** 259, 1676.
42. Curtis, B.M.; Catterall, W.A. *Biochemistry.* **1986,** 25, 3077.
43. Tanabe, T.; Takeshima, H.; Mikami, A.; Flockevzi, V.; Takahashi, H.; Kangawa, K.; Kojima, M.; Matsuo, H.; Hirose, T.; Numa, S. *Nature.* **1987,** 328, 313.
44. Henry, P.D. *J. Cardiol.* **1980,** 46, 1047.
45. Atlas, D.; Adler, M. *Proc. Natl. Acad. Sci. USA.* **1981,** 78, 1237.
46. Bregestovki, P.D.; Iljin, V.I. *J. Physiol. (Paris)* **1980,** 76, 515.
47. Miledi, R.; Parker, I. *Proc. R. Soc. Lond [Biol]* **1980,** 211, 143.
48. Bradbury, S.P.; Coats, J.R. *Rev. Environ. Contam. Toxicol.* **1989,** 108, 134.
49. Clark, J.M.; Nagy, S.R.; Edman, S.J.; Canhoto, A.; Houten, J.; Hecht, F. In *Pesticide/Environment: Molecular Biological Approaches;* Mitsui, T., Matsumura, F., Yamaguchi, I., Eds., Proc. of The First International Sym. on Pestic. Sci., Pestic. Sci. Soc. of Japan: Tokyo, Japan, **1993,** pp. 85-99.

RECEIVED January 23, 1994

LIGAND-GATED CHLORIDE CHANNELS

Chapter 12

Biophysical Analysis of a Single Amino Acid Replacement in the Resistance to Dieldrin γ-Aminobutyric Acid Receptor

Novel Dual Mechanism for Cyclodiene Insecticide Resistance

R. H. ffrench-Constant[1], H.-G. Zhang[2], and M. B. Jackson[2]

[1]Department of Entomology and [2]Department of Physiology, University of Wisconsin, Madison, WI 53706

Target site insensitivity to cyclodienes is associated with a novel insecticide insensitive γ-aminobutyric acid (GABA) receptor subunit gene termed *Resistance to dieldrin* (*Rdl*). To date, examination of this gene in resistant insects from three different orders has always shown the replacement of the same amino acid (alanine302) with either a serine or a glycine (*D. simulans* only) in the second membrane spanning region (M2) of the receptor. This provides the first direct evidence that cyclodienes and picrotoxinin bind within the M2 region of the receptor, the putative lining of the integral GABA gated chloride ion channel. However, it is unclear why resistance is *always* only associated with replacements of this single amino acid. We have shown, via a detailed biophysical analysis of the mutated GABA receptor, that the alanine302>serine replacement has a number of effects on channel function including: reduced sensitivity to picrotoxinin, lindane and TBPS; lower channel conductances; extended open times and shorter closed times and a marked reduction in the rate of GABA-induced receptor desensitization. Via a simple model for evaluating binding site versus allosteric changes, we propose that replacements of alanine302 are the *only* mutations that can *both* directly weaken cyclodiene binding to the antagonist favored (desensitized) conformation and indirectly destablilize the antagonist favored conformation through an additional allosteric mechanism. Thus a unique dual resistance mechanism achieves the high levels of target site insensitivity observed.

Despite their early introduction, the exact mode of action of cyclodienes has until recently remained obscure. Thus although the weight of pharmacological evidence has previously suggested that they block GABA gated chloride ion currents (1), it has remained unclear with *which* GABA receptor subunit(s), and precisely which amino acids on the receptor, they interact. As cyclodiene resistance has accounted for

0097–6156/95/0591–0192$12.00/0

approximately 60% of reported cases of insecticide resistance (2), we decided to employ a genetic approach to clone the gene conferring target site insensitivity and to examine the amino acid replacement(s) conferring resistance. Following field isolation of a mutant of the genetic model *Drosophila melanogaster* (3), and precise localization of the mutant gene on the polytene chromosome map (4), we cloned the gene responsible following a 200kb chromosomal walk (5). The resistance gene *Rdl* (*Resistance to dieldrin*) codes for a novel class of GABA receptor subunit (6). Here we describe the location of the single amino acid replaced in cyclodiene resistant strains and examine its role in resistance by a detailed biophysical analysis of the receptor. Although the cyclodienes themselves are no longer as widely used, the convulsant binding site(s) remain an important target for a number of classes of insecticidal compounds (7-9)(see also Deng, in this volume) and for important convulsant drugs such as picrotoxinin (PTX).

The Location of Alanine302 Within the Putative Ion Channel Pore

Nucleotide sequencing of the open reading frame of *Rdl* from a susceptible and resistant strain of *D. melanogaster* revealed a single amino acid replacement alanine302> serine within the proposed second membrane spanning domain of the receptor or M2 (10). Subsequent sequencing of other *D. melanogaster* strains, and also examination of this region with a diagnostic restriction enzyme site eliminated by the replacement, revealed that the same point mutation was found in all resistant strains of this species collected worldwide (10). Further, cloning of the M2 region by degenerate PCR from a number of other species from three different insect orders (11, 12) (Diptera: *Aedes aegypti, Musca domestica*; Dictyoptera: *Periplaneta americana* and Coleoptera: *Tribolium castaneum, Hypothenemus hampei*) revealed precisely the same change. Recently, the same replacement has also been reported in another cyclodiene resistant cockroach species *Blatella germanica* (K. Kaku and F. Matsumura, in this volume). In fact the only different mutation we have found has been in *D. simulans*, a close relative of *D. melanogaster*, where the *same* amino acid is replaced but with a glycine (alanine302> glycine) instead of a serine (10).

By site-directed mutagenesis of the susceptible *Rdl* cDNA and subsequent expression in *Xenopus* oocytes we have shown that the alanine302> serine replacement confers approximately 100 fold insensitivity to PTX and high levels of resistance to the cyclodiene dieldrin (13). The location of alanine302 within M2 is of particular interest in relation to the proposed mode of action of the cyclodienes and PTX. Work in the closely related nicotinic acetylcholine receptor involving serine> alanine replacements in this region has shown that a residue, close to the analogous position of alanine302 (Figure 1), interacts with elements within the channel pore (14). This led to the terming of this position as the "inner polar site" and the proposal that M2 actually forms the ion channel lining of this class of receptors. Further, alanine302 corresponds to the most cytoplasmic of the three small uncharged residues actually thought to be exposed to the channel lumen itself. This finding therefore represents the strongest evidence to date that cyclodienes and PTX actually bind within the channel pore.

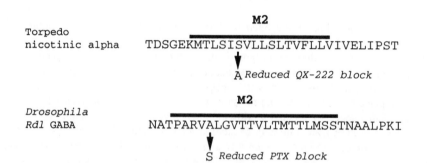

Figure 1. Alignment of the *Rdl* GABA receptor subunit sequence with that of the nicotinic acetylcholine receptor (nAChR) to illustrate that alanine302 occupies a similar position to the residue in the nAChR interacting with the channel blocker QX-222.

Biophysical Analysis of the Resistant Receptor

Despite the location of an amino acid conferring resistance to cyclodienes/PTX within the channel, physiological investigations have suggested that these drugs block by a mode of action different from open channel block. Newland and Cull-Candy in their extensive study of the kinetics of PTX action, have found that this drug binds preferentially to the desensitized conformation of the mammalian $GABA_A$ receptor (15). That is to say, that when in the desensitized state (not responsive to GABA) a higher affinity binding site for PTX is exposed. In order to examine the precise mechanism by which replacements of alanine302 confer resistance to cyclodienes/PTX, we carried out a detailed biophysical analysis of the mutant receptor in cultured *Drosophila* neurons using the patch clamp technique.

Response to GABA and Antagonists. GABA-activated whole cell current responses were recorded from patch-clamped cultured *Drosophila* neurons. Responses to brief pulses of GABA were identical in cells cultured from both susceptible and resistant flies. Current increased linearly with voltage and reversed near 0 mV, as expected for a chloride selective channel in symmetrical chloride solutions. GABA dose response curves from the two strains were also the same, with nearly identical apparent affinities for GABA and similar low cooperativities (with dissociation constants of 31 μM and 29 μM, and Hill coefficients of 1.2 and 0.83 for susceptible and resistant preparations respectively).

In contrast, widely differing sensitivities of GABA receptor antagonists were observed in susceptible and resistant receptors (Figure 2). Thus, PTX displayed 116 fold reduced sensitivity (K_i values of 224 nM and 26 μM for susceptible and resistant respectively) in resistant neurons, similar to the levels of resistance observed in oocytes expressing the mutant alanine302> serine cDNA (13). Lindane (or γ-HCH) exhibited extremely high levels (970 fold) of resistance (0.15 nM and 146 nM for susceptible and resistant respectively), in contrast to the low levels (9 fold) of cross resistance found in bioassays of adult cyclodiene resistant *Drosophila* (4). Whereas *t*-butylbicyclophosphorothionate or TPBS, an important vertebrate $GABA_A$ receptor antagonist, showed very low levels of resistance (only 9 fold) in the cultured neurons.

These findings raised two important questions that were difficult to reconcile. 1) Why does the alanine 302> serine mutant show such high levels of resistance to lindane in electrophysiological assays and such low levels of resistance in the whole insect? Is the primary site of action of this compound really, therefore, the same as that of the cyclodienes (i.e. a GABA gated chloride ion channel containing *Rdl* subunits)? 2) How can two different replacements of the same amino acid (alanine302> serine or glycine) confer similar levels of resistance in the whole insect if only *direct* differences in drug binding are involved? i.e. How can addition of a hydroxyl group (serine) or reduction of the amino acid side chain to only a hydrogen (glycine) have the same effect on resistance and therefore on drug binding to the receptor? To address the phenomena underlying these questions, we carried out further analysis of the channel properties of the wild type and mutant receptors by analysis of single channel recordings.

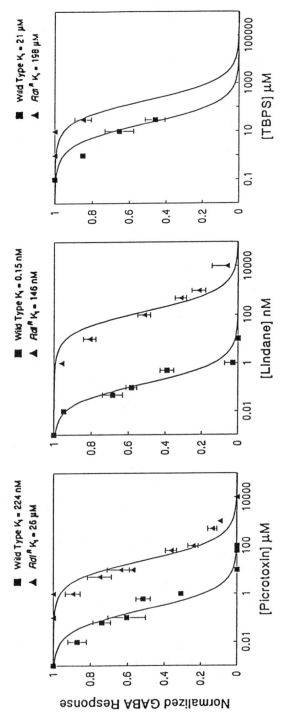

Figure 2. Dose-response curves for wild type and resistant cultured *Drosophila* neurons. The plots were fitted to the Hill equation to give apparent dissociation constants of 31 μM for wild type and 29 μM for resistant neurons. The Hill coefficients were 1.2 and 0.83 for wild type and resistant neurons respectively.

Mutant Channel Properties. Analysis of single channel recordings revealed several additional differences between susceptible and resistant channels. Firstly, the mutation reduced the single channel conductance by 5% for inward current and 17% for outward current (28 to 26.5 picosemens and 19 to 15.8 pS, for inward and outward conductance, respectively). Secondly, resistance was associated with significantly longer channel open times and shorter closed times, reflecting a net stabilization of the channel open state by approximately 5 fold. However, these changes were small and not likely to contribute to the drug resistance of the receptor. A much more important change was the marked reduction in the rate of GABA-induced desensitization (Figure 3), and a net destabilization of the desensitized conformation by a factor of 29 (current desensitized to a final level of 1.6% peak current in susceptible and 32% of peak current in resistant neurons). As PTX has been proposed to bind preferentially to the desensitized state of the GABA receptor (15), our results suggest that destabilization of the desensitized state could actually act in concert with the changes in the insecticide binding site, as a novel *dual* resistance mechanism.

Evaluation of Binding Site Alterations Versus Allosteric Effects

To form a framework for the evaluation of binding site versus allosteric changes in cyclodiene resistance we developed the following model, in which a receptor isomerizes between two conformations; activatable, A and inactivatable, N. Antagonists binding to these conformations therefore have affinities of K_A and K_N, respectively.

$$S_0 \left\Vert \begin{array}{c} A_0 \xrightarrow{\quad K_A \quad} A_1 \\ \\ N_0 \xrightarrow{\quad \quad \quad} N_1 \end{array} \right\Vert S_1$$

$$K_N$$

S_0 and S_1 represent equilibrium constants for interconversion between activatable and inactivatable states, in which the antagonist binding site is free or occupied, respectively. Similar models have proved very robust in the interpretation of ligand induced changes in the affinity of the acetylcholine receptor (16). The above model implies the following dependence for response to GABA as a function of concentration of antagonist, C (17).

$$R = \frac{R_{max}(1 + C/K_A)}{1 + S_0 + (1 + S_1)C/K_A}$$

(Eq. 1)

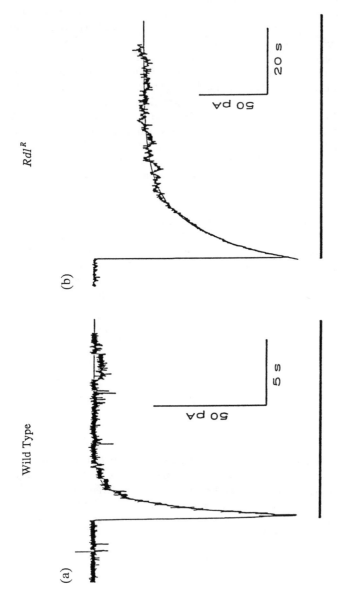

Figure 3. Differential desensitization of wild type (a) and resistant (b) GABA receptors by continuous application of GABA (solid bar). Traces were fitted to a single exponential giving mean (± SE) time constants of 1.2 ± 0.1 sec (n = 10) and 11.1 ± 0.9 sec (n = 9) for susceptible and resistant receptors, respectively. The current desensitized to a final level of 1.6% of the peak current in wild type and 32% of peak current in the resistant preparations.

With a few simplifying assumptions, the antagonist concentration at which response is half maximal can be reduced to:

$$K_i = \frac{K_N}{S_0}$$

(Eq. 2)

An analogous expression has been derived for agonist binding sites (18).

Equation 2 reduces the issue of resistance at the molecular level to two fundamentally different mechanisms. If a mutation alters the conformational equilibrium, then S_0 will change. If a mutation alters the binding site, then K_N will change. A change in either S_0 or K_N is sufficient to shift the dose-inhibition curve. However, when the change occurs in S_0 only, the mechanism will be *purely allosteric*. In such a case the sensitivity of the receptor to all antagonists of a given class will be shifted equally. In contrast, a change in K_N only is indicative of a mutation within the *binding site*. In such a case the change in sensitivity would vary in magnitude with the choice of ligand. Since we have in fact observed different levels of resistance to each of the three ligands tested, the mechanism of resistance cannot be purely allosteric. Resistance of the *Rdl* receptor must therefore be associated with a change in the antagonist binding site. Therefore, alanine302 makes a direct contribution to the binding site of cyclodienes/PTX.

Although the above reasoning argues that the binding site has changed, other lines of evidence indicate that allosteric processes are at work as well. PTX has been shown to bind preferentially to the desensitized conformation in vertebrate GABA receptors (15). This would suggest that the desensitized conformation corresponds to the conformation denoted as N in the model above. Since the mutation is accompanied by a change in desensitization, S_0 should change and make an allosteric contribution to resistance. Results from our whole cell patch clamp recordings have shown that resistance is associated with a 29 fold change in the open-desensitized equilibrium constant. The open-desensitized equilibrium constant differs from S_0 in the above model, in terms of the state of occupancy of the GABA binding sites. However, as the mutation is far from GABA binding sites (GABA and PTX do not compete in binding assays), it should destabilize the GABA-ligated and unligated forms of the receptor to similar degrees. Therefore, it can be assumed that the ratio of desensitization equilibrium constants for GABA induced desensitization can be used as an estimate of the factor by which S_0 was changed by the mutation. Thus, our desensitization results for *Rdl* imply a 29 fold change in S_0 above. This allows us to resolve the resistance ratios for the three ligands into their relative contributions from either resistance mechanism, namely a) destabilization of the desensitized conformation and b) changes in K_N (Table 1). Thus, for example within the 116 fold increase in the K_i of PTX, if destabilization of the desensitized conformation accounts for a factor of 29 the remaining 4 fold resistance represents an increase in K_N. Interestingly, this means that as the K_i ratio for TBPS is less than the factor by which desensitization is destabilized, one can infer that the mutation actually *enhances* binding of this particular ligand.

Table 1. The relative contributions of allosteric (destabilization of the desensitized conformation) and direct (changes in binding site affinity) effects on *Rdl* resistance conferred by alanine302> serine to three convulsants. The destabilization factor is assumed to be 29 fold throughout.

Compound	Change in sensitivity K_i	Change in binding K_N
1. PTX	116	4
2. Lindane	970	34
3. TBPS	9	0.31

In order to facilitate conceptualization of these direct and indirect effects on insecticide binding, we have included a diagrammatic model in Figure 4. Here, the receptor is visualized in either the open or desensitized state, in the presence or absence of antagonist. The key point to note is that passage to the desensitized state *increases* the affinity of binding between the receptor and the drug or insecticide. This is illustrated by the "tighter fit" of the antagonist into its binding site in the channel.

Conclusions and Future Directions

Although we now have a clearer understanding of the receptor with which cyclodienes interact and their mode of action, as with any scientific investigation, our studies have raised further important questions relevant both to insecticide resistance/mode of action and to GABA receptor structure/function. We will therefore briefly review our conclusions here and indicate future directions in the study of the cyclodiene receptor.

Binding site. By examining shifts in sensitivity within the framework of an allosteric theory, we were able to separate direct binding effects from allosteric effects. The unequal shifts in sensitivity to three different ligands at overlapping sites indicates that alanine 302 participates directly in drug binding. Since this residue is in the M2 region, and the M2 region is generally believed to form the pore, this provides strong evidence that picrotoxinin and cyclodienes bind to the pore region of the GABA receptor.

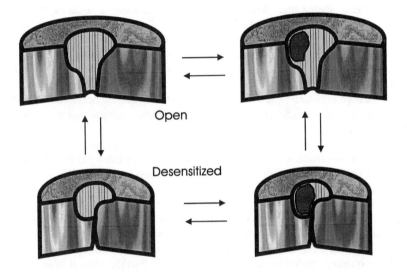

Open

Desensitized

Figure 4. Conceptual model of the interaction of the *Rdl* GABA receptor with the antagonists PTX or cyclodiene insecticides. The receptor in visualized in open (top) or desensitized (bottom) states, in the presence (right) or absence (left) of the antagonist. Note how the transition to the desensitized state increases the affinity of the receptor for the antagonist (bottom right).

Comparisons with other ligand-gated channels. The M2 regions of other ligand-gated channels have been studied extensively by site-directed mutagenesis. In the *Torperdo* α subunit of the nicotinic acetylcholine receptor the reciprocal substitution of a serine to alanine (four amino acids away from alanine 302) reduced block by the anesthetic QX-222 (14). Unfortunately, effects of this substitution on desensitization were not examined. However, other replacements in M2 regions were shown to alter desensitization dramatically; a leucine homologous to leucine 309 of *Rdl* influenced desensitization in both a neuronal nicotinic α subunit (19) and the 5-HT$_3$ receptor (20). Clearly, the M2 region has functions in important conformational transitions as well as ion permeation.

A Novel Dual Resistance Mechanism. Conventionally insecticide resistance conferred by target site insensitivity has been inferred to arise from a limited number of possibilities: 1) a change in binding site of the receptor, 2) a change in the number of receptors, 3) a change in the relative proportion of receptor subpopulations with differing affinities for insecticides and 4) changes in the membrane environment of the receptor (21). Here we report that cyclodiene resistance is conferred not only by a simple change in binding site affinity but *also* by an allosteric destabilization of the insecticide favored (desensitized) state. This therefore constitutes a novel *dual* resistance mechanism.

Is *Rdl* a Homomultimer? The finding that resistance ratios for PTX are similar in cultured neurons to those in heterologous expression systems (either in *Xenopus* oocytes (13) or *Spodoptera* cell lines (22)), and the similar pharmacology of the *in vivo* and *in vitro* receptors (13) (our unpublished results), raises the possibility that the cyclodiene GABA receptor is composed only of *Rdl* subunits. This would be in contrast to the complex heteromultimeric assemblies of vertebrate GABA$_A$ receptors (23-25). It should obviously be stressed that the inclusion of other subunits into the cyclodiene receptor cannot be excluded at this stage. Further, even if other subunits were present in the same receptor they may have no effect on PTX binding. However this is our working hypothesis at present. Thus, for example, it is presently unclear as to whether the *Drosophila* β GABA$_A$ receptor subunit homolog, recently cloned independently in two laboratories (26-27), may form part of the same receptor or not. In order to determine the subunit composition of the *Drosophila* cyclodiene receptor we will examine oocytes and transformed flies expressing mutants conferring lower channel conductances in the presence and absence of wild type subunits.

Future Directions. As two different replacements of the same amino acid (alanine302> serine or >glycine) are found in *Drosophila* species (10), which would be expected to have different effects on direct drug binding affinity, we might therefore also expect different relative contributions of allosteric effects to give the same net resistance levels observed in the whole fly. We will therefore directly test this new hypothesis on cyclodiene mode of action by examining the relative contributions of direct effects on binding site affinity and allosteric effects conferred by each of these two mutations.

Although this study leaves us with a clearer picture of cyclodiene and PTX mode of action it also suggests that lindane may have its primary site of action at a different receptor. We therefore hope to use this approach of isolating insecticide targets from resistant *Drosophila* mutants for lindane and for other (novel) compounds of interest. Finally, and interestingly, despite the demonstration of aberrant channel functions associated with resistance, there is little evidence that cyclodiene resistance in *Drosophila* is associated with a serious fitness disadvantage. Thus, cyclodiene resistance persists in *Drosophila* populations in the field at a frequency of approximately 1% despite the large scale withdrawal of most cyclodienes (28). Further, resistance frequencies remain unchanged in population cage studies run for over a year in the laboratory (our unpublished results). The only adverse phenotype associated with resistance, that we have been able to document, is paralysis at high temperatures (29). This study therefore not only illustrates a novel mechanism of insecticide resistance but also highlights how resistance genes can still persist in the absence of selection despite documented differences in susceptible and resistant receptor functions.

Literature Cited

1. Gant, D. B.; Eldefrawi, M. E.; Eldefrawi, A. T. *Toxicol. Appl. Pharmacol.* **1987**, *88*, 313-21.
2. Georghiou, G. P. In *Pesticide Resistance: Strategies and Tactics for Management*; National Academy of Sciences, Ed.; National Academy Press: Washington DC, 1986, pp 14-43.
3. ffrench-Constant, R. H.; Roush, R. T.; Mortlock, D.; Dively, G. P. *J. Econ. Entomol.* **1990**, *83*, 1733-1737.
4. ffrench-Constant, R. H.; Roush, R. T. *Genet. Res., Camb.* **1991**, *57*, 17-21.
5. ffrench-Constant, R. H.; Mortlock, D. P.; Shaffer, C. D.; MacIntyre, R. J.; Roush, R. T. *Proc. Natl. Acad. Sci. U.S.A.* **1991**, *88*, 7209-7213.
6. ffrench-Constant, R. H. *Comp. Biochem. Physiol.* **1993**, *104C*, 9-12.
7. Cole, L. M.; Nicholson, R. A.; Cassida, J. E. *Pestic. Biochem. Physiol.* **1993**, *46*, 47-54.
8. Deng, Y.; Casida, J. E. *Pestic. Biochem. Physiol.* **1992**, *43*, 116-122.
9. Deng, Y.; Palmer, C. J.; Casida, J. E. *Pestic. Biochem. Physiol.* **1991**, *41*, 60-65.
10. ffrench-Constant, R. H.; Steichen, J.; Rocheleau, T. A.; Aronstein, K.; Roush, R. T. *Proc. Natl. Acad. Sci. U.S.A.* **1993**, *90*, 1957-1961.
11. Thompson, M.; Shotkoski, F.; ffrench-Constant, R. *FEBS Lett.* **1993**, *325*, 187-190.
12. Thompson, M.; Steichen, J. C.; ffrench-Constant, R. H. *Ins. Mol. Bio.* **1993**, *2*, 149-154.
13. ffrench-Constant, R. H.; Rocheleau, T. A.; Steichen, J. C.; Chalmers, A. E. *Nature* **1993**, *363*, 449-451.
14. Leonard, R. J.; Labarca, C. G.; Charnet, P.; Davidson, N.; Lester, H. A. *Science* **1988**, *242*, 1578-1581.
15. Newland, C. F.; Cull-Candy, S. G. *J. Physiol.* **1992**, *447*, 191-213.
16. Jackson, M. B. (1993). *Thermodynamics of Membrane Receptors and Channels.* Boca Raton, FL. CRC Press.

17. Zhang, H.-G.; ffrench-Constant, R. H.; Jackson, M. B. *J. Physiol.* **1994**, in press.
18. Jackson, M. B. **1992**, *Biophys. J. 63*, 1443-1444.
19. Revah, F.; Bertrand, D.; Galzi, J.-L.; Devillers-Thiéry, A.; Mulle, C.; Hussy, N.; Bertrand, S.; Ballivet, M.; Changeux, J.-P. *Nature.* **1991**, *353*, 846-849.
20. Yakel, J. L.; Lagrutta, A.; Adelman, J. P.; North, R. A. *Proc. Natl. Acad. Sci. USA.* **1993**, *90*, 5030-5033.
21. Sattelle, D. B.; Leech, C. A.; Lummis, S. C. R.; Harrison, B. J.; Robinson, H. P. C.; Moores, G. D.; Devonshire, A. L. In *Neurotox '88: Molecular Basis of Drug and Pesticide Action*; G. G. Lunt, Ed.; Elsevier Science Publishers BV (Biomedical Division): 1988, pp 563-582.
22. Lee, H.-J.; Rocheleau, T.; Zhang, H.-G.; Jackson, M. B.; ffrench-Constant, R. H. *FEBS Lett.* **1993**, *335*, 315-318.
23. Olsen, R. W.; Tobin, A. J. *FASEB J.* **1990**, *4*, 1469-1480.
24. Betz, H. *Neuron* **1990**, *5*, 383-392.
25. Luddens, H.; Wisden, W. *Trends Pharmacolog. Sci.* **1991**, *12*, 49-51.
26. Sattelle, D. B.; Marshall, J.; Lummis, S. C. R.; Leech, C. A.; Miller, K. W. P.; Anthony, N. M. A.; Bai, D.; Wafford, K. A.; Harrison, J. B.; Chapaitis, L. A.; Watson, M. K.; Benner, E. A.; Vassallo, J. G.; Wong, J. F. H.; Rauh, J. J. In *Transmitter Amino Acid Receptors, Structures, Transduction and Models for Drug Development*; E. A. Barnard and E. Costa, Ed.; Thieme Medical Publishers: New York, 1992, pp 273-291.
27. Henderson, J. E.; Soderlund, D. M.; Knipple, D. C. *Biochem. Biophys. Res. Comm.* **1993**, *193*, 474-482.
28. Aronstein, K.; Ode, P.; ffrench-Constant, R. H. *Pestic. Biochem. Physiol.* **1994**, *48*, 229-233.
29. ffrench-Constant, R. H.; Steichen, J. C.; Ode, P. *Pestic. Biochem. Physiol.* **1993**, *46*, 73-77.

RECEIVED June 24, 1994

Chapter 13

Structural and Functional Characterization of Insect Genes Encoding Ligand-Gated Chloride-Channel Subunits

Douglas C. Knipple, Joseph E. Henderson, and David M. Soderlund

Department of Entomology, New York State Agricultural Experiment Station, Cornell University, Geneva, NY 14456

Ligand-gated chloride channels mediate synaptic inhibition in animal nervous systems and are sites of action of important drugs and toxins. Recent studies of vertebrate GABA- and glycine-gated chloride channels have revealed their heteromultimeric organization and the large number of genes encoding unique but structurally related subunits that comprise them. In insects, physiological studies have demonstrated the existence of diverse ligand-gated chloride channels, but their underlying structural and functional properties remain poorly characterized. This paper describes our isolation of members of this gene family from *Drosophila melanogaster* by PCR-based homology probing and summarizes investigations in this genetic model system of the only two insect subunit genes isolated to date to which functional properties have been ascribed. The implications of these findings for target-based insecticide discovery efforts are discussed in the context of the major paradigms established from the study of homologous vertebrate receptors.

Inhibitory neurotransmission plays a fundamental role in animal nervous systems by counterbalancing and modulating the excitatory inputs into postsynaptic cells. This process is mediated by amino acid neurotransmitter receptors, which, when activated, permit the selective flow of chloride ions across the postsynaptic membrane resulting in hyperpolarization of the postsynaptic cell (*1*). In mammals, glycine-gated chloride channels are the principal inhibitory receptors in the spinal cord and brainstem whereas GABA-gated chloride channels are the predominant inhibitory receptors elsewhere in the brain. These receptors are sites of action of important drugs and neurotoxins and have complex pharmacologies that are believed to reflect an underlying structural diversity owing to their heteromultimeric nature and variable subunit composition (reviewed in ref. *2*).

In contrast to the situation in vertebrates, the underlying structural basis of inhibitory neurotransmission in insects is poorly understood. Physiological studies provide evidence for the existence of several classes of functionally distinct chloride channels in insect nerve and muscle, but to date the complete structures of only two subunits having homology to vertebrate GABA and glycine receptors have been deduced (*3,4*). This paper provides an overview of the current state of knowledge of ligand-gated chloride channels in insects and discusses the implications of recent functional studies of vertebrate receptors for ongoing investigations of insect ligand-gated chloride channels.

Structural and Functional Properties of Vertebrate GABA and Glycine Receptors

The Ligand-Gated Chloride Channel Gene Family of Vertebrates. Molecular cloning and expression studies have revealed an unexpectedly large number of genes encoding GABA- and glycine-gated chloride channels. More than 20 individual members of the ligand-gated chloride channel gene family in mammals have been characterized to date, including five classes of GABA receptor subunits (α, β, γ, δ, and ρ) (5) and two classes of glycine receptor subunits (α and β) (6). The diversity of this family of genes is further increased by the existence of multiple subunit isoforms in some of these classes (e.g., $\alpha1$, $\alpha2$). GABA- and glycine-gated chloride channel subunit genes comprise a gene family that is itself a part of a larger superfamily of ligand-gated ion channel genes, which includes genes encoding acetylcholine receptor subunits (7-9). Gene products of this superfamily have a conserved structural organization characterized by four hydrophobic membrane-spanning domains that contribute to the formation of the ion channel and a large extracellular domain containing a postulated cysteine-cysteine bridge (Figure 1). Recent mutational analyses have identified a region of the second transmembrane domain that confers ionic selectivity (10) and two homologous and discontinuous segments of the extracellular domain between the cysteine loop and the first transmembrane domain that are implicated in neurotransmitter-binding (11,12).

Heteromultimeric Nature of Vertebrate Ligand-gated Chloride Channels. Biochemical purification and photoaffinity labelling experiments have provided direct evidence that mammalian GABA (13) and glycine (14) receptors are heteromultimeric. By analogy to the acetylcholine receptors, vertebrate GABA and glycine receptors have been thought to form pentameric assemblies (Figure 2). On the basis of the heteromultimeric nature of ligand-gated chloride channels and the large number of genes encoding individual subunits that have been identified, many subunit combinations are possible in native receptors. Various combinations of GABA receptor subunits have been examined using transient expression assays employing either *Xenopus* oocytes or transfected vertebrate cells in culture (15). Expression of various combinations of GABA receptor subunits results in the formation of heteromultimeric receptors with differing structures and pharmacology (reviewed in ref. 2). Furthermore, functional properties of expressed receptors can often be altered by isoform substitution (16-18). Thus the observed pharmacological diversity of vertebrate GABA and glycine receptors may be due primarily to the diversity of subunits that can coassemble to form functional heteromultimers.

Although some subunits expressed in these systems are capable of forming homomultimeric receptors, they typically exhibit biophysical and pharmacological properties that are unlike those of native receptors. A notable exception to this rule is provided by the GABA receptor ρ subunit, which forms a homomultimeric receptor that produces chloride currents in response to GABA that are comparable in magnitude to those produced by heteromultimeric receptors (19). Furthermore these expressed receptors exhibit the pharmacology of the $GABA_C$ receptors found in the rat retinal system (20,21) and the ρ subunit is highly expressed in this tissue (22). Thus it is possible that GABA receptors of the vertebrate visual system may, unlike most GABA receptors, form a homomultimeric receptor *in vivo*. It is important to note that whereas expression studies examining the properties of receptors comprising various subunit combinations have contributed greatly to our understanding of important

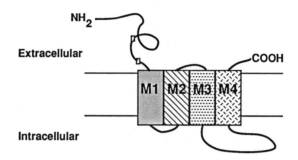

Figure 1. Conserved features of ligand-gated chloride channel subunits (adapted from ref. *34*). Depicted are the inferred transmembrane domains (M1-M4), hydrophilic domains (thin lines), cysteine loop in the extracellular domain (*9*), and regions implicated in neurotransmitter binding (*11,12*) (boxed).

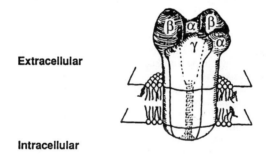

Figure 2. Postulated pentameric organization of vertebrate acetylcholine, GABA, and glycine receptors (adapted from ref. *39*). A hypothetical heteromultimeric GABA receptor containing α, β, and γ subunits is shown. This is one of many possible assemblies comprised of these three subunits and it must be stressed that the stoichiometry and organization of native GABA receptors are not known.

aspects of these pharmacological targets, particularly in regard to their benzodiazapine pharmacology, the subunit composition and stoichiometry of native receptors in specific neuroanatomical regions remains unknown (2,5).

Ligand-gated Chloride Channels in Insects

Electrophysiological studies provide evidence for a variety of of ligand-gated chloride channels in insect nerve and muscle (23,24). Pharmacologically distinct GABA receptor types are found in the insect central nervous system and at neuromuscular junctions (24). Histamine-gated chloride channels are found in arthropod visual systems, and have been shown to play a major role in the transduction and processing of visual information in the optic lobe of the blowfly (25). The sixth abdominal ganglion of the American cockroach contains taurine-gated chloride channels that are physiologically and pharmacologically distinct from the GABA-gated chloride channels in the same preparation (26). Finally, insect muscle contains extrasynaptic glutamate-gated chloride channels (27). These diverse receptors represent at least five distinct classes of ligand-gated chloride channels, and strongly suggest an underlying diversity of channel structures. If insect ligand-gated chloride channels are heteromultimeric like their vertebrate counterparts, then there may easily be 10-20 genes in the ligand-gated chloride channel gene family of insects.

Isolation of Insect Ligand-gated Chloride Channel Subunit Genes by Homology Probing. Despite the apparent diversity of ligand-gated chloride channels in insects, virtually nothing is known of their native structures because until very recently no genes encoding channel subunits had been isolated from insects. To address this problem, we implemented an approach that exploited the conserved amino acid sequence elements of the vertebrate GABA and glycine receptor gene family of vertebrates (4,28). The existence of a short stretch of invariant amino acids (TTVLTMTT) in the second inferred transmembrane domain (see Figure 1) of most reported vertebrate ligand-gated chloride channels (with the exceptions of the unusual β subunit of the glycine receptor (29) and the GABA receptor ρ subunit (22)) led us to hypothesize that this octapeptide sequence might be a "signature motif" for the family of ligand-gated chloride channel subunits, and therefore might be conserved in homologous ligand-gated chloride channels of insects. Unfortunately, the overall level of amino acid sequence identity between different vertebrate subunit classes is relatively low and the TTVLTMTT motif is the only stretch of invariant amino acid sequence encoded by this gene family that is of suitable length for PCR primer design. Consequently, we employed a "single-site" PCR-based homology screening of *D. melanogaster* genomic DNA with a degenerate oligonucleotide target primer that encodes the first seven amino acids of the octapeptide motif and either *Bgl*II- or *Bam*HI-cut and anchor-adapted genomic DNA (4). Sequencing of the several discrete amplification products obtained by this procedure revealed that two of them (designated LCCH1 and LCCH2) contained open reading frames that extended far enough to permit recognition of sequence similarity (about 40% identity) to conserved residues in the M2-M3 regions of vertebrate GABA and glycine receptor subunits. The other amplification products were either too short or their open reading frames were terminated a short distance downstream from the target priming site, precluding the facile determination of homology relationships.

   We also employed conventional PCR homology probing using two degenerate target primers specifying defined sequence elements of vertebrate α and β subunits. This strategy yielded a single product (designated LCCH3) that contained an extended open reading frame of predicted length delimited by the octapeptide signature motif in M2 and the conserved CFVFVF motif found in M3 of all GABA receptor β subunit sequences. The predicted amino acid sequence of this fragment exhibited greater than

40% identity to the corresponding sequences encoded by GABA receptor β subunit genes. In our effort to isolate gene segments with homology to GABA receptor α subunit-like sequences, we failed to obtain a specific amplification product using the degenerate signature motif primer in combination with the ATVNYFT amino acid sequence motif present in M3 of all vertebrate α subunits.

In order to more fully characterize the genes from which the LCCH1, LCCH2, and LCCH3 amplification products were initially obtained, we screened a *D. melanogaster* genomic library and isolated clones labeled by each probe. Southern analysis from each group of genomic clones identified restriction fragments that were subsequently subcloned and sequenced. The sequences of the regions flanking the original amplified segments of these DNAs were analyzed by deducing the coding and noncoding sequences and comparing the encoded amino acid sequences to vertebrate GABA and glycine receptors in order to identify the conserved structural features described above. Hydropathy plots of the predicted amino acid sequences deduced from the identified open reading frames of LCCH1, LCCH2, and LCCH3 were similar to those obtained for the corresponding regions of vertebrate GABA and glycine receptor subunits, and in particular exhibited the characteristic highly hydrophobic segments corresponding to the inferred membrane-spanning regions separated by hydrophilic segments. All three genes also contained other sequence elements that were conserved in some of the vertebrate GABA and glycine receptor subunit genes but were not invariant across the entire family. LCCH3 was found to have a high level of homology to previously characterized members of the β subunit class of GABA receptors with sequence identity of 79% to members of the vertebrate β subunit class in the core region encompassing domains M1 through M3. It exhibits a remarkable 98% identity over the same interval to a subunit encoded by a gene isolated from the pond snail *Lymnaea stagnalis*, which when coexpressed in *Xenopus* oocytes with vertebrate α1 produces functional heteromultimeric GABA receptors (*30*). On the basis of this structural analysis the LCCH1, LCCH2, and LCCH3 genes were positively identified as members of the ligand-gated chloride channel gene family. Further information on these genes and their encoded gene products is described separately below.

LCCH3: A Ligand-Gated Chloride Channel Gene That Maps Near the *slrp* Locus. The high degree of sequence identity between LCCH3 and known GABA receptor β subunits in the core region led us to give high priority to the isolation and complete sequence determination of the LCCH3 cDNA. Thus, the complete coding sequence of LCCH3, comprising a single open reading frame of 1488 nucleotides, was inferred from a 1.65 kb cDNA clone, which we isolated from a pupal cDNA library (*4*). Besides the conserved hydrophobic domains corresponding to the four membrane spanning segments, the 496-amino acid polypeptide encoded by this cDNA was found to contain the other structural features that are conserved in both relative position and amino acid sequence to vertebrate GABA and glycine receptor subunits (*7*). In addition two four-amino-acid domains (Tyr[177]-Gly-Tyr-Thr and Thr[220]-Gly-Val-Tyr) implicated in GABA binding in the extracellular domain between the cysteine loop and M1 (*11*) are identical to corresponding domains of the β subunit class (Table I).

In situ hybridization of LCCH3 to polytene chromosomes showed that this gene maps onto the same cytogenetic interval (13F-14A) (*28*) as the *slrp* (slow receptor potential) locus (*31*). The general hypoactivity and aberrant electroretinogram phenotypes of *slrp* mutants are consistent with defects in inhibitory neurotransmission (*31-33*). Thus the mapping of *slrp* and LCCH3 onto the same cytogenetic interval suggests that the defects exhibited in *slrp* flies could result from mutations in the LCCH3 gene.

Table I. Structures of Postulated Neurotransmitter Binding Domains of Ligand-Gated Chloride Channels and Effects of Amino Acid Substitutions on Neurotransmitter Recognition Properties

Vertebrate GABA receptor subunits[a]			
α subunit consensus	YAYT/P	TGEY	responsive to GABA
α1 subunit Y160S	SAYT	TGEY	no effect
β subunit consensus	YGYT	TGXY	responsive to GABA
β2 subunit Y157F	FGYT	TGSY	decreased GABA sensitivity
β2 subunit Y159F	YGFT	TGSY	"
β2 subunit Y205F	YGYT	TGSF	"
β2 subunit T160S	YGYS	TGSY	"
β2 subunit T202S	YGYT	SGSY	"
γ subunit consensus	YGYP	A/SGDY	responsive to GABA
γ2 subunit Y172S	SGYP	SGDY	no effect
Invertebrate subunits			
Drosophila LCCH3 subunit[b]	YGYT	TGVY	unresponsive to GABA
Lymnaea β subunit[c]	YGYT	TGDY	responsive to GABA
Drosophila Rdl subunit[d]	FGYT	TGNY	responsive to GABA
Vertebrate glycine receptor subunits[e]			
α1 subunit consensus	FGYT	TGKF	responsive to glycine
α1 subunit F159Y	YGYT	TGKF	increased sensitivity to diverse ligands
α1 subunit Y161F	FGFT	TGKF	"
α1 subunit F159Y, Y161F	YGFT	TGKF	"

[a] as heteromultimeric GABA receptors resulting from coexpression of α1, β2, and γ2 subunits after Amin and Weiss (11).
[b] as expressed as homomultimer and when coexpressed with rat α1 or *Rdl* GABA receptor subunits (this paper).
[c] as heteromultimer resulting from coexpression with rat α1 after Harvey et al. (30).
[d] as homomultimer after ffrench-Constant et al. (36) and our unpublished results (this paper).
[e] as homomultimeric α1 glycine receptor after Schmieden (12).

LCCH1: The *Rdl* Locus. Both DNA sequencing and *in situ* hybridization to polytene chromosome squashes (*28,34*) showed LCCH1 to be identical to the *Rdl* gene isolated independently using a molecular genetic approach (*3*). The investigations of Roush, ffrench-Constant, and co-workers to established the significance of *Rdl* as a target of cyclodiene insecticides and as a probable GABA receptor subunit are described in a recent review (*35*) and elsewhere in this volume.

LCCH2: A Novel Ligand-Gated Chloride Channel Gene? DNA sequencing of the LCCH2 gene showed that LCCH2 encodes the invariant landmarks of the ligand-gated chloride channel gene family with the notable exception that the octapeptide motif is truncated by the replacement of the terminal Thr with Phe (*28,34*). The inferred amino acid sequence of LCCH2 in the core region is equally divergent from all other known sequences in the gene family suggesting that LCCH2 may represent a novel class of ligand-gated chloride channel subunit genes. In contrast to LCCH1 and LCCH3, LCCH2 maps onto a cytogenetic region (75A) that has no mutations that might provide insight into the functional role of its gene product (*28*).

Analysis of Insect Ligand-gated Chloride Channel Structure and Function

Expression of LCCH3 and *Rdl* in *Xenopus* Oocytes. Heterologous expression studies in *Xenopus* oocytes serve two important functions. First, expression of mRNA obtained from *in vitro* transcription of cloned cDNAs provides essential confirmation that genes isolated by virtue of their structural homology to known ligand-gated chloride channel subunits actually encode such subunits. Second, the co-expression of different combinations of subunit genes permits us to test hypotheses about the functional assembly of individual subunits in heteromultimeric receptor complexes.

The expression of both the *Rdl* and LCCH3 gene products has been analyzed in this system (*36*, this paper). The *Rdl* subunit is capable of forming functional homomultimeric receptors that give robust (approaching 1 μA) chloride currents to GABA administered in the 10 μM range. However, the formation of GABA receptors comprised solely of *Rdl* subunits in this system requires an order of magnitude higher concentrations of RNA than that required to elicit a comparable response from heteromultimeric vertebrate α/β receptors. The latter finding suggests that the assembly of *Rdl* subunits into homomultimeric receptors is inefficient and that the homomultimeric structure may not be reflective of native receptors in the fly.

In contrast to the results obtained with the *Rdl* subunit, oocytes injected with LCCH3 RNA do not respond measurably to GABA or to a variety of other ligands, which suggests that other subunits must coassemble with LCCH3 to form functional receptors. The large GABA-gated currents we measured using expressed *Rdl* homomultimers and rat α/β heteromultimers as positive controls for methodology were similar in magnitude to values reported in the literature (*15,17,36*). The report of chimeric GABA receptors comprised of vertebrate α1 subunits and *L. stagnalis* β subunits that give robust responses to GABA, together with the high level of structural homology between the *L. stagnalis* β subunit and LCCH3, led us to hypothesize that LCCH3 might also form a functional chimeric GABA receptor with a vertebrate α1 subunit. However, we found that oocytes injected with rat α1 and LCCH3 RNAs yielded only very low level (3-5 nA) responses to GABA and no detectable responses to a variety of other neurotransmitters. These findings suggest either that the subunit interactions between LCCH3 and rat α1 are less strong than those resulting in the formation of chimeric rat α1/*L. stagnalis* β receptors, or that rat α1/LCCH3 chimeric receptors are formed but are insensitive to GABA.

In other experiments no responses to GABA were detected from oocytes coinjected with LCCH3 and *Rdl* RNAs. For these experiments, lower concentrations of RNA (<25 ng/oocyte) were used because of the robust GABA-gated currents that are obtained from homomultimeric *Rdl* receptors using higher RNA concentrations. These results are consistent with, but do not prove, the hypothesis that LCCH3 and *Rdl* do not assemble *in vivo* to form a functional receptor.

Structures of the Postulated Neurotransmitter Binding Domains of LCCH3 and *Rdl*. Recent studies of vertebrate GABA and glycine receptors have implicated two regions in the extracellular domain of ligand-gated chloride channel subunits that are involved in neurotransmitter binding *(11,12)* (Table 1). Each of these sites consists of four amino acids having aromatic residues in conserved positions. In one of these studies *(11)* the responsiveness to GABA of expressed heteromultimeric GABA receptors was assayed following systematic substitutions of the aromatic residues of these domains in α, β, and γ subunits. Table 1 summarizes the results obtained for various substitutions. Wheras no effect was observed when substitutions were made in the first domain of α and γ subunits, changes in the homologous domain of the β subunit significantly diminished the sensitivity of the expressed channels to GABA, strongly suggesting that activation of the native channel results from binding of GABA to the β subunit. In a separate study *(12)* substitutions of tyrosine for phenylalanine in the homologous domain of homomultimeric α1 glycine receptors produced a dramatic increase in sensitivity to diverse amino acid neurotransmitters, including significant responses to β-alanine and taurine.

Although the above studies imply an essential role of β subunits in neurotransmitter binding, the assignment of the ligand binding domain solely to the β subunit in native receptors is probably an oversimplification. For example, mutational studies by Sigel *et al.* *(37)* showed that substitutions of leu for phe[64] (between the amino terminus and the cysteine-loop) in the GABA receptor α subunit resulted in a significant decrease in sensitivity to GABA of heteromultimeric (α/β/γ) GABA receptors, whereas mutations in the homologous residue of β and γ subunits produced much less profound effects. This study points to the importance of allosteric interactions of the extracellular domains of dissimilar adjacent subunits in neurotransmitter recognition.

The regions of the LCCH3 and *Rdl* subunits corresponding to the domains of vertebrate ligand-gated chloride channel subunits implicated in neurotransmitter binding show strong conservation of the essential elements of these domains (Table 1). An intriguing aspect of the *Rdl* subunit is that phenylalanine is the aromatic amino acid in the first position of the first domain unlike all other GABA receptor subunits, which have tyrosine in this position. Furthermore substitution of phenylalanine for tyrosine in the vertebrate β subunit of receptors having native-like heteromultimeric subunit composition results in greatly decreased sensitivity to GABA. In the context of these findings, the lack of conservation of this residue in the *Rdl* subunit suggests the possibility that in native heteromultimeric receptors *Rdl* may confer an unusual pharmacological profile or instead may not be involved in GABA binding.

Conclusions

The identification of conserved structural motifs in vertebrate GABA and glycine receptor subunits provided the basis for our PCR-based search for ligand-gated chloride channel genes in *D. melanogaster*. Our isolation by this method of three genes encoding proteins having structural features common to all vertebrate ligand-gated chloride channel subunits demonstrates the conservation of these essential structural elements in inhibitory neurotransmitter receptor subunits of insects and provides technical validation of this experimental approach *(28,34)*. Physiological evidence for diverse ligand-gated chloride channels in insects (*23-27*) predicts a large and diverse gene family encoding the subunits that comprise these receptors. Because of the practical need to focus on the most promising DNA amplification products obtained in our initial screen, our search for insect homologues of this gene family was far from exhaustive and it is reasonable to assume that several additional members of this gene family remain to be identified.

Heterologous expression studies have shown that the majority of GABA and glycine receptors of vertebrates are heteromultimeric in native tissues and have suggested that the observed pharmacological diversity of these receptors is likely attributable to receptor subtypes comprised of different subunits combinations (*2,5,13-18*). Our finding that the LCCH3 gene product fails to form detectable chimeric GABA receptors when coexpressed in *Xenopus* oocytes with rat GABA receptor $\alpha 1$ subunits indicates that LCCH3 is dissimilar to the molluscan β subunit in this respect despite its high level of homology to the latter. This negative finding nevertheless implies that the LCCH3 subunit requires an additional and as yet unidentified subunit if it is, in fact, involved in the formation of functional receptors *in vivo* (as suggested by the coincidence of the cytogenetic localization the LCCH3 gene with the *slrp* locus). In contrast to the situation with LCCH3, functional analysis of the *Rdl* gene product has demonstrated that the *Rdl* subunit forms homomultimers in *Xenopus* oocytes that are responsive to GABA (*35,36*). Although this finding is consistent with the hypothesis that native receptors containing *Rdl* are homomultimeric, two pieces of circumstantial evidence, specifically the high RNA concentrations required to obtain detectable *Rdl* homomultimers in *Xenopus* oocytes and the unusual sequence of one of the two domains of the *Rdl* subunit implicated in GABA binding, suggest that the coassembly of other subunits with *Rdl* is required *in vivo*. Besides heterologous expression studies, the elucidation of the structures of native insect receptors will require the implementation of additional approaches such as the *in situ* localization of specific subunits in tissue sections using subunit-specific antibodies.

Although GABA receptors incorporating the *Rdl* subunit have been established as the primary targets of chlorinated cyclodiene insecticides (*35,36*), the diversity in insects of ligand-gated chloride channels and their heteromultimeric nature suggest that target-based insecticide discovery approaches focusing exclusively on this subunit protein will be unduly restricted in scope. This point is illustrated by evidence showing that the avermectins, a class of compounds with highly specific insecticidal properties, act on ligand-gated chloride channels in insects that are not gated by GABA (*39*). The number of unique physiological systems in insects in which inhibitory ligand-gated chloride channels play a role is significant, suggesting a diversity of potential target sites available for future development. At present efforts to identify additional receptor target sites that mediate inhibitory neurotransmission in insects are limited principally by the small number of genes encoding the subunit proteins of this family that have been cloned to date.

Acknowledgments

We thank R. Vasquez and R. Oswald (Cornell University), A. Chalmers and M. Wolff (Rhone-Poulenc Ag Company) for their technical and conceptual contributions to the expression studies summarized in this paper; P. Marsella-Herrick for technical assistance with molecular biology; R. Roush and R. J. MacIntyre (Cornell University) for providing data prior to publication and the *Rdl* cDNA clone; W. Pak for providing *slrp* mutants; and D. Pritchett (University of Pennsylvania) for providing vertebrate GABA receptor cDNA clones. Our investigations summarized in this paper were supported by grants from the Cornell Biotechnology Program, which is sponsored by the New York Science and Technology Foundation, a consortium of industries, the U.

S. Army Research Office, and the National Science Foundation; by a Training Grant (ES070520) from the National Institute of Environmental Health Sciences (to J. E. H.); and by a gift from the Rhone-Poulenc Ag Company, Research Triangle Park, NC.

Liturature Cited

(1)   Hille, B. *Ionic Channels of Excitable Membranes;* 2nd ed.; Sinauer: Sunderland,   MA, 1992, pp 607.
(2)   Macdonald, R. L.; Angelloti, T. P. *Cell Physiol. Biochem.* **1993**, *3*, 352-373.
(3)   ffrench-Constant, R. H.; Mortlock, D. P.; Shaffer, C. D.; MacIntyre, R. J.; Roush, R. T. *Proc. Natl. Acad. Sci. USA* **1991**, *88*, 7209-7213.
(4)   Henderson, J. E.; Soderlund, D. M.; Knipple, D. C. *Biochem. Biophys. Res. Commun.* **1993**, *193*, 474-482.
(5)   Burt, D. R.; Kamatchi, G. L. *Federation of the American Societies of Experimental Biology Journal* **1991**, *5*, 2916-2923.
(6)   Langosch, D.; Becker, C. M.; Betz, H. *Eur. J. Biochem.* **1990**, *194*, 1-8.
(7)   Barnard, E. A.; Darlison, M. G.; Seeburg, P. *Trends Neurosci.* **1987**, *10*, 502-509.
(8)   Grenningloh, G.; Rienitz, A.; Schmitt, B.; Methfessel, C.; Zensen, M.; Beyreuther, K.; Gundelfinger, E. D.; Betz, H. *Nature* **1987**, *328*, 215-220.
(9)   Schofield, P. R.; Darlison, M. G.; Fujita, N.; Burt, D. R.; Stephenson, F. A.; Rodriguez, H.; Rhee, L. M.; Ramachandran, J.; Reale, V.; Glencorse, T. A.; Seeburg, P. H.; Barnard, E. A. *Nature* **1987**, *328*, 221-227.
(10)   Galzi, J.-L.; Devillers-Thiéry, A.; Hussy, N.; Bertrand, S.; Changeux, J.-P.; Bertrand, D. *Nature* **1992**, *359*, 500-505.
(11)   Amin, J.; Weiss, D. *Nature* **1993**, *366*, 565-569.
(12)   Schmieden, V.; Kuhse, J.; Betz, H. *Science* **1993**, *262*, 256-258.
(13)   Sigel, E.; Barnard, E. A. *J. Biol. Chem.* **1984**, *259*, 7219-7223.
(14)   Graham, D.; Pfeiffer, F.; Simler, R.; Betz, H. *Biochem.* **1985**, *24*, 990-994.
(15)   Sigel, E.; Baur, R.; Trube, G.; Mohler, H.; Malherbe, P. *Neuron* **1990**, *5*, 703-711.
(16)   Levitan, E. S.; Schofield, P. R.; Burt, D. R.; Rhee, L. M.; Wisden, W.; Kohler, M.; Fujita, N.; Rodriguez, H. F.; Stephenson, A.; Darlison, M. G.; Barnard, E. A.; Seeburg, P. H. *Nature* **1988**, *335*, 76-79.
(17)   Ymer, S.; Schofield, P. R.; Draguhn, A.; Werner, P.; Kohler, M.; Seeburg, P. H. *EMBO J.* **1989**, *8*, 1665-1670.
(18)   Olsen, R. W.; Bureau, M. H.; Endo, S.; Smith, G. *Neurochemical Research* **1991**, *16*, 317-325.
(19)   Shimada, S.; Cutting, G.; Uhl, G. R. *Mol. Pharmacol.* **1992**, *41*, 683-687.
(20)   Woodward, R. M.; Polenzani, L.; Miledi, R. *Mol. Pharmacol.* **1992**, *42*, 165-173.
(21)   Feigenspan, A.; Wassle, H.; Bormann, J. *Nature* **1993**, *361*, 159-162.
(22)   Cutting, G. R.; Lu, L.; O'Hara, B. F.; Kasch, L. M.; Montrose-Rafizadeh, C.; Donovan, D. M.; Shimada, S.; Antonarakis, S. E.; Guggino, W. B.; Uhl, G. R.; H. H. Kazazian, J. *Proc. Natl. Acad. Sci. USA* **1991**, *88*, 2673-2677.
(23)   Lummis, S. C. R. *Comp. Biochem. Physiol.* **1990**, *95C*, 1-8.
(24)   Sattelle, D. B. *Adv. Insect Physiol.* **1990**, *22*, 1-112.
(25)   Hardie, R. C. *Nature* **1989**, *339*, 704-706.
(26)   Hue, B.; Pelhate, M.; Chanelet, J. *J. Canadien Sci. Neurol.* **1979**, *6*, 243-250.

(27) Dudel, J.; Franke, C.; Hatt, H.; Usherwood, P. N. R. *Brain Res.* **1989,** *481,* 215-220.
(28) Henderson, J. E.; Knipple, D. C.; Soderlund, D. M. *Insect Biochem. Mol. Biol.* **1994,** *24,* 363-371.
(29) Grenningloh, G.; Pribilla, I.; Prior, P.; Multhaup, G.; Beyreuther, K.; Taleb, O.; Betz, H. *Neuron* **1990,** *4,* 963-970.
(30) Harvey, R. J.; Vreugdenhil, E.; Zaman, S. H.; Bhandal, N. S.; Usherwood, P. N. R.; Barnard, E. A.; Darlison, M. G. *EMBO J.* **1991,** *10,* 3239-3245.
(31) Homyk, T. J.; Pye, Q. *J. Neurogenet.* **1989,** *5,* 37-48.
(32) Homyk, T. J.; Sheppard, D. E. *Genetics* **1977,** *87,* 95-104.
(33) *Mutations affecting the vision of Drosophila melanogaster*; Pak, W. L., Ed.; Plenum: New york, 1975; Vol. 3.
(34) Knipple, D. C.; Doyle, K. D.; Henderson, J. E.; Soderlund, D. M. In *Neurotox '91*; Duce, I. R., Ed.; Elsevier Science Publishers, Ltd.: London, 1992; pp 271-283.
(35) ffrench-Constant *Insect Biochem. Molec. Biol.* **1994,** *24,* 335-345.
(36) ffrench-Constant, R. H.; Rocheleau, T. A.; Steichen, J. C.; Chalmers, A. E. *Nature* **1993,** *363,* 449-451.
(37) Sigel, E.; Baur, R.; Kellenberger, S.; Malherbe, P. *EMBO J.* **1992,** *11,* 2017-2023.
(38) Bloomquist, J. R. *Comp. Biochem. Physiol.* **1993,** *106C,* 301-314.
(39) Stroud, R. M. In *Proteins of Excitable Membranes*; Hille, B. and Fambrough, D. M., Eds.; Society of General Physiologists Series; Wiley Interscience: New York, 1987, Vol. 41; p. 68.

RECEIVED July 19, 1994

Chapter 14

Chloride-Channel Gene Probes from Cyclodiene-Resistant and -Susceptible Strains of *Blattella germanica*

Koichiro Kaku[1] and Fumio Matsumura

Department of Environmental Toxicology and Department of Entomology, University of California, Davis, CA 95616

The DNA and amino acid sequence of the membrane spanning region of a GABA receptor of the German cockroach (*Blattella germanica*) has been identified, along with information on the nature and the specific site of mutation in a cyclodiene resistant strain (LPP strain). In this resistant strain the mutation has occurred at the most conserved, lower M2 cylinder region involving a *G* to *T* conversion, resulting in an amino acid change of alanine (GCC) residue to serine (TCC). The site, furthermore, coincides with the most conserved region of all GABA receptor subunits and the expected Cl⁻ transporting segment constituting the innermost surface of the channel opening. The deduced sequence of the German cockroach GABA receptor differs from that of the *Drosophila* mainly in the connecting region between M_3 and M_4.

It has been known for a long time that many insect species are capable of developing high levels of specific resistance to highly toxic cyclodiene insecticides (1-2). Dr. A.W.A. Brown, a founding father of insecticide resistance studies, has described the cyclodiene resistance problem as "a delight for geneticists and a nightmare for biochemists." Indeed, the mode of inheritance of the resistance gene was found to follow a straightforward Mendelian model, but the mechanism by which such a resistance phenomenon is phenotypically expressed remained a mystery for the long time period from the early 1960s to the 80s. Early efforts included studies on lipid differences, search on resistance antagonists, metabolism, tissue distribution and uptake studies as well as cross-resistance studies including that to DDT, organophosphates and carbamates. These studies have clearly established that this resistance mechanism is effective to most chemically defined cyclodiene insecticides and lindane, but not to any other types of commercially used pesticides. Since this group of chemicals includes very metabolically stable insecticides, such as heptachlor epoxide, and since metabolism and insect uptake studies could not

[1]Current address: K-I Chemical Research Institute Company Ltd., Shizuoka, Japan

reveal any difference between resistant and susceptible strains within the same species, the type of resistance developed here has been called "target insensitivity." Electrophysiological data clearly supported such a view as isolated, resistant nervous systems required higher concentrations of cyclodienes to show symptoms of excitation than control nerves. Interestingly, Brown (2) mentioned the anecdotal observation that pre-resistant populations among malaria mosquitoes were already present in Africa, even before dieldrin was used there. The level of pre-resistant population has been said to be in the order of several percent among Anopheles mosquito larvae. These observations suggest that a certain biochemical change must have taken place in the nervous system and that such a change is stable in a given population (i.e., is not giving the resistant individuals selective disadvantages against other stresses). The main question has been what the actual main target of these insecticides is. Without this knowledge we have no idea about the site of change in the resistant nervous system.

In 1982 Dr. Ghiassudin and I (3) formulated a hypothesis that the target of this group of insecticides could be the GABA receptor, based upon the observation that heptachlor epoxide and lindane could prevent GABA-induced increase in $^{36}Cl^-$ uptake by isolated nerve cords and coaxial muscles from the American cockroach. Also helpful in this regard was the incipient observation made in our laboratory that cyclodiene resistant German cockroaches showed cross-resistance to picrotoxin, whose action mechanism at that time was being established to act on the GABA receptors of the mammalian central nervous system. By using ^3H-dihydropicrotoxinin as an artificial ligand we were able to clearly establish that the picrotoxinin binding site in the resistant nervous system is less affected by dieldrin than that of the susceptible counterpart.

Subsequent mechanistic studies (3-8) have shown that these insecticides are specific antagonists of the GABA action on GABA receptors in both mammals and insects, and that their binding site within the receptor is probably identical to that of picrotoxinin, a well known naturally occurring antagonist of GABA. In the case of the German cockroach, cyclodiene resistance was shown to have evolved as a result of a very specific change in the biochemical properties of the GABA receptor itself (3, 6, 8). Such a change in this target site gives the homozygous individuals 10- to 100-fold resistance to these insecticides (6). The GABA receptors of several cyclodiene-resistant species have been shown to exhibit less binding affinity to cyclodienes (9, 10) or specific radioligands (11, 12) including ^3H-dihydropicrotoxinin (6) as compared to each of their susceptible counterparts. Therefore, these insects obviously offer us a unique opportunity to understand the site of action of these chemicals.

Recently ffrench-Constant et al. (13) have identified the DNA sequence of a GABA receptor from *Drosophila melanogaster* (14) from a cloned DNA. This GABA receptor differs considerably from the mammalian forms, and was named *Rdl*. Subsequently Henderson et al. (14) have also cloned and sequenced another GABA receptor gene, LCCH3 from *Drosophila melanogaster* cDNA libraries, which has been termed as a β-subunit based on the overall similarity to mammalian β subunits and that of *Lymnaea stagnalis*, the first GABA receptor identified in an invertebrate species (15). In more recent papers ffrench-Constant et al. (16) and Thompson et al. (17) have shown that a mutation occurred in cyclodiene resistant *Rdl* type GABA receptor to cause an Ala → Ser shift (at 302) in *Drosophila* and *Aedes aegypti*, respectively.

MATERIALS AND METHODS

The strains of German cockroaches (*Blattella germanica*) used for this work have been described elsewhere (9, 10). The cyclodiene-resistant strains were occasionally selected by using dieldrin to maintain the homozygous resistant individuals. The LPP strain was genetically selected from the original chlordane-resistant London strain (10) by back-crossing to the CSMA-susceptible strain for 8 generations, each time selecting for hetero-zygous resistance, and subsequent self crossing and selection for homozygous resistant individuals.

Isolation of poly A⁺ mRNA. Poly A⁺ mRNA was isolated (18) from adult German cockroaches (CSMA and LPP strain) (10). Heads and thoraxes were collected from 100 adult German cockroaches (50 males and 50 females) combined, and immediately frozen under liquid nitrogen. Approximately 3 g of tissues were ground in mortar with pestle under liquid nitrogen, and the slurries were transferred into 50 ml cell culture tubes. Immediately thereafter nitrogen was evaporated, and 40 ml lysis buffer [0.2 M NaCl, 0.2 M Tris-HCl (pH 7.5), 1.5 mM MgCl$_2$, 2% SDS, 200 µg/ml Proteinase K (Boehringer Mannheim) in DEPC-treated H$_2$O] was added, followed by immediate homogenization by Polytron® for 15-30 sec. The homogenates were incubated at 45°C for 2 hr with inter-mittent agitation. The residue in the lysate, mainly the cuticle of cockroaches, was separated by low speed centrifugation, and the supernatant was transferred into a sterile 50 ml cell culture tube. The NaCl concentration of the lysate (0.2 M) was adjusted to that of "binding buffer" (0.5 M NaCl) with 60 µl of 5 M NaCl in the same buffer per ml lysate. 80 mg of Oligo(dT) cellulose (GIBCO BRL/Life Technologies, Gaitherburg, MD) which was equilibrated to the same concentration of binding buffer [0.5 M NaCl, 0.01 M Tris-HCl (pH 7.5)] was mixed with the lysate, followed by incubation for 1 hr at room temper-ature with intermittent agitation. The treatment of poly A⁺ mRNA bound oligo(dT) cellulose and the elution scheme were identical to that of Badley et al. (18). The precipitated poly A⁺ mRNA was pelleted by centrifugation at 10,000 x g for 10 min, washed with ice-cold 75% ethanol, vacuum dried for 30 min and dissolved in 54 µl of DEPC-treated H$_2$O. Yields of extracted poly A⁺ mRNAs were approximately 42 µg both from CSMA and LPP strains.

Preparation of cDNA. Both mRNAs (38.5 µg) were reverse-transcribed into first strand cDNAs in 200 µl of reaction buffer containing 20 mM Tris-HCl (pH 8.4 at 25°C), 50 mM KCl, 5 mM MgCl$_2$, 0.01% gelatin [wt/vol.], 1 mM deoxynucleotide triphosphates (Pharmacia LKB), 200 units RNAsin (cloned, Promega), 3000 units M-MLV reverse transcriptase (GIBCO BRL/life Technologies) and 1 µg oligo(dT)$_{17}$-adapter primer (5'-GACTCGAGTCGACATCGATTTTT-TTTTTTTTTTTT-3') (Frohman et al., 1988). The reaction mixture was incubated at room temperature for 15 min, followed by incubation at 37°C for 1.5 hr and 95°C for 5 min. The mixture was then quickly chilled on ice and stored at -20°C until further use. All primers were synthesized on Model 391 DNA Synthesizer (Applied Biosystems, Foster City, CA) and purified by gel filtration using Sephadex-50.

Preparation of double stranded DNA by PCR. Each 1 µl cDNA mixture was combined in a 50 µl reaction mixture with 5 µl 10 X PCR buffer [100 mM Tris-HCl (pH 9.0 at 25°C), 500 mM KCl, 1.5 mM MgCl$_2$, 1% Triton X-100], 0.2 mM deoxynucleotide triphosphates (Pharmacia LKB Biotechnology), 1 unit Taq DNA polymerase (Promega), and 2.5 µg of both upstream and downstream degenerate primers or 5 pmol of specific primers. The mixture was overlaid with mineral oil. The PCR condition used was denaturation 95°C for 1 min, annealing 55°C for 2 min and extension 72°C for 3 min for 40 cycles in a DNA Thermal Cycler (Precision Scientific, Chicago, IL). 9 µl of each amplified product was run on a 2% agarose gel (FMC BioProducts, Rockland, ME) made with 1 X TBE (0.45 M Tris-HCl, 0.45 M boric acid, 10 mM EDTA) and stained with ethidium bromide. The desired PCR fragments were exercised from agarose gel and cut into pieces using a sterile blade, rinsed by ddH$_2$O twice, frozen and thawed in 100 µl ddH$_2$O (repeated twice), and left in a refrigerator overnight. 3 µl of the eluted DNA solution was re-amplified in 150 µl reaction mixture with 15 µl of 10 X PCR buffer, containing 0.2 mM deoxynucleotide triphosphates, 3 units Taq DNA polymerase, and 7.5 µg of both upstream and downstream degenerate primers or 15 pmol of specific primers. The PCR conditions were identical to the above except the annealing time was 1 min. In the second round PCR, three to six identical batches were amplified at the same time to prepare a large quantity of samples for DNA sequencing reaction or enzymatic restriction. The mass-produced PCR solution was collected in a 1.5 ml microcentrifuge tube and the mineral oil removed with chloroform. The water layer (450-900 µl) was concentrated to 35-40 µl by Centricon-100® (Amicon) at 1,000 x g for 30 min at 4-10°C. The concentrated DNA fragment was electrophoresed on 1% agarose gel made with 1 X TBE and excised from the gel, followed by purification by using QIAEX matrix (QIAGEN) according to the manufacturer's protocol. The DNA was eluted with ddH$_2$O and the concentration of the purified DNA was estimated by GelMarker® (Research Genetics, Huntsville, AL) as a quantitative standard on a 2% agarose gel. Usually 5-10 µg DNA was obtained from 450 µl reaction mixture under our standard experimental condition.

DNA sequencing reaction of PCR amplified double-stranded DNA. Sequenase® Version 2.0 System (United States Biochemical, Cleveland, OH) and DNA polymerase I (Klenow fragment) Promega) (20) were used for DNA sequencing reaction. DNA template-primer solution was made to 10 µl including the purified double-stranded DNA (approximately 0.5-2.0 pmol), primer (30 times the amount of DNA for a specific primer or approximately 1 µg for degenerate primer) and 2 µl of 5 X Sequenase Buffer [200 mM Tris-HCl (pH 7.5), 100 mM MgCl$_2$, 250 mM NaCl], which was boiled for 3 min and quickly cooled in dry ice–ethanol bath. The rest of DNA sequencing procedure was identical to that of Schuurman and Keulen (1991). Electrophoresis using 1 X TBE system was performed for 1.75 hr or 4.75 hr at 60 W on 7.4 M urea/6% Long Ranger gel (AT Biochem, Malvern, PA, 42 x 36 x 0.02 cm) with STS-45 Gel Electrophoresis Unit (IBI, New Haven, CT). The gel was dried at 60-80°C for 2 hr on Whatman 3MM paper under vacuum and exposed to Kodak XAR-5 film (Rochester, NY) for 8 to 72 hr.

Strategies for identification of the GABA receptor genes through PCR approach. The general flow diagram of the approach adopted is shown in Fig. 1. In brief, Step I. Primer A, B and C as shown were used to obtain DNA fragments designated as PCR-DNA A-C and B-C. Step II. Since the DNA fragment amplified by the combination of primer B and primer C (i.e., PCR-DNA- B-C) contained two kinds of subunits of GABA receptor genes, it was digested by several restriction endonucleases for analysis of DNA sequencing. One type possessed Rsa I recognition site but no Sfu I recognition site, whereas the other type possessed Sfu I recognition site but no Rsa I recognition site. The mass-produced PCR-DNA B-C (approximately 3 µg) was incubated at 37°C for 1 hr in

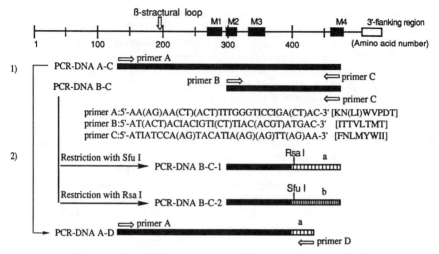

primer A:5'-AA(AG)AA(CT)(ACT)TTTGGGTTCCIGA(CT)AC-3' [KN(LI)WVPDT]
primer B:5'-AT(ACT)ACIACIGTI(CT)TIAC(ACGT)ATGAC-3' [ITTVLTMT]
primer C:5'-ATIATCCA(AG)TACACATIA(AG)(AG)TT(AG)AA-3' [FNLMYWII]

3) Adapter attached cDNA encoding GABA receptor genes which was concentrated by hybridization-selection

Fig. 1. Strategies for identification of GABA receptor gene of German cockroach
through polymerase chain reactions (PCR).The details of PCR strategies and the
structure of primer E, F, G, I, J and K are described in the Method section.
(Reproduced with permission from reference 47. Copyright 1994, Pergamon.)

40 μl reaction mixture containing 50 mM Tris-HC1 (pH 7.5, 37°C), 100 mM NaC1, 10 mM MgCl$_2$, 1 mM dithioerythritol and 10 units Sfu 1 (Boehringer Mannheim, Indianapolis, IN) or containing 10 mM Tris-HC1 (pH 7.5, 37°C), 10 mM MgCl$_2$, 1 mM dithioerythritol and 11 units Rsa I (Boehringer Mannheim). The reaction mixture was separated by using electrophoresis on 1% agarose gel made with 1 X TBE, and the unrestricted fragment (approximately 520 bps) was excised from the gel. The DNA was purified by QIAEX matrix (QIAGEN, Dusseldorf, Germany) according to the manufacturer's instruction and sequenced using primer B and primer C following the standard procedure. Two kinds of sequences (DNA-a and DNA-b type) around M4 membrane spanning region were revealed, and thereafter DNA sequencing was expanded to primer A region by using PCR-DNA A-D which had been prepared by primer A, primer D (5'-TGGGGTGGATGAGATGTT-3', corresponding to QHLIHPK) and PCR-DNA A-C as a template DNA. Step III. The method of Frohman et al. (19) was modified to extend PCR to 3'-flanking region. PCR-DNA A-D was used as the probe DNA. The purified probe-DNA (0.25 μg), which had been denatured at 95°C for 5 min, was applied on a strip of BIODYNE A 0.2 μm membrane (2.6 X 1.5 cm, Pall BioSupport, Glen Cove, NY) and washed with 1.5 M NaC1 and 1M Tris-HC1 (pH 7.4) followed by fixation by UV irradiation (120 mj). The membrane was prehybridized in a polypropylene tube (#2063 Falcon, Lincoln Park, NJ) with hybridization buffer [5 X SSC, 0.5% Block Reagent (Boehringer Mannheim), 0.1% sarkosyl, 0.02% SDS] at 68°C for 1 hr. cDNA (50 μl) was denatured by heating at 95°C for 5 min and quickly chilled on ice and mixed wit 600 μl of the hybridization buffer and hybridized with probe-DNA on the membrane at 68°C overnight. After removing the hybridization solution, the membrane was washed with 1 ml of 0.1 X SSC and 0.1% SDS solution at room temperature for 5 min, and incubated with 1 ml of 0.1 M NaOH at room temperature for 15 min to strip the GABA receptor cDNA from the probe DNA. This alkali solution was neutralized with 1M HCl on ice and desalted by Centricon-100 (Amicon, Beverly, MA) at 1,000 x g for 30 min with 1 ml ddH$_2$O three times. The retained solution was made up to 50 μl with ddH$_2$O and the first round PCR was performed with 1 μ of the hybridization-selected cDNA as a DNA template, adapter primer (5'GACTCGAGTCGACATCG-3') and primer E (5'-GCGCCCAAACAAACAGTA-3'), or primer F (5'-GCGCCCAAACAAACAGTT-3'). The primer E and F correspond to the amino acid sequence APKQTV of DNA-a and DNA-b type, respectively.   In the second round PCR, primer G (5-ACGTGCAGACGAAGAAGT-3', corresponding to GRADEEV) and primer H (5'-CAGCATCTCATTCACCCA-3', corresponding to QHLIHP) were used for confirmatory studies. Thus, obtained PCR products showed the two basic types of DNA sequence between M4 transmembrane region and 3'-flanking region (i.e., Rdl1, Rdl2-type vs. Rdl3, Rdl4-type) as shown in Fig.2.

RESULTS

We first made an effort to sequence the GABA receptor from the cDNA library of the susceptible German cockroach strain (CSMA strain) through a series of polymerase chain reactions (PCR), using various primers as shown in Fig. 1 and its caption. We chose the PCR approach, instead of DNA cloning.

The sequencing results shown in Fig. 2 revealed a few surprising features of the German cockroach GABA receptors. First, despite the fact that we employed primers representing the most common (conserved) regions (21-25) for all known subunits of mammalian and *Drosophila* GABA (13) receptors (i.e., α, β, γ and δ subunits), only one

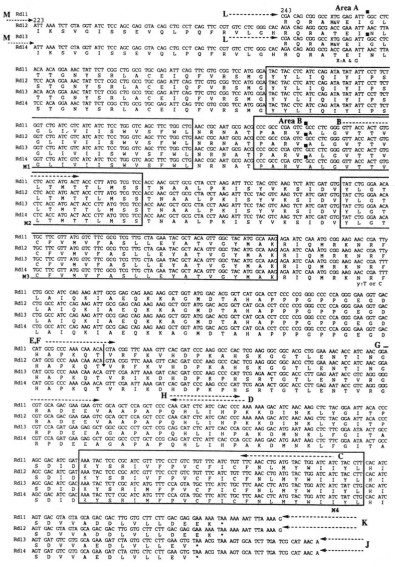

Fig. 2. The DNA and amino acid sequence of the German cockroach GABA receptor subunit, Rdl1, Rdl2, Rdl3 and Rdl4. The numbers at the beginning of Rdl1 and Rdl2 indicate the corresponding amino acid number of Drosophila Rdl-subunit (ffrench-Constant et al., 1991). The sites of mutation are indicated by solid squares. The DNA regions corresponding to PCR primers are shown by dotted lines with alphabetical designations directly above the region. The base shown at "X" was R= either A or G, and by "Y" was T or C. The site indicated by an arrow (at the tail end of primer E) was observed to contain a GAG (E) insertion in many cases. Rdl1 and Rdl2 were found to contain an equal amount of inserted vs. noninserted sequences, while Rdl3 and Rdl4 consist mostly of noninserted sequences according to the band intensities. (Reproduced with permission from reference 47. Copyright 1994, Pergamon.)

type resembling the *Rdl* sequence of *Drosophila* was picked up from the cDNA library of the German cockroaches. The overall features of this type of German cockroach GABA receptor sequences are similar to *Rdl*. Second, we found at least 4 basic types of strands of *Rdl* subunits in this species. Since both *Rdl*1 and *Rdl*2 possess a minor variant DNA base (Fig. 2 amino acid #13, shown as X meaning that it is either R=A or G at this position, yielding either methionine or valine), there are altogether 6 variants of these *Rdl* subunits. While the most variable region was in the connecting part between M3 and M4, variations were also found in other areas (Fig. 2). Third, these *Rdl*-subunits were much shorter than the *Rdl*-subunit of *Drosophila melanogaster* (13), the main difference being a shorter interconnecting region between M3 and M4.

For comparison, the cDNA library from the resistant strain (LPP strain) was analyzed through the same PCR and sequencing approach. The results showed that the sequences of all PCR fragments from this resistant strain were totally identical except at two DNA bases (see the solid squares, Fig. 2). In one of the sites (area B, Fig. 3) G to T mutation was detected in all six *Rdl*-subunits identified from this strain. Examples of the portion of the sequencing gel images indicating this mutational change are shown in Fig. 4 (area B). It must be noted that there was no sign of the presence, even at a trace level, of the guanine residue in the resistant DNA preparation at the position of mutation. In the case of the other mutations (area A) only one of the four basic types (*Rdl*3) of these *Rdl*-subunits showed a change of the amino acid composition (G to S). We could not detect any other difference between the susceptible and resistant strains with respect to the quantity or the quality of PCR products among various *Rdl*-subunits. Based on the observations that (a) this is the same site and the nature of mutation recently reported by others (16, 17), (b) the mutation occurring at area B involved all German cockroach *Rdl* subunits, (c) this is the most conserved region among all GABA subunits published so far, and (d) area B is within the membrane spanning region, whereas area A is not, we estimate that the mutation occurring at area B is the main factor conferring the cyclodiene resistance to this resistant strain.

DISCUSSION

A key question we must raise now is whether this mutational change occurring at such a conserved region (i.e. area B) of the GABA receptor would functionally allow the chloride channel to operate normally. A literature search of all sequences of known GABA receptor subunits has uncovered the presence of three mammalian subunits (22, 24) (δ, γ2 and rho2 human) which possess the same serine substitution for alanine configuration (Fig. 3 area B). Therefore, it is unlikely that such a mutation is a totally destructive one in terms of the functional aspect of GABA interactions and chloride ion transport.

What then is the evidence indicating that this site of mutation confers the resistance to cyclodiene-type chemicals? It must be pointed out, first, that there is enough evidence that the GABA receptors of the nervous systems of the resistant cockroaches (4, 6, 8-10) and other species (6, 11, 12) show lower binding affinities or a lower sensitivity (26) to cyclodiene-type ligands than do their susceptible counterparts. Second, there are now enough examples of the same type of mutations occurring in cyclodiene resistant insect species (16, 17). Therefore, one could conclude that the above mutation somehow changes the affinity of the assembled GABA receptor to cyclodiene-type chemicals. The main reasons to consider that the binding site of cyclodiene-type chemicals could also be within or close to this most conserved region where the mutation took place are: (a) it is located at the innermost surface of the chloride channel which is directly accessible to influxing

Area A

German cockroach (CSMA) *Rdl*1,*Rdl*3	R Q R A X E I G L T T	G N Y S R	L
German cockroach (LPP) *Rdl*1,*Rdl*3	R Q R A X E I G L T T	G N Y S R	L
	S		
German cockroach (CSMA) *Rdl*2,*Rdl*4	R Q R A T E I N L S T	G N Y S R	L
German cockroach (LPP) *Rdl*2,*Rdl*4	R Q R A T E I N L S T	G N Y S R	L
Drosophila (*Rdl, 13*)	R Q R A T E I N L T T	G N Y S R	L
Lymnaea (ß, *37*)	A T I N K I E E L L T	G D Y Q R	L
Mammal (ß1)*	K M V S K K V E F T T	G A Y P R	L
Mammal (ß2)†	K L I T K K V V F S T	G S Y P R	L
Bovine brain, Rat forebrain (ß3,*38*)	R L V S R N V V F A T	G A Y P R	L
Chicken (ß3,*42*)	R L V S K N V V F A T	G A Y P R	L
Mammal (α1)‡	T V D S G I V Q S S T	G E Y V V	M
Bovine brain (α2,*30*)	S I G K E T I K S S T	G E Y T V	M
Bovine brain (α3,*30*)	V V G T E I I R S S T	G E Y V V	M
Rat brain (α4,*40*)	T V G T E N I S T S T	G E Y T I	M
Rat brain (*22*),Human (*39*),Mouse (*43*) (γ2)	R N T.T E V V K T T S	G D Y V V	M
Chicken (γ2,*43*)	R N T T E V V K T T S	G D Y V V	M
Rat brain (δ,*22*)	R F T T E L M N F K S A G Q F P R		L
Human (ρ1,*44*)(rho1,*45*)	H T T T K L A F Y S S T G W Y N R		L
Human (rho2,*45*)	H T T S R L A F Y S S T G W Y N R		L

     X:Met or Val
     *:Bovine brain(*21*), Rat forebrain(*38*), Human(*39*)
     †:Bovine brain(*38*), Rat forebrain(*38*)
     ‡:Bovine brain(*21*), Rat brain(*40*), Mouse(*41*), Human(*39*)

Area B

German cockroach(CSMA) *Rdl*1-4	V S F W	L N R N A T P	A R	V A L G V	T	T V L T M	T	T
German cockroach(LPP) *Rdl*1-4	V S F W	L N R N A T P	A R	V **S** L G V	T	T V L T M	T	T
Drosophila (*Rdl, 13*)	V S F W	L N R N A T P	A R	V A L G V	T	T V L T M	T	T
Lymnaea (ß, *37*)	V S F W	I N H E A T S	A R	V A L G I	T	T V L T M	T	T
Mammal (ß)*	V S F W	I N Y D A S A	A R	V A L G I	T	T V L T M	T	T
Mammal (α)†	V S F W	L N R E S V P	A R	T V F G V	T	T V L T M	T	T
Mammal (γ2)‡	V S F W	I N K D A V P	A R	T S L G I	T	T V L T M	T	T
Rat brain (δ,*22*)	V S F W	I S Q A A V P	A R	V S L G I	T	T V L T M	T	T
Human (ρ1,*44*)(rho1,*45*)	V S F W	I D R R A V P	A R	V P L G I	T	T V L T M	S	T
Human (rho2,*45*)	V S F W	I D R R A V P	A R	V S L G I	T	T <u>V L T M</u>	T	T

                                        ━━━━━ M2 ━━━━━

* :Bovine brain(ß1,*21*), Rat forebrain(ß1,*38*), Human(ß1,*39*), Bovine brain(ß2,*38*),
    Rat forebrain(ß2,*38*), Bovine brain(ß3,*38*), Rat forebrain(ß3,*38*), Chicken(ß3,*42*)
† :Bovine brain(α1,*21*), Rat brain(α1,*40*), Mouse(α1,*41*), Human(α1,*39*),
    Bovine brain(α2,*30*), Bovine brain(α3,*30*), Rat brain(α4,*40*)
‡ :Rat brain(*22*), Human(*39*), Chicken(*43*), Mouse(*43*)

Fig. 3. Comparison of the amino acid sequences of two areas of GABA receptors, each containing a point mutation, in resistant *B. germanica* among several species. The locations of the mutations in the resistant German cockroach are shown by bold letters. The mutation in area A was found in only two of the *Rdl*-subunits, whereas that in area B was found in all of them.

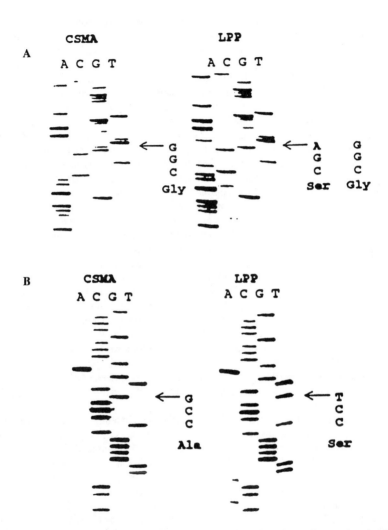

Fig. 4. Examples of DNA-sequencing gels showing the sites of mutation in the resistant strain (Lpp strain) in comparison with the susceptible counterpart (CSMA strain). The upper figures (panel A) show the area A where only one of the *Rdl* subunits (*Rdl4*) showed a mutation. The lower figures (panel B) show the center M$_2$ region where the major mutation responsible for cyclodiene resistance occurred. Note that the G to T conversion is complete in this homozygous R strain, resulting in an amino acid change of alanine to serine. The amino acid sequences of this area are highly conserved by being common among most GABA receptor subunits from invertebrates and vertebrate species.

(Reproduced with permission from reference 47. Copyright 1994, Pergamon.)

ligands through the pore, and (b) this region is the only common denominator for all hitherto identified mammalian and insect GABA receptor subunits as well as inhibitory glycine receptors (27) which also show sensitivities to picrotoxinin. Indeed, it has been shown by Pribilla et al. (28) that in the case of the chloride channel of inhibitory glycine receptor, the M_2 segment of the β-subunit is the action site of picrotoxinin. Judging by the fact that none of the mammalian subunits by themselves are capable of forming homo-polymers capable of responding to diazepines in the same *Xenopus* tests (29), it is evident that this conserved region does represent a specific picrotoxinin and barbiturate interaction site (28-35). On the other hand, in this work we have not experimentally shown that picrotoxinin type chemicals indeed bind to this region. Therefore, the precise binding site within this M2 region must be identified through vigorous experimental efforts in the future. Also, it would be of great interest to conduct functional tests on these mutated *Rdl* subunits. Unfortunately, this PCR approach is not suited to isolate a complete *Rdl* gene clone, which will be required for expression work.

The presence of at least 4 variants of this *Rdl* type GABA receptor sequence is note-worthy. Perhaps these are the result of alternative splicing, as in the case of *Drosophila* (36). However, much more work would be needed to confirm such a possibility.

Another unusual observation is that there is a site in *Rdl*1 and *Rdl*2 subunits where one extra amino acid (glutamic acid) equivalent of a DNA triplet (GAG) is inserted (Fig. 2). This is not an isolated incidence, as judged by the fact that as much as about 50% of the existing *Rdl*1 and *Rdl*2 mRNAs contained this insertion. Since *Rdl*3 and *Rdl*4 subunits did not show any sign of extra insertion, this phenomenon appears to indicate a fundamental difference in constructing *Rdl*1 and *Rdl*2 from *Rdl*3 and *Rdl*4 subunits. A close comparison of these units shows that there is a long stretch of conserved sequence from amino acid 253 through 401, the only difference being the position 366 indicated by "y". The triplet base insertion mentioned above occurred between 401 and 402 just at the end of the conserved region. This unusual occurrence of insertion of one amino acid may require special attention.

Also worth mentioning are the observed differences in length and amino acid sequen-ces of the M_3 - M_4 connecting region between the receptors of German cockroaches and those of *Drosophila melanogaster*. The former is only 75 amino acids long as opposed to 230 for *Rdl* type (13) and 130 for LCCH3 (14). Not only that, there is no similar segment of sequence series anywhere in this stretch (i.e., 377-555, according to the numbering scheme of ffrench-Constant et al. (13)). For instance, the characteristic proline-rich sequence, PPPGPPG, in the subunits from the German cockroach could not be located in any of the GABA subunits published so far. In this regard the GABA receptor sequence of the German cockroach is more homologous to that of *Aedes aegypti* (17) than to *Drosophila*.

To summarize this section, we have found the type of GABA receptor subunits in the German cockroach to be unique. There are at least 6 variants of this type of subunit in this species. The resistant counterparts showed two point mutations. One of them, occurring at the most conserved region, appears to be correlated to the resistance of this strain to cyclodiene-type chemicals.

The foregoing part of this article was published in a recent technical paper by us (47).

Having identified the site of mutation in the GABA receptor in *B. germanica*, which appears to be certain to confer the resistant individuals the selective advantage against the toxic action of cyclodiene-type chemicals, one can now ask what the toxicological significance is of such an accomplishment. Certainly, one outcome of this finding is that the action site of these chemicals within the GABA receptor is likely to be the location very close to the site of mutation. One could speculate that such a mutation makes the binding affinity of the resistant GABA receptor to these chemicals lower than that toward the wild type GABA receptor in the susceptible nervous system. There are several supporting pieces of evidence. As early as 1966, Dr. Hayashi and I (9) studied the binding behavior of ^{14}C-dieldrin among various nerve components from the central nervous system of the resistant (London strain) and susceptible (CSMA strain) German cockroach nerves. We could demonstrate then that the extent of dieldrin binding to components of the plasma membrane fraction was less in the London strain, the ancestral strain for the LPP strain used in the current study. A subsequent study by Telford and Matsumura, using an autoradiographic detection method on section specimen for electron microscopy prepared from ^{14}C-dieldrin treated nerve cords, also clearly showed that the number of ^{14}C grains, representing labeled dieldrin, observed to be present on the plasma membrane was much less in the preparation from the London strain, despite the fact that they were treated under identical incubation conditions (10). It is indeed very gratifying that after so many years we could finally demonstrate what causes such a differential binding behavior of dieldrin, as we had to maintain these strains for over 30 years, during which period we had to genetically purify and select them by using dieldrin residues. Other scientists have also shown in housefly strains and different radioligands that the resistant nervous system binds less with these GABA receptor interactors as compared to the susceptible counterparts (6, 11). However, it must be pointed out that the actual site of attachment of these insecticides within this region of the GABA receptor has not been determined. Therefore, the precise mechanism of cyclodiene interactions at the molecular level still remains to be elucidated.

Another beneficial outcome of this line of research is the establishment of the basic framework of the characteristics of one of the major insecticide target sites in the insect nerve. Certainly the basic structures of insect GABA receptors are similar to those found in mammalian species. However, thanks to several groups of insect toxicologists (e.g., 13, 14, 16, 46) we now know some significant structural differences between these two groups. While the origins of *Rdl*-type GABA receptors are not known, the ones found in *B. germanica* are very similar to those found in *Drosophila* and *Musca domestica*. Thus, now that it is possible to use very specific ligands or antibodies directed at particular sites of the receptor, it is possible to study the basis of selective toxicity (i.e., insect vs. mammalian systems) of toxicants at this target site.

The fact that the GABA receptor is one of the major insecticidal targets is now firmly established. A number of new insecticides are being developed to exploit this weakness of insects. Future research along this line should also include studies on the basic differences between insect and mammalian GABA receptors, to allow the rational design of selective insecticides based on these differences. A brief account of this paper, along with a similar alanine to serine mutation in dieldrin resistant red flour beetle (*Tribolium castenium*) *Rdl* gene, has been reported previously (48).

Acknowledgments. This work was supported by grants ES01963 and ES05707 from the National Institute of Environmental Health Sciences, Research Triangle Park, North Carolina.

LITERATURE CITED

1. Busvine, J.R. *Nature* **1954**, *174*, 783-785.
2. Brown, A.W.A. *Ann. Rev. Entomol.* **1960**, *5*, 301-326.
3. Ghiasuddin, S.M.; Matsumura, F. *Comp. Biochem. Physiol.* **1982**, *73C*, 141-144.
4. Matsumura, F.; Ghiasuddin, S.M. *J. Environ. Sci. Health.* **1983**, *B18*, 1-14.
5. Lawrence, L.J.; Casida, J.E. *Science* **1983**, *221*, 1399-1400.
6. Tanaka, K.; Matsumura, F. In *Membrane Receptors and Enzymes as Targets of Insecticide Action;* Clark, J.M. and Matsumura, F., Eds; Plenum Press: New York, NY, **1986**; pp. 33-49.
7. Eldefrawi, A.T.; Abalis, I.M.; Eldefrawi, M.E. In *Membrane Receptors and Enzymes as Targets of Insecticide Action;* Clark, J.M. and Matsumura, F., Eds.; Plenum Press: New York, NY, **1986**; pp. 107-124.
8. Matsumura, F. In *Pesticide Science and Biotechnology. Proceedings of the 6th IUPAC Congress of Pesticide Chemistry*; Greenhalgh, R. and Roberts, T.R., Eds.; Blackwell Scientific Publications: Oxford, 1987; pp. 151-159.
9. Matsumura, F.; Hayashi, M. *Science* **1966**, *153*, 757-759.
10. Telford, J.N; Matsumura, F. *J. Econ. Entomol.* **1970**, *63*, 795-800.
11. Deng, Y.; Palmer, C.J.; Casida, J.E. *Pestic. Biochem. Physiol.* **1991**, *41*, 60-65.
12. Anthony, N.M.; Benner, E.A.; Rauh, J.J.; Sattelle, D.B. *Neurochem Int.* **1992**, *21*, 215-2212.
13. ffrench-Constant, R.H.; Mortlock, D.P.; Shaffer, C.D.; MacIntyre, R.J.; Roush, R.T. *Proc. Natl. Acad. Sci. USA* **1991**, *88*, 7209-7213.
14. Henderson, J.E.; Soderlund, D.M.; Knipple, D.C. *Biochem. Biophys. Res. Commun.* **1993**, *193*, 474-482.
15. Harvey, R.J.; Vreugdenhil, E.; Zaman, S.H.; Bhandal, N.S.; Usherwood, P.M.; Barnard, E.A; Davlison, M.G. *EMBO J.* **1991**, *10*, 3239-3245.
16. ffrench-Constant, R.H.; Rocheleau, T.A.; Steichen, J.C.; Chalmers, A.E. *Nature* **1993**, *363*, 449-451.
17. Thompson, M.; Shotkoski, F.; ffrench-Constant, R.H. *FEBES Lett.* **1993**, *325*, 187-190.
18. Badley, J.E.; Bishop, G.A.;St. John, T.; Frelinger, J.A. *Biotechniques* **1988**, *6*, 114-116.
19. Frohman, M.A.; Dush, M.K.; Martin, G.G. *Proc. Natl. Acad. Sci. USA* **1988**, *85*, 8998-9002.
20. Schuurman, R.; Keulen, W. *Biotechniques* **1991**, *10*, 185.
21. Schofield, P.R.; Darlison, M.G.; Fujita, N.; Burt, D.R.; Stephenson, F.A.; Rodriguez, H.; Rhee, L.M.; Ramachandran, J.; Reale, V.; Glencorse, T.A.; Seeburg, P.H.; Barnard, E.A. *Nature* **1987**, *328*, 221-227.
22. Shivers, B.D.; Killisch, I.; Sprengel, R.; Sontheimer, H.; Kohler, M.; Schofield, P.R.; Seeburg, P.H. *Neuron* **1989**, *3*, 327-337.
23. Garrett, K.M.; Duman, R.S.; Saito, N.; Blume, A.J.; Vitek, M.P.; Tallman, J.F. *Biochem. Biophys. Res. Commun.* **1988**, *156*, 1039-1045.
24. Stephenson, F.A. *Biochem. J.* **1988**, *249*, 21-32.
25. Lunt, G.G.; Brown, M.C.S.; Riley, K.; Rutherford, D.M. In *Neurotox '88 Molecular Basis of Drug & Pesticide Action;* Lunt, G.G., Ed.; Excerpta Medica: Amsterdam, **1988**; pp. 185-192.

26. Bloomquist, J.R.; Roush, R.T.; ffrench-Constant, R.H. *Arch. Insect Biochem. Physiol.* **1992**, *19*, 17-25.
27. Betz, H. *Quart. Rev. Biophys.* **1992**, *25*, 381-394.
28. Pribilla, I.; Takagi, T.; Langosch, D.; Bormann, J.; Betz, H. *EMBO J.* **1992**, *11*, 4305-4311.
29. Olsen, R. W.; Tobin, A.J. *FASEB* **1990**, *4*, 1469-1480.
30. Levitan, E.S.; Schofield, P.R.; Burt, D.R.; Rhee, L.M.; Wisden, W.; Köhler, M., Fujita, N.; Rodriguez, H.; Stephenson, F.A.; Darlison, M.G.; Barnard, E.A.; Seeburg, P.H. *Nature* **1988**, *335*, 76-79.
31. Levitan, E.S.; Blair; L.A.C.; Dionne, V.E.; Barnard, E.A. *Neuron* **1988**, *1*, 773-781.
32. Blair, L.A.C.; Levitan, E.S.; Marshall, J.; Dionne, V.E.; Barnard, E.A. *Science* **1988**, *242*, 577-579.
33. Pritchett, D.B.; Sontheimer, H.; Gorman, C.M.; Kettenmann, H.; Seeburg, P.H.; Schofield, P.R. *Science* **1988**, *242*, 1306-1308.
34. Sigel, E.; Barnard, E.A. *J. Biol. Chem.* **1984**, *259*, 7219-7223.
35. Bormann, J.; Hamill, O.P.; Sakmann, B. *J. Physiol.* **1987**, *385*, 243-286.
36. ffrench-Constant, R.H.; Rocheleau, T.H. 1993. *J. Neurochem.* **60**, 2323-2326.
37. Harvey, R.J.; Vreudgenhil, E.; Barnard, E.A.; Davlison, M.G. *Biochem. Soc. Trans.* **1990**, *18*, 438-439.
38. Ymer, S.; Schofield, P.R.; Draguhn, A.; Werner, P.; Köhler, M.; Seeburg, P.H. *EMBO J.* **1989**, *8*, 1665-1670.
39. Pritchett, D.B.; Sontheimer, H.; Shivers, B.D.; Ymer, S.; Kettenmann, H.; Schonfield, P.R.; Seeburg, P.H. *Nature* **1989**, *338*, 582-585.
40. Khrestchaitisky, M.; MacLennan, A.J.; Chiang, M.; Xu, W.; Jackson, M.B.; Brecha, N.; Sternini, C.; Olsen, R.W; Tobin, A.J. *J. Neuron* **1989**, *3*, 745-753.
41. Keir, W.J.; Kozak, C.A.; Chakraborti, A.; Deitrich, R.A.; Sikela, J.M. *Genomics* **1991**, *9*, 390-395.
42. Bateson, A.N.; Harvey, R.J.; Bloks, C.C.M.; Darlison, M.G. *Nucleic Acids Res.* **1990**, *18*, 5557.
43. Kofuji, P.; Wang, J.B.; Moss, S.J.; Huganir, R.L.; Burt, D.R. *J. Neurochem.* **1991**, *56*, 713-715.
44. Glencorse, T.A.; Bateson, A.N.; Darlison, M.G. *Nucleic Acids Res.* **1990**, *18*, 7157.
45. Cutting, G.R.; Lu, L.; O'Hara, B.F.; Kasch, L.M.; Montrose-Rafizadeh, C.; Donovan, D.M.; Shimada, S.; Antonarakis, S.E.; Guggino, W.B.; Uhl, G.R.; Karazian Jr., H.H. *Proc. Natl. Acad. Sci. USA* **1991**, *88*, 2763-2677.
46. Cutting, G.R.; Curristin, S.; Zoghbi, H.; O'Hara, B.F.; Seldin, M.F.; Yhl, G.R. *Genomics* **1991**, *12*, 801-806.
47. Kaku, K.; Matsumura, F. *Comp. Biochem. Physiol.* **1994**, in press.
48. Matsumura, F.; Kaku, K.; Enan, E.; Charalambous, P.; Miyazaki, M.; Muralidhara; Inagaki, S. I *Pesticides/Environment: Molecular Biological Approaches*; Mitsui, T., Matsumura, F. And Yamaguchi, I., Eds.; Proceedings of the First International Symposium on Pesticide Science, Pesticide Science Society of Japan, Tokyo, **1993**, pp. 3-15.

RECEIVED December 5, 1994

Chapter 15

Insecticide Binding Sites in the House Fly Head γ-Aminobutyric Acid Gated Chloride-Channel Complex

Yanli Deng

FMC Corporation, 855 Parr Boulevard, Richmond, CA 94801

The insect γ-aminobutyric acid (GABA)-gated chloride channel complex is one of the most sensitive targets for insecticide action. The complex has been evaluated in house fly head membranes by radioligand binding studies with 1-(4-ethynylphenyl)-4-n-[2,3-^{3}H$_2$]propyl-2,6,7-trioxabicyclo[2.2.2]octane ([^{3}H]EBOB), [^{35}S]t-butylbicyclophosphorothionate ([^{35}S]TBPS) and [^{3}H]avermectin B$_{1a}$ ([^{3}H]AVM). The [^{3}H]EBOB binding site is identical to or overlaps that of five classes of structurally-diverse insecticides, $i.e.$ polychlorocycloalkanes, bicycloorthobenzoates, dithianes, silatranes, and picrotoxins. One mechanism of cyclodiene resistance in house flies is associated with a low-affinity binding site for [^{3}H]EBOB. The [^{35}S]TBPS binding site in brain is relevant in its toxic action in mammals but not house flies. The [^{3}H]AVM binding in house fly heads measures a site coupled to the EBOB site. Avermectin analogs (AVMs) are potent competitive inhibitors of [^{3}H]AVM binding and non-competitive inhibitors of [^{3}H]EBOB binding with the same structure-activity relationship for the binding sites and LD$_{50}$ assays. Findings on these binding sites lay the background for further studies on the GABA channel structure and function.

Gamma-aminobutyric acid (GABA)-gated chloride channel regulates nerve and neuromuscular functions in insects. The GABA-gated chloride channel complex has recently become the subject of intensive study for three reasons. First, the structure has been recognized as one of the major sites for insecticide action and insecticide resistance modification. The polychlorocycloalkane (PCCA) insecticides including lindane, toxaphenes and chlorinated cyclodienes (Figure 1), were once used worldwide with about three billions of pounds being applied over the past 50 years.

0097–6156/95/0591–0230$12.25/0

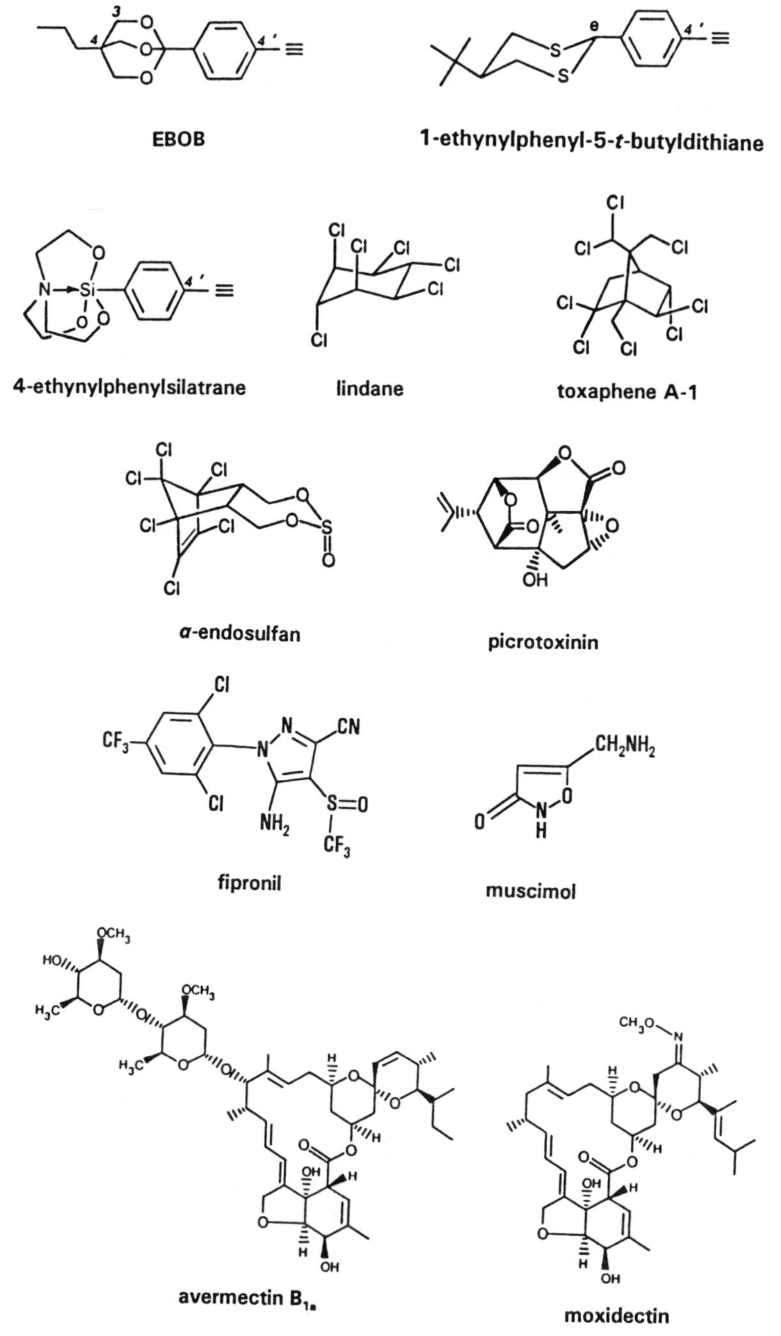

EBOB

1-ethynylphenyl-5-*t*-butyldithiane

4-ethynylphenylsilatrane   lindane   toxaphene A-1

α-endosulfan   picrotoxinin

fipronil   muscimol

avermectin B$_{1a}$

moxidectin

Figure 1. Structures of representative insecticides acting at the house fly head GABA-gated chloride channel complex.

The PCCAs were introduced decades ago and most of them have been banned due to their adverse environmental effects, yet their mode of action remained unclear until the 1980's when Matsumura and Giasuddin (1) proposed that these insecticides might act at the insect GABA-gated chloride channel. Later studies with improved techniques have confirmed their mode of action as directly blocking GABA-gated chloride channel (2,3). After prolonged use the efficacy of the PCCAs has diminished as evidenced by the 290 species of insects that have developed strong resistance to these insecticides (4). One mechanism for the resistance is reduction in receptor site sensitivity from modification on the GABA receptor (3,5,6,7-9). The second reason for intensive study is that an expanding number of new insecticides were discovered to affect the GABA-gated chloride channel, i.e. bicycloorthobenzoates (10,11), phenylpyrizoles (12) and avermectins (13-15) (Figure 1). Bicycloorthobenzoates and phenylpyrazoles induce similar hyperexcitation poisoning signs and share cross resistance with the PCCAs in insects, implicating some component(s) of the GABA receptor complex as their site of action. Avermectins cause strong sedation and share no resistance with PCCAs in insects, thus implying a different mechanism (e.g. opening of chloride channel) is involved (16). These new insecticides are more environmentally acceptable than PCCAs since they are used at very low rates and readily degrade in the environment (14,15,17) and, therefore, do not accumulate in biological systems. They may soon take over the role once played by the PCCAs as pest control agents and be more compatible with Integrated Pest Management programs. Once neglected in the past, the insect GABA receptor complex is being re-evaluated as a target for selecting potential insecticides and manipulating resistance problems. Third, in the last decade, a combination of biological, physiological, and molecular approaches has pushed our understanding of the mammalian $GABA_A$ receptor up to a level that has never been previously achieved (18,19). Logically, similar research tools are being used to explore the insect GABA receptor to elucidate the difference between the mammalian and insect GABA receptors to better design selective insect control agents.

Vertebrate GABA-gated Chloride Channel. Two types of GABA receptors are identified in vertebrates: the bicuculline-sensitive $GABA_A$ receptor regulating transient chloride ion conductance (20) and the baclofen-sensitive $GABA_B$ receptor mediating a variety of slow responses through receptor-G-protein-effector complexes, including regulation of the Ca^{++} channel (21). The GABA-gated chloride channel (or $GABA_A$ receptor) is estimated to make up to one third of all synapses in the nervous system of vertebrates (22). The mammalian $GABA_A$ receptor has been defined pharmacologically with a variety of specific agonists and antagonists. It possesses binding sites for GABA and agonist muscimol (23,24), benzodiazepines (25), convulsants such as picrotoxins (PTX) and t-butylbicyclophosphorothionate (TBPS) (26,27), modulatory sites for barbiturates (28,29) and barbiturate-behaving steroid hormones (30,31). Two distinct subunit types have been purified from bovine brain using photoaffinity labelling with specific radioligands: the 53 kDa α subunit and the 57 kDa β subunit (32-34). The α subunit is labeled by [³H]flunitrazepam and the β subunit by [³H]muscimol, suggesting that the α subunit possesses the benzobiazepine binding site and the β subunit carries the agonist binding site. Molecular cloning techniques have identified several subunits, α_{1-6}, β_{1-3}, γ_{1-2}, δ_1 and ρ_1 (35,36). They all belong to a gene superfamily of ligand-gated ion channels. There is a high degree of amino acid sequence homology between same subunit types (60-80%), a low level of sequence homology between different subunit types (20-40%) and even less amino acid homology with the nicotinic acetylcholine receptors and strychnine-sensitive glycine receptors (10-20%) (19). The distribution of mRNAs for different subunits varies considerably between brain regions, indicating receptor heterogeneity in mammals (19). Individual or co-expression of the subunit mRNAs in different combinations in *Xenopus* oocyte and mammalian cell lines has generated functional GABA receptors with unique pharmacological and electrophysiological profiles (20,37,38). The GABA$_A$ receptor is believed to be a hetero-oligomeric complex containing five glycoprotein subunits (38).

Insect GABA-gated chloride channel. Much less is known about the biology, physiology and molecular biology of the GABA-gated chloride channel in insects. GABA was first recognized as an inhibitory neurotransmitter in crustacea and later found to occur widely in invertebrates including insects (39-42). Ligands for mammalian GABA receptors have been less than satisfactory in characterizing GABA receptor in insects (43). Nevertheless, three different binding sites have been defined with these ligands in several insect preparations with similar characteristics to the vertebrate GABA$_A$ receptor. The GABA/agonist binding site was investigated with [³H]GABA binding assays in cockroach (44,45) and locust (46); and [³H]muscimol binding assays in house fly and locust (47) and honey bee heads (48). As in mammals, [³H]GABA and [³H]muscimol bindings are sodium-independent and inhibited by muscimol and GABA. Other mammalian GABA$_A$ receptor agonists such as isoguvacine, thiomuscimol, and 3-aminopropane sulfonic acid are much less effective in displacing the radioligands. The biggest difference between mammals and insects, however, may be the effect of bicuculline. This potent antagonist of the vertebrate GABA$_A$ receptor is not active in insect GABA receptor binding assays (49). The pharmacological profile of the insect binding sites suggests a GABA receptor rather than an uptake/transport site. One binding site is commonly detected for these ligands in insects with a few exceptions. In honey bee heads, for example, two sites with different affinities are distinguished for [³H]muscimol with the high-affinity site resembling the GABA binding site defined in other insect species (48). Interestingly, avermectin B$_{1a}$ is the most potent inhibitor to the site (2X as potent than muscimol itself) (48). Multiple binding sites for [³H]GABA are also reported in locust and *Drosophila* (50,51). The benzodiazepine site in insects has similar pharmacological characteristics to both central and peripheral benzodiazepine receptors in mammals (44,52). Two polypeptides (45 kDa and 59 kDa) are photoaffinity-labeled by [³H]flunitrazepam in locust ganglia (52). The convulsant site was first investigated with the radioligand [³H]dihydropicrotoxin ([³H]DHPTX) in cockroach nerve preparation to study cyclodiene resistance based on the observation that picrotoxinin shares cross resistance with chlorinated cyclodienes (5). Characterization of [³H]DHPTX binding sites of susceptible and resistant strains of cockroach provided early evidence on the association of the convulsant site with insect GABA receptor and its involvement with cyclodiene resistance (5,6,53). Due to its almost unacceptable level of non-specific binding (up to 90% of the total binding), however, [³H]DHPTX was quickly replaced by another radioligand [³⁵S]-*t*-

butylbicyclophosphorothionate ([^{35}S]TBPS) (27). [^{35}S]TBPS undergoes much higher specific binding to a number of insect preparations (40-75 % of the total binding) and its binding is inhibited by PCCA insecticides and trioxabicyclooctanes (TBOs) (45,54,55). Unlike its success in mammals, however, this binding assay seems toxicologically irrelevant in insects (16,56). No difference in [^{35}S]TBPS binding is observed between cyclodiene-resistant and -susceptible strains of house fly (57), further indicating that the radioligand is not suitable for insects. The effect of GABA on [^{35}S]TBPS binding is controversial, ranging from inhibitory (44), ineffective (55), to enhancing its binding (58). Picrotoxinin inhibits [^{35}S]TBPS binding noncompetitively in house fly (54) but is not active in locust (58). Hexobarbital inhibits [^{35}S]TBPS binding competitively in house fly (54) while pentobarbital enhances its binding in locust (58). These studies indicate that TBPS may not bind to the convulsant site of insect GABA receptor.

Radioligands. Introduction of the novel radioligand 4'-ethynyl-4-n-[^{3}H]propylbicycloorthobenzoate ([^{3}H]EBOB) has contributed significantly to our understanding of the insect GABA-gated chloride channel complex. This new bicycloorthobenzoate radioligand binds specifically to vertebrate and insect preparations with high affinity (specific binding 51-93 % of the total binding) (3,59). The binding assay has been optimized and studied in detail with house fly head membranes. Characterization of the binding site has shed light on the mode of action for a variety of structurally-diverse insecticides, i.e. PCCAs, dithianes, phenylsilatranes, trioxabicyclooctanes (TBOs), picrotoxinin, phenylpyrazoles, as well as the mechanism for cyclodiene resistance in house flies (3,60). [^{35}S]TBPS binds to the same site as [^{3}H]EBOB in mammals, but not in insects (3,16,59). The insect [^{35}S]TBPS site is toxicologically irrelevant for the insecticides examined (16). Binding assays of [^{3}H]EBOB in combination with [^{3}H]avermectin B$_{1a}$ ([^{3}H]AVM) have provided sound evidence that avermectin binding site is associated with the house fly head GABA-gated chloride channel (61). The [^{3}H]EBOB binding assay has also identified the action site of phenylpyrazole insecticides in house fly (60). The purpose of this paper is to summarize these findings on the binding sites in the house fly head GABA-gated chloride channel.

[^{3}H]EBOB Binding Site

4'-Ethynyl-4-n-propylbicycloorthobenzoate (EBOB) (Figure 1) was selected as a potential radioligand based on the structure-activity work with bicycloorthobenzoates suggesting that the 4'-ethynyl and 4-n-propyl substituents confer potent insecticidal activity (11) and ease of radiolabeling the 4-n-propyl substituent (62). The [^{3}H]EBOB binding studies with house fly head membranes indicate that the radioligand binds specifically with high affinity, exceeding the previous radioligands such as [^{3}H]DHPTX, [^{35}S]TBPS and t-[^{3}H]butylbicycloorthobenzoate ([^{3}H]TBOB) (63) in measuring insecticide toxicological relevance (3,64). The chloride-dependent [^{3}H]EBOB binding is moderately inhibited by GABA and muscimol and is strongly inhibited competitively by various cage convulsants such as PCCAs, bicycloorthobenzoates, dithianes (65-67), phenylsilatranes (68) and picrotoxinin. The pharmacological profile of [^{3}H]EBOB binding partly resembles that of [^{3}H]DHPTX binding to cockroach central nervous system (5). As established earlier (49), bicuculline does not affect ligand binding to the insect GABA receptor, including [³H]EBOB binding to house fly head membranes (3). [³H]EBOB binding is not sensitive to baclofen or strychnine, a specific inhibitor to mammalian GABA$_B$ and glycine receptors, respectively (21,69). These characteristics strongly suggest that EBOB binds to the convulsant site of the GABA receptor complex. [³H]EBOB binding is also strongly modulated by avermectins and phenylpyrazoles (60,61). In comparison to the mammalian TBPS site, the EBOB site in house fly is more sensitive to avermectins, equally sensitive to cage convulsants, less sensitive to GABA and its mimetics, insensitive to benzodiazepines, barbiturates and steroid hormones.

Scatchard analysis of [³H]EBOB binding reveals a single-affinity binding component for [³H]EBOB in house fly head membranes (K$_D$ = 1.4 nM and B$_{max}$ = 346 fmol/mg protein, Figure 2) (3). One binding site is commonly detected for ligands for insect GABA receptor complex (44,46,47,52,54,55,58). The structural heterogeneity of the insect GABA receptor still requires confirmation although a few groups have reported more than one binding site for [³H]muscimol or [³H]GABA in certain insect species (48,50,51). The sites with different binding affinity may reflect different conformational stages of the same receptor .

The affinity of the EBOB site is reduced four-fold in a cyclodiene-resistant strain of house fly (K$_D$ = 5.6 nM and B$_{max}$ = 398 fmol/mg protein, Figure 2), suggesting reduction in receptor sensitivity as one mechanism for cyclodiene resistance. This observation supports the early proposal that picrotoxinin resistance in insects is related to modified GABA receptors (6). Recent progress in molecular cloning in *Drosophila* and yellow-fever mosquito suggests that a single amino acid substitution may be the key modification in GABA receptor sensitivity toward cyclodiene resistance (8,9). Interestingly, alteration of the TBPS binding site is established as a mechanism of vertebrate resistance to cyclodiene insecticides (71).

[³H]EBOB binding measures a toxicologically relevant site in house fly heads for five classes of insecticides that other radioligands fail to do (3,43,72). The potency of PCCA insecticides, bicycloorthobenzoates, dithianes, phenylsilatranes, and picrotoxins in inhibiting [³H]EBOB binding correlates with their insecticidal activity (Figure 3). The correlation further indicates that these classes of insecticides act at the EBOB site and block the GABA-gated chloride channel as their mode of action. Not surprisingly, these insecticides inhibit [³H]EBOB binding competitively and share resistance in the cyclodiene-resistance house fly strain (3,73). Similar *in vitro* and *in vivo* relationships were also established for these insecticides in inhibiting mammalian GABA$_A$ receptor (74,75). The pharmacological profile of [³H]EBOB binding is similar to [³⁵S]TBPS binding in all the vertebrates examined (59,76). The profile of the [³H]EBOB binding site varies considerably between vertebrates and insects, paralleling with the species sensitivity to EBOB as a toxicant (59). Thus, the receptor site specificity may contribute to the species selectivity.

[³⁵S]TBPS Binding Site

The [³⁵S]TBPS binding assay has helped to characterize the mammalian GABA$_A$ receptor to a greater extent (27). Based on the correlation between the potency of

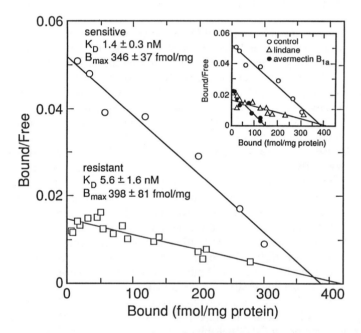

Figure 2.   Scatchard plots showing [³H]EBOB binding to house fly head membranes of a sensitive and a cyclodiene-resistant strain and effect of lindane at 10 nM and avermectin B₁ₐ at 5 nM on [³H]EBOB binding. (Reproduced with permission from ref. 3.   Copyright 1991 Academic Press, Inc.)

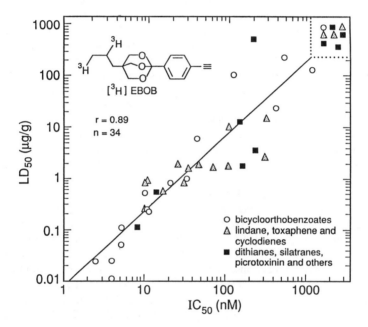

Figure 3. Correlation between potency of five classes of insecticides in inhibiting [³H]EBOB binding to house fly head membranes of a sensitive strain and their toxicity to the house fly (pretreated with piperonyl butixide). Inactive compounds (IC_{50} > 2000 nM and topical LD_{50} > 500 $\mu g/g$) are shown in the upper right box. (Reproduced with permission from ref. 3. Copyright 1991 Academic Press, Inc.)

inhibitors of [^{35}S]TBPS binding and chloride uptake and their toxicity to mice, two types of action have been classified: type A for PCCAs, picrotoxinin and trioxabicyclooctanes (TBOs) with large 1-substituents; type B for TBOs with smaller 1-substituents and bicyclophosphorus esters (75). Type A compounds are generally 300 to 500-fold more potent inhibitors than the type B compounds and tend to be more insecticidal. Enough evidence has been gathered to believe that type A and B actions also exist in insects. The type A and B compounds in insects are differentiated based on the correlation between the potency of the compounds in inhibiting [^{3}H]EBOB and [^{35}S]TBPS binding assays and their toxicity to flies, differences in poisoning signs, temperature coefficient of poisoning, and cross-resistance in a cyclodiene-resistant house fly strain (16). As in mammals, type A action in house fly involves the EBOB site with inhibitors like PCCAs, bicycloorthobenzoates, 2-aryl-dithianes, phenylsilatranes and picrotoxinin. Their action is characterized with hyperexcitatory poisoning signs, a positive temperature coefficient, and cross-resistance with cyclodienes. Type B action is believed to involve the [^{35}S]TBPS site for some of the bicyclophosphorus esters, trithiabicyclooctanes (16) and dithianes with smaller 1-substituent, even though the insecticidal relevance of [^{35}S]TBPS binding assay has never been established (Figure 4) (16,57). The type B action is characterized with slightly different hyperexcitatory poisoning signs, a negative temperature coefficient, and no cross-resistance to cyclodienes.

[^{35}S]TBPS and [^{3}H]EBOB appear to bind to the same site in the mammalian brain GABA$_A$ receptor (59), yet their interaction with insect GABA receptor varies considerably (3,54,55,59). [^{3}H]EBOB acts at the convulsant site of house fly head GABA receptor complex. [^{35}S]TBPS binds to a single saturable site in the same preparation with a much different B$_{max}$ and pharmacological profile with respect to the effect of benzodiazepines and insecticides examined (16). The density of the binding site (B$_{max}$ = 2.4 pmol/mg protein) is 7-times higher than that of [^{3}H]EBOB (B$_{max}$ = 0.35 pmol/mg protein) (3,16). It is likely that the EBOB site or the GABA receptor is a subtype of TBPS sites in house fly. The rest of TBPS sites may not be associated with the GABA-gated chloride channel in the insect. Abalis et al. (77) has shown that not all [^{35}S]TBPS binding sites in vertebrates are associated with GABA$_A$ receptor from their research on the electric ray *Torpedo nobiliana* electric organ, a tissue lacking GABA receptors. Voltage-dependent chloride ion channels have been proposed as possible TBPS binding sites in the organ (77). Alternatively, ligands like [^{3}H]EBOB may preferentially measure the GABA receptor at one particular conformational stage, thus giving a lower estimate of true receptor density. [^{35}S]TBPS may bind to GABA receptor at all stages, as do the PCCA insecticides. A PCCA radioligand should therefore measure a receptor density similar to TBPS site. Other ligands like the bicyclophosphates may selectively bind to other conformational stages of the GABA receptor inactive to EBOB, and are thus not measurable by [^{3}H]EBOB binding, or vise versa. Both theories should explain the dilemma that TBPS appears to be a competitive inhibitor of [^{3}H]EBOB binding while EBOB has no effect on [^{35}S]TBPS binding (16).

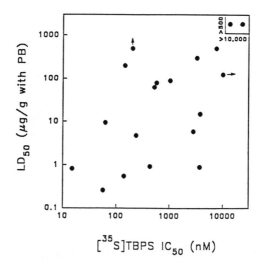

Figure 4. Lack of correlation between potency of insecticides in inhibiting [^{35}S]TBPS binding to house fly head membranes (the sensitive strain) and their toxicity to the house fly pretreated with piperonyl butoxide. Arrows designate IC_{50}s > 10,000 nM or LD_{50}s > 500 μg/g. Inactive compounds (IC_{50} > 10,000 nM and LD_{50} > 500 μg/g) are shown in the upper right box. (Reproduced with permission from ref. 16. Copyright 1993 Academic Press, Inc.)

[³H]AVM Binding Site

Avermectins are a mixture of antibiotics isolated from the soil bacterium *Streptomyces avermitilis* (*78,79*), with potent anthelmintic (*80*) and insecticidal activities (*13,15*). Structurally avermectins are a family of macrocyclic lactones possessing a 16-membered lactone ring with a spiroacetal system (C-17 to C-25) consisting of two six-membered rings and an α-L-oleandrosyl-α-L-oleandrosyloxy disaccharide substituent at the C-14 position (Figure 1). They consist of four closely-related major (A_{1a}, A_{2a}, B_{1a}, B_{2a}) components and four homologous minor (A_{1b}, A_{2b}, B_{1b}, B_{2b}) components, of which the B series are generally more biologically active (*15*). The mode of action of avermectins is partially understood from electrophysiological studies and radioligand binding assays with [³H]AVM and its dihydro analog [³H]ivermectin ([³H]IVM). Electrophysiological experiments indicate that avermectin B_{1a} acts on the GABAergic synapses of crustacea, a nematode and a mammal. Fritz *et al.*, (*81*) and Mellin *et al.* (*82*) have demonstrated that perfusion of lobster stretcher muscle with avermectin B_{1a} irreversibly eliminated inhibitory postsynaptic potentials as a result of increased chloride ion conductance. This stimulating effect on membrane conductance was inhibited by PTX, suggesting a close-coupling between the action of AVM and the GABA-gated chloride channel. Kass *et al.* (*83*) reported that avermectins acted as GABA agonists or stimulated GABA release from presynaptic inhibitory membranes to stimulate the inhibitory neuronal input in *Ascaris lumbricoides*. Similar observations were made with mammalian dorsal root ganglion neurons that avermectin B_{1a} at 10 μM activated GABA-sensitive chloride channel and was antagonized by bicuculline or PTX (*84*). Contrarily, avermectins were also found to block non-competitively GABA-stimulated chloride uptake in mammalian brains (*85,86*), to block GABA-induced chloride conductance in locust (*87*) and in *Ascaris* (*88*), and to block GABA-gated chloride channel conductance by acting as GABA-antagonists in *Ascaris* (*89*). Schaeffer and Haines (*90*) defined a specific binding site for [³H]IVM in a crude membrane preparation from the nematode *Caenorhabditis elegans*. Avermectin analogs inhibited [³H]IVM binding proportionally to their *in vivo* effect on *C. elegens* mobility, indicating a toxicologically relevant IVM site. The binding site is insensitive to GABA, muscimol, bicuculline, PTX, diazepam and several putative neurotransmitters (*90*). Cochlioquinone A, a yellow pigment isolated from a parasitic mold of rice, is the only known non-avermectin inhibitor to avermectin binding (*91*). Several other binding reports suggest a close coupling of the AVM site with the GABA-benzodiazepine receptor complex. Avermectins have been found to modulate the specific binding of GABA (*92,93*), a benzodiazepine (*93,94*), and TBPS (*77*) in mammalian systems. In insects, avermectin B_{1a} affects binding of muscimol (*48*) and TBPS (*95*). The affinity of avermectins for retinol binding proteins has also been proposed as a possible mechanism of action for avermectins in filarial parasites (*96*).

The development of the [³H]EBOB and [³H]AVM binding assays has provided three pieces of evidence supporting that the AVM site is associated with the GABA-gated chloride channel in house fly. First, avermectin B_{1a} and moxidectin [a milbamycin analog with an O-methyloxime substituent at the 23-position (*97*, Figure 1)], are extremely potent noncompetitive inhibitors to [³H]EBOB binding (Figure 2) and competitive inhibitors to [³H]AVM binding, suggesting that the AVM and EBOB sites are different yet closely-coupled (*3,61*). Second, eleven avermectin analogs inhibited both [³H]EBOB and [³H]AVM binding proportionally to their insecticidal activities (Figure 5). The potency for one assay parallels with that for the other (*61*). Thus, both [³H]EBOB and [³H]AVM assays measure toxicologically relevant sites for AVM analogs in house fly. Third, the inhibition curves of avermectin analogs for [³H]EBOB binding is sigmoidal, reaching a compound-dependent maximum (varying from 45-50% inhibition for the least potent compounds to 72-80% for the most active ones) (Figure 6), further indicating an allosteric relationship between the AVM and EBOB sites (*61*). [³H]AVM binds specifically to the house fly head membranes with high affinity (K_D = 1.9 nM) (*61*), about ten-fold higher than [³H]IVM binding to rat brain (*90*). This may account for the relatively high insecticidal activity of avermectins. As in [³H]IVM binding with nematode membranes, [³H]AVM binding in house fly is also not affected by compounds other than avermectin analogs themselves (*61*). [³H]AVM binds specifically to the thorax membrane of house fly and a decrease in B_{max} is observed for an AVM-resistant strain (*98*). The avermectin binding density tends to be high, ranging from 0.11 pmol/mg protein in house fly thorax (*98*) to 9.6 pmol/mg protein in house fly head (*61*). The B_{max} of 9.6 pmol/mg protein for [³H]AVM is 27-fold higher than that of [³H]EBOB site (B_{max} = 0.35 pmol/mg protein) in the same house fly head preparation. The reasons for the difference are not yet known. One possibility is that [³H]AVM may bind to sites in addition to GABA-gated chloride channel. The sites for IVM binding in locust muscle, for example, are concentration-dependent, acting at the GABA-gated chloride channel at low levels and at sites including the glutamate H-receptor in high concentrations (*87,99*). [³H]AVM may act similarly in house fly, but the sites for AVM at high concentrations may be less relevant to their insecticidal activity.

The avermectin binding sites have been solubilized from *C. elegans* membranes with a nonionic detergent (*100*). A slightly different [³H]IVM binding was observed with the detergent-solubilized extract (K_D = 0.20 nM and B_{max} = 0.66 pmol/mg protein) from the non-treated membranes (K_D = 0.11 nM and B_{max} = 0.54 pmol/mg protein) (*100*). Three polypeptide subunits (53, 47, and 8 kDa) were photoaffinity-labeled with an azido-[³H]AVM analog in *C. elegans* membranes (*101*). The same azido ligand labels a single polypeptide (47 kDa) from *Drosophila melanogester* head membranes, implying that a single binding protein for IVM may be present in the insect (*101*). Cloning and structure determination of avermectin site(s) may provide further understanding of the mode of action of these novel insecticides.

Phenylpyrazole Binding Site

Phenylpyrazoles are a recent class of compounds with herbicidal and insecticidal activities (*12*). The herbicidal action involves induction of porphyrin accumulation by interfering with porphyrin biosynthesis, causing rapid wilting of treated plants (*12*). The insecticidal phenylpyrazoles induce salivation, rigor, and convulsion in mammals as well as hyperexcitation and knockdown in house flies, with poisoning signs similar to that of pyrethroids (*102*) and trioxabicyclooctanes (*60,103*). Phenylpyrazole insecticides induce membrane depolarization not associated with voltage-dependent Na^+ channel or GABA-gated chloride channel in frog nerve and

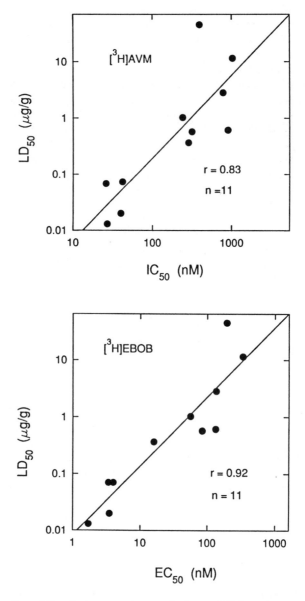

Figure 5. Correlation between potency of eleven AVM analogs as inhibitors of [3H]AVM or [3H]EBOB binding to house fly head membranes (the sensitive strain) and their toxicity to the house fly pretreated with piperonyl butoxide. (Reproduced with permission from ref. 61. Copyright 1992 Academic Press, Inc.)

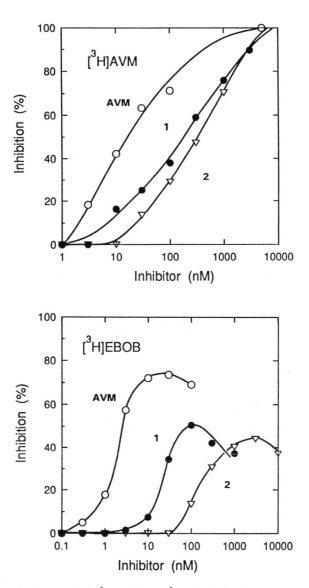

Figure 6. Displacement of [³H]AVM or [³H]EBOB binding to house fly head membranes by representative AVM analogs: AVM, avermectin B_{1a}; **1**, 22,23-dihydro-AVM B_{1a} 4"-O-phosphate Na^+ salt; **2**, 22,23-dihydro-AVM B_{1a} monosaccharide. (Reproduced with permission from ref. 61. Copyright 1992 Academic Press, Inc.)

sense organ, and acetylcholine release in frog neuromuscular junction, explaining in part the *in vivo* excitatory effect of these compounds on animals (*102,104*). In isolated locust nerve cell, one phenylpyrazole has been demonstrated to potently antagonize the GABA-activated Cl⁻ currents, suggesting that this compound may act at the insect GABA receptor (*103*). Cole *et al.* (*60*) reported that phenylpyrazole analogs inhibited [³H]EBOB binding to house fly head membranes with the insecticides exhibiting higher potency than the herbicides. The insecticides inhibit [³H]EBOB binding by reducing the B_{max} but not the K_D (Figure 7), indicating that they bind irreversibly at the EBOB site or at a site closely-coupled in the house fly head GABA receptor complex. A cyclodiene-resistant strain of house fly is also tolerant to these compounds (*60*). Phenylpyrazoles do not inhibit [³H]AVM binding to house fly head membranes. Also, selected phenylpyrazoles block GABA-stimulated chloride uptake in mouse brain microvesicles (*60*). These studies suggest that phenylpyrazole insecticides may act as a unique type at insect GABA-gated chloride channel complex.

Other Insecticide Binding Sites

Muscimol, a natural product isolated from *Amanita muscaria* as well as *Amanita pantherina*, possesses moderate insecticidal activity (*105,106*). Muscimol and its synthetic analogs are well-established GABA receptor agonists in mammals (*24,107-109*) and in insects (*44-48,50,51*). Muscimol and dihydromuscimol inhibit [³H]EBOB binding moderately in house fly head membranes (*3*), indicating that the muscimol binding site may reside closely to the EBOB site in the GABA receptor complex. The muscimol site may be coupled to the AVM site in insects as well. Avermectin B_{1a} blocks [³H]muscimol binding in honey bee heads at extremely low level (3 nM) (*48*) and modulates the [³H]GABA binding in cockroach CNS by enhancing its binding at 1 μM and inhibiting its binding at 10 μM (*110*).

A new experimental insecticide, spirosultam LY219048 or 3-(4-chlorophenyl)-4-methyl-2-thia-1-azaspiro[4,5]dec-3-ene 2,2-dioxide, may act at the mammalian $GABA_A$ receptor as it suppresses the GABA-stimulated chloride ion uptake and inhibits [³H]TBOB binding competitively in bovine brain preparations (*111,112*). The benzodiazepine site and GABA agonist site show little affinity for this compound. A dieldrin-resistant *Drosophila* strain is tolerant to LY219048 (*111,112*). It will be very interesting to see how the compound affects [³H]EBOB binding in house flies.

Conclusions and Prospects

Only recently has the insect GABAergic system been recognized as one of the most sensitive targets for a variety of structurally-diverse insecticides. [³H]EBOB binding assays have effectively differentiated four types of insecticide binding sites in the house fly head GABA receptor complex. The [³H]EBOB site is identical to or overlaps with the convulsant site characterized earlier with [³H]DHPTX and is the site of action for five classes of insecticides, *i.e.* polychlorocycloalkanes, bicycloorthobenzoates, selected dithianes, phenylsilatranes, and picrotoxins. Cross-resistance in a cyclodiene-tolerant house fly strain extends to these insecticides. Their action is characteristic with hyperexcitation signs and a positive temperature

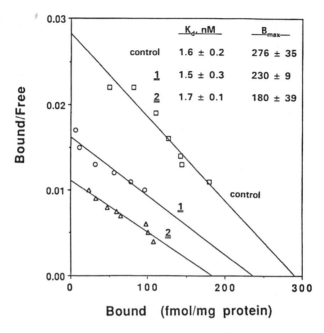

Bound (fmol/mg protein)

Figure 7. Effect of phenylpyrazoles at 2-4 nM on [³H]EBOB binding to house fly head membranes (the sensitive strain). **1**, fipronil; **2**, SLA 4454. Reproduced with permission from ref. 60. Copyright 1993 Academic Press, Inc.

coefficient, and may be classified as type A. [³⁵S]TBPS binds with low affinity to sites that may include [³H]EBOB site. [³H]EBOB and [³⁵S]TBPS assays share PCCAs and some TBOs as common inhibitors while [³H]EBOB assay generally predicts their *in vivo* potency for house fly and [³⁵S]TBPS does not. TBOs with smaller terminal substituents presumably act at the TBPS sites with different poisoning signs from type A, a negative temperature coefficient, and no cross-resistance with type A compounds, and the action is classified as type B. The action of phenylpyrazole insecticides is coupled with type A insecticides with similar poisoning signs and cross-resistance, yet these insecticides inhibit [³H]EBOB binding non-competitively or irreversibly, implying a different site for them, and therefore, their action is designated as type C. Types A-C compounds presumably act as non-competitive blockers to the GABA receptor by disrupting GABA-activated chloride conductance. The AVM site is coupled to the EBOB site since the AVM analogs are potent non-competitive inhibitors to [³H]EBOB binding. AVM sites can be assayed with either [³H]EBOB or [³H]AVM assays since both assays predict their *in vivo* activity. Their action is entirely different from the types A-C with respect to their poisoning signs and cross-resistance with cyclodienes (no resistance to AVMs for the house fly strain) and is classified as type D. Like [³⁵S]TBPS, [³H]AVM may also bind to sites not associated with the insect GABAergic system. Muscimol and its analogs are GABA receptor agonists and insecticides. Their action at the GABA binding site may be classified as type E. Types D-E are similar in that both types are chloride channel openers.

Subunit heterogeneity has been demonstrated for $GABA_A$ receptor in vertebrates. Little information, however, is available for the subunit organization of the GABA receptor in insects. Further examination of insecticide action with [³H]EBOB and new probes may improve our knowledge of the biochemistry and molecular toxicology of the insect GABAergic system. Molecular cloning of GABA receptor genes from different species and strains of insects will help us understand at molecular level the insecticide binding sites at the insect GABA-gated chloride channel and the mechanism underlying insecticide resistance. GABA receptor genes have already been cloned from susceptible and cyclodiene-resistant strains of *Drosophila* and yellow-fever mosquito, and a single point mutation (from Ala to Ser) within the second transmembrane domain in the resistant strain may confer the cyclodiene resistance (*7-9*).

With the demise of most PCCAs, the GABAergic system is for now an underutilized target for insect control. This situation soon may change if the increased understanding of the insect and mammalian GABA receptor complex leads to discovery of new insecticides with improved selective toxicity and better utilization of the current insecticides.

Acknowledgments

The author thanks Professor John E. Casida of University of California at Berkeley, Dr. Kirk Gohre and Vince R. Hebert of FMC Corporation for reviewing the manuscript and helpful suggestions, and John W. Stearns of FMC Corporation for computer assistance.

Literature Cited

1.   Matsumura, F.; Ghiasuddin, S. M. *J. Environ. Sci. Health* **1983**, *B18*, 1-14.
2.   Gant, D. B.; Eldefrawi, M. E.; Eldefrawi, A. T. *Toxicol. Applied Pharmacol.* **1987**, *88*, 313-321.
3.   Deng, Y.; Palmer, C. J.; Casida, J. E. *Pestic. Biochem. Physiol.* **1991**, *41*, 60-65
4.   Georghiou, G. G.; Mellon, R. In *Pest Resistance to Pesticides;* Georghiou, G. P. Saito, T., Eds.; Plenum: New York, NY, 1983. pp. 1-46
5.   Kadous, A. A.; Ghiasuddin, S. M.; Matsumura, F.; Scott, J. G.; Tanaka, K. *Pestic. Biocehm. Physiol.* **1983**, *19*, 157-166.
6.   Tanaka, K.; Matsumura, F. In *Membrane Receptors and Enzymes as Targets of Insecticidal Action;* Clark, J. M.; Matsumura, F., Eds., Plenum: New York, NY, 1986, pp 33-49.
7.   ffrench-Constant, R. H.; Mortlock, D. P.; Shaffer, C. D.; MacIntyre, R. J.; Roush, R. T. *Proc. Natl. Acad. Sci. USA* **1991**, *88*, 7209-7213.
8.   ffrench-Constant, R. H.; Rocheleau, T. A.; Steichen, J. C.; Chalmers, A. E. *Nature* **1993**, *363*, 449-451.
9.   Thompson, M.; Shotkoski, F.; ffrench-Constant, R. *FEBS Lett.* **1993**, *325* *(3)*, 187-190.

10. Palmer, C. J.; Casida, J. E. *J. Agric. Food Chem.* **1985**, *33*, 976-980.
11. Palmer, C. J.; Casida, J. E. *J. Agric. Food Chem.* **1989**, *37*, 213-216.
12. Yanase, D.; Andoh, A. *Pestic. Biochem. Physiol.* **1989**, *35*, 70-80
13. Ostlind, D. A.; Cifelli, S.; Lang, R. *Vet. Rec.* **1979**, *105*, 168.
14. Lasota, J. A.; Dybas, R. A. *Acta Leidensia* **1990**, *59*, 217-225.
15. Lasota, J. A.; Dybas, R. A. *Annu. Rev. Entomol.* **1991**, *36*, 91-117.
16. Deng, Y.; Palmer, C. J.; Casida, J. E. *Pestic. Biochem. Physiol.* **1993**, *47*, 98-112.
17. Cole, L. M.; Sanders, M.; Palmer, C. J.; Casida, J. E. *J. Agric. Food Chem.* **1991**, *39*, 560-565.
18. Stephenson, F. A. *Biochem. J.* **1988**, *249*, 21-32.
19. Olsen, R. W.; Tobin, A. J. *FASEB J.* **1990**, *4*, 1469-1480.
20. Schofield, P. R.; Darlison, M. G.; Fujita, N.; Burt, D. R.; Stephenson, F. A.; Rodriguez, H.; Rhee, L. M.; Ramachandran, J.; Reale, V.; Glencorse, T. A.; Seeburg, P. H.; Barnard, E. A. *Nature* **1987**, *328*, 221-227.
21. Ogata, N. *Gen. Pharmac.* **1990**, *21*, 395-402.
22. Bloom, F. E.; Iversen, L. L. *Nature* **1971**, *229*, 628-630.
23. Enna, S. J.; Snyder, S. H. *Brain Res.* **1975**, *100*, 81-97.
24. Enna, S. J.; Snyder, S. H. *Mol. Pharmacol.* **1977**, *13*, 442-453.
25. *Benzodiazepine/GABA Receptors and Chloride Channels: Structural and Functional Properties;* Olsen, R. W.; Venter, J. C., Eds., Alan R. Liss: New York, NY, 1986.
26. Ticku, M. K.; Ban, M.; Olsen, R. W. *Mol. Pharmacol.* **1978**, *14*, 391-402.
27. Squires, R. F.; Casida, J. E.; Richardson, M.; Saederup, E. *Mol. Pharmacol.* **1983**, *23*, 326-336.
28. Ticku, M. K.; Maksay, G. *Life Sci.* **1983**, *33*, 2363-2375.
29. Ramanjaneyulu, R.; Ticku, M. K. *Neurosci. Abs.* **1983**, *9*, 403.
30. Majewska, M. D.; Harrison, N. L.; Schwartz, R. D.; Barker, J. L.; Paul, S. M. *Science* **1986**, *232*, 1004-1007.
31. Harrison, N. L.; Majewska, M. D.; Harrington, J. W.; Barker, J. L. *J. Pharmacol. Exp. Ther.* **1987**, *241*, 346-353.
32. Siegal, E.; Stephenson, F. A.; Mamalaki, C.; Barnard, E. A. *J. Biol. Chem.* **1983**, *258*, 6965-6971.
33. Casalotti, S. O.; Stephenson, F. A.; Barnard, E. A. *J. Biol. Chem.* **1986**, *261*, 15013-15016.
34. Mamalaki, C.; Stephenson, F. A.; Barnard, E. A. *EMBO J.* **1987**, *6*, 561-565.
35. Seeburg, P. H.; Wisden, W.; Verdoorn, T. A.; Pritchett, D. B.; Werner, P.; Herb, A.; Luddens, H.; Sprengel, R.; Sakmann, B. *Cold Spring Harbor Symposium on Quantitative Biology* **1990**, *LV*, 29-40.
36. Cutting, G. R.; Lu, L.; O'Hara, B. F.; Kasch, L. M.; Montrose-Rafizadeh, C.; Donovan, D. M.; Shimada, S.; Antonarakis, S. E.; Guggino, W. B.; Uhl, G. R.; Kazazian, Jr. H. H. *Proc. Natl. Acad. Sci. USA* **1991**, *88*, 2673-2677.
37. Pritchett, D. B.; Sontheimer, H.; Gorman, C. M.; Kettenmann, H.; Seeburg, P. H.; Schofield, P. R. *Science* **1988**, *242*, 1306-1308.

38.   Burt, D. R.; Kamatchi, G. L. *FASEB J.* **1991**, *5*, 2916-2923.
39.   Dudel, J.; Gryder, R.; Kaji, A.; Kuffler, S. W.; Potter, D. D. *J. Neurophysiol.* **1963**, *26*, 721-728.
40.   Takeuchi, A.; Takeuchi, N. *J. Physiol.* **1965**, *177*, 225-238.
41.   Takeuchi, A.; Takeuchi, N. *J. Physiol.* **1969**, *205*, 377-391.
42.   Roberts, E. In *Benzodiazepine/GABA Receptors and Chloride Channel: Structural and Functional Properties;* Olsen, R. W.; Venter, J. C., Eds.; Alan R. Liss: New York, NY, 1986, pp 1-40.
43.   Casida, J. E.; Nicholson, R. A.; Palmer, C. J. In *Neurotox'88: Molecular Basis of Drug and Pesticide Action*; Lunt, G. G., Ed.; Elsevier: Amsterdam, Netherlands, 1988, pp 125-144.
44.   Lummis, S. C. R.; Sattelle, D. B. *Neurochem. Int.* **1986**, *9*, 287-293.
45.   Sattelle, D. B.; Lummis, S. C. R.; Wong, J. F. H.; Rauh, J. J. *Neurochem. Res.* **1991**, *16*, 363-374
46.   Rutherford, D. M.; Jeffery, D.; Lunt, G. G.; Weitzman, P. D. *J. Neurosci. Lett.* **1987**, *79*, 337-340
47.   Lunt, G. G.; Robinson, T. N.; Miller, T.; Knowles, W. P.; Olsen, R. W. *Neurochem. Int.* **1985**, *7*, 751-754.
48.   Abalis, I. M.; Eldefrawi, A. T. *Pestic. Biochem. Physiol.* **1986**, *25*, 279-287.
49.   Lummis, S. C. R. *Comp. Biochem. Physiol.* **1990**, *95C*, 1-8.
50.   Breer, H.; Heilgenberg, H. *J. Comp. Physiol A.* **1985**, *157*, 343-354.
51.   Rosario, P.; Barat, A.; Ramirez, G. *Neurochem. Int.* **1989**, *15*, 115-120.
52.   Robinson, T.; MacAllan, D.; Lunt, G.; Battersby, M. *J. Neurochem.* **1986**, *47*, 1955-1962.
53.   Tanaka, K.; Scott, J. G.; Matsumura, F. *Pestic. Biochem. Physiol.* **1984**, *22*, 117-127.
54.   Cohen, E.; Casida, J. E. *Pestic. Biochem. Physiol.* **1986**, *25*, 63-72.
55.   Olsen, R. W.; Szamraj, O.; Miller, T. *J. Neurochem.* **1989**, *52*, 1311-1318.
56.   Ozoe, Y.; Takayama, T.; Sawada, Y.; Mochida, K.; Nakamura, T.; Matsumura, F. *J. Agric. Food Chem.* **1993**, *41*, 2135-2141.
57.   Anthony, N. M.; Benner, E. A.; Rauh, J. J.; Sattelle, D. B. *Pestic. Sci.* **1991**, *33*, 223-230.
58.   Brown, M. C. S.; Lunt, G. G.; Stapleton, A. *Comp. Biochem. Physiol.* **1989**, *92C*, 9-13.
59.   Cole, L. M.; Casida, J. E. *Pestic. Biochem. Physiol.* **1992**, *44*, 1-8.
60.   Cole, L. M.; Nicholson, R. A.; Casida, J. E. *Pestic. Biochem. Physiol.* **1993**, *46*, 47-54.
61.   Deng, Y.; Casida, J. E. *Pestic. Biochem. Physiol.* **1992**, *43*, 116-122.
62.   Palmer, C. J.; Casida, J. E. *J. Labelled Compd and Radiopharm.* **1991**, *29*, 829-839.
63.   Lawrence, L. J.; Palmer, C. J.; Gee, K. W.; Wang, X.; Yamamura, H. I.; Casida, J. E. *J. Neurochem.* **1985**, *45*, 798-804.
64.   Deng, Y.; Palmer, C. J.; Casida, J. E. *Abstracts in Seventh International Congress of Pesticide Chemistry at Hamburg, Germany*; **1990**, Abstract 04C-07.
65.   Elliott, M.; Pulman, D. A.; Casida, J. E. *Abstracts in Seventh International Congress of Pesticide Chemistry, Hamburg, Germany*, **1990**, Abstract 01A-17.

66.   Palmer, C. J.; Casida, J. E. *J. Agric. Food Chem.* **1992**, *40*, 492-496.
67.   Wacher, V. J.; Toia, R. F.; Casida, J. E. *J. Agric. Food Chem.* **1992**, *40*, 497-505.
68.   Horsham, M. A.; Palmer, C. J.; Cole, L. M.; Casida, J. E. *J. Agric. Food Chem.* **1990**, *38*, 1734-1738.
69.   Gynther, B. D.; Curtis, D. R. *Neurosci. Lett.* **1986**, *68*, 211-215.
70.   Ozoe, Y.; Mochida, K.; Nakamura, T.; Yoyama, A.; Matsumura, F. *Comp. Biochem. Physiol.* **1987**, *87C*, 187-191.
71.   Bonner, J. C.; Yarbrough, J. D. *Pestic. Biochem. Physiol.* **1987**, *29*, 260-265.
72.   Casida, J. E.; Cole, L. M.; Hawkinson, J. E.; Palmer, C. J. In *Recent Advances in the Chemistry of Insect Control II;* Crombie, L., Ed.; Spec. Publ. 79; Royal Society of Chemistry: Cambridge, England, 1990, pp 212-234.
73.   Deng, Y. Ph.D. Theses, Univ. California, Berkeley, CA; **1991**.
74.   Casida, J. E.; Palmer, C. J.; Cole, L. M. *Mol. Pharmacol.* **1985**, *28*, 246-253.
75.   Palmer, C. J.; Casida, J. E. *Toxicol. Lett.* **1988**, *42*, 117-122.
76.   Cole, L. M.; Lawrence, L. J; Casida, J. E. *Life Sci.* **1984**, *35*, 1755-1762.
77.   Abalis, I. M.; Eldefrawi, M. E.; Eldefrawi, A. T. *Biochem. Pharmacol.* **1985**, *34*, 2579-2582.
78.   Miller, T. W.; Chaiet, L.; Cole, D. J.; Cole, L. J.; Flor, J. E.; Goegelman, R. T.; Gullo, V. P.; Joshua, H.; Kempf, A. J.; Krellwitz, W. R.; Monaghan, R. L.; Ormond, R. E.; Wilson, K. E.; Albers-Schonberg, G.; Putter, I. *Antimicrob. Agents Chemother.* **1979**, *15*, 368-371.
79.   Burg, R. W.; Miller, B. M.; Baker, E. E.; Birnbaum, J.; Currie, S. A.; Hartman, R.; Kong, Y.; Monaghan, R. L.; Olsen, G.; Putter, I.; Tunac, J. B.; Wallick, H.; Stapley, E. O.; Oiwa, R.; Omura, S.; *Antimicrob. Agents Chemother.* **1979**, *15*, 361-367.
80.   Egerton, J. R.; Ostlind, D. A.; Blair, L. S.; Eary, C. H.; Suhayda, D.; Cifelli, S.; Riek, R. F.; Campbell, W. G. *Antimicrob. Agents Chemother.* **1979**, *15*, 372-378.
81.   Fritz, L. C.; Wang, C. C.; Gorio, A. *Proc. Natl. Acad. Sci. USA* **1979**, *76*, 2062-2066.
82.   Mellin, T. N.; Busch, R. D.; Wang, C. C. *Neuropharmacol.* **1983**, *22*, 89-96.
83.   Kass, I. S.; Wong, C. C.; Walrond, J. P.; Stretton, A. O. W. *Proc. Natl. Acad. Sci. USA* **1980**, *77*, 6211-6215.
84.   Robertson, B. *Br. J. Pharmacol.* **1989**, *98*, 167-176.
85.   Soderlund, D. M.; Adams, P. M.; Bloomquist, J. R. *Biochem. Biophys. Res. Comm.* **1987**, *146*, 692-698
86.   Payne, G. T.; Soderlund, D. M. *Pestic. Biochem. Physiol.* **1993**, *47*, 178-184.
87.   Duce, I. R.; Scott, R. H. *Br. J. Pharmacol.* **1985**, *85*, 395-401.
88.   Martin, R. J.; Pennington, A. J. *Br. J. Pharmacol.* **1989**, *98*, 747-756.
89.   Holden-Dye, L.; Walker, R. J. *Parasitology* **1990**, *101*, 265-271.
90.   Schaeffer, J. M.; Haines, H. W. *Biochem. Pharmacol.* **1989**, *38*, 2329-2338.

91. Schaeffer, J. M.; Frazier, E. G.; Bergstrom, A. R.; Williamson, J. M.; Liesch, J. M.; Goetz, M. A. *J. Antibiotics* **1990,** *XLIII,* 1179-1182.
92. Pong, S. S.; Wang, C. C. *J. Neurochem.* **1982,** *38,* 375-379.
93. Olsen, R. W.; Snowman, A. M. *J. Neurochem.* **1985,** *44,* 1074-1082.
94. Drexler, G.; Sieghart, W. *Eur. J. Pharmacol.* **1984,** *101,* 201-207.
95. Bermudez, I.; Hawkins, C. A.; Tayler, A. M.; Beadle, D.J. *J. Receptor Res.* **1991,** *11,* 221-232.
96. Sani, B. P.; Vaid, A. *Biochem. J.* **1988,** *249,* 929-932.
97. Asato, G.; France, D. J. European Patent Application EP 259,779 16 March 1988, *Chem. Abstr* 109, 190134m, **1988.**
98. Konno, Y.; Scott, J. G. *Pestic. Biochem. Physiol.* **1991,** *41,* 21-28.
99. Scott, R. H.; Duce, I. R. *Pestic. Sci.* **1985,** *16,* 599-604.
100. Cully, D. F.; Paress, P. S. *Mol. Pharmacol.* **1991,** *40,* 326-332.
101. Rohrer, S. P.; Meinke, P. T.; Hayes, E. C.; Mrozik, H.; Schaeffer, J. M. *Proc. Natl. Acad. Sci. USA* **1992,** *89,* 4168-4172.
102. Klis, S. F. L.; Vijverberg, H. P. M.; van den Bercken, J. *Pestic. Biochem. Physiol.* **1991a,** *39,* 210-218.
103. von Keyserlingk, H. C.; Willis, R. J. In *Neurotox'91: Molecular Basis of Drug and Pesticide Action;* Duce, I. R., Ed.; Elsevier: New York, NY, 1992, pp 79-104.
104. Klis, S. F. L.; Nijman, N. J., Vijverberg, H. P. M.; van den Bercken, J. *Pestic. Sci.* **1991b,** *33,* 213-222.
105. Onda, M.; Fukushima, H.; Akagawa, M.; *Chem. Pharm. Bull.* **1964,** *12,* 751.
106. Eugster, C. H.; Muller, G. F. R.; Good, R. *Tetrahedron Lett.* **1965,** *23,* 1813-1815.
107. Olsen, R. W. *J. Neurochem.* **1981,** *37,* 1-13.
108. Krogsgaard-Larsen, P.; Larsen, A. L. N.; Thyssen, K. *Acta Chem. Scand. B* **1978,** *32,* 469-477.
109. Krogsgaard-Larsen, P.; Nielsen, L.; Falch, E.; Curtis, D. R. *J. Med. Chem.* **1985,** *28,* 1612-1617.
110. Lummis, S. C. R.; Sattelle, D. B. *Neurosci. Lett.* **1985,** *60,* 13-18.
111. Gajewski, R. P.; Jackson, J. L.; Karr, L. L.; Bloomquist, J. R. In *Abstracts Neurotox'91: An International Symposium on the Molecular Basis of Drug and Pesticide Action;* University of Southampton, 1991, pp 61-62.
112. Bloomquist, J. R.; Jackson, J. L.; Karr, L. L.; Ferguson, H. J.; Gajewski, R. P. *Pestic. Sci.* **1993,** *39,* 185-192.

RECEIVED October 15, 1994

Chapter 16

Effects of Ivermectin on γ-Aminobutyric Acid and Glutamate-Gated Chloride Conductance in Arthropod Skeletal Muscle

Ian R. Duce, Narotam S. Bhandal, Roderick H. Scott[1], and Timothy M. Norris[2]

Department of Life Science, University of Nottingham, University Park, Nottingham NG7 2RD, United Kingdom

Ivermectin at low concentrations produced reversible and irreversible increases in Cl⁻ conductance in locust skeletal muscle, but had little effect on tick muscle. Two electrode current clamp was used to monitor input conductance in both GABA-sensitive and GABA-insensitive muscle fibres in locust extensor tibiae muscle. In GABA-sensitive muscle fibres IVM (100pM-1nM) reversibly increased Cl⁻ conductance and antagonised GABA responses, however higher IVM concentrations (10nM-1μM) induced irreversible increases in Cl⁻ conductance which continued to increase during washing. GABA-insensitive muscle fibres produced only irreversible responses to IVM. IVM also inhibited Cl⁻ conductance gated by ibotenate-sensitive glutamate receptors in locust muscle fibres. Tick muscle fibres showed little sensitivity to IVM and were insensitive to ibotenate. We conclude that a major component of the IVM response in arthropod muscle is mediated via ibotenate-sensitive Cl⁻ channels.

Arthropod pests have an enormous impact on mankind. As vectors of viral, bacterial, protozoal and nematode pathogens, blood sucking insects and arachnids are a major threat to human and animal health. Damage by insects and mites to crops and stored products has massive effects on economic welfare and food avilability for humans and domestic stock. These pressures have inevitably lead to strenuous attempts to control arthropod pests.

Effective control of both free-living and ectoparasitic arthropods is heavily dependent on chemical control, the scale of which can be seen from sales figures which are projected to reach 7 billion $US by 1995 (1). Although the existing chemical agents in use are valuable both in terms of their ability to boost

[1]Current address: Department of Physiology, St. George's Hospital Medical School, Cranmer Terrace, London SW17 0RE, United Kingdom
[2]Current address: Department of Entomology, University of California, Riverside, CA 92521

agricultural output and to improve public and animal health, a heavy reliance is placed on two major chemical groups: the pyrethroids and the organophosphates. These compounds are neurotoxins affecting respectively: neuronal voltage-sensitive sodium channels, and acetylcholinesterase. However the heavy reliance on actions at these two target sites and the intensive application of these molecules   has resulted in a significant problem of resistance in many parts of the world. Historically older commercial pesticides such as DDT and the chlorcycloalkanes, although effective in terms of arthropod control, achieved notoriety as a result of the persistence and toxicity of residues in the natural environment. Thus the search for safer, effective pesticides preferably working at different target sites is driven on by these combined needs.

Avermectins. It was in this context that the announcement of a new class of pesticides, the Avermectins derived from a microorganism *Streptomyces avermitilis*(2), generated considerable interest. These compounds were reported to be very active nematicides(3) insecticides(4,5) and acaricides (5) and it was apparent from early work that the avermectins did not share the same mode of action as existing pesticides(6). The site of action was generally ascribed to 4-aminobutyric acid (GABA) receptors, however Turner and Schaeffer(6) in reviewing this topic point to the difficulties in interpreting the many studies carried out in widely different systems using a baffling array of experimental methods. In this chapter we shall review our attempts, using electrophysiological techniques, to determine the mechanism(s) of action of 22,23 dihydroavermectin B_{1a} (Ivermectin IVM) on the skeletal muscle of two arthropod species: an insect, the locust *Schistocerca gregaria*; and an acarine ectoparasite, the cattle tick *Amblyomma hebraeum*.

Insect Skeletal Muscle

The Locust Extensor Tibiae (ET) Muscle The ET muscle in the metathoracic (jumping) leg of grasshoppers and locusts consists of many bundles of muscle fibres.
    The complex postural movements and powerful jumping response mediated by this muscle are under the control of "fast excitatory", "slow excitatory" and "inhibitory" motorneurons situated in the CNS of the insect. The pattern of innervation and nomenclature of these muscle bundles was described by Hoyle(7), who showed that the smaller proximal and distal muscle bundles received both slow excitatory and inhibitory innervation whereas the large medial bundles were only supplied by fast excitatory axons (Figure 1). The excitatory and inhibitory transmitters are respectively L-glutamate and GABA(8).

Pharmacology of (ET) Muscle Bundles The actions of drugs and neurotransmitters on the mechanical and electrophysiological responses of locust ET have been extensively studied over the last 25 years (8,9). It is now generally agreed that the principal excitatory and inhibitory neurotransmitters are L-glutamate and GABA respectively. Ionophoresis of L-glutamate demonstrated that receptors occurred not only at neuromuscular junctions but were distributed across the extrajunctional sarcolemma (10); furthermore the extrajunctional receptors were of two types: Depolarising glutamate receptors which were sensitive to the glutamate analogue quisqualate and gated cationic channels (qGluR) and hyperpolarising glutamate receptors which were sensitive to ibotenate and gated chloride channels (iGluR) (11). As one would predict the excitatory junctional receptors responded to L-glutamate by gating cationic channels resulting in depolarisation.

Inhibitory post-synaptic potentials and ionophoretic GABA responses in ET muscle bundles were both found to activate chloride (Cl⁻) channels with the same reversal potential (12) and responses were antagonised by picrotoxin providing strong evidence that GABA was the inhibitory neuromuscular transmitter.

ET muscles have also been shown to be innnervated by neurons releasing octopamine (13) and proctolin (14,15) which modulate the contractility of ET muscle fibres, probably via second messenger systems.

Two-Electrode Current Clamp When we embarked on our studies of the actions of IVM on insect muscle we were aware of the problematic properties of this pesticide for electrophysiology, it is highly lipophilic, relatively insoluble in water, active at low concentrations and its actions were reported to involve irreversible changes in Cl⁻ permeability. The method we employed was to measure the input conductance of muscle cells using a two-electrode current clamp whilst delivering IVM and other drugs via bath perfusion or local microperfusion from a micropipette. Constant current pulses were injected into the muscle through a microelectrode whilst monitoring the membrane potential via a second microelectrode. Alterations in input conductance resulting frrom the opening or closing of ion-channels in the cell membrane result in changes in the size of the electrotonic potential resulting from the injected current pulses (Figure 2). The recording system is very stable permitting experiments to be carried out over several hours without change in the resting input conductance. The bathing physiological saline contained 2% dimethyl sulfoxide (DMSO) to keep IVM in solution. This vehicle is frequently used in studies on IVM and other lipophilic pesticides and its effects on cell physiology are often discounted, however in our studies we found that DMSO altered potassium permeability (16) and in all experiments muscles were equilibrated with saline containing 2% DMSO which was present in all subsequent bathing solutions and drug applications.

GABA-Sensitive Distal Muscle Bundles The distal muscle bundles 32-34 (Figure 1) are innervated by the inhibitory motor nerve(7). Application of GABA to these muscle fibres resulted in dose-dependent increases in input condcutance (17, 18) (Figure 2) which were abolished in Cl⁻-free saline. The pharmacology of the GABA response was characterised using a range of agonists and antagonists as well as modulators of GABA-gated Cl⁻ conductance such as pentobarbitone (19). The pharmacology of these locust ET muscle GABA receptors resembled that of the vertebrate GABA_A receptor in several respects, however locust ET muscle GABA receptors were insensitive to bicuculline.

The actions of IVM on these muscle fibres were complex (20). At low concentrations (100pM - 10nM) microperusion of IVM induced dose dependent increases in Cl⁻ conductance which were fully reversible on washing (Figure 3). Application of IVM (100pM - 10nM) during a GABA activated increase in conductance resulted in a decrease in the GABA response (Table I). The GABA antagonism was reversible and non-competitive (Table I).

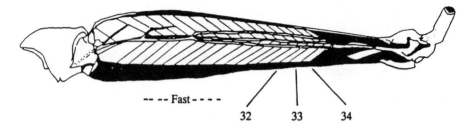

Figure 1. Diagram of the femur of a locust Schistocerca gregaria metathoracic ("jumping") leg showing the position of the GABA-sensitive distal muscle bundles 31-33 and the GABA-insensitive "Fast" muscle bundles of the extensor tibiae muscle.

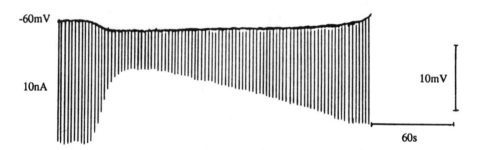

Figure 2. Current-clamp recording showing the change in input conductance in muscle bundle 33 due to perfusion of 1mM GABA. Downward deflections represent hyperpolarising electrotonic potentials resulting from the injection of 10nA pulses of current. GABA produces a hyperpolarisation of the membrane potential and an increase in input conductance.

Table I. Microperfusion of IVM Induces Reversible Conductance Increase and Inhibition of GABA-induced Conductance In Muscle Bundle 33 of Locust ET Muscle

Dose IVM	Conductance Increase (nS) ± S.E.M. of (n) observations	% Inhibition of Conductance produced by 1mM GABA ± S.E.M.(n)
75pM	16 ± 6 (5)	17.3 ± 5.1 (3)
100pM	50 ± 7 (6)	63.5 ± 3.3 (7)
1nM	105 ± 9 (5)	68.3 ± 1.5 (5)
7.5nM	258 ± 35 (6)	82.0 ± 3.7 (5)

When higher concentrations of IVM (10nM - 1μM) were perfused over muscle bundles 31-34 an increase in input conductance was produced which was both irreversible on washing with saline containing DMSO and continued to increase for 60 mins after IVM was washed from the bath(Figure 4; Table II). During these irreversible IVM-induced conductances, application of GABA produced a potentiated response (Table II).

Table II. Microperfusion of IVM Induces Irreversible Conductance Increases Which Continue for 60 Mins After Washing, and Potentiates GABA-induced Conductance in Locust ET Muscle Bundle 33

Dose IVM (μM)	Irreversible Increase in Conductance (μS) ± S.E.M. (n)		% Increase in GABA Response Compared with 1mM GABA ± S.E.M (n)
	3-5 Minutes After IVM	60 Minutes After IVM	
10 nM	0.63 ± 0.14	1.79 ± 0.45 (6)	60.8 ± 8.1 (25)
100nM	3.33 ± 0.69	5.62 ± 0.97 (7)	170.4 ± 24 (13)
500nM	4.2 ± 0.58	8.25 ± 0.65 (3)	Not Tested
1μM	9.11 ± 0.78	36.0 ± 5.9 (8)	334.1 ± 79.7 (16)

However the picture was further complicated by the following experiment: IVM (100nM) was applied and induced an irreversible conductance increase; it was then washed from the bath and GABA (1mM) was perfused over the muscle and as expected the GABA response was potentiated. If IVM (100nM) was now reapplied

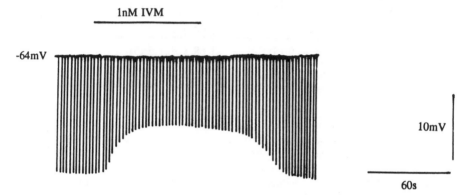

Figure 3. *Reversible increase in input conductance in muscle bundle 33 produced by microperfusion of 1nM IVM.*

Figure 4. *Irreversible increase in input conductance in muscle bundle 33 produced by microperfusion of 10nM IVM . The conductance continues to increase after IVM application ends.*

during the GABA perfusion, the potentiated GABA response was antagonised. We interpret these complex responses to IVM on GABA sensitive muscle fibres in locust ET as being due to effects on both GABA-gated Cl^- channels and other Cl^- channels which are not associated with GABA receptors. This hypothesis could be readily tested using the ET muscle where the majority of the fast muscle bundles (Figure 1) do not respond to GABA.

GABA-Insensitive Fast Muscle Bundles Microperfusion of IVM (1nM-1μM) onto fibres which were insensitive to GABA evoked an irreversible increase in Cl^- permeability which continued to develop for up to 60 mins following washing of the pesticide from the bath (20) (Table III).

Table III. Microperfusion of IVM onto GABA-insensitive Muscle Bundles of Locust ET Muscle Induces Irreversible Conductance Increases Which Continue for 60 Minutes Following Washing

Dose IVM	Irreversible Increase in Conductance (μS) $\pm$ S.E.M. (n)	
	3-5 Minutes After IVM	60 Minutes After IVM
1nM	0.18 $\pm$ 0.05	0.65 $\pm$ 0.15 (5)
7.5nM	0.47 $\pm$ 0.13	1.31 $\pm$ 0.04 (3)
10nM	1.36 $\pm$ 0.33	4.74 $\pm$ 0.99 (5)
100nM	2.81 $\pm$ 0.34	7.66 $\pm$ 0.25 (3)
1μM	11.1 $\pm$ 1.49	47.4 $\pm$ 8.44 (3)

If these responses are not due to the action of IVM on GABA receptor Cl^- channels the obvious question is what type of Cl^- might they be? In *Ascaris* muscle patch-clamp studies (21) showed that IVM depressed GABA-activated channel currents, but also opened non-GABA-activated Cl^- channels with long average open times. In locust ET muscle ionophoresis (11) and patch-clamping (22) revealed that iGluR are widely distributed across the sarcolemma (Figure 5). They gate Cl^- currents and can be selectively activated by the glutamate analogue [2-(3-hydroxyisoxazol-5yl)glycine] (ibotenic acid) (23,11). Using two-electrode current-clamp we found that perfusion of ibotenic acid on to ET muscle bundles (Figure 6) produced a dose-dependent increase in Cl^- conductance over the range 1μM - 1mM (24). Responses to ibotenic acid (100μM) were reduced by IVM (500pM-100nM) (Table IV).

These results demonstrate that IVM does interact at low concentrations with iGluR, however to gain a more precise understanding of the interaction of IVM with iGluR it will be necessary to use a method with a higher resolution. Recent developments in methodology have now enabled recordings of membrane currents resulting from the activation of small populations of iGluR using a liquid filament switch to apply brief pulses of ibotenate to an outside-out patch of membrane excised from ET muscles in a patch-clamp micropipette (Figure 7). This method

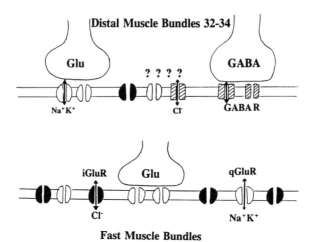

Figure 5. Cartoon of the putative receptors for neurotransmitter amino acids on ET muscle fibres. The presence of extrajunctional qGluR and GABA receptors on distal muscle fibres is likely but has not been shown experimentally.

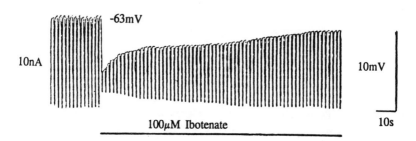

Figure 6. Perfusion of 100μM ibotenic acid induces an increase in input conductance in a GABA-insensitive fast muscle bundle. The response desensitises during ibotenate application.

should allow us to gain a fuller understanding of the actions of IVM on locust muscle.

Table IV. IVM Inhibits Increases in Input Conductance Produced by Perfusion of Ibotenic Acid (100μM) in Non-GABA Sensitive Muscle Bundles of Locust ET Muscle

Dose IVM (μM)	Mean % Reduction in Response to Ibotenic Acid (100μM) Caused by Microperfusion of IVM $\pm$ SEM (n)
500pM	51 $\pm$ 7.8 (4)
1nM	59 $\pm$ 4.1 (13)
10nM	100 (3)

Tick Skeletal Muscles

Treatment of domestic stock using IVM has been shown to control a range of ectoparsites including cattle ticks of the genus *Boophilus* (24). Injection of IVM into the cattle tick *Amblyomma hebraeum* caused paralysis which affected movement and feeding (25). Harrow *et al* (26) found that IVM at concentrations in excess of 1μM produced an increase in muscle membrane Cl⁻ conductance and inhibited spontaneous activity recorded from the synganglion (CNS) of *Amblyomma hebraeum*. Our findings on the actions of IVM on insect muscle Cl⁻ channels encouraged us to extend our studies to examine the skeletal muscles of the cattle tick *Amblyomma hebraeum*.

The size and anatomy of cattle ticks restricts the range of skeletal muscles which are accessible to electrophysiological studies. For the work described here an incision was made around the dorsal/lateral edge of the cuticle (shield). The shield was folded anteriorly and the viscera were dissected away to reveal four sets of coxal muscles. The preparation was continuously perfused with tick saline (mM: NaCl 203, KCl 6, CaCl$_2$ 10, HEPES 10, MgCl$_2$ 1; pH 6.9) and drugs were applied via the bath perfusion or microperfused onto the target muscle. Coxal muscle fibres were impaled with two microelectrodes and two electrode current-clamp was used to investigate changes in input conductance produced by drug perfusion.

Pharmacology of tick coxal muscle Coxal muscle fibres responded to GABA (1μM - 1mM) with dose-dependent increases in Cl⁻ permeability (Figure 8). Muscimol was a more potent agonist with an EC$_{50}$ approximately 10 fold lower than that for GABA. GABA responses were antagonised by picrotoxin but not by bicuculline and were potentiated in the presence of pentobarbitone and the benzodiazepine flurazepam (27). This pharmacological profile closely resembles that described above for locust skeletal muscle GABA receptors.

Application of L-glutamate resulted in depolarisation accompanied by a dose dependent increase in input conductance (Figure 9) which is consistent with glutamate induced increases in cation permeability described in other arthropod

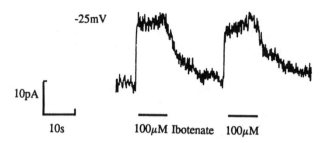

Figure 7. Pulses of ibotenate applied using the liquid filament technique activate transmembrane Cl⁻ currents in outside-out patches of membrane excised from locust fast muscle fibres.

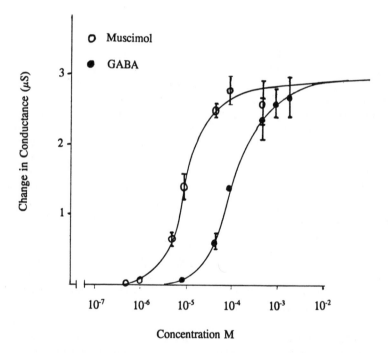

Figure 8. Dose-response curves for GABA and Muscimol application to tick coxal muscle fibres. Each point is the mean change in conduct-ance ± S.E.M (n=3).

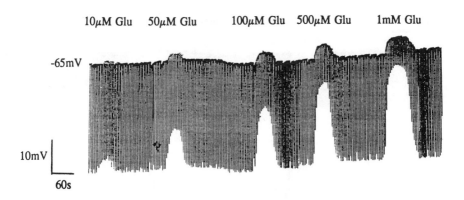

Figure 9. L-glutamate induces dose-dependent depolarisation and increased input conductance in tick coxal muscle fibres.

skeletal muscles (28). Glutamate has also been shown to produce muscle contractions in ticks (29).

The actions of GABA and Glutamate were closely parallel to those found in locust muscle, however in 85% of tick muscles tested perfusion of IVM (1nM - 50μM) had no effect on either the input conductance or membrane potential. Very high doses of IVM (>100μM) produced irreversible increases in conductance. 15% of fibres tested produced small reversible increases in input conductance at concentrations above 50 nM. These results were surprising considering the high sensitivity of insect muscle and the knowledge that IVM is a potent paralytic acaricide.

In view of the evidence presented above that IVM may interact with ibotenate-sensitive glutamate-gated Cl⁻ channels in insect muscle and the recent findings of Arena *et al* and Rohrer *et al* (this volume) that the high affinity binding site for IVM in nematodes and insect CNS is also an ibotenate sensitive glutamate-gated Cl⁻ channel, we decided to test whether tick muscle was sensitive to ibotenate. Figure 10 shows that perfusion of GABA 100μM over tick muscle produced a clear increase in input conductance, whereas the same muscle fibre is insensitive to both ibotenate and IVM. This strongly suggests that the sensitivity of arthropod skeletal muscles is associated significantly with iGluR channels.

The acaricidal actions of IVM are probably mediated through sites in the CNS. We obtained electropphysiological data confirming the findings of Harrow *et al* (26) that IVM potently inhibits spontaneous electrical activity in the CNS of *Amblyomma hebraeum*.

Conclusion

Our results show that IVM has several sites of action and more than one pharmacological effect in arthropod muscles. The actions on insect muscle appear to involve Cl⁻ channels gated by both GABA and iGluR. It is likely that the widespread distribution of iGluR, estimated from patch-clamp studies to be 100 per μm² (22), is responsibility for the high sensitivity of locust muscle to IVM. The low sensitivity of tick muscle to both ibotenate and IVM presumably implies a paucity

100μM GABA   100μM Ibotenate   100nM IVM

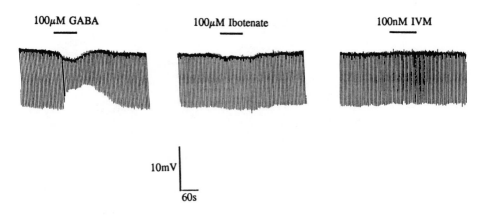

10mV

60s

Figure 10. Perfusion of 100μM GABA produces hyperpolarisation and a large increase in input conductance in a tick coxal muscle fibre. However the same muscle fibre is insensitive to 100μM ibotenate and 100nM IVM.

of glutamate-gated Cl⁻ channels in tick muscles, however without single channel studies this will remain speculative.

IVM was found to produce conductance increases in tick muscle at very high concentrations, however this result may a non-specific effect of IVM as at μM concentrations it has been shown to produce unitary currents in planar lipid bilayers which are cation specific (30). Although a picture is emerging from studies on nematodes and insects of IVM opening Cl⁻ channels associated with amino acid receptors, the precise action of IVM on these molecules is unknown and awaits detailed molecular and electrophysiological studies.

Acknowledgments

IRD thanks the SERC for grant support. NSB thanks Pfizer Central Research (UK) for support through the SERC CASE studentship scheme.

Literature Cited

(1) Voss,G.; Neumann,R. In *Neurotox '91: Molecular Basis of Drug and Pesticide Action*; Duce,I.R. Ed.; Elsevier Applied Science: London and New York, 1992; pp vii-xx

(2) Burg,R.W.; Miller,B.M.; Baker,E.E.; Birnbaum,J.; Currie,J.A.; Hartman,R.; Kong,Y-L; Monaghan,R.L.; Olson,G.; Putter,I.; Tunac,J.P.; Wallick,H.; Stapley,E.O.; Oiwa,R.; Omura,S.; *Antimicrob. Agents Chemother.* **1979**, *15*, 361-367.

(3) Egerton,J.R.; Ostlind,D.A.; Blair,L.S.; Eary,C.H.; Suhayda,D.; Cifelli, S.; Riek,R.F.; Campbell,W.C.; *Antimicrob.Agents Chemother.* **1979**, *15*, 372-378.

(4) Ostlind,D.A.; Cifelli,S.; Lang,R.; *Vet. Record* **1979**, *105*, 168.

(5)   Putter,I.;   MacConnell,J.G.;   Preiser,F.A.;   Haidri,A.A.;   Ristich,S.S.; Dybas,R.A.; *Experientia* **1981**, *37*, 963-964.

(6) Turner,M.J.; Schaeffer, J.M. In *Ivermectin and Abamectin;* Campbell,W.C. Ed.; Springer-Verlag: New York, 1989; pp 73-88

(7) Hoyle,G.; *J. Exp. Biol.* **1978**, *73*, 205-233.

(8) Usherwood,P.N.R.; Cull-Candy,S.G. In *Insect Muscle;* Usherwood,P.N.R. Ed.; Academic Press: New York, 1975, pp207-280.

(9) Piek,T. In *Comprehensive Insect Physiology, Biochemistry and Pharmacology;* Kerkut,G.A.; Gilbert,L.I. Eds.; Pergamon Press: Oxford, 1985, Vol 11;pp 55-118.

(10) Cull-Candy,S.G.; Usherwood,P.N.R.; *Nature New Biol.* **1973**, *246*, 62-64.

(11) Cull-Candy,S.G.; *J. Physiol.* **1976**, *255*, 449-464.

(12) Usherwood,P.N.R.; Grundfest,H.; *J.Neurophysiol.* **1965**, *28*, 497-518.

(13) Evans,P.D.;O'Shea,M.; *Nature (Lond.)* **1977**, *270*, 257-259.

(14) Piek,T.;Mantel,P.; *J.Insect Physiol.* **1977**, *23*, 321-325.

(15) O'shea,M; Adams,M.E.; *Adv.Insect Physiol.* **1986**, *19*, 1-27.

(16) Scott,R.H.; Duce,I.R.; *Pesticide Sci.* **1985**, *16*, 695.

(17) Duce,I.R.; Scott,R.H.; *J.Physiol (Lond.)* **1983**, *343*, 31-32p.

(18) Scott,R.H.; Duce,.I.R.; *J.Insect Physiol.* **1987**, *33*, 183-189.

(19) Scott,R.H.; Duce,I.R.; *Comp. Biochem. Physiol.* **1987**, *86C*, 305-311.

(20) Duce,I.R.; Scott, R.H.; *Brit. J. Pharmacol.* **1985**, *85*, 395-401.

(21) Martin,R.J.; Pennington,A.J.; *Brit. J. Pharmacol.* **1989**, *98*, 747-756.

(22) Dudel,J; Franke,Ch.; Hatt, H.; Usherwood,P.N.R.; *Brain Res.* **1989**, *481*, 215-220.

(23) Lea,T.J.; Usherwood,P.N.R.; *Comp. Gen. Pharmacol.* **1973**, *4*, 333-350.

(24) Benz,G.W.; Roncalli,R.A.; Gross S.J. In *Ivermectin and Abamectin;* Campbell,W.C. Ed.; Springer-Verlag; New York, 1989; pp 215-229.

(25) Kaufman,W.R.; Ungarian,S.G.; Noga,A.E.; *Exp. Applied. Acarology* **1986**, *2*, 1-18.

(26) Harrow,I.D.; Gration,K.A.F.; Evans, N.A.; *Parasitology* **1991**, *102*, S59-S69.

(27) Bhandal,N.S.; Duce,I.R.; *Pesticide Sci.* **1991**, *32*, 509-510.

(28)   Anis,N.A.;   Clark,R.B.;   Gration,K.A.F.;   Usherwood,P.N.R.; *J.Physiol.(Lond.)* **1981**, *312*, 345-364.

(29) Hart,R.J.; Potter,C.; Wright,R.A.; Lea,P.J.; *Physiological Entomol.* **1978**, *3*, 289-295.

(30) Bhandal,N.S.; Duce,I.R.; *Pesticide Sci.* **1991**, *33*, 509-510.

RECEIVED August 15, 1994

Chapter 17

Ivermectin Interactions with Invertebrate Ion Channels

Susan P. Rohrer and Joseph P. Arena

Merck Research Laboratories, Department of Cell Biochemistry and
Physiology, P.O. Box 2000, Rahway, NJ 07065

Ivermectin is a semi-synthetic analog of avermectin B_{1a}. It is
a potent anthelmintic/insecticide effective against a broad
range of parasites. Ivermectin binding sites from nematodes
and insects have been characterized by employing a ^{3}H-
ivermectin binding assay and subsequently identified with an
^{125}I-azido-avermectin analog. Avermectin binding proteins
from the free living nematode, *Caenorhabditis elegans*, and
from the dipteran, *Drosophila melanogaster*, have been
purified by immuno-affinity chromatography using a
monoclonal antibody against the ligand, avermectin.
Electrophysiological and pharmacological characterization of
the chloride ion channel target of ivermectin action in *C.
elegans* has been accomplished by injecting *C. elegans* mRNA
into *Xenopus* oocytes and studying the expressed chloride ion
channel. The endogenous ligand for the avermectin
sensitive chloride channel expressed in oocytes is glutamate.
At low concentrations, ivermectin potentiates the effect of
glutamate on the channel, while at high concentrations,
ivermectin causes irreversible opening of the channel.

The avermectins are a family of macrocyclic lactones synthesized as
natural fermentation products by the bacterium, *Streptomyces avermitilis*.
The producing organism was isolated from a soil sample collected in Japan
as part of a collaborative agreement between Merck Research Laboratories
and Kitasato Institute in Tokyo (*1*). Potent anthelmintic activity of the
avermectins was discovered when a fermentation broth generated from
the soil sample was introduced into the diet of mice infected with the
intestinal nematode parasite, *Nematospiroides dubius* (*2,3*). Subsequent
investigations revealed that this class of compounds also possessed potent
insecticidal activity but lacked antibacterial or antifungal properties (*4,5*) .

0097–6156/95/0591–0264$12.00/0
© 1995 American Chemical Society

Ivermectin (22, 23 dihydroavermectin B_{1a}, Figure 1), a semi-synthetic avermectin analog (6), was introduced commercially in 1981 and rapidly became the drug of choice for treating a broad spectrum of conditions caused by nematode and arthropod parasites (7,8). Ivermectin has been marketed for use against *Dirofilaria immitis*, the causative agent of heartworm disease in dogs (9). In humans, ivermectin (Mectizan®) was shown to be effective in the treatment and prevention of onchocerciasis (or river blindness) in West Africa, and Central and South America (10,11). Abamectin (avermectin B_{1a}), developed for use as an insecticide in crop protection programs, was introduced in 1985 and has been shown to be effective against mites, leafminers, and lepidopterans (12).

With respect to the general mechanism of action, the avermectins are not unlike the anthelmintic compounds that preceded them. The organophosphates (13), pyrantel and morantel (14,15), levamisole and tetramisole (16) and the avermectins (17,18) all act by interrupting neuromuscular transmission at some level. However, the avermectins are noteworthy for their extraordinary potency against a broad spectrum of endoparasites and ectoparasites combined with a lack of toxicity to mammals, features which have contributed to the superior therapeutic index displayed by the avermectins over previously discovered anthelmintics.

Ivermectin mode of action

Early studies showed that ivermectin induced paralysis in invertebrates was associated with increased cell permeability to chloride (for review see (19,20)). This suggested that the target of drug action was a chloride channel. The additional finding that picrotoxin blocked avermectin-sensitive increases in chloride conductance led to the suggestion that ivermectin interacted with GABA-gated chloride channels (17,21,22). However, Duce and Scott (23) showed that avermectin increased the chloride conductance of locust leg muscle bundles which did not contain GABA-sensitive chloride channels. It was subsequently demonstrated that avermectins activated a glutamate-sensitive chloride conductance in locust leg muscles (24). Its noteworthy that glutamate-gated chloride channels (or H-receptors) are also sensitive to picrotoxin (24,25,26). Moreover, Zufall et al. (27) showed that avermectins activated a picrotoxin-sensitive multitransmitter-gated (glutamate, GABA, acetylcholine) chloride channel on crayfish muscle. These studies supported the hypothesis that avermectins modulate ligand-gated chloride channels.

Identification of the target in nematodes and insects

Isolation and characterization of the nematode specific ivermectin channel from the free living nematode, *Caenorhabditis elegans* has been the

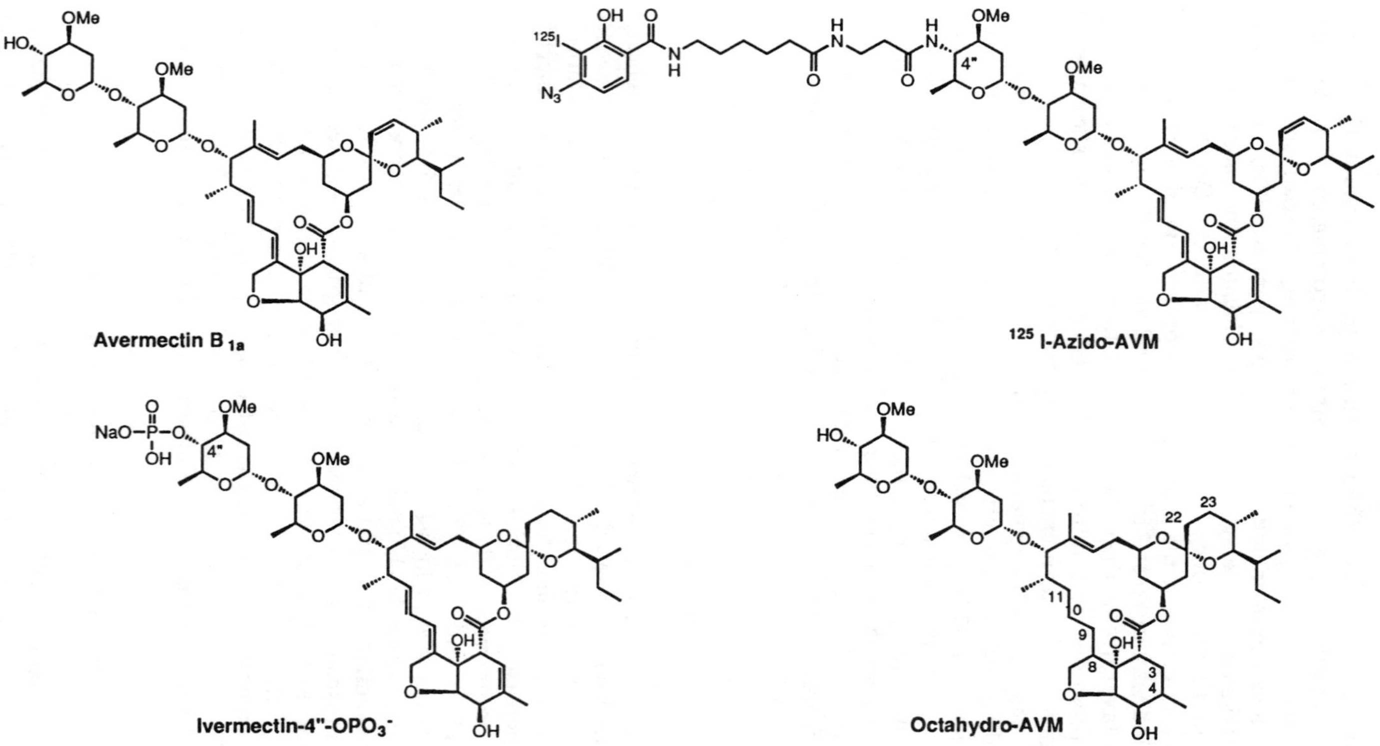

Figure 1. Structures of four different avermectin analogs used in the described studies. Azido-avermectin and ivermectin-phosphate are both biologically active. Octahydroavermectin is biologically inactive.

primary focus of our research efforts. We selected this animal as our model system because *C. elegans* is highly sensitive to avermectin (*28*). The ease with which the organism can be maintained in the laboratory and the feasibility of culturing massive quantities of worms for large scale experiments were additional reasons for choosing *C. elegans* over parasitic nematodes. Much of the biochemistry and molecular biology to be described (*29,30,31,32*) was facilitated by our development of a protocol for cultivating *C. elegans* worms in 500 gm quantities (*33*). In addition, the neuromuscular system of *C. elegans* has been well characterized developmentally, biochemically, and physiologically (*34*) and the neuronal "wiring diagram" is similar to that of parasitic nematodes, such as *Ascaris suum* (*35,36*).

Specific, high affinity ivermectin binding sites have been identified and thoroughly characterized in membrane preparations isolated from *C. elegans* (*37*). The affinities of a series of avermectin analogs for this binding site were determined and compared to the biological activity of each compound in a *C. elegans* motility assay (Figure 2). The motility assay is performed in liquid media and involves overnight incubation of *C. elegans* worms in the absence or presence of drug (*38*). The strong correlation between the binding affinity and biological activity suggest that binding of ivermectin to this site mediates neuromuscular paralysis.

High affinity ivermectin binding sites also have been identified in membranes prepared from several different arthropods including *Drosophila melanogaster, Schistocerca americana* (Figure 3), *Spodoptera frugiperda, Heliothis zea* and *Liriomyza sativae* (data not shown). All tissues tested were similar with respect to affinity for ^{3}H-ivermectin but receptor density was variable. The density of binding sites in *Drosophila* head membrane preparations was ten fold greater when compared to membranes from whole worm homogenates of *C. elegans* and the metathoracic ganglia neuronal membranes from *Schistocerca americana* were enriched 100-fold over *C. elegans*. (*39*).

Although resistance to ivermectin has not presented itself as a problem in the field thus far, the question of target site involvement in the development of resistance could be addressed by comparing wild type and resistant organisms in the ^{3}H-ivermectin binding assay if, and when, it arises in any of the arthropods listed above. Such a study was performed by comparing ivermectin sensitive and ivermectin resistant strains of the parasitic nematode, *Haemonchus contortus* (*40*). Membranes were prepared from L3 larvae of the wild type susceptible worms and ivermectin resistant worms. Both tissue preparations exhibited high affinity avermectin binding sites and the number of sites per mg of protein was the same indicating that resistance in this particular strain was not due to a change in the affinity of the receptor for the ligand or in the number of receptors present on the membranes. This experiment also demonstrated that the membrane receptor from a parasitic target organism, was nearly identical to the *C. elegans* membrane receptor with

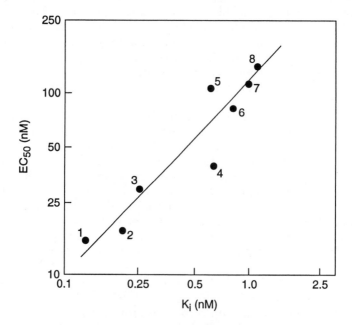

Figure 2. Correlation between binding affinities for a series of avermectin analogs and their efficacy in the *C. elegans* motility assay. [1] ivermectin [2] avermectin B_{2a} [3] avermectin B_{1a} 4"-O-phosphate [4] ivermectin-aglycone [5] ivermectin-monosaccharide [6] avermectin B_1-monosaccharide [7] 2-dehydro-4-hydro-α-2,3-avermectin B_1 [8] avermectin B_{1a}-5-ketone. (Reproduced with permission from ref. 37. Copyright 1989 Elsevier Science Ltd.)

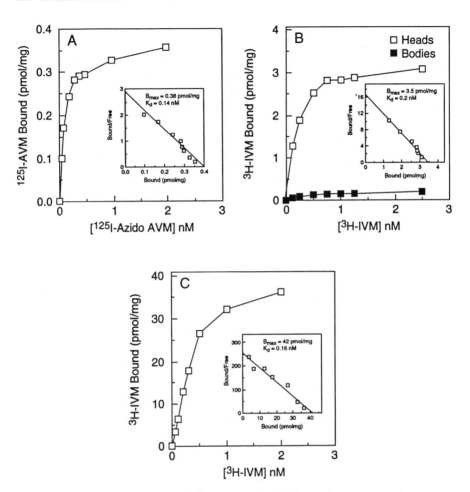

Figure 3. Characterization of avermectin binding sites on membranes from *C. elegans, Drosophila melanogaster,* and *Schistocerca americana.* [A] Membranes were prepared by homogenizing whole *C. elegans* worms and isolating a 28,000 x g membrane preparation. Specific binding of $125I$-azido-AVM was measured in the dark to avoid non-specific crosslinking of the ligand to other proteins. [B] Binding of $3H$-ivermectin to *D. melanogaster.* Membranes were prepared by homogenizing heads of adult flies and isolating a 28,000 x g membrane preparation. [C] $3H$-ivermectin binding to adult *S. americana.* Metathoracic ganglia were dissected from adult *Schistocerca americana.* Neuronal membranes were homogenized and a 28,000 x g pellet prepared. (Reproduced with permission from ref. 32. Copyright 1992 S.P. Rohrer and from ref. 39. Copyright 1994 Insect Biochemistry and Molecular Biology).

respect to affinity for ivermectin (K_d) and receptor density or number of binding sites per mg of membrane protein (B_{max}).

Crude *C. elegans* and *Drosophila* membrane preparations have been useful for characterizing many aspects of the interaction of ivermectin with the nematode and insect binding sites respectively. However, purification and cloning of the nematode or insect receptors could lead to a more precise understanding of the mechanism of action and facilitate the establishment of new mechanism based screens for identification of novel compounds which interact with the same ion channel proteins. Because of the inherent difficulties associated with isolation and cloning of ion channel proteins, independent biochemical and molecular biological approaches toward obtaining the *C. elegans* and *Drosophila* ivermectin-sensitive chloride channel have been taken.

Biochemical isolation of ivermectin receptors from insect and nematode tissues

The purification of the invertebrate ivermectin receptor was facilitated by the use of an azido-AVM analog as a photoaffinity probe (32). The compound shown in Figure 1 was synthesized (41) and found to be biologically active in the *C. elegans* motility assay as well as the *C. elegans* ivermectin binding assay (Figure 3A) in spite of the addition of the large substituent at the 4"-position. The ivermectin receptor was solubilized from *C. elegans* membranes with Triton X-100 and then incubated in the presence of the photoaffinity probe at room temperature in the dark. After a one hour incubation, dextran coated, activated charcoal was added in order to adsorb any unbound ^{125}I-azido-AVM. The affinity ligand was crosslinked to the receptor by exposure to UV light and the result analyzed by autoradiography of an SDS-PAGE gel (Figure 4A). Three *C. elegans* proteins with molecular weights of 53, 47, and 8 kDa were radiolabeled. Increasing concentrations of unlabeled ivermectin (lanes 2-5) resulted in elimination of the affinity labeling pattern. The low concentrations (10^{-8} M) of ivermectin required to block the labeling pattern suggested that all three proteins were associated with the high affinity drug binding site. It is unknown whether the three labeled proteins represent non-identical subunits of a multi-subunit receptor, metabolic breakdown products of a larger precursor, or tissue specific forms of the receptor.

The *Drosophila* proteins were affinity labeled while still bound to the membrane (Figure 4B). Photoaffinity labeling of the *Drosophila* head membranes resulted in identification of an apparent doublet in the 45 kDa size range. Labeling was blocked by ivermectin at low concentrations, indicating that these two proteins are part of the high affinity drug binding site. A biologically inactive avermectin analog (3,4,8,9,10,11,22,23-octahydroavermectin B_{1a}), did not block the labeling pattern, consistent with the interpretation that the two proteins at 45 kDa were specific labeling products.

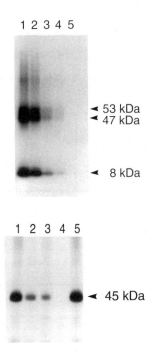

Figure 4. [Panel A] Photoaffinity labeling of the *C. elegans* avermectin binding site. Autoradiography of a 5-20% polyacrylamide gel. Each lane contained 200 mg of Triton X-100 soluble membrane protein. *C. elegans* proteins were labeled with ^{125}I-azido-AVM in the presence of 0.0, 0.2, 0.8, 2, or 20 nM ivermectin (lanes 1-5).
[Panel B] Photoaffinity labeling of the *Drosophila* avermectin binding proteins. Autoradiography of a 5-20% gradient gel. In the presence of increasing concentrations of unlabeled ivermectin (0.1 nM, 1 nM, and 10 nM in lanes 2-4 respectively) the labeling pattern was progressively blocked, indicating that the 45 kDa labeling product was associated with the high affinity site. The inactive analog, octahydroavermectin added at a concentration of 1 nM did not block labeling (Lane 5) indicating specificity of the labeling result shown in Lane 1. (Reproduced with permission from ref. 32. Copyright 1992 S.P. Rohrer and from ref. 39. Copyright 1994 Insect Biochemistry and Molecular Biology).

Purification of the photoaffinity labeled *C. elegans* and *Drosophila melanogaster* proteins was accomplished using a monoclonal antibody against avermectin B_{1a} to capture the ^{125}I-azido-AVM labeled proteins (42). Preparative scale labeling and purification of the *Drosophila* proteins led to the recovery of picomole amounts of pure protein (Figure 5 and Table I) which may prove to be adequate for obtaining internal amino acid sequence information.

TABLE I. Purification of the photoaffinity labeled *Drosophila melanogaster* protein

Purification Step	Total Protein (mg)	^{125}I- Azido AVM Receptor (pmol)	Specific Act. (pmol/mg)	Purification (fold)	Recovery (%)
Head Membranes	927	324[a]	0.35	1	100
X-Link and Resuspend in Tris/SDS-DTT	ND	324[a]	ND	ND	ND
S-300 Gel Filtration	125	293	2.34	7	90
Protein A-3A61 Antibody Affinity	.023[b]	78	3391	9688	24

a pmol of receptor covalently crosslinked to ^{125}I-Azido-AVM. Approximately 10% of total ligand bound became covalently crosslinked upon photolysis.
b Based on Coomassie Blue staining of radiolabeled protein in 10% SDS-PAGE gels. (Reprinted with permission from ref. 42. Copyright 1994 The Biochemical Society and Portland Press.).

Electrophysiology, pharmacology and expression cloning of the ivermectin receptor

Electrophysiological evaluation of *C. elegans* was not feasible because the tough cuticle prevents penetration with microelectrodes and procedures for patch clamping of isolated neurons or muscle cells have not been established. Therefore, the *Xenopus* oocyte system was explored as a surrogate for expression of the *C. elegans* ivermectin gated channel. *C. elegans* poly (A)$^+$ RNA (mRNA) was isolated and injected into *Xenopus*

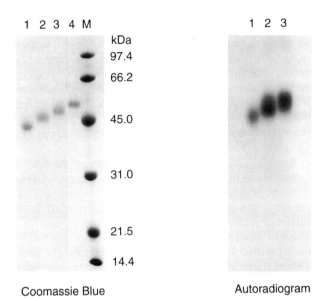

Figure 5. Purification of the photoaffinity labeled avermectin binding proteins from *Drosophila melanogaster*. Two major proteins (Lanes 2 and 3) with approximate molecular weights of 47 kDa and 49 kDa were obtained. A third protein with a molecular weight of 45 kDa is less heavily radiolabeled. The protein shown in lane 4 is mouse immunoglobulin heavy chain which co-elutes from the antibody affinity column along with the avermectin binding proteins as a result of leaching from the column. (Reproduced with permission from ref. 42. Copyright 1994 The Biochemical Society and Portland Press.).

oocytes (20,29,30). Within 48-72 hours the protein translational and processing machinery of the oocyte expressed a functional ion channel from the nematode mRNA. Channel activity was measured using a standard two microelectrode voltage clamp procedure (29,30). Application of ivermectin, or a water soluble derivative, 22,23-dihydroavermectin B_{1a}-4"-O-phosphate (IVMPO$_4$), induces a slowly activating, essentially irreversible increase in membrane current (Figure 6). The irreversible nature of the response is consistent with the slow rate of dissociation for ivermectin observed in binding studies (31,37). The ivermectin-sensitive membrane current displays a reversal potential, sensitivity to extracellular chloride, and current/voltage relationship consistent with the flow of chloride through chloride channels. In addition, the current is blocked by high concentrations of picrotoxin (29). The ability of a series of avermectin analogs to activate current in oocytes has been shown to be directly correlated with nematocidal efficacy and membrane binding affinity (manuscript submitted). Ivermectin has no effect on non-injected or water injected oocytes (Figure 6), and octahydroavermectin, the biologically inactive analog, does not activate the chloride channel expressed in oocytes. These data strongly suggest that the channel expressed in oocytes represents the target of neuromuscular paralysis in the worm.

Oocytes injected with *C. elegans* mRNA also respond to the neurotransmitter glutamate (Figure 7). In contrast to the response to ivermectin, the glutamate-sensitive current activates rapidly, desensitizes in the continued presence of glutamate, and is rapidly and completely reversible. The glutamate-sensitive current is also carried by chloride and is blocked with low affinity by picrotoxin (30). Ibotenate, a structural analog of glutamate activates current in oocytes expressing *C. elegans* mRNA (Figure 7). Ibotenate cross-desensitizes with glutamate, and is approximately four-fold more potent (EC$_{50}$ 70 μM compared to 300 μM for glutamate). Specific agonists for glutamate-gated cation channels such as kainate, N-methyl-D-aspartate (NMDA), α-amino-3-hydroxyl-5-methyl-4-isoxazole propionic acid (AMPA), and quisqualate have no effect on the chloride channel expressed from *C. elegans* mRNA (30).

Several lines of experimental data demonstrate that glutamate/ibotenate and ivermectin are acting on the same chloride channel. When the concentrations of glutamate and IVMPO$_4$ that evoke maximal responses are coapplied, the response is only slightly larger than when either is applied alone (Figure 7). In addition, low concentrations of IVMPO$_4$ (< 10 nM) that have no direct effect on membrane current, potentiate the response to sub-maximal concentrations of glutamate (Figure 7). This potentiation is due to a reduction in the EC$_{50}$ and the Hill coefficient of the glutamate dose-response curve (Figure 8). IVMPO$_4$ has the additional action of reducing the population of channels that desensitize in the presence of glutamate (Figure 8). Finally, both glutamate and IVMPO$_4$ responses are blocked with similar concentration of

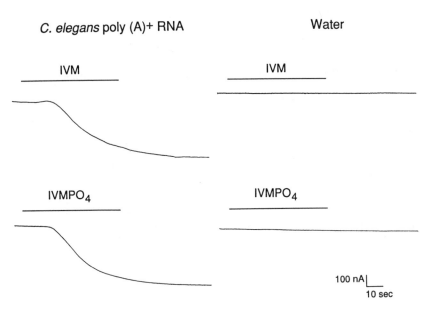

Figure 6. Ivermectin activates a membrane current in *Xenopus* oocytes expressing *C. elegans* mRNA. Membrane currents were recorded from oocytes injected with *C. elegans* mRNA (left) and with water (right). Solid lines above current traces indicate the time of drug application. Concentrations of ivermectin and IVMPO$_4$ applied were 1 µM. Ivermectin or IVMPO$_4$ activated current in >95 % of the oocytes injected with *C. elegans* mRNA, but not in oocytes injected with water. (Reproduced with permission from ref. 29. Copyright 1991 Waverly).

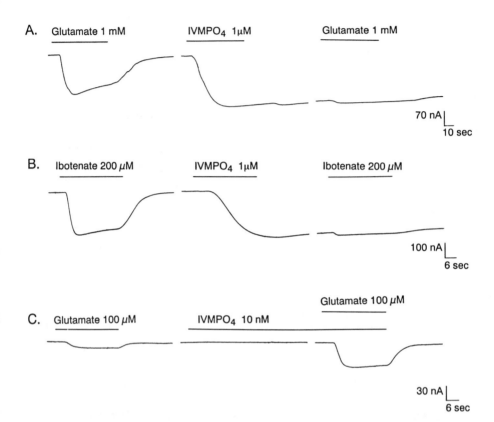

Figure 7. IVMPO4, glutamate, and ibotenate activate the same chloride current in oocytes injected with *C. elegans* mRNA. A. Response of an oocyte to maximal concentrations of glutamate, IVMPO4, and re-application of glutamate (last trace). B. Response of an oocyte to maximal concentrations of ibotenate, IVMPO4, and re-application of ibotenate (last trace). C. After the control response to a submaximal concentration of glutamate (left), the oocyte was pretreated with a concentration of IVMPO4 that failed to activate current (middle trace) for 3 min. Re-application of glutamate in the presence of IVMPO4 resulted in potentiation of the glutamate-sensitive current (last trace). (Reproduced with permission from ref. 30. Copyright 1992 Elsevier Science.).

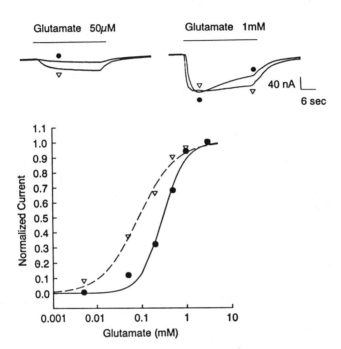

Figure 8. IVMPO4 shifts the concentration-response curve for glutamate. The plot shows the normalized glutamate-sensitive concentration response curve in the absence (filled circles) and presence (open triangles) of 2 nM IVMPO4. The smooth curves represent fits to a modified Michaelis-Menten equation (*30*). IVMPO4 decreased the EC50 for glutamate from 300 to 85 µM, and changed the Hill coefficient from 1.8 to 1.0 (smooth curves). The inset shows superimposed current traces in the absence (filled circles) and presence of IVMPO4 (open triangles). (Reproduced with permission from ref. 30. Copyright 1992 Elsevier Science.).

picrotoxin, and the same size range of the mRNA encodes for glutamate and ivermectin responses (29,30,43).

The activity of ibotenate on the glutamate-gated chloride channels expressed from C. elegans mRNA suggests that the channel is similar to glutamate-gated chloride channels reported in arthropods. Ibotenate gates glutamate-sensitive chloride channels present in locust muscle (H-receptors), locust neurons, cockroach neurons, and lobster muscle (25,26,44,45,46,47). Half-maximal activation of these channels by glutamate/ibotenate is in the range of 10-500 µM, similar to that for activation in C. elegans-injected oocytes. Finally, it appears that the glutamate-gated chloride channels found on locust muscle are gated by avermectins (24). These comparative studies support the hypothesis that avermectins target glutamate-gated chloride channels.

Obtaining electrophysiological responses in Xenopus oocytes has enabled us to take an expression cloning approach toward isolating cDNAs for the C. elegans avermectin receptor (43,48,49). C. elegans mRNA was size-fractionated and then injected into Xenopus oocytes. Glutamate- and ivermectin-sensitive currents were expressed from mRNA in the 1.8-2.0 kb size range. A cDNA library was synthesized from the active fraction. RNA has been synthesized in vitro from the pools of recombinant cDNAs and are being screened in Xenopus oocytes for expression of an ivermectin-sensitive glutamate-gated chloride channel.

Interactions of avermectins with other ligand-gated chloride channels

Avermectins are highly selective for invertebrate receptors (37). However they have been shown to act on GABA-gated chloride channels in vertebrate brain (37,50,51,52,53). We have investigated the response to ivermectin using oocytes injected with rat brain mRNA ((54), Figure 9). There was no direct activation of GABA-sensitive chloride current with ivermectin. However, ivermectin potentiated the response to low concentrations of GABA, and reduced the fraction of current that desensitized during application of maximal GABA concentrations. Except for the lack of direct activation, the interaction of ivermectin with mammalian $GABA_A$ receptors resembles the response observed in oocytes injected with C. elegans mRNA. In oocytes expressing C. elegans mRNA, ivermectin potentiated the effect of glutamate whereas in oocytes expressing rat brain, chick brain, or cloned $GABA_A$ receptor mRNA, ivermectin potentiates the effect of GABA (51,54). In both systems ivermectin causes a reduction in the EC_{50} and the Hill coefficient of the ligand dose-response curve and reduces the fraction of channels that become desensitized.

Avermectins do not readily cross the blood brain barrier, so only very low concentrations ever come in contact with the vertebrate receptors. The combination of lower affinity for the vertebrate receptor (37) and compartmentalization of ivermectin may account for the low incidence of host toxicity with this class of compounds. It has recently

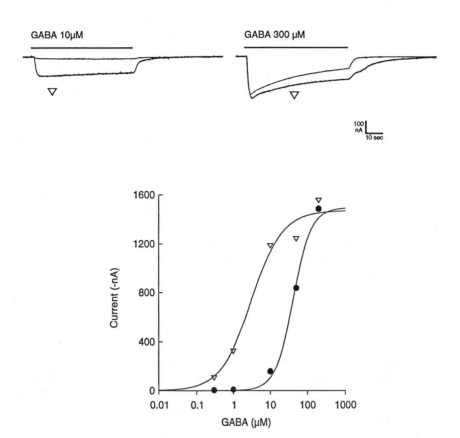

Figure 9. IVMPO4 shifts the concentration-response curve for GABA. Oocytes were injected with mRNA isolated from rat cerebral cortex, and recordings were made 2 days after injection. The inset shows superimposed current traces in the absence and presence of IVMPO4 (open triangles). The plot shows the GABA dose response curve in the absence (filled circles) and presence (open triangles) of 1 μM IVMPO4. The smooth curves represent fits to a modified Michaelis-Menten equation (30). IVMPO4 decreased the EC50 for GABA from 42 to 3 μM, and changed the Hill coefficient from 1.8 to 1.0 (smooth curves).

been shown that transgenic mice with a disrupted mdr1a gene are 100 times more sensitive to ivermectin than are wild type mice (55). The mrd1a gene encodes for the major drug transporting P-glycoprotein in the blood brain barrier (55). This result suggests that normal individuals expressing a functional mdr1a gene, are not subject to toxic side effects of ivermectin because they possess the ability to prevent drug accumulation in brain tissue by virtue of the P-glycoprotein efflux pump.

Avermectins also have been shown to interact with other members of the ligand gated chloride channel family. They inhibit strychnine binding to the mammalian glycine-gated chloride channels in a noncompetitive fashion (56). In crayfish stomach muscle, avermectins activate a multi-transmitter-(glutamate, acetylcholine, GABA) gated chloride channel (27). Moreover, there is evidence for both potentiation and blockade of GABA-gated chloride channels in arthropods (23,57), and blockade of the *Ascaris* body wall muscle GABA-gated chloride channels (58,59). Based on these accounts, avermectins appear to be a promiscuous class of compounds interacting with many of the known ligand-gated chloride channels. It remains to be determined whether the effects of direct channel activation, blockade, and/or potentiation of ligand-gated responses can be correlated with the biological activity of the avermectins.

Future studies

The biochemical isolation and expression cloning approaches outlined earlier are being applied in parallel in order to obtain the C. *elegans* and *Drosophila* avermectin receptor genes. Cloning of the genes that encode chloride channels gated by glutamate from both nematodes and insects will broaden our understanding of the physiological importance of these channels and the similarity of these channels to other ligand-gated chloride channels. It will also allow us to determine the native configuration and conformation of the receptor as well as its exact anatomical location in nematodes and insects. Cloning of homologous ion channel protein genes from parasitic nematodes and arthropods will be facilitated as will studies of the developmental regulation of the expression of these genes. Ultimately, we intend to use the genes in order to establish new mechanism based screens for novel anthelmintic and insecticidal compounds. If the avermectin gated chloride channels are similar to other ligand-gated chloride channel family members, then it is likely that additional drug binding sites distinct from the ivermectin binding site, will be present on the receptor molecule. It is our goal to discover and exploit these additional drug binding sites in our search for the next generation of anthelmintic/insecticidal chemical entities.

During preparation of this manuscript the expression cloning approach was successful and two cDNAs that encode for subunits of a C. *elegans* avermectin-sensitive glutamate-gated chloride channel were isolated (60). The subunits, termed GluClα and GluClβ, form homomeric chloride channels with unique pharmacological and electrophysiological properties. When coexpressed the subunits assemble heteromeric chloride channels with properties that are distinct from the homomeric channels. The characteristics of heteromeric channel responses to glutamate and ivermectin resemble those observed in oocytes injected with *C. elegans* mRNA.

Acknowledgments

Ed Hayes, Ethel Jacobson and Elizabeth Birzin collaborated with Susan Rohrer on the affinity labeling and purification experiments. Doris Cully, Philip Paress and Ken Liu carried out the characterization of the *C. elegans* receptor expressed in *Xenopus* oocytes in collaboration with Joe Arena. Helmut Mrozik, Peter Meinke and Tom Shih provided chemistry support for the projects described here and synthesized numerous other avermectin analogs. Kodzo Gbewonyo, Leonard Lister, and Bruce Burgess provided *C. elegans* tissue from large scale preparations. Scott Costa contributed large quantities of *Drosophila* flies needed for receptor purification. We would also like to acknowledge the support of Jim Schaeffer, Merv Turner, Roy Smith and Mike Fisher.

Literature Cited

1. Burg, R. W.; Stapley, E. O. In *Ivermectin and Abamectin*. W. C. Campbell, Eds.; Springer-Verlag: New York, 1989; pp 24-32.
2. Burg, R. W.; Miller, B. M.; Baker, E. E.; Birnbaum, J.; Currie, J. A.; Harman, R.; Kong, V. L.; Monaghan, R. L.; Olson, G.; Putter, I.; Tunac, J. P.; Wallick, H.; Stapley, E. O.; Oiwa, R.; Omura, S. *Antimicrob. Agents Chemother.* **1979**, *15*, 361-367.
3. Egerton, J. R.; Ostlind, D. A.; Blair, L. S.; Eary, D. H.; Suhayda, D.; Cifelli, S.; Riek, R. F.; Campbell, W. C. *Antimicrob. Agents Chemother.* **1979**, *15*, 372-378.
4. Ostlind, D. A.; Cifelli, S.; Lang, R. *Vet. Rec.* **1979**, *105*, 168.
5. Putter, I.; MacConnell, J. G.; Preiser, F. A.; Haidri, A. A.; Ristich, S. S.; Dybas, R. A. *Experentia.* **1981**, *37*, 963-964.
6. Chabala, J.; Mrozik, H.; Tolman, R.; Eskola, P.; Lusi, A.; Peterson, L.; Woods, M.; Fisher, M. *J. Med. Chem.* **1980**, *23*, 1134-1136.
7. Campbell, W. *Parasitology Today.* **1985**, *1*, 10-16.
8. Campbell, W. C. *Medicinal Research Reviews.* **1993**, *13*, 61-79.
9. Campbell, W. *Sem. Vet. Med. & Surg.* **1987**, *2*, 48-55.
10. *Ivermectin and Abamectin.*; W. C. Campbell, Eds.; Springer-Verlag: New York, 1989.
11. Aziz, M. A.; Diallo, S.; Diop, I. M.; Lariviere, M.; Porta, M. *Lancet.* **1982**, *2*, 171-173.
12. Lasota, J. A.; Dybas, R. A. *Annu. Rev. Entomol.* **1991**, *36*, 91-117.
13. Knowles, C. O.; Casida, J. E. *J. Agric. Food Chem.* **1966**, *14*, 566-572.
14. Eyre, P. *J. Pharm> Pharmacol.* **1970**, *22*, 26-36.

15. Forbes, L. *Southeast Asian J. Trop. Med. Publ. Health.* **1972**, *3*, 235-241.
16. Lewis, J.; Wu, J.; Levine, J.; Berg, H. *Neuroscience.* **1980**, *5*, 967-989.
17. Kass, I. S.; Wang, C. C.; Walrond, J. P.; Stretton, A. O. W. *Proc. Natl. Acad. Sci. USA.* **1980**, *77*, 6211-6215.
18. Kass, I. S.; Larsen, D. A.; Wang, C. C.; Stretton, A. O. W. *Exp. Parasitol.* **1982**, *54*, 166-174.
19. Turner, M. J.; Schaeffer, J. M. In *Ivermectin and Abamectin.* W. C. Campbell, Eds.; Springer-Verlag: New York, 1989; pp 73-88.
20. Arena, J. P. *Parasitology Today.* **1994**, *10*, 35-37.
21. Fritz, L. C.; Wang, C. C.; Gorio, A. *Proc. Natl. Acad. Sci. USA.* **1979**, *76*, 2062-2066.
22. Mellin, T. N.; Busch, R. D.; Wang, C. C. *Neuropharm.* **1983**, *22*, 89-96.
23. Duce, I. R.; Scott, R. H. *Brit. J. Pharmacol.* **1985**, *85*, 395-401.
24. Scott, R. H.; Duce, I. R. *Pestic. Sci.* **1985**, *16*, 599-604.
25. Lea, T. J.; Usherwood, P. N. R. *Comp. Gen. Pharmacol.* **1973**, *4*, 351-363.
26. Lea, T. J.; Usherwood, P. N. R. *Comp. Gen. Pharmacol.* **1973**, *4*, 333-350.
27. Zufall, F.; Franke, C.; Hatt, H. *J. Exp. Biol.* **1989**, *142*, 191-205.
28. Simpkin, K. D.; Coles, G. C. *J. Chem. Technol. Biotechnol.* **1981**, *31*, 66-69.
29. Arena, J. P.; Liu, K. K.; Paress, P. S.; Cully, D. F. *Mol. Pharmacol.* **1991**, *40*, 368-374.
30. Arena, J. P.; Liu, K. K.; Paress, P. S.; Schaeffer, J. M.; Cully, D. F. *Mol. Brain Res.* **1992**, *15*, 339-348.
31. Cully, D. F.; Paress, P. S. *Mol. Pharmacol.* **1991**, *40*, 326-332.
32. Rohrer, S. P.; Meinke, P. T.; Hayes, E. C.; Mrozik, H.; Schaeffer, J. M. *Proc. Natl. Acad. Sci.* **1992**, *89*, 4168-4172.
33. Gbewonyo, K.; Rohrer, S. P.; Lister, L.; Burgess, B.; Cully, D.; Buckland, B. *Biotechnology.* **1994**, *12*, 51-54.
34. Chalfie, M.; White, J. In *The Nematode Caenorhabdtitis elegans.* W. B. Wood, Eds.; Cold Spring Harbor Laboratory: Cold Spring Harbor, 1988; pp 337-391.
35. White, J. G.; Southgate, J. N.; Thomson, J. N.; Brenner, S. *Philos. Trans. R. Soc. Lond. B. Biol. Sci.* **1986**, *314*, 1-340.
36. Stretton, A. O. W.; Davis, R. E.; Angstadt, J. D.; Donmoyer, J. E.; Johnson, C. D. *Trends in Neurosci.* **1985**, *8*, 294-300.
37. Schaeffer, J. M.; Haines, H. W. *Biochem. Pharm.* **1989**, *38*, 2329-2338.
38. Schaeffer, J. M.; Fraizer, E. G.; Bergstrom, A. R.; Williamson, J. M.; Liesch, J. M.; Goetz, M. A. *J. Antibiotics.* **1990**, *9*, 1179-1182.
39. Rohrer, S. P.; Birzin, E. T.; Costa, S. D.; Arena, J. P.; Hayes, E. C.; Schaeffer, J. M. *Insect Biochem. and Mol. Biol.* **1994**, in press,
40. Rohrer, S. P.; Birzin, E.; Eary, C.; Schaeffer, J. M.; Shoop, W. *J. Parasitology.* **1994**, *80*, 493-497.
41. Meinke, P. T.; Rohrer, S. P.; Hayes, E. C.; Schaeffer, J. M.; Fisher, M. H.; Mrozik, H. *J. Medicinal Chemistry.* **1992**, *35*, 3879-3884.
42. Rohrer, S. P.; Jacobson, E. B.; Hayes, E. C.; Birzin, E.; Schaeffer, J. M. *Biochem. J.* **1994**, in press,
43. Cully, D. F.; Paress, P. S.; Liu, K. K.; Arena, J. P. *J. Cell Biochem.* **1993**, *17C*, 115.

44. Cull-Candy, S. G. *J. Physiol.* **1976**, *255*, 449-464.
45. Horseman, B. G.; Seymour, C.; Bermudez, I.; Beadle, D. J. *Neurosci. Lett.* **1988**, *85*, 65-70.
46. Lingle, C.; Marder, E. *Brain Res.* **1981**, *212*, 481-488.
47. Wafford, K. A.; Sattelle, D. B. *J. Exp. Biol.* **1989**, *144*, 449-462.
48. Sigel, E. *J. Membrane Biol.* **1990**, *117*, 201-221.
49. Lester, H. A. *Science.* **1988**, *241*, 1057-1063.
50. Wang, C.; Pong, S. *Prog. Clin. Biol. Res.* **1982**, *97*, 373-395.
51. Sigel, E.; Baur, R. *Mol. Pharmacol.* **1987**, *32*, 749-752.
52. Drexler, G.; Sieghart, W. *Neurosci. Letters.* **1984**, *50*, 273-277.
53. Drexler, G.; Sieghart, W. *Eur. J. Pharmacol.* **1984**, *101*, 201-207.
54. Arena, J. P.; Whiting, P. J.; Liu, K. K.; McGurk, J. F.; Paress, P. S.; Cully, D. F. *Biophys. J.* **1993**, *64*, a325.
55. Schinkel, A. H.; Smit, J. J. M.; van Tellingen, O.; Beijnen, J. H.; Wagenaar, E.; van Deemter, L.; Mol, C. A. A. M.; van der Valk, M. A.; Robanus-Maandag, E. C.; te Riele, H. P. J.; Berns, A. J. M.; Borst, P. *Cell.* **1994**, *77*, 491-502.
56. Graham, D.; Pfeiffer, F.; Betz, H. *Neurosci. Letters.* **1982**, *29*, 173-176.
57. Bermudez, I.; Hawkins, C. A.; Taylor, A. M.; Beadle, D. J. *J. Recept. Res.* **1991**, *11*, 221-232.
58. Holden-Dye, L.; Walker, R. J. *Parasitology.* **1990**, *101*, 265-271.
59. Martin, R. J.; Pennington, A. J. *Br. J. Pharmacol.* **1989**, *98*, 747-756.
60. Cully, D. F.; Vassilatis, D. K.; Liu, K. K.; Paress, P. S.; Van der Ploeg, L. H. T.; Schaeffer, J. M.; Arena, J. P. *Nature.* **1994**, *371*, 707-711.

RECEIVED November 1, 1994

Chapter 18

Resistance to Avermectins in the House Fly, *Musca domestica*

Jeffrey G. Scott

Department of Entomology, Cornell University, Comstock Hall, Ithaca, NY 14853

Resistance to abamectin (avermectin B_{1a}) has been studied in the house fly, *Musca domestica*. Cross-resistance was first detected in 1986 and appeared due to decreased penetration and a piperonyl butoxide suppressible factor (i.e. monooxygenase-mediated metabolism). Selection with abamectin resulted in a strain, called AVER, with $>60,000$-fold resistance by topical application. However, the level of resistance varied between technical and formulated material as well as by the method of bioassay. Resistance in the AVER strain is recessive, autosomal and appears to be due to decreased cuticular penetration and target site insensitivity. Insensitivity of the nervous system was associated with a small decrease in the number of binding sites per mg protein in non-neural tissue in the house fly thorax. The impact of resistance on the future use of abamectin is discussed.

Insecticide resistance continues to be a problem that plagues pest control efforts. As new insecticides are developed it is imperative that we move to understand how resistance develops in order to determine the optimum strategies for delaying the evolution of resistance.

Avermectins are a mixture of natural products produced by the soil actinomycete *Streptomyces avermitilis* that have insecticidal, acaricidal and nematocidal activity. The eight individual components produced by *S. avermitilis* vary in toxicity to different pest species. For insect species, avermectin B_{1a} generally had the greatest toxicity and has been developed as an insecticide with the common name of abamectin (*1*).

In this chapter I shall review what is known about abamectin resistance in the house fly. From this information, strategies for delaying the development of resistance in the field is discussed.

0097–6156/95/0591–0284$12.00/0

Detection of Cross-resistance

The first case of cross-resistance to abamectin was noted in 1986 in the Learn Pyrethroid-Resistant (LPR) strain of house fly (2). This strain was selected with permethrin (3) and is highly resistant to pyrethroid insecticides due to increased metabolic detoxication mediated by the microsomal cytochrome P450 monooxygenases (2,4), decreased sensitivity of the nervous system (i.e. kdr) and decreased cuticular penetration (2). Abamectin cross-resistance in LPR could be partially suppressed by the monooxygenase inhibitor piperonyl butoxide (PBO) and was associated with genes on chromosomes 2 and 3 (5). Due to the fact that kdr resistant house flies were not cross-resistant to abamectin (6), it was proposed that cross-resistance to abamectin in the LPR strain was due to increased oxidative metabolism and decreased cuticular penetration (5). Cross-resistance has subsequently been shown in field collected strains from New York dairies using a variety of bioassay methods (5,7,8). These results suggest that resistance might develop rapidly if abamectin was introduced for house fly control.

Selection and Levels of Resistance

To determine how rapidly resistance to abamectin could occur, house flies were collected from several New York dairies and selected in the laboratory by a residual exposure using AVID (i.e. formulated abamectin) (9). Resistance developed rapidly (9) and to a high level (see below). This highly resistant strain was named AVER. This strain is somewhat unusual in that the resistance appears to be unstable and cannot be maintained without frequent selection (unpublished observation).
    A comparison of abamectin toxicity using different bioassay methods is shown in Table I. Resistance ranged from 5.9-fold for a field collected strain (Dairy), to 25-fold for LPR, to >60,000-fold for AVER by topical application of technical abamectin (Table I). Topical application of formulated material could not be carried out because the formulation blank was toxic to susceptible flies (Table I). Technical abamectin was slightly more toxic than formulated material to the susceptible strain by residual exposure, although the level of resistance in the AVER strain was greater to technical material by this route of exposure (Table I).

Inheritance of Abamectin Resistance

Reciprocal crosses between AVER and the susceptible aabys strain produced F_1 flies with nearly identical abamectin dose-response lines, indicating that resistance was not sex linked or due to cytoplasmic factors (Table II, 10). A similar result was obtained for the LPR strain (5). Abamectin resistance was highly recessive in the AVER strain, having only 3.5- to 5.4-fold resistance in the F_1 compared to >60,000-fold for the AVER strain (Table II). While other recessive resistance mechanisms (such as kdr) are known, the tremendous change in abamectin resistance between the F_1 and parental AVER strain is quite remarkable.

Table I.   Toxicity of Technical and Formulated Abamectin
to House Flies

Material	Method[i]	Strain	LD_{50}[a]	RR[b]
Technical[c]	T	NAIDM	5.0	---
Technical[c]	T	LPR	38	7.6
Technical[d]	T	aabys	1.3	---
Technical[d]	T	Dairy	7.6	5.9
Technical[d]	T	LPR	33	25
Technical[d]	T	F_1(LPR x aabys)	5.9	16
Technical[e]	T	S+	1.7	---
Technical[e]	T	AVER	> 100,000	> 60,000
Technical[g]	T	aabys	0.79	---
Technical[g]	T	aabys (male)	0.50	---
Technical[g]	T	AVER	> 100,000	> 100,00
Technical[g]	T	AVER (male)	~ 100,000	~ 200,00
Technical[g]	T	S+	3.1	---
Technical[g]	I	S+	1.2	---
Technical[g]	I	AVER	43	35
Technical[e]	R	S+	0.033[h]	--
Technical[e]	R	AVER	> 6.3[h]	> 190
Formulated[e]	T	S+	ND[f]	---
Formulated[e]	R	S+	0.05[h]	--
Formulated[e]	R	AVER	1.8[h]	36

[a] In units of ng/fly. Females tested unless stated otherwise.
[b] LD_{50} resistant strain/LD_{50} susceptible strain.
[c] From Scott and Georghiou 1986.
[d] From Scott 1989.
[e] From Scott et al. 1991.
[f] Not determined. Formulation blank toxic to S+.
[g] From Konno and Scott 1991.
[h] LC_{50} in units of $\mu g/cm^2$.
[i] Bioassay method: T = topical, I = injection, R = residual.

Inheritance of abamectin resistance (recessive) is different than inheritance of permethrin resistance (co-dominant, 16) in this strain. Males generally had lower LD_{50} values than females (Table I), probably due to the slightly smaller size of

Table II. Toxicity of Technical Abamectin With and Without
Synergists, to Susceptible and Resistant House Flies

Strain	Synergist	LD_{50}[a]	SR[b]	RR[c]
aabys[d]	---	1.3	---	---
aabys[d]	PBO	0.36	3.6	---
Dairy[d]	---	7.6	---	5.9
Dairy[d]	PBO	0.89	8.5	2.5
LPR[d]	---	33	---	25
LPR[d]	PBO	5.9	5.6	16
F_1 (aabys x LPR)[d]	---	4.6	---	3.5
S+[e]	---	1.7	---	---
S+[e]	PBO	1.5	1.1	---
S+[e]	DEF	1.5	1.1	---
S+[e]	DEM	1.7	1.0	---
AVER[e]	---	>100,000	---	>60,000
AVER[e]	PBO	>100,000	ND[f]	>60,000
AVER[e]	DEF	>100,000	ND	>60,000
AVER[e]	DEM	>100,000	ND	>60,000

[a] In units of ng/fly.
[b] Synergistic Ratio = LD_{50} with synergist.
[c] Resistance Ratio = LD_{50} resistant strain/LD_{50} susceptible strain.
[d] From Scott 1989.
[e] From Scott et al. 1991.
[f] Could not be determined.

the males. The S+ (susceptible) strain was slightly less sensitive to abamectin compared to the aabys strain. This difference is commonly noted between these strains and is most likely due to the slightly larger size of the S+ flies.

Bioassays with Synergists. The effect of the oxidative synergist piperonyl butoxide (PBO) on abamectin toxicity in the aabys, LPR and Dairy strains is shown in Table II. PBO synergized the toxicity of abamectin 3.6-fold in the aabys strain, suggesting that abamectin is oxidatively metabolized, to at least a small degree, in this strain. PBO was also effective in suppressing, but not completely abolishing, resistance to abamectin in LPR and Dairy. This suggests the involvement of the monooxygenase system, as well as another mechanism that is insensitive to PBO.

The synergists PBO, DEF, and DEM were ineffective in substantially altering the toxicity of abamectin in either the S+ or AVER fly strains (Table II), suggesting abamectin is not rapidly metabolized by monooxygenases, hydrolases, or glutathione S-transferases in these strains. The lack in measurable reduction of the resistance ratio in the presence of the synergists suggests that metabolism is not the major mechanism of resistance in the AVER strain.

Pharmacokinetic Studies. The rate of penetration of radiolabeled abamectin was evaluated in susceptible and AVER strains of house flies (*10*). Penetration of radiolabeled abamectin was relatively slow, reaching only 53% after 8 hours in the S+ strain. The rate constant for penetration of radiolabel was 2.4-fold slower in AVER than in S+ with values of 1.37 x 10^{-2} hr^{-1} and 3.28 x 10^{-2} hr^{-1}, respectively. This suggests that decreased cuticular penetration is one of the mechanisms of resistance in the AVER strain (*10*). A comparison of the penetration rate constants with those obtained using [^{14}C]permethrin in a previous study (*2*) on pyrethroid resistance in the LPR strain indicates that abamectin penetrates about 2-fold more slowly than permethrin in both susceptible and resistant strains. Additionally, the difference between the resistant and susceptible strains was greater in the case of abamectin (2.4-fold) compared to permethrin (1.4-fold). These results suggest that decreased penetration might confer greater resistance to abamectin than this mechanism is normally thought to confer to other types of insecticides.

Injection of technical abamectin did not significantly enhance the toxicity to susceptible house flies, but it did change the LD_{50} from > 100,000 to 43 ng/fly in the AVER strain (Table I) suggesting that decreased cuticular penetration was a mechanism of resistance in this strain and agreeing with the penetration studies mentioned above. Penetration is known to have a multiplicative interaction with other resistance mechanisms (*14,15*), however, the nearly 3000-fold difference between injected and topical application resistance ratios is quite remarkable.

Metabolism Studies. No difference in the *in vivo* metabolism of [^{3}H]abamectin was noted between susceptible and AVER strains of house fly (*10*). Almost 90% of the radiolabel was recovered from the body homogenates while less than 11% was recovered from the holding vials even 8 hr after treatment (both strains). In body homogenates, the amount of [^{3}H]abamectin, radiolabel from the water-soluble fraction and the unknown metabolites in the ethyl acetate-soluble fraction were either not different or were higher in the S+ strain than in the AVER strain. The radiolabel from the unextractable materials was also higher in the S+ strain compared to the AVER strain. These data suggest that increased metabolism is not a mechanism of abamectin resistance in the AVER strain (*10*).

Receptor Binding. To investigate if changes in the abamectin receptor were associated with resistance in the AVER strain, receptor binding studies were carried out (*10*). Specific [^{3}H]-abamectin binding to the membrane preparations was saturable with increasing concentrations of [^{3}H]abamectin in both strains. The equilibrium dissociation constant (mean $\pm$ SD) was not significantly different between the AVER (K_D = 0.74 nM) and S+ (K_D = 0.72 nM) strains (*10*). However, the maximum number of binding sites (B_{max}) was significantly different between the two strains, with B_{max} for the S+ strain (0.113 $\pm$ 0.008 pmol/mg protein) being 1.5 times higher than that of the AVER strain (0.077 $\pm$ 0.006 pmol/mg protein). Nonspecific binding increased linearly, and was less than 7% of total binding at 3.0 nM [^{3}H]abamectin in both strains (Konno and Scott 1991).

To examine if the receptor binding differences correlated with the inheritance of resistance (i.e. recessive), Konno and Scott (*10*) investigated the specific binding of [^{3}H]abamectin in susceptible (aabys), resistant (AVER) and F_1 (aabys x AVER) house flies. The K_D values were 0.75, 0.74 and 0.76 nM for aabys, AVER and the F_1, respectively. The B_{max} values were 0.125, 0.077, and 0.113 pmol/mg protein for aabys, AVER and the F_1, respectively. Thus, both the decrease in B_{max} and the resistance are recessive traits. These results support the idea that part of the abamectin resistance in the AVER strain is associated with a decreased number of binding sites per mg protein (i.e., B_{max}).

Binding of [^{3}H]abamectin varies between tissues in house flies. In 1992, Deng and Casida (*11*) reported higher K_D and B_{max} values associated with head compared to thoraces plus abdomens. Recently, Hatano and Scott (*12*) reported a significant difference in the binding of [^{3}H]abamectin between thoraces without ganglia and the thoracic ganglia. The receptors in ganglia had a lower affinity (higher K_D) and higher number of binding sites per mg protein (B_{max}) compared to the receptors in thoraces without ganglia in both strains. The K_D and B_{max} of heads and thoracic ganglia were similar. These results suggest that there may be abamectin receptors with a lower affinity and increased abundance in preparations from neural tissue (i.e. head and ganglia) compared with a preparation of mainly muscle (i.e. thoraces without ganglia). In all cases the Hill coefficient was not significantly different from 1 suggesting no cooperativity between binding sites (*12*).

The only tissue specific difference in [^{3}H]abamectin binding between the resistant strain compared to the susceptible strain was a small, but significant, decrease in the B_{max} found in thoraces without ganglia. As there was no difference in [^{3}H]abamectin binding from preparations made mainly from neural tissue, this suggests that the lower B_{max} associated with resistance is due to a receptor not found in the brain or ganglia.

Although the small change in abamectin binding sites is clearly associated with resistance, this does not easily explain the putative 40-fold difference in target site sensitivity that might be expected from bioassay data (injection). A similar situation has been noted for *kdr*-type resistant German cockroaches where the *para*-homologous sodium channel gene has been linked to the resistance (*17*). In this case, pyrethroids and batrachotoxin are thought to act at similar sites on the sodium channel and these insects are resistant to pyrethroids and batrachotoxin (*18*). However, [3H]batrachotoxinin A-20-α-benzoate binding studies showed no difference in the binding between resistant and susceptible strains (*19*). It was suggested that the receptor may not be different between the strains, but that it could be the coupling of the binding to the toxic action that differed between resistant and susceptible strains (*19*). Therefore, although a decrease in B_{max} is associated with resistance, its actual role in abamectin resistance remains uncertain. Clearly, identification of the target site and use of functional assays will be needed to more fully understand abamectin resistance in the house fly.

[^{3}H]abamectin binding is not displaced by neurotoxins known to act at the GABA receptor/ionophore complex. Dieldrin, muscimol, Ro5-4864 (*12*), picrotoxinin, bicycloorthobenzoates (*11,12*), GABA, and lindane (*11*) did not produce dose dependent displacement of [^{3}H]abamectin in preparations of membranes from house flies. This suggests that the binding site of abamectin is not the same as the binding site of these other well characterized ligands in the house fly. These studies agree with the lack of cross-resistance observed in the AVER strain of house fly (see above).

Collectively these results indicate that decreased cuticular penetration and an altered target site are likely the mechanisms of abamectin resistance in the AVER strain. It appears that the decreased penetration mechanism provides an unusually large enhancement of resistance, as the topical and injected resistance ratios are > 60,000 and 35-fold, respectively. The protection conferred by one or both of these mechanisms is decreased by the use of formulated abamectin.

Cross-Resistance. The AVER strain was found to be > 4,000-fold cross-resistant to abamectin oxide with LD_{50} values of 0.02 and > 100 μg/fly for the S+ (susceptible) and resistant AVER strains, respectively (*10*). MK-243 was reasonably toxic to S+ house flies with an LD_{50} of 0.03 μg/fly. However, the AVER strain was very heterogeneous in its response to this compound having a dose-response line that was very flat, with an approximate LD_{50} of 0.4 μg/fly (*10*). Although the resistance ratio for this compound ($\approx$ 13) is much less than found toward abamectin oxide (> 4,000), the heterogeneity of response in the AVER strain suggests that high levels of MK-243 resistance could be selected for.

Following selection for abamectin resistance, the AVER strain was bioassayed to determine if the selection had brought about cross-resistance to other insecticides. The high level of abamectin resistance was not associated with cross-resistance to bicycloorthobenzoate (4-*tert*-butyl-1-(4-ethynylphenyl)trioxabicyclo[2.2.2]octane, *12*), organophosphate (crotoxyphos, dimethoate, tetrachlorvinphos or dichlorvos), pyrethroid (permethrin) or cyclodiene (lindane and dieldrin) insecticides (*9*). The lack of cross-resistance between abamectin and bicycloorthobenzoate or cyclodiene insecticides suggests that the change(s) in the target site causing insensitivity to abamectin does not alter the effectiveness of these other insecticides.

Conclusions

House flies are a serious pest capable of transmitting many animal and human pathogens including viruses and bacteria (*13*). Abamectin is potent and could become a highly useful insecticide for use against house flies. A major question remains: will resistance develop rapidly under field conditions? It is impossible to extrapolate laboratory results, such as those reviewed in this chapter, directly to field conditions. However, a few points are worthy of consideration. First, field populations of house flies are clearly not as sensitive to abamectin as laboratory susceptible strains (5,7,8) suggesting that efficacy could be less than expected by laboratory assays. Second, we were able to select for high levels of resistance quite rapidly (9) suggesting that genes for expression of high level resistance are present. On the other hand, resistance was recessive (10), apparently unstable (unpublished observation), did not confer resistance to other common insecticides (9) and resistance to formulated material was much less than to technical material (Table I). These factors suggest that resistance might not develop quickly under field conditions. Overall, it is clear that abamectin has potential for use against house flies, but that it must be used judiciously if the development of resistance is to be delayed.

Acknowledgements

The critical review of this manuscript by Dr. R. M. Hollingworth was greatly appreciated. Support for this research was provided by Hatch project 139414 and Merck Sharp and Dohme.

Literature Cited

1. Lasota, J. A.; Dybas, R. A. *Ann. Rev. Entomol.* **1991**, *36*, 91-117.
2. Scott, J. G.; Georghiou, G. P. *Pestic. Sci.* **1986**, *17*, 195-206.
3. Scott, J. G.; Georghiou, G. P. *Econ. Entomol.* **1985**, *78*, 316-19.
4. Scott, J. G.; Wheelock, G. D. In *Molecular Mechanisms of Insecticide Resistance: Diversity Among Insects*; Mullin, C. A.; Scott, J. G. Eds.; ACS Symp. Ser. No. 505; American Chemical Society; Washington, DC, 1992, pp 16-30.
5. Scott, J. G. *Pestic. Biochem. Physiol.* **1989**, *34*, 27-31.
6. Roush, R. T.; Wright, J. E. *J. Econ. Entomol.* **1986**, *79*, 562-564.
7. Geden, C. J.; Rutz, D. A.; Scott, J. G.; Long, S. J. *J. Econ. Entomol.* **1992**, *85*, 435-440.
8. Geden, C. J.; Steinkraus, D. C.; Long, S. J.; Rutz, D. A.; Shoop, W. L. *J. Econ. Entomol.* **1990**, *83*, 1935-1939.
9. Scott, J. G.; Roush, R. T.; Liu, N. *Experientia* **1991**, *47*, 288-291.
10. Konno, Y.; Scott, J. G. *Pestic. Biochem. Physiol.* **1991**, *41*, 21-28.
11. Deng, Y.; Casida, J. E. *Pestic. Biochem. Physiol.* **1992**, *43*, 116-122.
12. Hatano, R.; Scott, J. G. *J. Pestic. Sci.* **1993**, *18*, 281-284.
13. Kettle, D. S. In *Medical and Veterinary Entomology; Kettle*, D. S., Ed.; Croom Helin, London, 1984, pp 229-231.
14. Oppenoorth, F. J. In *Comprehensive Insect Physiology, Biochemistry and Pharmacology*, Kerkut, G. A.; Gilbert, L. I., Eds.; Pergamon Press, Oxford, 1985, pp 731-773.

15. Scott, J. G. In *Handbook of Pest Management*, Vol. 2, Pimentel, D., Ed.; CRC Press, Boca Raton, FL, 1991, pp 663-677.
16. Scott, J. G.; Shono, T.; Georghiou, G. P. *Experientia* **1984**, *40*, 1416-1418.
17. Dong, K.; Scott, J. G. *Pestic. Biochem. Physiol.* **1991**, *41*, 159-169.
18. Dong, K; Scott, J. G.; Weiland, G. A. *Pestic. Biochem. Physiol.* **1993**, *46*, 141-148.
19. Dong, K; Scott, J. G. *Insect Biochem. Molec. Biol.* **1994**, (In press).

RECEIVED December 8, 1994

BACILLUS THURINGIENSIS INDUCED CATION CHANNELS

Chapter 19

Permeability of *Bacillus thuringiensis* CryI Toxin Channels

Michael G. Wolfersberger

Department of Biology, Temple University, Philadelphia, PA 19122

During sporulation *Bacillus thuringiensis* produces parasporal inclusions with insecticidal activity. These parasporal inclusions consist of one or more Cry proteins. Type-I Cry proteins are active only against the larvae of lepidopteran insects. Their mode of action includes binding to specific components of the brush border membrane of larval midgut columnar cells and culminates with formation of a pore in the cell membrane. The resulting increase in cell membrane permeability leads eventually to cell lysis, disruption of gut integrity, and finally death of the insect. Recently we have been able to reconstitute both CryIA(c) toxin proteins and toxin binding proteins from the brush border membrane of insect midgut cells into planar lipid bilayers and determine the conductance of the resulting pores. The results of these determinations are compared with previous estimates of pore size obtained by less direct methods.

The species *Bacillus thuringiensis* consists of numerous strains of gram-positive rod-shaped bacteria which during sporulation produce crystalline proteinaceous parasporal bodies. The parasporal body proteins produced by many strains of *B. thuringiensis* are toxic to certain insect larvae. Therefore, they are frequently called either delta-endotoxins (older literature) or insecticidal crystal proteins (more recent literature). For many years *B. thuringiensis* strains have been classified primarily on the basis of flagellar antigens (*1*). This classification system brought some order to the huge collection of strains but proved to be of very limited utility in predicting the spectrum of larvicidal activity of the various strains. The application of modern molecular genetic methods to *B. thuringiensis* revealed that genetic information for most insecticidal parasporal body proteins was encoded on transmissible plasmids rather than on the bacterial chromosomes. Furthermore, it was determined that a single plasmid could carry the information for more than one insecticidal parasporal body protein and that a single bacterium could contain more than one plasmid (*2*).

0097–6156/95/0591–0294$12.00/0

These findings explained why seemingly closely related strains could have rather different spectra of larvicidal activity. When a critical mass of information, including primary structure, about the genes encoding *B. thuringiensis* insecticidal parasporal body proteins had accumulated a classification system was proposed (*2*). This system grouped the proteins on the basis of their similarities in both primary structure and insecticidal spectra. The rapid adoption of this classification system by the *B. thuringiensis* research community has made it possible to compare and combine the more recent results from various laboratories and in some cases even make sense of the older literature.

Relatedness of *B. thuringiensis* Crystal Proteins

Briefly, the *B. thuringiensis* insecticidal crystal protein (ICP) classification system divided all parasporal body proteins into five groups. Four of these groups contain Cry proteins that are encoded by *cry* genes. The fifth group contains Cyt proteins that are encoded by *cyt* genes. Unlike Cry proteins, Cyt proteins are cytolytic to a variety of non-insect cells. Cyt proteins seem to be genetically unrelated to Cry proteins. All Cry proteins seem to be genetically related. Cry proteins in the first and by far the largest group (CryI) have greater than 50% amino acid sequence identity and are active against lepidopteran larvae. CryII proteins show activity against lepidopteran and/or dipteran larvae. Although they are often found together in the same bacterium, CryII proteins are rather distant relatives of CryI proteins. CryIII proteins are more closely related genetically to CryI proteins than are CryII proteins. However, CryIII proteins are toxic only to certain coleopteran larvae. A major breakthrough in *B. thuringiensis* research occurred when the crystal structure of a CryIII toxin was determined at 2.5 Å resolution (*3*). Members of the fourth class of Cry proteins (CryIV) often occur together with one another as well as Cyt proteins in the parasporal bodies of *B. thuringiensis* strains that show larvicidal activity against certain dipteran insects. The CryIV class of ICPs is a comparatively diverse group. CryIVA and CryIVB are about as closely related to one another as are the proteins in the CryI class. CryIVA and CryIVB proteins are also related, about equally, to CryI and CryIII ICPs. However, on the basis of amino acid sequence, CryIVC proteins are only slightly more closely related to CryIVA and CryIVB proteins than to CryI or CryIII proteins. Finally, CryIVD proteins are related most closely to CryII ICPs.

A somewhat controversial update of the original classification system introduced two new classes of Cry proteins (*4*). ICPs in one of these new proposed classes were said to be completely unrelated to any other Cry proteins. These authors also proposed that CryIVC proteins are much more closely related to CryIVA and CryIVB proteins than to any other ICPs. Hopefully, a consensus update of the 1989 classification will be fourthcomming following the Second International Conference on *Bacillus thuringiensis* during the summer of 1994.

Processing of Cry Proteins

Most Cry ICPs are regarded to be protoxins because full larvicidal activity is contained within a portion of the ICP. Ingested IPCs are partially digested in the midguts of susceptible insect larvae. In the case of CryIVA and CryIVB as well as all

CryI ICPs the "toxin" portion of the ICP resides in the N-terminal half of each molecule, the entire C-terminal half being dispensible for toxicity if not stability of the molecules (5). CryII, CryIII and CryIVC ICPs occur naturally C-terminally truncated. Removal of only a few amino acids from the C-termini of these ICPs results in loss of larvicidal activity. However, at least the CryIII ICPs, like CryI ICPs, undergo limited N-terminal digestion in the midguts of susceptible insect larvae (6).

Crystal Structure of CryIIIA Toxin

The sequence of about a dozen amino acids near the C-termini of CryI, CryIII, CryIVA, CryIVB, and CryIVC toxins is extremely similar. This is one of five "blocks" of amino acid sequences shared by all of these toxins (2). As mentioned previously, the crystal structure of CryIIIA toxin has been determined. It was found to consist of three domains. Domain I consists of a bundle of seven α-helixes connected by short loops. Amino acids in the conserved sequence block closest to the N-terminus of all Cry toxins, including CryII and CryIVD, are involved in forming hydrophobic helix 5 in the center of the seven helix bundle. Amino acids in conserved sequence block 2 participate in forming the sixth and seventh helixes of the bundle. Domain II consists of three antiparallel ß sheets stacked roughly parallel to the helix bundle of domain I. The only conserved sequence block found in domain II consists of several amino acids at the C-terminal end of block 2 which connect the end of helix 7 of domain I with the first sheet of domain II. Amino acids of conserved sequence block 2 that are involved in forming helix 7 of domain I are also in contact with one of the sheets in domain II. Domain III again consists of a group of antiparallel ß sheets laying atop domain II at its junction with domain I. Three of the five conserved amino acid sequence blocks are located in domain III. As mentioned above, block 5 amino acids are found at the C-terminus of the toxin. Amino acids of block 3 are found in loops connecting domain II with domain III. Amino acids of block 4 constitute a ß strand adjacent to the strand containing block 5 amino acids. Both of these strands are buried within the molecule at the junction of all three of its structural domains. Since nearly all conserved amino acid sequences are found at sites within the CryIIIA toxin molecule involved in stabilizing its three dimensional structure, it is thought likely that other molecules containing these sequence blocks would adopt a similar structure (3).

Mode of Action of CryI Toxins

Studies of the mode of action of *B. thuringiensis* delta-endotoxins have been in progress since the 1950s. Nearly all of the early studies were conducted on whole larvae and by far the most widely used experimental method was microscopic examination of tissues removed from the larvae and fixed at various times after they had ingested a delta-endotoxin preparation. These studies were invariably conducted with lepidopteran larvae and CryI delta-endotoxin preparations. The first CryIV and Cyt ICP producing strain of *B. thuringiensis* was discovered in 1977 (7) and the first CryIII ICP producing strain of *B. thuringiensis* was discovered in 1983 (8). We now know that the delta-endotoxin preparations used in these early studies often contained more than one CryI ICP and were sometimes contaminated with CryII proteins.

Nonetheless, as seen in the summary by Luethy and Ebersold (*9*), a surprisingly complete picture of critical steps in the mode of action of CryI ICPs emerged from these studies. The primary site of toxin action was clearly the larval midgut. The importance of high lumen pH for solubilization of parasporal bodies as well as the role of lumen proteinases in cleaving ICPs to toxins were recognized. The potential consequences of disrupting the large pH gradient between midgut lumen and midgut cell cytoplasm were discussed. The primary effect of the toxins seemed to be on cell membrane permeability and evidence favored the primary site of toxin action being the brush border membrane of midgut columnar cells. A requirement for specific toxin binding proteins (receptors) in the brush border membrane was not mentioned explicitly but was implied strongly in the Luethy and Ebersold (*9*) review.

The authors of a recent article on the mode of action of *B. thuringiensis* Cry proteins (*10*) endorse the idea, based on the presence of conserved blocks of amino acid sequences in most Cry toxins and strengthened by the locations of these blocks in the structure of the CryIIIA toxin, that the mode of action of all Cry toxins is basically similar. They summarize the mode of action of Cry toxins in four steps: 1) Ingestion, 2) Solubilisation and proteolytic activation, 3) Receptor binding, 4) Formation of toxic lesion.

Studies with Insect Cell Lines. It may seem that advances in understanding the mode of action of CryI toxins between 1981 and 1993 were limited. If this is the case, it might be due in part to the introduction of insect cell lines into *B. thuringiensis* research (*11*). The prospects of eliminating or at least reducing insect rearing plus convenient access to cells *in vitro* lured many investigators into working with insect cell lines. These cell lines were neither derived from nor resembled larval midgut columnar cells. Furthermore, they were 50 to 100 times less sensitive to Cry toxins. Nonetheless, working with these cell lines was not without some rewards. Some of the first evidence for the existence and concerning the nature of CryI toxin binding cell membrane components (*12*) as well as the concept of colloid-osmotic lysis as a general mechanism for the larvicidal action of *B. thuringiensis* delta-endotoxins (*13*) came from studies using insect cell lines. However, quantitative characterization of binding between Cry toxins and membranes of cultured insect cells has never been achieved and the prospects for isolation of toxin binding proteins from insect cell lines continue to appear slim. The use of cultured insect cell lines in *B. thuringiensis* research seemed to decrease considerably after publication of a study showing a lack of correlation between the toxicity of several CryI toxins to spruce budworm larvae and a cell line derived from spruce budworm larvae (*14*). Following the introduction of larval midgut brush border membrane vesicles (BBMV) to CryI mode of action studies (*15*) and especially after the quantitative demonstration of specific high affinity binding between BBMV and CryI toxins (*16; 17*), the amount of effort expended by all but a few laboratories on studies of the effects of *B. thuringiensis* toxins on established insect cell lines seems to have decreased even further.

Molecular Mode of Action. Although the overall course of a successful intoxication of a lepidopteran larva by a CryI *B. thuringiensis* toxin seems to be well established at the level of the major steps involved, many questions remain unanswered at a molecular level. The concentration and affinity of toxin binding sites on its midgut brush border membranes often but not always correlates positively with the susceptibility of a larva to a toxin (18). The irreversible step that follows binding of toxin to the brush border membrane is thought to be associated with insertion of all or part of the toxin into the membrane. This insertion of toxin molecules into the membrane is believed to be necessary for formation of a pore which mediates the potentially lethal increase in membrane permeability (10). However, the size and composition of this membrane pore is unknown.

Membrane pores formed by CryI toxin proteins in the presence of insect receptor proteins could consist of toxin molecules alone or they could consist of some combination of toxin and receptor molecules (10). Pores consisting of toxin molecules alone might be expected to exhibit the same permeability properties whether or not receptor molecules are also present. However, pores composed of both receptor molecules and toxin molecules seem likely to differ significantly in their permeability properties from pores formed by toxin alone. The permeability properties of pores formed by CryIA(c) and CryIC toxins in planar phospholipid bilayers have been determined quantitatively (19; 20). Several studies of the effects of CryI toxins on the permeability of lepidopteran insect tissue culture cells (13; 21) and brush border membrane vesicles prepared from midguts of lepidopteran larvae (15; 22-25) have been published. For reasons discussed above the results of studies with tissue culture cells must be interpreted with caution. Midgut BBMV have the advantage of being the true target of the toxins and contain authentic toxin binding proteins. Although Hendrickx and associates (22) attempted to measure the effect of toxin on alanine permeability, all but the most recent of the studies with BBMV cited above used indirect methods which limited them to detecting changes in membrane permeability only for certain inorganic ions. With their light-scattering assay, Carroll and Ellar (25) were able to study the effects of CryIA(c) toxin on the permeability of larval *Manduca sexta* midgut BBMV to a variety of solutes. However, they were able to draw only qualitative conclusions about relative changes in permeability for the different solutes. By incorporating larval *M. sexta* midgut BBMV into planar phospholipid bilayers we have been able to measure directly and quantitatively the current flow through pores formed by CryIA(c) toxin in the presence of insect midgut proteins that interact specifically with this bacterial protein (Martin, F. G.; Wolfersberger, M. G. *J Exp. Biol.*, in press).

Toxin Pores in Hybrid Bilayers. Fusions of BBMV with planar bilayer lipid membranes (BLMs) were detected as small step increases in membrane current. After evidence of one or more BLM-BBMV fusions had been observed, addition of a small amount (final [toxin] < 2 nM) of CryIA(c) toxin to the chamber supporting the BLM bathed by pH 9.6 KCl solution resulted in one or more large step increases in membrane current. The smallest step increase recorded corresponded to an increase in membrane conductance of 13 nS. However, current increases corresponding to changes in membrane conductance of approximately 26 nS or 39 nS were most common. Membrane current never decreased following one of these step increases; any subsequent increases simply added to the total membrane current. Similar increases in membrane current were never recorded from BLMs exposed to only

BBMV. CryIA(c) toxin at the concentrations used in these experiments had no affect on BLMs that had not been exposed to BBMV. The quantal nature of changes in membrane conductance following addition of toxin to the solutions bathing BLMs that showed evidence of having fused with BBMV (all changes in membrane conductance were within experimental error some multiple of 13 nS) lead us to propose that 13 nS was most likely to be the conductance of a single pore. A cylindrical pore spaning the insulating portion of the bilayer filled with the buffer used in these studies with a conductance of 13 nS would have a diameter of 22 Å. This is approximately twice the diameter of the pore required to allow passage of the largest solutes shown by light scattering measurements to enter CryIA(c) toxin treated BBMV prepared from larval *M. sexta* midguts (*25*). However, it is within the range of diameters of the larger of the two most likely six toxin molecule model pores constructed by Hodgman and Ellar (*26*).

pH Sensitivity of Pores. All light scattering studies of the effects of CryI toxins on the permeability of BBMV have been conducted at pH 7.5. We have not attempted to measure the effects of CryIA(c) toxin on BBMV containing BLMs at this pH. However, a few experiments conducted at pH 8.8 gave results that were qualitatively similar to but quantitatively different from those obtained at pH 9.6. At pH 8.8 the seemingly irreversible stepwise increases in membrane current that occured after addition of toxin were never more than 25% as large as the smallest increases observed at pH 9.6. Pores with conductances in this range would be expected to have diameters of about 9 Å and solute permeabilies very similar to those seen in the light scattering studies at pH 7.5. The pore diameter of the smaller of the two most likely six toxin molecule pore models constructed by Hodgman and Ellar (*26*) brackets this caluclated pore diameter.

The model pores constructed by Hodgman & Ellar (*26*) consisted only of conserved amino acid sequences found in Cry toxins that fulfilled certain criteria for potentially forming membrane spanning helixes. pH was not considered explicitly in model construction. Slatin and associates (*19*) were the first to demonstrate that Cry toxins alone actually could form pores in phospholipid bilayers. Neither CryIA(c) nor CryIIIA toxin pores were observed at pH 7 but both toxins formed pores at pH $\geq$ 9.5. The pores formed by CryIA(c) toxin at pH 9.7 were highly cation-selective and showed two major conductance states. The higher and more frequently observed "single channel" conductance was 600 pS. The other major state had a conductance of 200 pS. Both states showed the same 25:1 P_K:P_{Cl} ion selectivity. Both high and low conductance channels alternated, in what appeared to be a completely independent manner, between open and closed on a time scale of seconds.

Pore Composition. We have independently confirmed the conductance and gating characteristics reported by Slatin and associates (*19*) for pores formed by CryIA(c) toxin in planar phospholipid bilayers (Martin F. G.; Wolfersberger, M. G., unpublished data). However, as mentioned above, when BBMV prepared from the midguts of *M. sexta* larvae were introduced into an otherwise identical system, very different results were obtained. The toxin concentration required for pore formation was at least two orders of magnitude lower, the pores formed were much larger, and after forming or opening they have never been observed to close. The great differences between the properties of CryIA(c) toxin pores formed in phospholipid bilayers in the presence and absence of specific toxin-binding components of target insect cell membranes favors insect cell membrane components playing an active role in pore formation.

Knowles and Dow (*10*) discuss two ways in which insect membrane receptors might interact with Cry toxins to form pores different from those formed in their absence. In one case the receptor molecules, which are presumed to be integral membrane proteins, combine with the toxin to form the walls of the pore. The resulting hybrid pores would differ in structure and almost certainly also in functional properties from pores composed of toxin molecules alone. In their second model, the pore is lined only with toxin molecules. However, binding between toxin and receptor triggers a change in the conformation of the toxin so that the portions of the toxin that insert into the membrane and line the pore are ones that are buried within the structure of the toxin molecules in aqueous solution. In this case, although the pore is lined only by portions of toxin molecules, the portions of toxin molecules lining the pore could be different from those lining pores formed in the absence of receptors. Knowles and Dow (*10*) seem to favor their second model. It is certainly not without appeal and fits well with the results of molecular modeling (*26*). However, they suggest that high pH and/or proximity to a hydrophobic membrane might have similar effects on toxin conformation as receptor binding. Proximity to a hydrophobic membrane is necessarily part of all studies of the effects of toxins on natural or artificial membranes. pH clearly affects toxin pore formation as well as the properties of the resulting pores (*20*) but not nearly as much as the presence or absence of toxin binding proteins in the membrane under study.

Acknowledgments

My research on the mode of action of *Bacillus thuringiensis* toxins has been supported by grants from the United States Department of Agriculture and the National Institutes of Health.

Literature Cited

1. deBarjac, H.; Bonnefoi, A. *Entomophaga* **1973**, *18*, 5-17.
2. Hoefte, H.; Whiteley, H. R. *Microbiol. Rev.* **1989**, *53*, 242-255.
3. Li, J.; Carroll, J.; Ellar, D.J. *Nature* **1991**, *353*, 815-821.
4. Feitelson, J.S.; Payne, J.; Kim, L. *Bio/Technol.* **1992**, *10*, 271-275.
5. Aronson, A. I.; Beckman, W.; Dunn, P. *Microbiol. Rev.* **1986**, *50*, 1-12.
6. Gill, S.S.; Cowles, E.A.; Pietrantonio, P.V. *Ann. Rev. Entomol.* **1992**, *37*, 615-636.
7. Goldberg, L.; Margalit, J. *Mosquito News* **1977**, *37*, 355-358.
8. Krieg, A.; Huger, A.M.; Langenbruch, G.A.; Schnetter, W. *Z. Angew. Entomol.* **1983**, *96*, 500-508.
9. Luethy, P.; Ebersold, H.R. In *Pathogenesis of Invertebrate Microbial Diseases*; Davidson, E.W. Ed.; Allanheld Osmun:Totowa, NJ, 1981; pp 235-267.
10. Knowles B.H.; Dow, J.A.T. *BioEssays* **1993**, *15*, 469-476.
11. Murphy, D.W.; Sohi, S.S.; Fast, P.G. *Science*, **1976** *194*, 954-956.
12. Knowles, B.H.; Thomas, W.E.; Ellar, D.J. *FEBS Lett.* **1984**, *168*, 197-202.

13. Knowles, B.H.; Ellar, D.J. *Biochim. Biophys. Acta* **1987**, *924*, 509-518.

14. Witt, D.P.; Carson, H.; Hodgdon, J.C. In *Fundamental and Applied Aspects of Invertebrate Pathology*; Samson, R.A.; Vlak, J.M.; Peters, D. Eds.; Foundation of the Fourth International Colloquium of Invertebrate Pathology: Wageningen, 1986; pp. 3-6.

15. Sacchi, V.F.; Parenti, P.; Hanozet, G.M.; Giordana, B.; Luethy, P.; Wolfersberger, M.G. *FEBS Lett.* **1986**, *204*, 213-218.

16. Hofmann, C.; Luethy, P.; Huetter, R.; Pliska, V. *Eur. J. Biochem.* **1988**, *173*, 85-91.

17. Hofmann, C.; Vanderbruggen, H.; Hoefte, H.; VanRie, J.; Jansens, S.; VanMellaert, H. *Proc. Natl. Acad. Sci. USA* **1988**, *85*, 7844-7848.

18. Wolfersberger, M.G. *Experientia* **1990**, *46*, 475-477.

19. Slatin, S.L.; Abrams, C.K.; English, L. *Biochem. Biophys. Res. Commun.* **1990**, *169*, 765-772.

20. Schwartz, J-L.; Garneau, L.; Savaria, D.; Masson, L.; Brousseau, R. *J. Membrane Biol.* **1993**, *132*, 53-62.

21. Schwartz, J-L.; Garneau, L.; Masson, L.; Brousseau, R. *Biochim. Biophys. Acta* **1991**, *1065*, 250-260.

22. Hendrickx, K.; deLoof, A.; VanMellaert, H. *Comp. Biochem. Physiol.* **1989**, *95C*, 241-245.

23. Wolfersberger, M.G. *Arch. Insect Biochem. Physiol.* **1989**, *12*, 267-277.

24. Uemura, T.; Ihara, H.; Wadano, A.; Himeno, M. *Biosci. Biotech. Biochem.* **1992**, *56*, 1976-1979.

25. Carroll, J.; Ellar, D.J. *Eur. J. Biochem.* **1993**, *214*, 771-778.

26. Hodgman, T.C.; Ellar, D.J. *DNA Sequence* **1990**, *1*, 97-106.

RECEIVED January 12, 1995

Chapter 20

Modulation of δ-Endotoxin Ion Channels

L. English[1], F. Walters[1,3], M. A. Von Tersch[1], and S. Slatin[2]

[1]Ecogen, Inc., 2005 Cabot Boulevard West, Langhorne, PA 19047
[2]Albert Einstein College of Medicine, Department of Physiology and
Biophysics, 1300 Morris Park Avenue, Bronx, NY 10437

The importance of Delta-endotoxin-formed ion channels to
the mode of action of these toxins is not fully understood,
but there are no examples of toxicity without this function.
Not all toxins have equivalent channel activity in planar lipid
bilayers and the impact of receptor proteins in modulating
channel function is just beginning to be evaluated. The
activity of a single ion channel-producing toxin has not been
rigorously evaluated. Channel conductance, frequency of
gating, the frequency distribution of numerous possible
conductant states, as well, and the duration of the on-state
are important variables that require more careful analysis
before the toxicity of this specific step in the mode of action
is understood. It is important to emphasize that the mode
of action of the delta endotoxins consists of several
sequential events; so the relative strength of any one step
can be modulated by the prior or subsequent events.

We have evaluated the ion channels formed by several delta endotoxins
in planar lipid bilayers and phospholipid vesicles. All of these evaluations
examined, only superficially, the channel forming ability of each toxin. A
thorough analysis of a delta endotoxin might include a description of the
different conductance states, the frequency of their occurrence, and an
examination of the factors that modulate the appearance of one
conductance state over another. In addition, the frequency of gating and
the probability of observing the ion channels in the open state need to be
determined for each conductance state of the toxin.

In our initial analysis of CryIA(c) we observed highly cation-
selective channels with conductance states from 150-600 pS (*1*). These
channels were observed infrequently, but had open states between 0.5-1

[3]Current address: Department of Biology, Geneva College, Beaver Falls, PA 15010

second. In subsequent observations of CryIA(c) and CryIIIA we also observed very low conductances between 5-10 pS as well as conductances of 1000-2000 pS that do not resolve into "well-behaved" ion channels; that is, they do not have well-resolved open states, but flutter on and off. There is no reason to exclude these phenomena as events that are less characteristic of the behavior of CryIA(c) and CryIIIA. Unfortunately, it is not clear what regulates the behavior of these different events.

Modulating Factors

Some insight into the variable behavior of CryIA(c) was gained in our analysis of Domain 1, the putative α helical N-terminal half of the active toxin molecule. This region of the toxin was sufficient to form well-resolved channels (2); but also produced "noisy" conductances. Unlike the full-length toxin, however, Domain 1 was significantly less cation selective and the net $^{86}Rb^+$ leak through phospholipid vesicles was significantly greater than CryIA(c). These data suggest that Domain 1 contains the necessary structural information to form the well-resolved ion channels and the noisy conductance states, but perhaps the ion selectivity is modulated by residues in other domains of either the same toxin or other toxin molecules that form a higher order quarternary structure.

We also cloned and expressed the α helical Domain 1 from CryIIIB2 (3), one of two delta endotoxins with a well-resolved (2.4 Å) tertiary structure (Cody *et al*, in preparation) containing three domains similar to CryIIIA (4). As with the putative α helical domain from CryIA(c), this Domain 1 did not have the identical ion channel activity shown by the full length CryIIIB2 toxin. In this instance the channels formed by Domain 1 were smaller than the parent toxin, and the concentration-dependent leak from phospholipid vesicles was also reduced. Unlike Domain 1 from CryIA(c) Domain 1 from CryIIIB2 had similar cation-selectivity when compared with the native toxin. Once again, these data suggest that Domain 1 contains sufficient information to form ion channels, but Domains 2-3 can modulate this activity.

We also observed a qualitative improvement in the frequency of CryIA(c) and CryIIIA-induced conductance by using the neutral lipid phosphatidyl choline in the bilayer instead of the mixture of lipids found in asolectin (a mixture of soybean phospholipids). Even though CryIA(c) binds to several different phospholipids, and certainly not all lipid combinations have been evaluated, this qualitative discovery has permitted more rapid evaluation of other *Bt* toxins in the bilayer system.

EDTA increased the frequency of events observed in several experiments with CryIA(c). Ca^{2+} (5 mM) is generally present in all of the buffers in the bilayer experiments, along with 1 mM EDTA to chelate any heavy metals that might contaminate the toxin. Addition of sufficient EDTA to the bilayer to chelate most Ca^{2+} increased CryIA(c) conductance. Once again, more detailed analyses of this phenomena are required. Crawford and Harvey (5) observed that Ba^{2+} and Ca^{2+} can reduce the toxin-induced inhibition of the transepithelial short circuit across the midgut epithelium, but these ions did not affect the ability of a delta endotoxin to inhibit Na^+ or K^+-dependent amino acid transport (6). These data simply do not prove or disprove the possibility of a direct interaction of metals with the toxin in a bilayer.

Our analysis of CryIIA provided evidence for another modulatory force that can alter the behavior of delta endotoxins in the bilayer and in the biological membrane. Unlike CryIA(c), CryIIIA and CryIIIB2, CryIIA conductance demonstrated voltage-dependence in squalene films (7). The sign of the voltage was such that it would turn channels on under physiological conditions, that is, the negative membrane potential would open CryIIA channels. The voltage-dependence of CryIIA was not observed in thicker black lipid membranes prepared according to the method of Mueller et al (8). The squalene films are thinner and perhaps more physiologically representative than the black lipid membranes. Whatever the case, the data suggested that CryIIA can take advantage of the membrane potential across the midgut epithelium. If this is true then the membrane potential might govern insect sensitivity to CryIIA to some degree. The voltage-dependence of CryIIA in Montal films also suggests that other delta endotoxins might also display voltage-dependence under some conditions. Our experience, however, suggests that CryIA(c) and CryIIIA are not voltage-sensitive channels.

Schwartz et al. (9) reported that pH can modulate the behavior of CryIC and actually change the ion selectivity of the CryIC channels. A pH-dependent change in ion selectivity is certainly possible as charged groups that line the ion pore formed by the delta endotoxin can change depending on the pH. At low pH one can imagine that carboxyl groups lining the channel might be titrated to create a more positively charged electrostatic surface and thereby change the selectivity from cations to anions. In addition, because delta endotoxins have multiple conductance states, a change in pH might selectively favor one or more channel conformations. Of course, in designing experiments to evaluate ion selectivity, the specificity of the channel for cations or anions of specific dimensions should be considered. The effect of ion size on the reversal potential has not been evaluated. All of our measurements were made under alkaline conditions primarily to insure that the toxins remained in solution. Therefore, we have not determined selectivity except under alkaline conditions and where KCl was the current-carrying salt.

The single greatest modulation of ion channel activity is likely to be the receptor for these toxins. In experiments where brush border membrane was reconstituted into phospholipid vesicles, the effective concentration of toxin was reduced 1000-fold (*10*). Similarly, when the binding proteins from *Manduca sexta* were reconstituted into phospholipid vesicles, the effective toxin concentration was reduced 1000-fold (*11*). It is reasonable to propose that the binding proteins serve to increase the local concentration of toxin at the membrane, but it is also possible that the binding proteins change the channels formed by the toxin. Most likely a combination of these events occurs. By increasing the local concentration of toxin on the membrane, the membrane-bound toxin can form higher order structures resulting in increased conductance. We conjecture that selectivity also might diminish based on the results of experiments to characterize the toxin pore size (*12*). In these experiments where osmotic pressure is used to move solutes, delta endotoxins induce membrane changes where sugars can leak out of vesicles.

A central question in the analysis of delta endotoxin-formed channels is: How many toxin molecules are required to form a channel? While we have determined that the quarternary structure of CryIIIA and CryIIIB2 is a dimer in solution (*13*), it is not at all obvious that this dimer corresponds to a form of the toxin capable of spontaneously penetrating a bilayer and forming a channel. All that can be said is that the dimer, in these instances, is the most likely form of the toxin in solution and therefore, the minimal unit to initially associate with the bilayer. Our observation of numerous conductance states for CryIA(c), CryIIA, CryIIIA and CryIIIB2 suggests that several arrangements of the toxin in the bilayer are possible. Naively, one might consider the smallest association of toxin molecules to be responsible for the small conductance channels. Increasing conductance might then be the result of larger arrays of toxins that gate together. The high cation selectivity of the CryIA(c) and CryIIIA-induced channels is consistent with this hypothesis. In these instances, the 4000 pS channels of CryIIIA and the 600 pS channels of CryIA(c) were still cation selective suggesting that the large conductance was the result of numerous smaller channels gating synchronously. We have since observed very small cation-selective CryIIIA channels on the order of 8-10 pS. These may be the smallest channels formed by this toxin, but even so, we have not determined how many toxin molecules form this channel.

Our data suggest that at least two types of toxin-toxin interactions occur to modulate the ion channel function. There are toxin-toxin interactions that occur to stabilize the formation of the dimer in solution, and then, there are toxin-toxin associations in the membrane that favor the formation of larger arrays of toxins that gate together and yet retain ion selectivity.

Ion Channel Activity and Toxin Strategies

The formation of ion channels and less specific leaks in the insect midgut are believed to be the end-point toxic events in the mode of action of *Bt* delta endotoxins. The modulation of these channels by the physiology of the midgut membrane may play an important role in producing specific and lethal activities for these toxins. But, it cannot be overstated that the diverse channels formed by delta endotoxins are only one of several events used by delta endotoxins in the overall toxicological strategy. In fact, the most severe modulation of toxin activity is the loss of toxin molecules along the way to the membrane. From the initial ingestion step and following, these insecticidal agents have taken advantage of specific chemical interactions to produce a toxic response. No one has yet demonstrated that insects taste proteins (*14*). Specifically, no protein-dependent electrical discharges have been recorded from insect taste cells or whole sensilla, so taste-dependent avoidance of the *Bt* toxin is unlikely. Solubility of *Bt* toxins is quite variable which makes early or premature sensing of the toxin less likely for CryIIA than for CryIA(c) (*14*). The poor solubility of CryIIIA, however, may be so severely restricted that most of the toxin molecules never form channels (*16*).

All delta endotoxins have remarkable structural stability permitting residence as soluble proteins in the insect digestive track, though some proteins appear to be more stable than others. For example, CryIA(c) (55 kDa) and CryIIA have half lives in digestive fluids that are greater than 20 minutes, but CryIIIA has a 14-minute half life in the Colorado potato beetle (*17*).

We also know that binding to brush border epithelium can range from highly specific with few saturable binding proteins to non-specific, nonsaturable binding seen with CryIIA (*15*). Collectively, the biochemical and biophysical properties of individual toxins in the host insect are strategically associated to successfully pass toxin equivalents to the membrane. Weaknesses in one step may be compensated in another. For example, the poor solubility of CryIIIA may be compensated by the high conductance channels produced by this toxin.

The reduced solubility of CryIIA, when compared with CryIA(c) might be compensated by the voltage-dependence of CryIIA. The very few but highly specific binding for CryI toxins in larval Lepidoptera may be just as effective as the less specific but more abundant binding sites for CryIII toxins in larval Coleoptera. In other words, *Bt* toxins might achieve potency by several different compensatory mechanisms.

Conclusions

The channel forming ability of delta endotoxins is a versatile multi-component step in the overall mode of action of these toxins. The type of channels, their frequency and conductance properties are not well-characterized but each toxin that has been examined is different. Factors modulating the channel function include the brush border membrane and associated biophysical properties of that membrane (Dj, V, pH, binding proteins, lipids, digestive detergents). In addition, steps in the mode of action prior to channel formation will dramatically modify the behavior of the toxins (solubility, stability, binding). All together, these parameters offer the biophysical chemist an enormous variety of opportunities to redesign *Bt* toxins to improve the performance of this natural resource.

Literature Cited

1. Slatin, S. L., Abrams, C. K., and English, L., 1990. *Biochem. Biophys. Res. Commun.* 169, 765-772.
2. Walters., F., Slatin, S. L., Kulesza, C. A., and English, L. H., 1993. *Biochem. Biophys. Res. Commun.* 196, 921-926.
3. VonTersch, M., Slatin, S. L., Kulesza, C. A. and English, L. H., 1994. *Appl. Environ. Microbiol.* In Press.
4. Li, J., Carroll, J., and Ellar, D. J., 1991. *Nature.* 353, 815-821.
5. Crawford, D. N. and Harvey, W. R., 1988. *J. Exp. Biol.* 137, 277-286.
6. Wolfersberger, M. G., 1989. *Arch. Insect Biochem. Physiol.* 12, 267-277.
7. Montal, M. C., 1974. *Methods Enzymol.* 32, 545-554.
8. Mueller, P., Rudin, D. O., Tein, H. T., and Wescott, W. C., 1963. *J. Phys. Chem.* 67, 534-535.
9. Schwartz, J. L., Garneau, L., Masson, L., Brousseau, R., and Rousseau, E., 1993. *J. Memb. Biol.*
10. English, L. H., Readdy, T. L., and Bastian, A. E., 1991. *Insect Biochem.* 21, 177-184.
11. Sangadala, S., Walters, F. W., English, L. H., and Adang, M. J., 1994. *J. Biol. Chem.* 269, 10088-10092.
12. Carroll, J., and Ellar, D., 1993. *Eur. J. Biochem.* 214, 771-778.
13. Walters, F., Phillips, A., Kulesza, C. A., and English, L., 1994. *Insect Biochem. Molec. Biol.* In Press.
14. Schoonhoven, L. M., 1987. In *Perspectives in Chemoreception and Behavior*, Chapman, R. F., Bernays, E. A., Stoffolano, J. G., Springer-Verlag, London, 1987.
15. English, L., Robbins, H. L., Von Tersch, M. A. Kulesza, C., Avé, D., Coyle, D., Jany, C. S., and Slatin, S. L., 1994. *Insect Biochem. Molec. Biol.* In Press.
16. Koller, C. N., Bauer, L. S., Hollingsworth, R. M., 1992. *Biochem. Biophys. Res. Commun.* 692-699.
17. Slaney, A. C., Robbins, H. L., and English, L., 1992. *Insect Biochem. Molec. Biol.* 22, 9-18.

RECEIVED August 23, 1994

Chapter 21

Mechanism of Action of *Bacillus thuringiensis* CytA and CryIVD Mosquitocidal Toxins

Sarjeet S. Gill[1,2], Cheng Chang[1-3], and Edward Chow[1,4]

[1]Department of Entomology and [2]Environmental Toxicology Graduate Program, University of California, Riverside, CA 92521

Both the CytA and CryIVD mosquitocidal toxins have structural domains that have different functionality. The CytA toxin has distinct domains involved in cell binding and pore formation. Monoclonal antibodies developed against the native CytA toxin identify a putative cell binding region of the CytA toxin. These antibodies can inhibit the cytolytic activity of the CytA toxin. Similarly the N- and C-terminal fragments of the CryIVD toxin appear to have distinct functions, with the N-terminal region retaining the pore forming ability. Both of these toxins act synergistically to exert their mosquitocidal activity. This synergistic action is probably critical in delaying the onset of mosquito resistant to wild-type *B. thuringiensis* subsp. *israelensis* parasporal inclusions which are known to contain multiple toxins.

Bacillus thuringiensis subsp. *israelensis*, an isolate highly insecticidal to mosquitoes and blackflies (1), produces parasporal inclusions containing proteins of 27-, 72-, 125- and 135 kDa, the CytA, CryIVD, CryIVB, and CryIVA proteins, respectively (2). The amino acid sequences of these proteins have been deduced from their gene nucleotide sequences. Three of these proteins, the CryIVD, CryIVB, and CryIVA, are structurally related and are primarily mosquitocidal, while the fourth, a structurally unrelated protein, CytA, has cytolytic activity in addition to mosquitocidal activity (2,3).

Although the primary structures of these proteins has been elucidated there is significantly less information on the mechanisms by which these toxins exert their mosquitocidal and/or cytolytic effects. The CytA is generally regarded as non-

[3]Current address: Department of Biochemical Genetics, City of Hope Medical Center, Duarte, CA 91010
[4]Current address: Ciba-Geigy Company, 400 Farmington Avenue, Farmington, CT 06032

specific, although recent evidence suggests that in vivo it demonstrates selective action (4). The focus of this chapter is to highlight some of the recent findings regarding these mosquitocidal toxins. The primary emphasis of this chapter is on the cytolytic CytA and the mosquitocidal CryIVD toxin.

Mechanism of CytA toxin action

CytA phospholipid interaction. Early studies demonstrated that the CytA protein interacts with cell membrane phospholipids, with better toxin binding to zwitterionic phospholipids (3). Incubation of the activated CytA toxin with dioleolyl phosphatidylcholine decreases its cytolytic activity (3). This neutralizing ability is characteristic of phosphatidylcholine having unsaturated acyl residues. In fact, unsaturation only at the *syn*-2 position of the phosphatidylcholine is sufficient to inhibit the hemolytic activity of the toxin (5). Identical lipids with saturated acyl residues did not neutralize the toxin The phospholipid and not the triglyceride moiety are key in this inactivation because monooleolyl glycerol, dioleolyl glycerol or triolein did not inactivate the CytA toxins (3,6).

More recently elegant studies performed by Gazit and Shai (7) demonstrated that selected parts of the CytA toxin could also interact with phospholipids. They synthesized the peptides, helix-1 and helix-2 (Table I) encompassing putative α-helix 1 and α-helix 2 of the CytA toxin (8) and showed that these peptides could form amphipathic helices in phospholipid vesicles (7). Both peptides could bind zwitterionic phospholipids with high affinity; the affinity of these peptides being similar to that of other membrane-permeating peptides such as *Staphylococcus* δ-toxin (9). Moreover, the N-termini of both peptides is located within the vesicle lipid environment. However, while helix-2 can form large aggregates within the lipid membrane at low peptide:lipid molar ratios, helix-1 is unable to. It is likely hence that CytA α-helix 2 plays a role in the pore formed by CytA aggregates.

Table I. Membrane interaction and assembly of synthetic hydrophobic segments of the CytA toxin

Peptide	Sequence	Activity observed
Helix-1	NYILQAIMLANAF QNALVPTST	Binds zwitterionic phospholipid
		N-termini located in lipid environment
		Does not form large aggregates
Helix-2	NQVSVMINKLVE VLKTVLGVAL	Binds zwitterionic phospholipid
		N-termini located in lipid environment
		Forms large aggregates

Data from Gazit and Shai (7)

Cell membrane binding. In most cases when an insect ingests a *B. thuringiensis* toxin, the toxin is solubilized by midgut proteases and an activated toxin is obtained. This activated toxin then binds in a competitive manner a receptor on the insect midgut epithelium. Elegant work initially performed by a number of investigators (10-12) shows a receptor specific to a particular toxin can be identified in the midgut of most insects. With a number of Cry toxins this binding has been well characterized, and other chapters in this book address the characterization and isolation of a putative *B. thuringiensis* toxin receptor. Although no kinetic data has been obtained, the CytA toxin also binds to selective regions of the mosquito midgut (4). There is, however, greater difficulty in determining the kinetics of CytA toxin binding. High toxin concentrations are needed to effectively saturate the available CytA toxin sites (13), and with living cells it is not possible to achieve these levels because the inherent cytolytic activity of these toxins causes cell lysis. At toxin concentrations tolerable by living cells the concentration of bound toxin increases proportionally with toxin concentration in solution, with no saturation observed. Similarly, increased incubation times increase the bound toxin (13).

Analysis of the cell membrane bound toxin has been performed (13). To determine the nature of this bound toxin, radiolabeled CytA toxin treated *Aedes albopictus* cells were extracted with Triton X-100, and the detergent soluble fractions were separated by ultracentrifugation on a 5-20% sucrose density gradient. These experiments were performed with varying CytA toxin concentrations. At low toxin concentrations used (< 0.25μg/ml) and for an incubation time of 30 min, the CytA toxin was present in its monomeric form. At toxin concentrations of 0.3 - 0.4μg/ml, a second peak of about 300-400 kDa molecular weight was observed. At even higher toxin concentrations, 0.7 - 4.0μg/ml, this second peak became pronounced (Fig. 1). Although the aggregate peak size increased with toxin concentration, there is no apparent change in the size or the shape of the monomeric peak. No aggregates of intermediate size were observed. Further, the size of the aggregate peak remained constant for each cell type used. For example, for *Aedes* (C6/36) and *Choristoneura* (CF1) cells the aggregate corresponded to about 400 kDa in molecular weight, while for human erythrocytes the aggregate was about 170 kDa (13).

Toxin aggregation is dependent on the concentration of cell membrane bound toxins. At low toxin concentrations in solution and at short incubation times, the cell membrane toxin concentrations are low and no toxin aggregates are observed. However, when higher toxin concentrations are used membrane toxin concentration increases and toxin aggregates are formed. Hence aggregate formation occurs when a threshold membrane toxin concentration is attained. At high membrane toxin concentration the CytA toxin exists predominantly as an aggregate. Similar results have been observed in erythrocytes with *Staphylococcus* α-toxin, which exists mostly in an aggregate form (14). This *Staphylococcus* α-toxin spontaneously assembles into a dodecameric form under appropriate conditions.

Aggregate formation dependence on membrane toxin concentration demonstrates probable interactions between toxin molecules and implies the presence of more than one CytA toxin in each aggregate. An aggregate of 400 kDa, as observed with *Aedes* or CF1 cells , suggests that if it is composed solely of the CytA

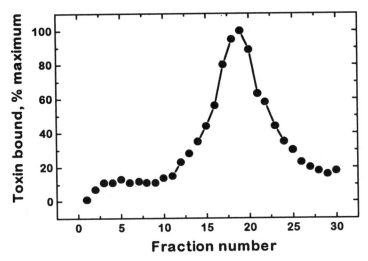

Fig. 1.   Isolation of CytA toxin aggregates from *Aedes albopictus* cell membranes using 2.0 μg/ml toxin at 27°C for 30 min.   Following incubation the cells were washed, solubilized with Triton X-100 and fractionated on a sucrose density gradient.   Individual fractions were analyzed for bound toxin.   The peak in bound toxin is about 400 kDa. Marker enzymes (not shown) were used as standards.
(Reproduced with permission from reference 13.   Copyright 1989 American Society for Microbiology.)

the aggregate consists of sixteen CytA molecules.   It is more likely, however, that the aggregate contains a small number of CytA molecules because the cell membrane aggregate isolated probably also contains other membrane components.   Therefore difference in aggregate size between different cell types probably reflects the presence of these membrane components.

CytA inhibition by monoclonal antibodies.   To evaluate whether specific toxin domains played a role in either binding or aggregation, anti-CytA mouse monoclonal antibodies were developed against both denatured and native toxin.   Mice injected with the native CytA became tolerant after continued injection of this toxin suggesting the production of inhibitory antibodies.

A number of monoclonal antibodies were isolated and analyzed for their ability to inhibit toxin binding to insect cells.   All antibodies capable of blocking toxin binding to cells were also observed to abolished its cytolytic effect.   Only antibodies obtained from mice injected with native toxins were able to block the binding of radiolabeled CytA toxin to cells.   Some of these antibodies recognized the native but not the denatured CytA, however, others recognized both the native and denatured toxin.   None of the monoclonal antibodies obtained from mice treated with heat denatured toxin showed any significant inhibition of binding or aggregation of the CytA toxin on cell membranes.

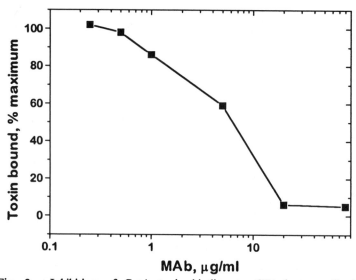

Fig. 2. Inhibition of CytA toxin binding to CF1 insect cells by monoclonal antibody 10B9. Trypsin-activated CytA, 1.0 µg/ml, was incubated with increasing antibody concentrations, and the mixture then incubated with CF1 cells. CF1 bound toxin concentrations were then measured. (Reproduced with permission from reference 13. Copyright 1989 American Society for Microbiology.)

Anti-native CytA monoclonal antibodies could inhibit 80-97% of labeled toxin binding to cells and protect them from lysis by the toxin (Fig. 2). The inhibition of toxin binding to cells was dependent on antibody concentrations. Approximately 10µg/ml antibody could abolish most of the CytA toxin binding (13).

To determine the toxin fragments recognized by these antibodies the CytA protein was cleaved with various proteases, and/or cyanogen bromide. The protein fragments were then separated by SDS-PAGE and transferred to nitrocellulose or Immobilon paper and then probed with these monoclonal antibodies. Bands recognized were subjected to N-terminal sequencing. Three fragments, of 6-, 15- and 17 kDa, were sequenced. Sequence information showed that the 15 kDa band was from the N-terminal of the trypsin activated CytA protein, with an N-terminus identical to that previously reported (5,15). The 6- and 17 kDa fragments were internal sequences of the CytA protein (Fig. 3). Preliminary findings from these mapping studies using one of these antibodies showed that the C-terminal fragments (Fig. 3, B and C) contained the antibody epitope(s). The inability of antibodies, that inhibit binding, to recognize the N-terminal half CytA toxin suggests that this part of the toxin is probably not involved in cell binding.

Summary of CytA action. Three defined steps appear necessary for CytA action (Fig 4). The first involves cell binding, followed by toxin insertion, and finally oligomerization resulting in pore formation. Toxin insertion and formation of aggregates accounts for the irreversibility of the CytA toxin binding and its detergent-like action (3,13).

CytA

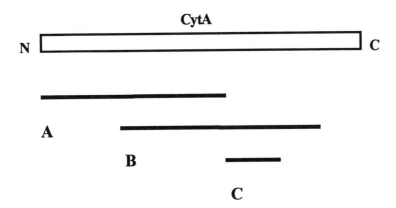

Fig. 3. Proteolytic fragments of the activated CytA toxin. Fragments A, B and C were isolated and sequenced. The epitope for the monoclonal antibody 10B4 is in fragments B and C but not A. The location of the C-termini of these fragments is estimated. The hydrophobic domain which overlaps with the C fragment also contains α helices 3 and 4 based on a predicted secondary structure. (Adapted from reference 8.)

Key for CytA toxicity is the initial binding to the cell membrane. The CytA binds irreversibly to cell membranes in a concentration dependent manner It appears that a specific toxin domain is required for this binding to occur. CytA toxicity inhibition by monoclonal antibodies directed towards the C-terminus region suggests that this region potentially plays an important role in this initial cell membrane recognition. This domain also contains a cysteine residue which when modified affects CytA toxicity (5). This cysteine at position 190 is in a major hydrophobic region from amino acid 170 to 210; the cysteine is adjacent to α-helices 3 and 4 (8). The cysteine itself is probably not involved in toxin binding, however, its modification does disrupt the predicted coil/turn at this site and/or affecting the cell recognition motif nearby. This C-terminal domain also contains the epitope for an cell binding inhibitory monoclonal antibody. Therefore this domain probably is important for CytA binding and critical for toxicity.

The binding of the CytA toxin to cell membranes facilitates its insertion into membranes. It is likely that α-helix 2, amino acids 110-131 of the CytA toxin, forms part of the toxin molecule that is inserted into the membrane. Synthetic peptides encompassing this region are not only able to insert into the membrane but are also able to form aggregates (7). In contrast the putative α-helix 1 (8) while able to interact with the lipid vesicles is apparently unable to form aggregates (7).

The third step in toxin interaction with insect cell membranes is the formation of toxin aggregates. There is evidence that the CytA toxin monomers insert into the membrane (7,13,16). Formation of CytA aggregates probably occurs in the cell membrane rather than in solution, and limited studies performed with the CytA toxin

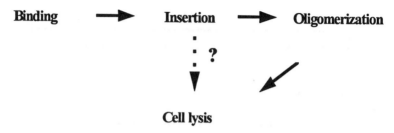

Fig. 4. Putative steps in the cytolytic activity of the CytA toxin.

support toxin oligomerization in the cell membrane (13,16). Two potential models of oligomerization are possible. In the first toxin molecules bind randomly to the membrane, then insert and diffuse laterally in the membrane where they form toxin oligomers. Alternatively, cell membrane inserted toxin molecules facilitate additional toxin insertions thereby leading to the formation of an aggregate. Potentially if this were to occur toxin aggregates of varying size will probably be obtained from cell membranes, however, experimental evidence, i.e. isolation of cell membrane toxin aggregates (13) does not support this hypothesis. It is possible that this toxin-toxin interaction is rapid and not measurable by the techniques used in this latter experiment.

Nevertheless it is clear that toxin aggregates are formed. These aggregates lead to the formation of a pore, which has been estimated to be 1nm in radius (17). The pore size was estimated at a toxin concentration of 10 μg/ml at which the cell membrane bound toxin would exist as aggregates. Pores of similar sizes have been reported for *Staphylococcus* α-toxin and *E. coli* aerolysin (18,19) but substantially smaller than the >30nm pore size estimated for streptolysin O (20). The CytA toxin pore formed is cation selective (21), thereby facilitating disruption of osmotic balance and ultimately cell lysis.

Mechanism of CryIVD toxin action

Pore formation by CryIVD. Investigations by a number of laboratories demonstrate that *B. thuringiensis* toxins insert into membranes and form pores that are permeable to small ions and molecules (21-23). Using a variety of techniques with or without insect cell membrane proteins, these studies demonstrate the capacity of *B. thuringiensis* toxins to form pores that have the capacity to conduct ions. The channels formed are voltage independent cation pores. Although the conductances obtained vary depending on the toxin used in most cases these conductances are large (21-23).

To determine whether the mosquitocidal toxins can form pores the CryIVD toxin was used in a lipid bilayer to monitor ion conductances. These experiments illustrate that like other proteins the CryIVD can form cation channels (Fig. 5) (Gill et al., unpublished results). The CryIVD formed channel conductance is relatively large, approximately 2nS. Further studies to determine the role of the N- and C-terminal regions, i.e. the role of the α-helix rich region and the cell membrane binding

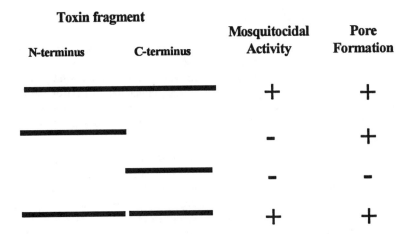

Fig. 5. Mosquitocidal and pore forming ability of the N- and C-terminal fragments of the CryIVD toxin from *B. thuringiensis* subsp. *israelensis*.

domains, respectively, were also performed. Therefore the N- and C-terminal domains individually and together were expressed in *E. coli*, the proteins purified, and channel formation determined. Analysis reveal that the N-terminal retains most of the pore forming ability. This fragment, is however, not mosquitocidal. The C-terminal does not have channel forming activity nor does it posses any mosquitocidal activity. In contrast when both the N and C terminal fragments are utilized together, both pore forming ability and mosquitocidal activity are observed. These studies demonstrate that the N-terminal half by itself can form pore that are functionally similar to the whole CryIVD toxin.

These studies and those in other chapters in this book demonstrate the increasing evidence of channel like activity of the *B. thuringiensis* toxins. Most evidence suggests the probable Cry toxins pore forming amphipathic α-helix is helix 5. Nevertheless it is probable that other parts of the toxin could contribute to pore characteristics. The Cry toxin pore radii has been measure to be about 0.6 nm (17), which is large than the crystal radii of K^+ and Na^+, the ions that are selectively conducted. Moreover large conductances, in the nS range, have bean measured by many investigators. Hence the basis for the ion selectivity that is observed by Cry and Cyt toxins is not known. Potentially the large conductances observed represent population of channels with potentially different ionic selectivities, or the same channel with many different conductance substates.

Expression of mosquitocidal proteins in *Bacillus thuringiensis*. *B. thuringiensis* subsp. *israelensis* and *B. thuringiensis* subsp. *morrisoni* (PG-14) produce four mosquitocidal toxins all of which are essential to obtain the full complement of activity observed in the wild-type parasporal inclusions. These toxins, encoded by a large plasmid, have been cloned and expressed in both *E. coli* and in *B. thuringiensis*

(24-26). The isolate *B. thuringiensis* subsp. *morrisoni* (PG-14) also contains a lepidopteran toxin encoded by another plasmid in this strain. Expression of the CytA toxins in *E. coli* requires the 20 kDa toxin (27,28), however, high level expression of the CytA in *B. thuringiensis* does not require the specific presence of the 20 kDa protein (Fig. 6), but cell viability is enhanced with presence of a second protein in the cell (29). The presence of an additional protein, either the 20 kDa or the CryIVD, enhances the stability of the CytA protein preventing proteolytic cleavage, and hence increasing cell viability (29). In contrast, the CryIVA, CryIVB and CryIVD proteins are highly expressed by themselves (24,25,29).

The presence of multiple proteins in *B. thuringiensis* subsp. *israelensis* suggests that these proteins could potentially interact synergistically, i.e. the mosquitocidal toxicity of each toxin is lower than the toxicity when more than one toxin is present. Indeed a number of investigators over the years have demonstrated that various combinations of these *B. thuringiensis* subsp. *israelensis* toxins act synergistically (30-32). To better estimate the level of synergism obtained with these toxins we expressed the CytA and CryIVD toxins in various combinations to determine the level of this interaction (Fig. 6). When both these toxins are expressed as in the construct CG8, which produces 70% CytA and 30% CryIVD, the LC_{50} value obtained towards *Culex quinquifasciatus* is 26 ng/ ml. As the CytA protein has little mosquitocidal activity, this LC_{50} value results from about 11ng/ml of CryIVD

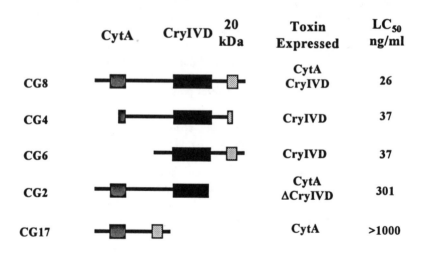

	Cyta	CryIVD	20 kDa	Toxin Expressed	LC_{50} ng/ml
CG8				CytA CryIVD	26
CG4				CryIVD	37
CG6				CryIVD	37
CG2				CytA ΔCryIVD	301
CG17				CytA	>1000

Fig. 6. Construction of combinations of the CytA and CryIVD toxins in the *B. thuringiensis/E. coli* expression vector pHT3101 (33) for expression in *B. thuringiensis*. The toxins expressed were detected by SDS-PAGE and immunoblotting with antibodies raised against either the CytA or CryIVD proteins. Mosquitocidal activity was monitored against *Culex quinquifasciatus* fourth instar larvae and values represent 24-hr LC_{50} values.

protein. When only the CryIVD toxin is expressed, as in CG4 and CG6, the LC_{50} value is 37 ng/ml, and if only the CytA toxin is expressed the LC_{50} value is > 1000 ng/ml. Although the LC_{50} value for the CryIVD/CytA combination is 26 ng/ml, and for CryIVD alone is 37 ng/ml, there is a 4-5 fold synergism between these two proteins even, though the CytA itself has a relatively low toxicity itself (29). Nevertheless the toxicity observed with both the CytA and CryIVD proteins is an order of magnitude lower than that observed with intact *B. thuringiensis* subsp. *israelensis* parasporal inclusions, LC_{50} of 4ng/ml. Synergism has also been observed between the CryIVA and CryIVB and the CytA proteins, however, the precise fold synergism observed has not been established. It is also likely that synergism among other *B. thuringiensis* toxins will be detected.

Effect of multiple toxins and their synergistic effects. Although high resistance levels, both in the laboratory and in the field, have been demonstrated for a number of lepidopteran *B. thuringiensis* toxins (34-39) resistance to *B. thuringiensis* subsp. *israelensis* has been difficult to attain (2,40). Only low resistance levels have been reported. However, selection of mosquitoes with single *B. thuringiensis* subsp. *israelensis* toxin, namely the CryIVD, it is possible to develop resistance more rapidly (2). In part this inability of mosquitoes to develop high resistance levels to *B. thuringiensis* subsp. *israelensis* toxins is due to the interaction of the multiple toxins that are present in this isolate. Consequently, synergism that has been demonstrated between the CytA and CryIVD toxins, albeit of low level, plays an important if not crucial role in deferring resistance development.

Acknowledgment

Supported in part by grants from the NIH (ES03298), and the University of California Mosquito Research

References

1. Goldberg, L. J. & Margalit, J. (1977) *Mosq. News* **37**, 355-358.

2. Gill, S. S., Cowles, E. A. & Pietrantonio, P. V. (1992) *Ann. Rev. Entomol.* **37**, 615-636.

3. Thomas, W. E. & Ellar, D. J. (1983) *FEBS Lett.* **154**, 362-368.

4. Ravoahangimalala, O., Charles, J. F. & Schoeller-Raccaud, J. (1993) *Res. Microbiol.* **144**, 271-278.

5. Gill, S. S., Singh, G. J. P. & Hornung, J. M. (1987) *Infect. Immun.* **55**, 1300-1308.

6. Thomas, W. E. & Ellar, D. J. (1983) *J. Cell Sci.* **60**, 181-197.

7. Gazit, E. & Shai, Y. (1993) *Biochemistry* **32**, 12363-12371.

8. Ward, E. S., Ellar, D. J. & Chilcott, C. N. (1988) *J. Mol. Biol.* **202**, 527-535.

9. Thiaudiere, E. , Siffert, O. , Talbot, J. -C., Bolard, J. , Alouf, J. E. & Dufourcq, J. (1991) *Eur. J. Biochem.* **195**, 203-213.

10. Hofmann, C. , Vanderbruggen, H. , Höfte, H. , Van Rie, J. , Jansens, S. & Van Mellaert, H. (1988) *Proc. Natl. Acad. Sci. USA* **85**, 7844-7848.

11. Hofmann, C. , Luthy, P., Hutter, R. & Pliska, V. (1988) *Eur. J. Biochem.* **173**, 85-91.

12. Van Rie, J. , Jansens, S. , Höfte, H. , Degheele, D. & Van Mellaert, H. (1989) *Eur. J. Biochem.* **186**, 239-247.

13. Chow, E. , Singh, G. J. P. & Gill, S. S. (1989) *Appl. Environ. Microbiol.* **55**, 2779-2788.

14. Bhakdi, S. & Tranum-Jensen, J. (1991) *Microbiol. Rev.* **55**, 733-751.

15. Armstrong, J. L., Rohrmann, G. F. & Beaudreau, G. S. (1985) *J. Bacteriol.* **161**, 39-46.

16. Maddrell, S. H. P., Overton, J. A., Ellar, D. J. & Knowles, B. H. (1989) *J. Cell Sci.* **94**, 601-608.

17. Knowles, B. H. & Ellar, D. J. (1987) *Biochim. Biophys. Acta* **924**, 509-518.

18. Palmer, M. , Weller, U. , Messner, M. & Bhakdi, S. (1993) *J. Biol. Chem.* **268**, 11963-11967.

19. Howard, S. P. & Buckley, J. T. (1982) *Biochem.* **21**, 1662-1667.

20. Buckingham, L. & Duncan, J. L. (1983) *Biochim. Biophys. Acta* **729**, 115-122.

21. Knowles, B. H., Blatt, M. R., Tester, M. , Horsnell, J. M., Carroll, J., Menestrina, G. & Ellar, D. J. (1989) *FEBS Lett.* **244**, 259-262.

22. Slatin, S. L., Abrams, C. K. & English, L. H. (1990) *Biochem. Biophys. Res. Comm.* **169**, 765-772.

23. Schwartz, J. , Garneau, L. , Savaria, D. , Masson, L. , Brousseau, R. & Rousseau, E. (1993) *J. Membr. Biol.* **132**, 53-62.

24. Donovan, W. P., Dankocsik, C. & Gilbert, M. P. (1988) *J. Bacteriol.* **170**, 4732-4738.

25. Ward, E. S. & Ellar, D. J. (1988) *J. Bacteriol.* **170**, 727-735.

26. Waalwijk, C. , Dullemans, A. M., Workum, M. E. S. & Visser, B. (1985) *Nucl. Acids Res.* **13**, 8207-8217.

27. Adams, L. F., Visick, J. E. & Whiteley, H. R. (1989) *J. Bacteriol.* **171**, 521-530.

28. Visick, J. E. & Whiteley, H. R. (1991) *J. Bacteriol.* **173**, 1748-1756.

29. Chang, C. , Yu, Y. -M., Dai, S. -M., Law, S K. & Gill, S. S. (1993) *Appl. Environ, Microbiol.* **59**, 815-821.

30. Wu, D. & Chang, F. N. (1985) *FEBS Lett.* **190**, 232-236.

31. Angsuthanasombat, C. , Crickmore, N. & Ellar, D. J. (1992) *FEMS Microbiol. Lett.* **94**, 63-68.

32. Yu, Y. M., Ohba, M. & Aizawa, K. (1987) *J. Gen. Appl. Microbiol.* **33**, 459-462.

33. Lereclus, D. , Arantès, O. , Chaufaux, J. & Lecadet, M. -M. (1989) *FEMS Microbiol. Lett.* **60**, 211-217.

34. Stone, T. B., Sims, S. R. & Marrone, P. G. (1989) *J. Invertebr. Pathol.* **53**, 228-234.

35. McGaughey, W. H. (1985) *Science* **229**, 193-196.

36. Ferre, J. , Real, M. D., Van Rie, J. , Jansens, S. & Peferoen, M. (1991) *Proc. Natl. Acad. Sci. USA* **88**, 5119-5123.

37. Gould, F. , Martinez-Ramirez, A. , Anderson, A. , Ferre, J. , Silva, F. J. & Moar, W. J. (1992) *Proc. Natl. Acad. Sci. USA* **89**, 7986-7990.

38. McGaughey, W. H. & Johnson, D. E. (1992) *J.Econ.Entomol.* **85**, 1594-1600.

39. Tabashnik, B. E. (1994) *Annu. Rev. Entomol.* **39**, 47-79.

40. Goldman, I. F., Arnold, J. & Carlton, B. C. (1986) *J. Invertebr. Pathol.* **47**, 317-324.

RECEIVED December 6, 1994

Chapter 22

Identification and Functional Characterization of the *Bacillus thuringiensis* CryIA(c) δ-Endotoxin Receptor in *Manduca sexta*

M. J. Adang, S. M. Paskewitz[1], S. F. Garczynski, and S. Sangadala

Department of Entomology, University of Georgia,
Athens, GA 30602–2603

Binding sites for insecticidal δ-endotoxins of *Bacillus thuringiensis* are located in the midgut brush border. We report the purification and functional characterization of 120- and 65-kDa proteins from *Manduca sexta* that bind CryIA(c) toxin. The 120-kDa protein, the major CryIA(c) binding protein, was identified as aminopeptidase N (EC 3.4.11.2). A mixture containing 120- and 65-kDa proteins was obtained by isoelectric focusing and affinity chromatography. When reconstituted into phospholipid membrane vesicles, the 120/65-kDa proteins enhanced toxin binding and catalyzed $^{86}Rb^+$ release. The presence of the 120/65-kDa proteins allows channels to form at nanomolar concentrations of toxin. The structure that attaches the aminopeptidase to the *M. sexta* brush border was also determined. In contrast to vertebrate aminopeptidase N, the 120-kDa aminopeptidase in *M. sexta* is anchored by a glycosyl-phosphatidylinositol moiety.

The bacterium *Bacillus thuringiensis* (*Bt*) produces insecticidal crystals that are important for control of crop pests (1). When susceptible lepidopteran larvae ingest CryI crystals, 130- to 140-kDa protoxins are released in the alkaline midgut lumen. Midgut proteases process the protoxin molecule to form a stable 55- to 70-kDa toxin core. Activated toxin subsequently binds to epithelial cells lining the midgut. Bound toxin molecules form ion channels, causing cell lysis and insect mortality (1, 2).

Brush border membrane vesicles (BBMV) provide a method for presenting membrane surface molecules to CryI toxins *in vitro* for the purpose of characterizing toxin binding. Studies using ^{125}I-labeled toxins and BBMV have correlated high-affinity binding sites in the midgut with susceptibility to CryI toxins (3). Binding of

---

[1]Current address: Department of Entomology, University of Wisconsin, Madison, WI 53706

0097–6156/95/0591–0320$12.00/0

the three CryIA toxins and CryIC toxin has been studied in the greatest detail. The CryIA toxins share overlapping populations of binding sites in susceptible insects, while CryIC toxin clearly does not compete for CryIA binding sites (3 and Ke Luo unpublished observations). The impact of these distinct populations of binding sites on *Bt* activity is best illustrated by *Plodia interpunctella* (Indian meal moth) and *Plutella xylostella* (diamondback moth) larvae that have acquired resistance to the CryIA toxins. Both species acquired resistance to CryIA toxins by reducing the number of CryI binding sites in the midgut, while toxicity and binding of CryIC toxin was relatively unchanged (4, 5). In *P. xylostella*, the first insect with *Bt*-resistant field populations (5), resistance is reversible (6). Reversal of resistance was correlated with the return of CryIA toxin binding sites (6). Unfortunately, exceptions to the correlation between binding and toxicity are reported for natural (7, 8) and acquired resistance (9) to *Bt* toxins.

Molecular Sizes of CryIA(c) Binding Proteins

Lepidopteran insects have single or, more often, multiple toxin-binding proteins. To identify toxin-binding proteins we have employed a protein blot technique called ligand blot analysis. After sodium dodecyl sulfate polyacrylamide gel electrophoresis (SDS-PAGE), membrane vesicle proteins are transferred to nitrocellulose membranes. Candidate binding proteins are identified by hybridization with ^{125}I-labeled ligand. Alternatively, toxin binding is detected using anti-toxin antibodies and enzyme-conjugated secondary antibodies (10). The validity of the ligand blot technique depends on the relevance of binding determinants that survive the conditions of SDS-PAGE and protein blotting. Figure 1 is a ligand blot displaying ^{125}I-CryIA(c) binding to four insect species. *Manduca sexta* had a major CryIA(c) binding protein at 120 kDa and *Spodoptera frugiperda* had only one band at 148 kDa. *Heliothis virescens* and *Helicoverpa zea*, displayed a more complex pattern of CryIA(c) binding proteins ranging in size from 63 to 155 kDa in size. These ligand blot results agreed with the studies of binding of ^{125}I-CryIA(c) toxin to brush border vesicles from larval midguts. We focused attention towards a more detailed characterization of the *M. sexta* 120-kDa protein. Following is a summary of experiments described in Sangadala et al. (11) that identify the 120-kDa protein as aminopeptidase N (3.4.11.2) and demonstrate its function as a toxin receptor.

The 120-kDa CryIA(c) Toxin Binding Protein is Aminopeptidase N

The approach was to purify the 120-kDa protein that binds CryIA(c) toxin on blots by preparative SDS-PAGE using a Bio-Rad prep cell. We collected 60 fractions after the dye front had migrated to the bottom of the gel, then analyzed the samples by SDS-PAGE and ligand blotting. Fraction 33 contained a single protein of approximately 120 kDa that bound ^{125}I-CryIA(c) toxin. This protein was separated a second time by SDS-PAGE, blotted onto membrane, and subjected to N-terminal amino acid sequencing. The N-terminal amino acid sequence YRLPTT from PSYRLPTTTG matched residues 70-75 predicted from the porcine aminopeptidase N gene (12). Knight et al. (13) also purified the 120-kDa protein from *M. sexta* that binds CryIA(c) toxin. The N-terminal sequence of the 120-kDa protein obtained by Knight *et al.* (13) agrees with the sequence we reported (11).

Brush border aminopeptidase is an ectoenzyme that has most of the protein mass protruding external to the cell membrane and a portion anchored in the membrane (14). In mammals, aminopeptidase N is anchored in the membrane via a peptide stalk of approximately 60 amino acids (14). The 120-kDa N-terminal sequence indicates that the CryIA(c)-binding aminopeptidase does not have an N-terminal peptide stalk. A recent study (15) suggested an alternative anchor for membrane attachment of *Bombyx mori* aminopeptidase. Takesue *et al.* (15) reported that at least some of the brush border aminopeptidase is anchored to the epithelial membrane by a glycosyl-phosphatidylinositol (GPI) linkage. Garczynski and Adang (16) concluded that the 120-kDa protein has a GPI membrane anchor based on susceptibility to glycosyl-phosphatidylinositol-specific phospholipase C. The structure of the *M. sexta* aminopeptidase is unique in several respects; the protein has a membrane anchor different from that found in other animal taxa, and most importantly, binds CryIA(c) δ-endotoxin.

Purification of the 120-kDa Protein for Reconstitution and Enzymatic Analyses

For purposes of membrane reconstitution and enzymatic analyses we needed to purify the 120-kDa protein using a method that did not utilize detergents that denature proteins or interfere with membrane reconstitution. Affinity-purification using immobilized CryIA(c) toxin and standard chromatographic methods were unsuccessful (data not shown). A combination of preparative isoelectric focusing followed by immunoaffinity selection yielded a mixture consisting of 120- and 65-kDa proteins.

We prepared monoclonal antibodies (mAbs) against the 120-kDa protein. We used mice and a specialized immunization technique (17) because only a limited amount of purified 120-kDa protein (approximately 10 to 20 μg) was available. Briefly, BBMV proteins were separated on large preparative SDS-gels. Each gel was lightly stained with Coomassie blue in distilled H_2O and the band at 120 kDa excised. Protein was electroeluted from the gel slice and checked for purity by SDS-PAGE and toxin binding by ligand blot analysis. About 10 μg of 120-kDa protein was spotted onto a square of nitrocellulose and implanted intrasplenically in mice as described in Sangadala *et al.* (11). An antibody screening strategy was devised that circumvented the lack of purified antigen. We screened for anti-120 mAbs by western blot analysis using small strips (2 x 20 mm) of blotted BBMV protein. A multi-well tray was devised from a Styrofoam microcentrifuge tube box (Sarstedt) by removing the divider between two holes then sealing the tray with melted paraffin. This simple apparatus made it convenient to screen 50 culture supernatants at a time. From approximately 500 supernatants we selected 5 hybridoma cell lines for further study. Each mAb was able to detect a 120-kDa-protein on a western blot of *M. sexta* BBMV (data not shown).

A pool of three mAbs (2B3, 8G1, and 12B8), collectively called anti-120 mAbs, was used as affinity ligands to purify toxin binding proteins from solubilized BBMV. Polypeptides of 120, 95, 65, and 45 kDa were detected by SDS-PAGE and silver staining (Figure 2). When protein blots were probed with either ^{125}I-CryIA(c) or ^{125}I-mAb, proteins at 120 and 65 kDa showed visible signals (Figure 2). Note, the 120-kDa protein signal on the autoradiogram was much stronger with both probes. Also, a third faint band was seen at 60 kDa. This pattern of bands was identical to total BBMV.

Preparative isoelectric focusing was performed on BBMV proteins solubilized in octyl glucoside using pH 3-10 ampholytes in a Rotofor (Bio-Rad) . The 120-kDa protein along with 80% of the total protein partitioned into two fractions of pH 5 and 6. When the isoelectric focusing step preceded the immunoselection step, the result was a protein fraction containing just the 120- and 65-kDa proteins (Figure 3). Based on molecular size and recognition by CryIA(c) and mAbs, the 120-kDa protein appears to be the toxin-specific protein seen previously on blots of *M. sexta* BBMV.

The 120/65-kDa Proteins Catalyze Toxin-Induced Pore Formation

Using the reconstitution method of English *et al.* (18), phospholipid vesicles were prepared with or without the 120/65-kDa proteins in the presence of [86]Rb+. Dried phosphatidylcholine was resuspended in assay buffer then purified brush border proteins and [86]Rb+ were reconstituted into vesicles by freeze-thaw sonication. Efflux from vesicles was measured by trapping external [86]Rb+ on a cation exchange column while internal [86]Rb+ passes through and is counted. We found that when external CryIA(c) approaches μM concentrations it induces channel formation in membranes even without binding proteins (11). The spontaneous release of [86]Rb+ was minimal below 100 nM. When the 120/65-kDa proteins were incorporated, the concentration of toxin required to induce [86]Rb+ efflux was reduced at least 1000-fold (Figure 4A). Also, per cent of efflux corresponded to the amount of 120/65-kDa proteins in the reconstituted vesicles (Figure 4B). These reconstitution experiments provided the first evidence that proteins which bind toxin on ligand blots actually function as toxin receptors.

CryIA(c) Toxin-Enzyme Interactions

Two enzymes were identified in the 120- and 65-kDa toxin binding protein fraction. Immunoaffinity beads carrying 120/65-kDa proteins were directly assayed for enzyme activities.   The 120-kDa protein had the activity profile characteristic of aminopeptidase N. The Zn^{++} chelating agents 2,2'-dipyridyl and 1,10 phenanthroline inhibited activity, as did the selective aminopeptidase inhibitors; actinonin, amastatin and bestatin.   An unexpected result was the increase in relative aminopeptidase activity in the presence of CryIA(c) toxin up to 20 nM (Figure 5).

A phosphatase was also detected in the 120/65-kDa protein fraction. The 65-kDa *M. sexta* protein is similar in molecular size to the 70-kDa brush border alkaline phosphatase in *H. virescens* (19) and the 58-kDa enzyme of *Bombyx mori* (20). *Bt* CryIA(c) was previously shown to bind and inhibit this enzyme (19). The phosphatase activity in the 120/65-kDa protein fraction was reduced to 50% of the maximum level by the presence of 1 μM CryIA(c) (Figure 5). CryIA(c) did not change the activity of bovine alkaline phosphatase (11).

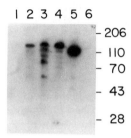

Figure 1. Ligand blot identification of brush border membrane vesicle-binding proteins specific to CryIA(c) toxin [SDS-PAGE of membrane vesicles (20 µg), incubated with ^{125}I-CryIA(c) toxin]. Lane 1, rat; lane 2, *S. frugiperda*; lane 3, *H. zea*; lane 4, *H. virescens*; lane 5, *M. sexta*. Lines indicate the positions of M_r standards (kDa).
(Reproduced with permisson from ref. 8. Copyright 1991 American Society for Microbiology.)

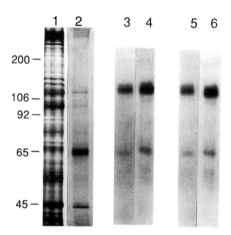

Figure 2. SDS-PAGE of BBMV proteins bound to the mAb affinity column. Lanes 1 and 2 were visualized by silver staining. Protein blots were probed with ^{125}I-CryIA(c) (lanes 3 and 4) or ^{125}I-mAbs 2B3, 8G1 and 12B8 (lanes 5 and 6). Lanes 1, 3 and 5 contained 5 µg of *M. sexta* BBMV protein. Lanes 2, 4, and 6 contained 0.5 µg of immuno-affinity selected proteins.
(Reproduced with permisson from ref. 11. Copyright 1994 American Society for Biochemistry and Molecular Biology, Inc.)

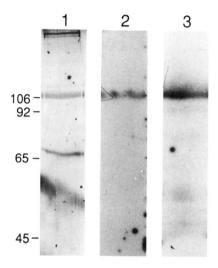

Figure 3.   SDS-PAGE and protein blot analyses of the proteins obtained by preparative isoelectric focusing and mAb chromatography.   Each lane contains 0.5 μg of protein.   Lane 1 was the silver stained gel, lane 2 was incubated with ^{125}I-CryIA(c) and lane 3 ^{125}I-mAbs.
(Reproduced with permisson from ref. 11.   Copyright 1994 American Society for Biochemistry and Molecular Biology, Inc.)

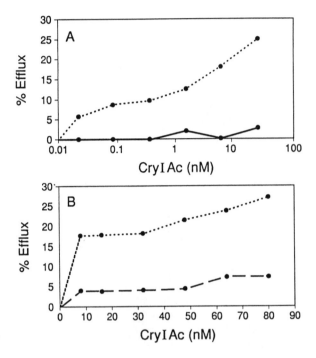

Figure 4. CryIA(c) induced release of [86]Rb+ from phospholipid vesicles containing either phospholipid-only or phospholipid plus proteins purified from BBMV. Varying amounts of CryIA(c) were added externally and [86]Rb+ release measured. In Panel A, the vesicle reconstitution mixture contained either 2.5 μg (•••●•••) of 120/65-kDa proteins isolated by preparative isoelectric focusing followed by immuno-affinity chromatography, or with phospholipid-alone (▬▬●▬▬). In Panel B, the vesicle reconstitution mixture contained either 2.5 μg (•••●•••) or 0.5 μg (▬▬●▬▬) of 120/65-kDa proteins. Values are expressed as a percentage of counts retained in vesicles in the absence of toxin treatments. Points in Panels A and B are the mean of duplicate samples. The standard deviation for all points in Panels A and B was less than 1.4%. (Adapted from ref. 11.)

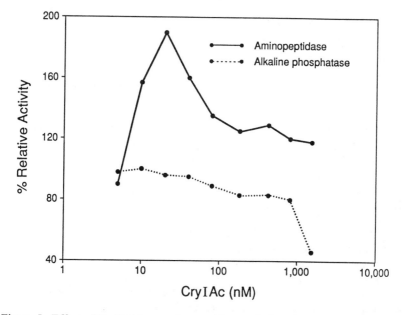

Figure 5. Effect of CryIA(c) on aminopeptidase and phosphatase activities in the 120/65 kDa protein fraction. The 120/65 kDa protein fraction was assayed in the presence of various toxin concentrations. Relative activity is expressed in terms of a control containing no CryIA(c). Values are the means of triplicate determinations.
(Adapted from ref. 11.)

Summary

Our data support a bifunctional role for 120-kDa protein as a *Bt* CryIA(c) receptor and aminopeptidase N. Aminopeptidase N in mammals exhibits several biological functions in addition to its hydrolysis of peptides in the intestine. Aminopeptidase N has a role in Na^+-dependent amino acid transport in bovine renal brush border membranes (21, 22). Sacchi *et al.* (23) demonstrated that toxin inhibits K^+-dependent amino acid cotransport into membrane vesicles. While speculative, it is possible that toxin shut-down of amino acid transport results from a direct interaction with toxin and receptor. It is also possible that shut-down of amino acid transport is a secondary effect caused by disruption of the K^+ gradient. In addition to normal physiological functions, viruses target this class of proteins to gain entry into cells. Aminopeptidase N acts as a receptor for certain coronaviruses which infect intestinal cells in pigs (24) and humans (25).

Our investigations of the interaction between CryIA(c) toxin and its target protein show that toxin binds to a 120-kDa aminopeptidase in the brush border membrane of *M. sexta* and causes ion channels to form at very low toxin concentrations. This result supports and refines the model developed by Knowles and Ellar (26), and English and Slatin (2). They proposed that toxin binding proteins act by increasing the toxin concentration at the membrane surface microenvironment resulting in pore formation at extremely low concentrations of toxin.

In mammals the cell surface aminopeptidase is anchored in the membrane via an N-terminal peptide stalk of approximately 60 amino acids (21). Since the N-terminal amino acid sequence from the 120-kDa *M. sexta* midgut aminopeptidase N aligns with pig microsomal aminopeptidase N starting at amino acid 69 (11, 13), it appears that the CryIA(c) binding aminopeptidase does not have an N-terminal stalk typical of the mammalian enzyme anchor. Our recent results demonstrate that the 120-kDa aminopeptidase in the *M. sexta* brush border membrane possesses a GPI-anchor (16). While the major function of the GPI-anchor is to attach the protein to the plasma membrane, GPI-anchors have been implicated in a variety of other structural and physiological functions.

Future studies will be directed at defining the structures on the aminopeptidase which specify CryIA(c) binding. We are also investigating how insects modify binding proteins to acquire resistance to *Bt* toxins.

Acknowledgments

This research was supported through grants from the USDA-CSRS. We thank Drs. Leigh English and Fred Walters (Ecogen, Inc.) for providing the training that made the reconstitution studies possible.

Literature Cited

1. Adang, M. J. In *Biotechnology for Biological Control of Pests and Vectors*; Maramorosch, K., Ed.; CRC Press: Boca Raton, **1991**; pp 3-24.

2. English, L.; Slatin, S. L. *Insect Biochem. Mol. Biol.* **1992**, *22*, 1-7.
3. Van Rie, J.; Jansens, S.; Höfte, H.; Degheele, D.; Van Mellaert, H. *Appl. Environ. Microbiol.* **1990**, *56*, 1378-1385.
4. Van Rie, J.; McGaughey, W. H.; Johnson, D. E.; Barnett, B. D.; Van Mellaert, H. *Science* **1990**, *247*, 72-74.
5. Ferré, J.; Real, M. D.; Van Rie, J.; Jansens, S.; Peferoen, M. *Proc. Natl. Acad. Sci. USA* **1991**, *88*, 5119-5123.
6. Tabashnik, B. E.; Finson, N.; Groeters, F. R.; Moar, W. J.; Johnson, M. W.; Luo, K.; Adang, M. J. *Proc. Natl. Acad. Sci. USA* **1994**, *91*, 4120-4124.
7. Wolfersberger, M. G. *Experientia* **1990**, *46*, 475-477.
8. Garczynski, S. F.; Crim, J.W. Adang, M. J. *Appl. Environ. Microbiol.* **1991**, *57*, 2816-2820.
9. Gould, F.; Martinez-Ramirez, A.; Anderson, A.; Ferré, J.; Silva, F. J.; Moar, W. J. *Proc. Natl. Acad. Sci. USA* **1992**, *89*, 7986-7990.
10. Knowles, B. H.; Knight, P. J.; Ellar, D. J. *Proc. R. Soc. Lond. B* **1991**, *3*, 31-35.
11. Sangadala, S.; Walters, F.; English, L. H.; Adang, M. J. *J. Biol. Chem.* **1994**, *269*, 10088-10092.
12. Olsen, J.; Sjostrom, H.; Noren, O. *FEBS Letters* **1989**, *251*, 275-281.
13. Knight, P. J. K.; Crickmore, N.; Ellar, D. J. *Mol. Microbiol.* **1994**, *11*, 429-436.
14. Taylor, A. *FASEB J.* **1993**, *7*, 290-298.
15. Takesue, S.; Yokota, K.; Miyajima, S.; Taguchi, R.; Ikezawa, H.; Takesue, Y. *Comp. Biochem. Physiol.* **1992**, *102B*, 7-11.
16. Garczynski, S. F., Adang, M. J. *Insect Biochem. Mol. Biol.* in press.
17. Nilsson, B. O.; Svalander, P. C.; Larsson, A. *J. Immunol. Methods* **1987**, *99*, 67-75.
18. English, L.; Readdy, T. L.; Bastian, A. E. *Insect Biochem.* **1991**, *21*, 177-184.
19. English, L.; Readdy, T. L. *Insect Biochem.* **1989**, *19*, 145-152.
20. Takeda, S.; Azuma, M.; Itoh, M.; Eguchi, M. *Comp. Biochem. Physiol.* **1993**, *104B*, 81-89.
21. Kenny, A. J.; Stephenson, S. L.; Turner, A. J.In *Mammalian ectoenzymes*; Kenny, A. J., Turner, A. J., Eds.; Elsevier: Amsterdam, **1987**; pp 169-210.
22. Plakidou-dymock, S.; Tanner, M. J.; McGivan, J. D. *Biochem. J.* **1993**, *290*, 59-65.
23. Sacchi, V. F.; Parenti, P.; Hanozet, G. M.; Giordana, B.; Lüthy, P.; Wolfersberger, M. G. *FEBS Letters* **1986**, *204*, 213-218.
24. Delmas, B.; Gelfi, J.; L'Haridon, R.; Vogel, L. K.; Sjostrom, H.; Noren, O.; Laude, H. *Nature* **1992**, *357*, 417-419.
25. Yeager, C. L.; Ashmun, R. A.; Williams, R. K.; Cardellichio, C. B.; Shapiro, L. H.; Look, A. T.; Holmes, K. V. *Nature* **1992**, *357*, 420-422.
26. Knowles, B. H.; Ellar, D. J. *Biochem. Biophys. Acta* **1987**, *924*, 509-518.

RECEIVED November 17, 1994

Chapter 23

Mode of Action of δ-Endotoxin from *Bacillus thuringiensis* var. *aizawai*

M. Himeno and H. Ihara

Department of Applied Biological Chemistry, College of Agriculture, University of Osaka Prefecture, Sakai, Osaka 593, Japan

CryIA(a) and CryIA(b), δ-Endotoxins from *Bacillus thuringiensis* Var. *aizawai*, were found to interact with a brush border membrane vesicle (BBMV) prepared from silkworm, *Bombyx mori* L. The irreversibly bound toxin was resistant to proteinase K, suggesting that it was inserted into lipid bilayer of BBMV. Cultured cabbage looper TN-368 cells swelled during treatment with δ-endotoxin. The cell swelling induced by the toxin was selectively caused with monovalent cations. Effects of pyrethroids on the toxin-induced cell swelling and the effect of various drugs that affect the Na^+ or K^+ channel suggested that the δ-endotoxin stimulated the Na^+ channel of the cell in isotonic NaCl solution and also Na^+, K^+-ATPase in KCl solution. The content of cGMP in the swollen cells did not increase.

Bacillus thuringiensis is a gram-positive spore-forming bacterium that forms a crystalline inclusion body containing proteinaceous δ-endotoxin, which is simply called the crystal. The δ-endotoxins are only toxic to insects and are harmless to vertebrates and plants. As such, they have been widely used as microbial insecticides for pest control. Höfte and Whiteley classified the δ-endotoxin into 4 *cry* groups (*1*). After ingestion by susceptible larvae, crystals dissolve into 130-140 kDa proteins (pro-toxin) in the alkaline conditions of the insect midgut. Proteases in the midgut cleave the pro-toxins to yield 55-70 kDa active toxins. Active toxins bind to high-affinity binding sites on the midgut epithelial membrane, which decide the insecticidal spectrum. Then the epithelium cells swell and their microvilli disappear and finally the cells lyse. The flux of the potassium ion into the blood induces paralysis and death of the insect. The mode of action of these toxins was described in several reviews (*1-5*) and various models of the mechanism of action have been proposed (6). Nevertheless, the exact mode of action remains unknown.

In this paper, we describe the binding of δ-endotoxin to a positive receptor protein in brush border membrane vesicles (BBMV) from midgut epithelium of *Bombyx mori* L. and the activation and other disturbance of monovalent cation-selective channels induced by the toxin in an insect-cultured cell system.

0097–6156/95/0591–0330$12.00/0
© 1995 American Chemical Society

MATERIALS AND METHODS

Cells. The TN-368 cells from cabbage looper were cultured in TC199-MK medium containing 3~5% fetal bovine serum and 50 μg/ml kanamycin sulfate at 27°C. The cells were cultured on a Falcon Micro-Test plate 3040 and used to determine the toxic activity of various agents as described by Himeno.*et al* (7).
Purification of Crystals and Dissociation of the δ-Endotoxin. Crystals from *B. thuringiensis* var. *aizawai* I45 and var. *israelensis* 4Q1 were purified by the flotation procedures by Yamamoto (8) and by the density gradient centrifugation of Renographin. CryIA(a) and CryIA(b), which constitute the crystal of *B. thuringiensis* var. *aizawai* I45, were individually purified from *E. coli* JM103 carrying *cryIA(a)* or *cryIA(b)* gene and described in previous paper (9). Binding experiments were performed with CryIA(a) and CryIA(b) from recombinant *E. coli* and the others were done with crystal protein from *B. thuringiensis* var. *aizawai* I45.
Preparation of BBMV. The preparation of BBMV from midgut cells of *Bombyx mori* follows the precipitation methods using Mg-EGTA described by Wolfersberger *et al*. (10).
Iodination of Activated Toxins and Binding Assay. Iodination of activated toxin was done by the chloramine-T method (11). The binding assay of the [^{125}I]-toxin with the BBMV was performed in the presence or absence of excess unlabeled toxin in phosphate buffer saline (PBS)-bovine serum albumin (BSA). Bound and free toxins separated by fast ultrafiltration through a nitrocellulose filter. Each filter was rapidly washed with ice cold PBS-BSA. Radioactivity on the filter was measured in a gamma counter (12).
Protease Digestion of Irreversibly Bound d-Endotoxin.
[^{125}I]-Labeled CryIA(a) and BBMV were incubated for 2 hr at 30 °C. The mixture was diluted 10-fold in the buffer containing 0.1 μM unlabeled toxin and incubated for 2 hr at 30 °C. Binding of labeled toxin to BBMV was in irreversible component by this procedure. The mixture was centrifuged at 40,000 x g for 30 min at 2 °C to remove free toxin. The pellet was suspended in the buffer containing several concentration of proteinase K and was incubated for 2hr at 30 °C. The BBMV were separated from enzymes and degradation products by rapid filtration method. Each filter was rapidly washed and the radioactivity was measured with a gamma counter.
Effect of Monovalent Cation on Cell Swelling. The swelling activity induced by the toxin on the TN-368 cells and effects of the some drugs on the toxin-treated cells were determined by the procedures described in previous paper (7). The cells were incubated in 160 mM sucrose solution containing 20 mM concentration of the following monovalent cations (KCl, LiCl, NaCl, RbCl, and CsCl) and the cells were treated with the toxin (18 μg/ml) prepared from the *aizawai* crystal (7). Also the cells were incubated in 140 mM sucrose solution containing 20 mM MgCl$_2$ or CaCl$_2$, and 160 mM sucrose solution containing 20 mM NH$_4$Cl, 0.8% NaCl, and 10 mM phosphate buffer saline, pH 6.3, without MgCl$_2$ or CaCl$_2$. The cell swelling was counted after suitable incubation time at room temperature.
Determination of cGMP Content in Swollen Cells. The confluent monolayers of TN-368 cells were washed 3 times with 150 mM NaCl or 230 mM KCl solution and treated with 1 ml of the enzyme-digested δ-endotoxin preparation (3.7 μg/ml) (7). The monolayer cells were treated with the isotonic NaCl solution containing silkworm digestive enzyme as a control. After incubation for an appropriate period, the cells were treated with 6% TCA at 0°C and homogenized. The homogenates were centrifuged for 5,000 rpm for 15 min at 4 °C and supernatant saved. The pellets were again extracted with cold 6% TCA and recentrifuged.

The combined supernatants were extracted with equal volumes of ether to remove TCA. After repeating the ether extraction 3 times, the aqueous fraction was dried. The dried samples were assayed for cGMP levels with the YAMASA cGMP assay kit according to the manufacture's instructions.

RESULTS

Binding of δ-Endotoxin to Brush Border Membrane Vesicles (BBMV). Proteolytically activated δ-endotoxins primarily bind to specific receptors on the midgut epithelial cell membranes and that event is important in the action of the toxin and for the determination of its specificity (3). We have studied the binding of CryIA(a) and CryIA(b) to BBMV (12). CryIA(a) and CryIA(b) were found to specifically bind to crude membranes and BBMV from larval *B. mori* midguts. Neither toxins bound to the BBMVs from rabbit kidney cortex and small intestine. This correlates to the lack of toxicity of this toxin in rabbit. The binding of both toxins was dependent on membrane protein concentrations. The specific activity of the binding to BBMV increased approximately three-fold relatives to that of the crude membranes, indicating that the binding sites of both toxins are localized on BBMV. The binding of each toxin to BBMV was saturable with increasing concentrations of toxins. Scatchard plot analysis indicated a single binding site for each toxin. The dissociation constant and the binding capacity for CryIA(a) binding to BBMV were 0.89 nM and 7.68 pmole/mg vesicle protein, respectively (Figure 1). For CryIA(b) the dissociation constant was 1.46 nM and the binding capacity was 5.22 pmole/mg vesicle protein (Figure 1). Competition experiments showed that binding of each toxin was displaced by unlabeled toxin in the nanomolar concentration range, suggesting that both toxins compete for the same high affinity binding site.

Irreversible Binding of δ-Endotoxin to BBMV. Investigators have indicated that the mechanism of toxin binding to BBMV is complicated (13-16). Association of toxin apparently obeys simple bimolecular kinetics, but the dissociation process is biphasic with fast and very slow stages. This suggests that the binding of toxins proceeds through reversible and apparently irreversible stages. The two-step binding appears to be a common feature of CryI toxins. We studied the dissociation kinetics of the binding of CryIA(a) and CryIA(b) to BBMV (12). The dissociation curves of both toxins followed a biphasic process with a fast and a very slow state, suggesting that the binding of toxins proceeds through reversible and apparently irreversible states. The ratio of the reversible state to the irreversible state was different for these two toxins. For CryIA(a), the ratio was approximately 2 : 8 and for CryIA(b) it was approximately 9 : 1. Interestingly, we found that the differences in their toxicities to *B. mori* were better correlated with the reversible : irreversible ratio, rather than just to binding ability. Therefore the irreversible binding state is likely to be a physiologically important factor.
　　　　　Investigators have studied effects of δ-endotoxin on artificial lipid bilayers (17-21). The toxins apparently are involved in forming pores or channels in the lipid bilayers. Because there are conserved hydrophobic regions near the N-terminus of the δ-endotoxins, it has been predicted that the toxins may interact with the membrane of midgut epithelial cells (1, 22). The irreversible state of the toxin binding may be due to the insertion of the toxin into lipid bilayer (13-16, 23), but direct experimental evidence for such interaction has not been reported. To determined the fraction of the toxin which was irreversibly bound to BBMV, the irreversibly bound toxin was digested with several concentration of proteinase K (Figure 2). As the enzyme concentration increased, the radioactivity from [^{125}I]-toxin retained on BBMV decreased to 40% of that without enzyme-treatment. The level of radioactivity was not further reduced even when

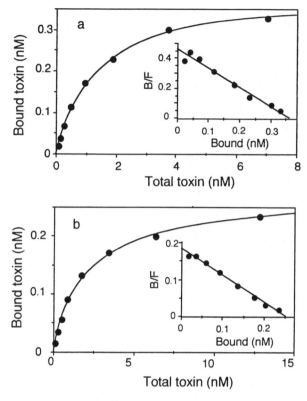

Figure 1. Binding of [^{125}I]-labeled CryIA(a) [panel a] and CryIA(b) [panel b] to B. mori BBMV.

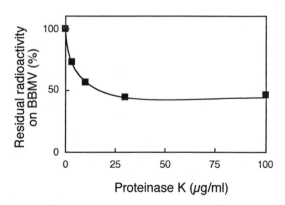

Figure 2. Digestion of [^{125}I]-labeled CryIA(a) bound irreversibly to BBMV by proteinase K.

the reaction time was extended. This result indicated that approximately 40% of the toxin bound irreversibly to BBMV was resistant to the enzyme treatment. It is reasonable to assume that the membrane partially protects the toxin from the attack of the enzyme and that the toxin is likely inserted into the membrane.

Effect of Monovalent Cation on the Cell Swelling. It was reported previously that TN-368 cell swelled by treatment with enzyme-digested δ-endotoxin of *B. thuringiensis* var. *aizawai* I45 and var. *israelensis* 4Q1 (2). The cytotoxic swelling was dependent upon the concentration of NaCl or KCl in the isotonic sucrose solution (Figure 3). The toxin of *B. thuringiensis* var. *israelensis* as well as var. *aizawai* induced the cell swelling in NaCl and KCl isotonic solution, but did not in the isotonic sucrose solution.

The effects of various monovalent cations were examined with the toxin from *aizawai* (Figure 4). Monovalent cations were required for cell swelling induced by the δ-endotoxin (18 μg/ml). However, divalent cations (Ca^{++}, Mg^{++}) were not effective cation replacements. Monovalent cations differed in their ability to effect cell swelling in following order: K^+ > Li^+ > Na^+ > Rb^+ > Cs^+. Cell swelling was not apparent in the presence of NH_4Cl and sucrose solutions. Cell swelling by monovalent cations was concentration dependent. In figure 3, the isotonic NaCl solution was more effective than the isotonic KCl solution (7) but K^+ ion was more effective than Na^+ ion in the isotonic sucrose solution containing 20 mM K^+ or Na^+ (Figure 4).

Effect of 4-AP and Ouabain on the Cell Swelling. It has been reported that cell swelling induced by δ-endotoxin is inhibited by tetrodotoxin (TTX) and cAMP in isotonic NaCl solution (7). However, 4-AP, which is a selective blocker of the K^+ channels, stimulated cell swelling caused by the toxin in isotonic KCl solution (Figure 5). The concentration of 4-AP (400 μM) by itself did not result in significant cell swelling in either isotonic solution. That is, 4-AP inhibited an efflux of K^+ leak channels.

Ouabain (686 μg/ml), an inhibitor of Na^+, K^+-ATPase, stimulated cell swelling caused by the δ-endotoxin (8 μg/ml) in isotonic NaCl solution but inhibited the swelling in the isotonic KCl solution (Figure 6).

When both of 4-AP (600 μM) and ouabain (686 μg/ml) were added to cells in the presence of δ-endotoxin (7 μg/ml), the swelling was stimulated in isotonic NaCl solution. The swelling activity caused by the toxin and 4-AP in isotonic KCl solution was reduced by the addition of ouabain (Figure 7). In isotonic NaCl solution, ouabain did not reduce the swelling activity caused by the 4-AP in the presence of the toxin. However, in the isotonic KCl solution, cell swelling in the presence of toxin and 4-AP was reduced by ouabain. Although the toxin stimulated the K^+ influx through the Na^+, K^+-ATPase pump, 4-AP blocked the K^+ efflux through the K^+ leak channel and stimulated the cell swelling. Then the cell swelling by the toxin and 4-AP was deduced by the addition of ouabain in the KCl isotonic solution.

Effect of Pyrethroids on the Cell Swelling. The site of action of pyrethroid insecticides is likely ion channel, especially sodium channel on nerve, muscle, and other excitable tissues (24, 25). The effects of five pyrethroids (allethrin, fenvalerate, fenpropathrin, permethrin, and tetramethrin : Each compound is 60 μM) were investigated (Figure 8 and Figure 9). Permethrin and tetramethrin stimulated cell swelling induced by the toxin (9 μg/ml) in isotonic NaCl solution but were inhibitory to cell swelling in isotonic KCl solution (Figure 9). Fenpropathrin stimulated the cell swelling in isotonic NaCl or KCl solution (Figure 8). Fenpropathrin is the same a blocker of the K^+ channel just as 4-AP and permethrin and tetramethrin the same an inhibitor of Na^+, K^+-ATPase just as

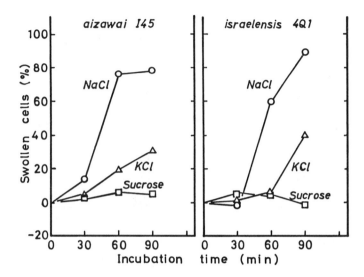

Figure 3.   Effects of monovalent cations on the cell swelling by the δ-endotoxin from var. *aizawai*.

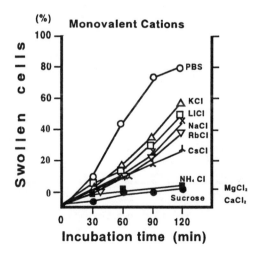

Figure 4.   Time course of cytolytic response to the toxin from var. *aizawai*.

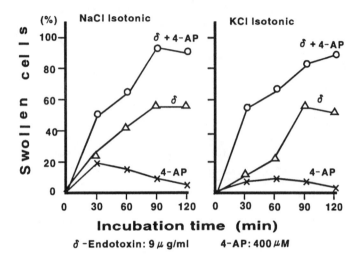

Figure 5.   Stimulation of the cell swelling by 4-AP the δ-endotoxin.

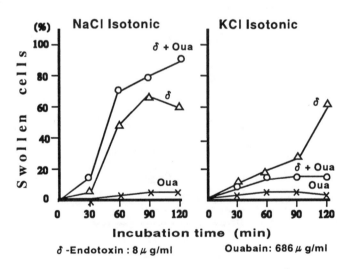

Figure 6.   Effects of ouabain on the cell swelling with δ-endotoxin.

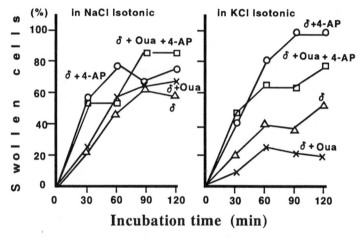

δ -Endotoxin : 7 μ g/ml   Ouabain:686 μ g/ml   4-AP:600 μM

Figure 7.   Effects of ouabain and 4-AP on the cell swelling of the δ-endotoxin.

NaCl Solution

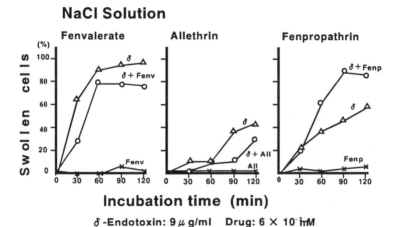

δ-Endotoxin: 9 μg/ml   Drug: 6 × 10⁻¹ mM

KCl Solution

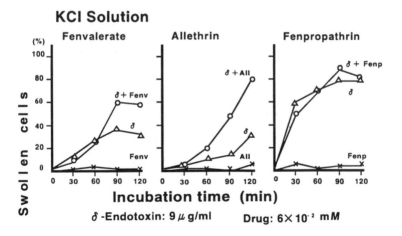

δ-Endotoxin: 9 μg/ml   Drug: 6×10⁻² mM

Figure 8.   Effect of allethrin, fenvalerate and fenpropathrin on the cell swelling.

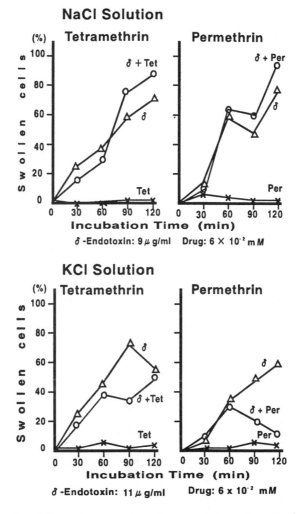

Figure 9.   Effect of permethrin and tetramethrin on the cell swelling

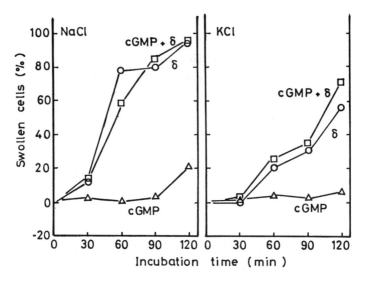

Figure 10.   Stimulation of cell swelling by cGMP.

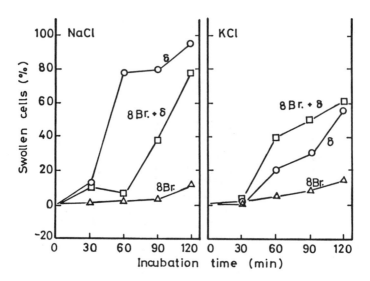

Figure 11.   Effect of 8-Br-cGMP on the cell swelling.

ouabain. Allethrin and fenvalerate inhibited the cell swelling caused by toxin in isotonic NaCl solution. However they stimulated the cell swelling by the toxin (9 μg/ml) in isotonic KCl solution (Figure 8).

The Effect of cGMP or 8-Br-cGMP on Cell Swelling. Cyclic nucleotides, such as cAMP or cGMP, act as intracellular second messengers and cAMP inhibited cell swelling induced by δ-endotoxin (5.2 μg/ml) (7). In view of this, the action of cGMP (1 mM) and 8-Br-cGMP (1 mM) were similarly examined (Figure 10 and Figure 11, respectively). cGMP did not affect cell swelling in isotonic NaCl solution but resulted in slight stimulation in isotonic KCl solution. 8-Br-cGMP, an unnatural cGMP derivative inhibited cell swelling induced by the δ-endotoxin in isotonic NaCl solution but was stimulated in isotonic KCl solution.

cGMP Content in Cells treated with δ-Endotoxin. There was an increase of approximately 18% in the cGMP content in cells treated with the δ-endotoxin (3.7μg/ml) than in untreated cells in isotonic NaCl solution, and an increase of approximately 30% in isotonic KCl solution (Figure 12). It seems that the slight increase of the cGMP content is observed in the swelling cells. It is obscure whether or not a few increment of cGMP content is effective amounts.

DISCUSSION

The δ-endotoxin from *Bacillus thuringiensis* var. *aizawai* attacked brush border membrane of larval midgut cells. The brush border membrane vesicle prepared from midgut epithelium of *Bombyx mori* L. is highly susceptible to the toxin, CryIA(a) and CryIA(b), which purified from *E. coli* carrying a *cryIA(a)* or *cryIA(b)* gene cloned from *B. thuringiensis* var. *aizawai*. The dissociation constants and binding capacities for [^{125}I]-CryIA(a) and [^{125}I]-CryIA(b) to BBMV were determined. The toxins bound to BBMV through reversible and apparently irreversible process. The toxin, which is irreversibly bound to BBMV, is resistant to proteinase K activity. This suggests that the bound toxin may insert itself into the lipid bilayer. The presence of monovalent cations is an essential component for cell swelling caused by the enzyme-activated δ-endotoxin. Cell swelling induced by the toxin may be due to the stimulation of Na$^+$ influx through Na$^+$ channels of the cells, because the swelling is inhibited by addition of TTX to isotonic NaCl solution described in previous paper (7). Also, ouabain inhibited the swelling only in isotonic KCl solution and did not in isotonic NaCl solution. Additionally, 4-AP stimulated the swelling by blocking K$^+$ efflux through K$^+$ leak channels of the insect cells.

Effects of pyrethroid on cell swelling induced by δ-endotoxin are summarized in table I. The activity of the pyrethroids can be divided to three groups. First, fenpropathrin stimulated the toxin-induced swelling activity in isotonic NaCl or KCl solution. Second, permethrin and tetramethrin stimulated the toxin-induced swelling activity in NaCl solution but inhibited the activity in KCl solution. Third, fenvalerate and allethrin stimulated the swelling activity induced by the toxin in the KCl solution but inhibited the activity in the NaCl solution. Thus, this action is not related to the presence or absence of an alpha cyano grouping.

The protein δ-endotoxin appears to insert into the plasma membrane of brush border cell, and bind with ion channel proteins or some unknown proteins, forming a monovalent cation-selective ion channel that seems to be not NH$_4^+$ or Cl$^-$ permeable. The cGMP content of cells treated with toxin did not drastically change. it is obscure wheter or not levels of cyclic nucreotides do function as a second messenger in these aspect.

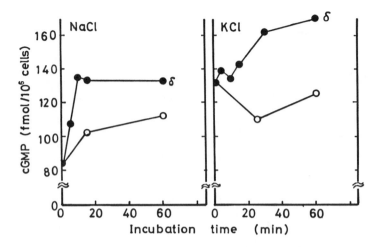

Figure 12.   cGMP contents in the cells treated with δ-endotoxin.

TABLE I.   Effects of various drugs on the cell swelling induced by the δ-endotoxin

Drug	the NaCl Isotonic Solution	KCl Isotonic Solution
4-AP	+ [1]	+
App(NH)p	+	+
Nitroprusside	+	+
Fenpropathrin	+	+
Permethrin	+	− [2]
Tetramethrin	+	−
Ouabain	+	−
TEA	± [3]	±
Fenvalerate	−	+
Allethrin	−	+
Indomethacin	−	+
Gpp(NH)p	−	+
8-Br-cGMP	−	+
TTX	−	−
cAMP	−	−
AMP	−	−

(1) : +; The drug stimulated the cell swelling.
(2) : −; The drug inhibited the cell swelling.
(3) : ±; The drug did not affected on the cell swelling.

Acknowledgments
We wish to thank the Biotechnology Laboratory, Takarazuka Research Center Sumitomo Chemical Co., Ltd., for kindly providing pyrethroids and cloned δ-endotoxins, CryIA(a) and CryIA(b), from *Bacillus thuringiensis* var. *aizawai*.

Literature Cited
1. Höfte, H.; Whiteley, H. R. *Microbiol. Rev*. **1989**, *53*, 242.
2. Himeno, M.; *J. Toxicol. Toxin Rev*. **1987**, *6*, 45.
3. Gill, S. S.; Cowles, E. A.; Pietrantonio, P. V. *Annu. Rev. Entomol*. **1992**, *37*, 615.
4. English, L.; Spatin, S. L. *Insect Biochem. Molec. Biol*. **1992**, *22*, 1.
5. Adang, M. J.; In *Biotechnology for Biological Control of Pests and Vectors*; Maramorosch, K., Ed.; CRC Press Inc.: Boca Raton, Florida, 1991, pp.3-24.
6. Knowles, B. H.; Dow, J. A. T. *BioEssays* **1993**, *15*, 469.
7. Himeno, M.; Koyama, N.; Funato, J.; Komano, T. *Agric. Biol. Chem*. **1985**, *49*, 1461.
8. Yamamoto, T.; *J. Gen. Microbiol*. **1983**, *129*, 2595.
9. Oeda, K.; Inoue, K.; Ibuchi, Y.; Oshie, K.; Shimizu, M.; Nakamura, K.; Nishioka, R.; Takada, Y.; Ohokawa, H. *J. Bacteriol*. **1987**, *171*, 3568.
10. Wolferberger, M.; Lüthy, P. A.; Maurer, P.; Parent, P.; Sacchi, F. V.; Giordana, B.; Hanozet, G. M. *Comp. Biochem. Physiol*. **1987**, *87A*, 301.
11. Hunter, W. M.; Geenwood, F. C. *Nature* **1962**, *194*, 495.
12. Ihara, H.; Kuroda, E.; Wadano, A.; Himeno, M. *Biosci. Biotech. Biochem*. **1993**, *57*, 200.
13. Hofmann, C.; Vanderbruggen, H.; Höfte, H.; Van Rie, J.; Jansens, S.; Van Mellaret, H. *Proc. Natl. Acad. Sci. USA* **1988**, *85*, 7844.
14. Van Rie, J.; Jansens, S.; Höfte, H.; Degheele, D.; Van Mellaert, H. *Eur. J. Biochem*. **1989**, *186*, 239
15. Van Rie, J.; Jansens, S.; Höfte, H.; Degheele, D.; Van Mellaert, H. *Appl. Environ. Microbiol*. **1990**, *56*, 1378.
16. MacIntosh, S. C.; Stone, T. B.; Jokerst, R. S.; Fuchs, R. L. *Proc. Natl. Acad. Sci. USA* **1991**, *88*, 8930.
17. Younovitz, H.; Yawetz, A. *FEBS Letts*. **1988**, *23*, 195.
18. Harder, M. Z.; Ellar, D. J. *Biochem. Biophys. Acta* **1989**, *978*, 216.
19. English, L. H.; Readdy, T. L.; Bastian, A. E. *Insect Biochem*. **1991**, *21*, 177.
20. Slatin, S. L.; Abrams, C. K.; English, L. H. *Biochem. Biophys. Res. Commun*. **1990**, *169*, 765.
21. Schwartz, J-L.; Garneau, L.; Savaria, D.; Masson, L.; Brousseau, R.; Rousseau, E. *J. membrane Biol*. **1993**, *132*, 53.
22. Nakamura, K.; Murai-Nishioka, R.; Shimizu, M.; Oshie, K.; Mikitani, K.; Oeda, K.; Ohokawa, H. *Biosci. Biotech. Biochem*. **1992**, *56*, 1.
23. Wolfersberger, M. G. *Experientia* **1990**, 46, 475.
24. Narahashi, T.; *Pysiol. Rev*. **1974**, *54*, 813.
25. Wouters, W.; Van den Berchen, J. *Gen. Pharmacol.*, **1978**, *9*, 387.
26. Narahashi, T.; In *Neuro tox '88; Molecular Basis of Drug and Pesticide Action*. Lunt, G. G. Ed.; Elservier Science Pub., **1988**, pp269-288.
27. Himeno, M.; In *Biotechnology in Invertebrate Pathology and Cell Culture* Maramorosh, K.Ed., Academic Press Inc. San Digo, **1987**, pp29-43.

RECEIVED October 25, 1994

Author Index

Affiliation Index

Subject Index

A

Production: Meg Marshall
Indexing: Deborah H. Steiner
Acquisition: Anne Wilson
Cover design: Cornithia Harris

Printed and bound by Maple Press, York, PA

Highlights from ACS Books

Bestsellers from ACS Books

The ACS Style Guide: A Manual for Authors and Editors
Edited by Janet S. Dodd
264 pp; clothbound ISBN 0–8412–0917–0; paperback ISBN 0–8412–0943–X

Understanding Chemical Patents: A Guide for the Inventor
By John T. Maynard and Howard M. Peters
184 pp; clothbound ISBN 0–8412–1997–4; paperback ISBN 0–8412–1998–2

Chemical Activities (student and teacher editions)
By Christie L. Borgford and Lee R. Summerlin
330 pp; spiralbound ISBN 0–8412–1417–4; teacher ed. ISBN 0–8412–1416–6

Chemical Demonstrations: A Sourcebook for Teachers,
Volumes 1 and 2, Second Edition
Volume 1 by Lee R. Summerlin and James L. Ealy, Jr.;
Vol. 1, 198 pp; spiralbound ISBN 0–8412–1481–6;
Volume 2 by Lee R. Summerlin, Christie L. Borgford, and Julie B. Ealy
Vol. 2, 234 pp; spiralbound ISBN 0–8412–1535–9

Chemistry and Crime: From Sherlock Holmes to Today's Courtroom
Edited by Samuel M. Gerber
135 pp; clothbound ISBN 0–8412–0784–4; paperback ISBN 0–8412–0785–2

Writing the Laboratory Notebook
By Howard M. Kanare
145 pp; clothbound ISBN 0–8412–0906–5; paperback ISBN 0–8412–0933–2

Developing a Chemical Hygiene Plan
By Jay A. Young, Warren K. Kingsley, and George H. Wahl, Jr.
paperback ISBN 0–8412–1876–5

Introduction to Microwave Sample Preparation: Theory and Practice
Edited by H. M. Kingston and Lois B. Jassie
263 pp; clothbound ISBN 0–8412–1450–6

Principles of Environmental Sampling
Edited by Lawrence H. Keith
ACS Professional Reference Book; 458 pp;
clothbound ISBN 0–8412–1173–6; paperback ISBN 0–8412–1437–9

Biotechnology and Materials Science: Chemistry for the Future
Edited by Mary L. Good (Jacqueline K. Barton, Associate Editor)
135 pp; clothbound ISBN 0–8412–1472–7; paperback ISBN 0–8412–1473–5

For further information and a free catalog of ACS books, contact:
American Chemical Society
Product Services Office
1155 16th Street, NW, Washington, DC 20036
Telephone 800–227–5558

Lady's Choice

*Also by Jayne Ann Krentz
in Large Print:*

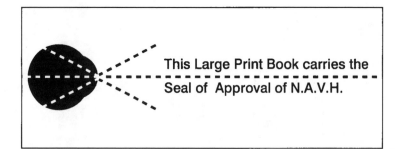

This Large Print Book carries the
Seal of Approval of N.A.V.H.

Lady's Choice

Jayne Ann Krentz

Published in 2002 by arrangement with Harlequin Books S.A.

Wheeler Romance Series.

The text of this Large Print edition is unabridged.
Other aspects of the book may vary from the original edition.

Cover design by Thorndike Press Staff.

Set in 16 pt. Plantin by Christina S. Huff.

Printed in the United States on permanent paper.

Library of Congress Control Number: 2002105932
ISBN 1-58724-266-4 (lg. print : hc : alk. paper)

Lady's Choice

One

"I love you, Travis. Hold me, hold me."

"Juliana."

Travis Sawyer heard his own muffled shout as he shuddered heavily over the flame-haired woman in his arms. The last of his white-hot passion spent itself in a blinding, driving storm of pure release. He lost himself in his lover's arms, surrendering to her fire even as he exulted in his victory. She clung to him with all her sleek strength, drawing him into her until he felt as though he'd stepped into another universe.

It had never been this good with anyone else. Travis Sawyer was thirty-eight years old. He'd never been a womanizer but he considered he'd lived long enough to make the judgment. This was special. Nothing had ever been this good before in his life.

It was everything he had instinctively sensed it would be with this woman. Hot, wild, powerful. He had never felt so alive, so strong. Satisfaction swept through him in the wake of the slowly dissipating passion.

She was his now. Reluctantly he disengaged himself and rolled to one side, his hand trailing heavily over the gentle curve of Juliana's breast. She smiled up at him from the pillow, the expression as dazzling as always, even here in the darkness of her bedroom.

The thick, untamed mass of her hair was an elegant, pagan crown framing her vivid features. Travis stared down at her, captivated by huge, long-lashed eyes, a noble nose, an arrogant, yet surprisingly delicate chin and a luscious mouth. Her long leg slid between his in a languidly sensual movement. Then she closed her eyes and snuggled into his warmth.

He had done it, Travis thought triumphantly as his arm tightened around her. He had claimed his red-haired, topaz-eyed queen.

And then, in the next moment, reality settled back into place around him. What the hell was he doing here, holding her like this? He'd never intended to take his revenge this far; never intended to wind up in bed with Juliana Grant.

He stared deeply into the shadows of the bedroom, searching for answers that weren't there. He felt dazed now that the fiery passion had receded.

Vengeance led a man down strange paths. Juliana Grant had been an unexpected detour in the long road he'd been walking for the past five years. But he could not, would not allow the detour, no matter how exotic, to deflect him from his chosen course. He had come too far. There was no turning back now, even if he wanted to do so.

Travis Sawyer was very good at what he did, and when he had set out to orchestrate his revenge he had left no loopholes. There was no escape for anyone, not even for himself.

The clear, bright California sunshine danced across the bay and slammed cheerfully through the condominium's bedroom windows. Juliana opened her eyes slowly and watched the early-spring light as it bounced around the dramatic white-on-white room. It sparkled on the thick white carpet, bounded off the white walls, struck the chrome and white leather chair and tap-danced over the gleaming white lacquer dressing table. It sizzled when it struck the only color accent in the bedroom, an egg-yolk-yellow abstract painting that hung on the wall over the chrome and white bed.

Mesmerized, Juliana followed the trail of sunlight as it ricocheted between the mirror and the painting and splashed across the foaming white sheets of the rumpled bed. There, in a final burst of dazzling brilliance, the morning sun revealed the alien male being who had invaded her room last night.

A man in her bed. That, in and of itself, was a rare enough event to excite wonder and curiosity, but in this case it was an even more notable occurrence. Juliana hugged herself with her secret knowledge.

Because she knew beyond a shadow of a doubt that this particular man — this hard, lean, sexy man named Travis Sawyer — was *the* man. The right man. The one she'd been waiting for all her life.

She savored the delicious secret and held herself very still so as not to awaken the exotic creature lying next to her. She wanted a moment to luxuriate in the thrilling certainty that she had finally encountered her true mate.

He was not exactly as she had fondly imagined over the years when she had indulged in a little harmless fantasizing. He wasn't quite as tall as he should have been, for one thing. She, herself, was just a sliver under six feet and she had always envisioned her true mate as being somewhere in the neighborhood of six feet, four inches or so. Tall enough so that she could wear high heels comfortably around him. Travis was barely an inch over six feet. In two-inch heels she was eyeball to eyeball with him. In two-and-a-half-inch heels she was taller than he was.

But whatever he lacked in height, he more than compensated for in build, Juliana assured herself cheerfully. Travis was sleekly muscular and as solid as a chunk of granite. Last night she had been in no doubt of his strength. The masculine power in him had been totally controlled and all the more exciting for that sense of control. This was a man who exercised a sure command over himself, a man who had learned the techniques of self-mastery. She admired that kind of control in a man. It gave a woman a sense of security — an old-fashioned, primitive assurance that his greater physical strength need never be feared but could be relied upon for protection.

Travis did not quite match Juliana's inner image of her perfect man in a few other minor respects, either. His eyes were the wrong color. Juliana preferred sensual, warm brown or hazel eyes in those of the male persuasion. Travis had cool, crystal gray eyes that did not betray his emotions except in the most intense situations. Last night had been intense, however, she recalled with delight. She'd seen the passion blazing in his eyes and it had sent shivers of excitement through her.

This morning she was quite prepared to drop her old standards regarding eye color in view of the fact that Travis's eyes were not only capable of reflecting his passion, but also an intelligence that complemented her own and a rare sense of humor that delighted her when it showed itself.

His hair was a bit off, too, unfortunately. It was a far darker shade than she'd fantasized. Juliana had always liked men with tawny-colored or blond hair, but she had to admit that Travis's severely trimmed, night-black hair seemed to suit him. The hint of silver at the temples was not at all unattractive.

There were a few other minor discrepancies between the real Travis Sawyer and Juliana's fantasy version of her true mate. If she were inclined to be picky, for example, she could have carped about the undeniable fact that his rough, grim looks would probably forever keep him from gracing the cover of *Gentlemen's Quarterly* magazine. Ah, well, it was GQ's loss, she told herself. He looked perfect here in her bed.

Then, too, there was Travis's apparent total lack of interest in style and clothes. She had known him for almost one whole month now and she had never seen him in anything but a pair of dark trousers, an austere white shirt, a conservatively striped tie and wing-tip shoes. His jackets were all muted shades of gray. But Juliana figured she could fix the problem. After all, she had more than enough style for both of them, she told herself. She glanced at her closet and smiled as she pictured the rack of expensive, high-fashion clothes and the boxes of shoes inside. Shopping was high on her list of hobbies.

All in all, Juliana was more than willing to make allowances for the few areas in which Travis Sawyer fell short of her idealized image of Mr. Right. She was used to working for what she wanted, and she was quite prepared to put in whatever time and effort was required to polish her very special diamond in the rough. Last night she had received ample assurance that the effort would be worth it. She still tingled from head to toe with the hot memories.

Having finished her perusal, Juliana stretched slowly, deliberately stroking one toe down the length of Travis's muscular calf. When there was no response, she sighed and accepted the

12

fact that the man probably needed his sleep after last night.

Juliana grinned with amused regret, pushed back the sheet and got to her feet. She was mildly startled to discover she ached pleasantly all over. Travis had been a demanding as well as a bold and generous lover. He'd taken everything he could get but he had given back passion with equal intensity. If she closed her eyes she could still feel his strong, sensitive hands on her this morning. She felt as if she'd been imprinted with his touch.

Standing in the middle of her bright white room Juliana allowed herself one last, fond gaze at the man in her bed and then she headed for the bathroom with a long, exuberant stride.

She would welcome her true mate with a proper display of feminine domesticity, she decided. Might as well give the man a little foretaste of the wonders that were in store for him.

Half an hour later, showered, her mass of red curls caught up in a dramatically cascading ponytail and dressed in a pair of fashionably cut, high-waisted slacks and wide-sleeved painter's shirt, Juliana made her way back into the bedroom. She was carrying a black enameled tea tray. Perched on the elegant tray was an art deco teapot and two cunningly designed, bright red cups.

"Good morning." She smiled brilliantly when she saw that Travis was awake. He sprawled on his back, watching her through half-closed eyes.

"Good morning." His voice was husky with sleep and very sexy.

"Beautiful day, isn't it? But, then, it always seems to be beautiful here in Jewel Harbor. That's one of the things I had trouble getting used to when I first moved here four years ago. It's the perfect California seaside town, and perfection always makes a person a little suspicious, doesn't it?" Juliana busied herself with the tray. "Even the fog, when it shows up, is different here than it is anywhere else. Soft and romantic and eerie. You don't take milk or sugar in your tea, do you?"

"Uh, no. No, I don't." Travis sat up slowly against the pillows.

"Didn't think so. You're not the type."

"There's a type?" He watched her, as if deeply intrigued by the whole process of pouring tea.

"Oh, definitely. But I knew you wouldn't be one of those." She handed him a red cup. "Just as I knew the day you walked into my shop that you drank just plain coffee, not espresso or latte or cappuccino."

Seemingly bemused, he glanced down at the strong, dark tea and then up to meet her expectant gaze. "No offense, but it is a little surprising to discover that the queen of the local coffee empire serves tea in bed."

Juliana laughed and helped herself to the second cup. "I'll let you in on a little secret," she said as she sat down in the white leather and chrome chair. "I really don't like coffee, especially all those fancy French and Italian variations I serve at the shop. The stuff upsets my stomach."

Travis's mouth curved faintly. "I know most of your secrets but you've hidden this one well. I would never have guessed you're a closet tea drinker. What would the patrons of Charisma Espresso say if they knew?"

"I don't intend for them to ever find out. Until, that is, I get ready to open up a chain of tea shops."

Travis frowned, shaking his head in an automatic, negative gesture. "Forget the idea of tea shops. Your goal is to expand Charisma, remember? There are a lot more coffee drinkers than tea drinkers around here."

"Never mind about my tea shop idea. I don't really want to talk about it this morning, anyway." Juliana eyed him with great interest. "Did you think you knew all my secrets just because you've been looking into my business affairs for the past couple of weeks?"

"Most of them." Travis shrugged, his bare, bronzed shoulders moving with masculine grace against the white satin pillow. "I'm a business consultant, remember? I'm good at what I do. And I've learned that once you know someone's financial secrets, you usually know all the rest of his or her secrets, too."

"Sounds ominous." Juliana shuddered elegantly and took a sip of her Darjeeling. "I'm

15

glad that in our case there are still a few surprises left. More fun that way, don't you think?"

"Not all surprises are pleasant ones."

The warning was soft. And, predictably enough, it went unheeded. Juliana figured Travis was still a bit sleepy.

"Oh, in our case I'm sure the surprises will all be at least interesting, if not downright pleasant," she said with assurance. "I'm looking forward to each and every one." A rush of happiness sizzled through her as she studied him. He looked so good lying there in her bed. She loved that mat of dark hair on his broad chest. To think she had ever wasted time fantasizing about fair-haired men. She shook her head in disbelief at the recollection of her own foolishness.

"Something wrong?" Travis asked.

"Not in the least."

"I thought you might be having a few regrets —" he paused carefully, his eyes meeting hers "— about last night."

Juliana's stared at him in astonishment. "Of course not. If I'd worried about having regrets, I wouldn't have gone to bed with you in the first place. I knew exactly what I was doing."

"Did you?"

"Absolutely. I'm sure you did, too."

"Yes," he said, looking contemplative. "I knew what I was doing. You look pleased with yourself this morning, Juliana."

"I am." She smiled widely, vastly pleased indeed, with him, life and the world.

"I'm glad you weren't disappointed."

"Disappointed?" She was shocked. "How could I possibly have been disappointed? It was glorious. Perfect. Everything I imagined. You are a fabulous lover, Travis Sawyer. Magnificent."

An unexpected telltale red stained his high cheekbones. For a few seconds Juliana could have sworn Travis looked embarrassed by the expansive praise. She was instantly touched by his lack of ego in that particular department.

"No," Travis said, concentrating on his tea, "I don't think it was anything special I did. We just sort of clicked, I guess. It happens that way sometimes. Two people meet, find each other attractive and, well, things work out in bed."

Juliana's brows rose and she pursed her lips thoughtfully. "Has it happened that way a lot for you in the past?"

Travis blinked and his crystal eyes gleamed behind his lashes. "No," he admitted quietly. "It hasn't happened that way a lot for me."

Juliana relaxed immediately, satisfied to know for certain that the emotions in this situation were not one-sided. "Good. I knew last night was special."

"I take it that it hasn't happened that way a lot for you, either?" The question was reluctant, as if Travis did not want to know the answer but was unable to stop himself from asking the intimate question.

"Never in my whole life," she assured him with perfect honesty.

He grinned slightly. "Maybe you just haven't had enough experience to judge."

"I'm thirty-two years old and along the way I've had to kiss a few frogs to find my prince."

"But you haven't slept with too many of those frogs have you?"

"Of course not. Frogs can be very slimy, you know. A woman has to be cautious."

"Ummm. These days it works both ways."

"I'm aware of that. And I know you're not the kind of man who jumps into bed with any willing body that comes along." Juliana wrinkled her nose with distaste. "I could never fall for a man who didn't have enough sense to be extremely discriminating when it came to his sex life. More tea?"

"All right." He held out the red cup, watching with amusement as she leaped out of the chair to pour from the stunning little art deco pot. "I could get accustomed to this kind of service."

She laughed. "You've caught me in a good mood. Either that or the novelty of the situation has inspired me." She handed him back the cup, enjoying the brush of his fingers against hers. "Well?" she asked, barely able to conceal her impatience. "Would you like to talk about it now or later?"

"Talk about what?"

"Our future, of course." She was aware of Travis going very still against the pillows, but she ignored the implications. "When did you plan to ask me? I wouldn't want to spoil the surprise, but it would be helpful to know what date you had in mind so that I could start making plans. So much to do, you know. I want everything to be just perfect."

Travis stared at her, his tea forgotten. "Plans? What the devil are you talking about?"

"Are you always this dense in the morning?" She smiled at him indulgently. "I'm talking about our marriage plans, of course."

"Our *what?*" The red tea cup slipped from Travis's hand, spilling its contents on the white sheets. It rolled off the edge of the bed and landed on the white carpet with a soft thud.

"Oh, dear. I'd better get something on that right away. Tea stains, you know." Juliana jumped to her feet again and scurried into the white-tiled bathroom to fling open a cupboard.

"Juliana, wait. Come back here. What the hell did you mean a minute ago? Who said anything about marriage?"

She turned around, sponge and carpet cleaner in her hands and marveled at the sight of Travis standing fully nude in the doorway. For a few seconds she forgot about the risk of tea stains in the other room. "Who cares if you're a little on the short side?" she asked softly. "You're just perfect."

"Short?" He scowled at her. "I'm not short. You're too tall, that's the problem."

"It's not a problem. We can work it out. I'll wear flats or short heels most of the time," she vowed. Her eyes traveled wistfully down the length of him. "And certain parts of you are not very short at all."

"Juliana, for Pete's sake." Travis snatched a white towel off the nearest rack and wrapped it around his lean hips.

"You're blushing. I didn't know men could do that."

"Juliana, put down that sponge and start talking sense. What did you mean about getting married?"

She remembered the carpet with a start. "Hang on a second, I have to get that tea out right away. That's the trouble with a white carpet. You can't let anything sit on it very long or the stain sets." She brushed past him and hurried over to where the brown stain was slowly sinking into the beautiful white fibers. "This thing is supposed to be stain resistant, but unfortunately that doesn't mean it will take just any sort of abuse. I probably shouldn't have put in white but it looked so terrific in here I couldn't resist."

Travis stalked slowly across the room to stand towering over her while she knelt and began to scrub industriously. "Damn it, Juliana, I'm trying to talk to you."

"Oh, right. About our marriage. Well, I've been thinking about it and I've decided there's no real reason to wait, is there? I mean, we're hardly kids."

"No, we are not kids," Travis shot back. "We

20

are adults. Which means we don't have to start talking marriage just because we went to bed together on one occasion."

"You know me well enough by now to realize I like to get going on things right away," she reminded him airily. "Once my mind is made up, there's no stopping me. Just ask anyone."

"Juliana, stop scrubbing that damned carpet and pay attention."

"But while it's true there's no real reason to wait, maybe we shouldn't move too quickly on this." Juliana chuckled. "I mean, you're always telling me that I should take my time and plan major moves carefully, right?"

"Damn right."

"And I do want to plan this wedding carefully. I think I'd like to have a big, splashy one with all the trimmings including an engagement party first. After all, I only intend to go through this once, you know? I'd like to do it right. I'd want the whole family present, naturally. My cousin Elly and her husband live just a few miles down the coast and my parents will be able to come down from San Francisco quite easily. Uncle Tony lives in San Diego so that's no problem, either. Then there are all my friends here in Jewel Harbor. There are also several good Charisma customers I'd like to invite."

"Juliana . . ."

"We could use that lovely little chapel that overlooks the harbor."

"Does it occur to you, Juliana," Travis interrupted in a grim voice, "that you're rushing this a bit?"

She paused in her scrubbing and looked up curiously. "Rushing?"

"Yes, rushing." He seemed gratified to have her full attention at last. "I recall everything that happened last night as well as every detail of everything we've discussed for the past three and a half weeks and I know for a fact that nothing, I repeat *nothing*, was ever said about marriage."

"Oh, dear. I've gone and ruined your big proposal scene, haven't I? You were probably planning a romantic evening with wine and caviar topped off by a stroll along the harbor front and a formal proposal of marriage." Juliana bit her lip contritely as she got to her feet. "I'm sorry, Travis. But there's really nothing to worry about. We can still do all that tonight or tomorrow night. That restaurant where we ate last night, The Treasure House, makes a great location for marriage proposals. We can go back there this evening."

"How would you know it makes such a great location for proposals? Oh, hell, what am I saying? Forget it." Travis's gray eyes glittered with anger and exasperation. "Damn it, Juliana, I have no intention of asking you to marry me."

There was a heartbeat of silence while Juliana absorbed that information. For a moment she was convinced she must have misunderstood. "I beg your pardon?"

"You heard me," he said, rubbing the back of

his neck in a gesture of irritation and frustration.

"But I thought . . . I assumed . . ." Juliana ran out of words, a stunningly unusual event. Vaguely she waved her hand, the one with the sponge in it, in a helpless little gesture. "I mean, last night we . . ."

Travis's mouth twisted wryly. "You think that just because we'd been dating for the past couple of weeks and we went to bed together that it automatically followed that I intended to marry you? Come on, Juliana, you're not that naive. In fact, you can be pretty damn savvy when you choose. You're one heck of a sharp businesswoman. You know how to take care of yourself and as you pointed out, you're thirty-two years old and you've kissed a few frogs. So don't give me that wounded doe look."

The accusation stung. Juliana instantly narrowed her eyes as a slow anger began to simmer through her veins. "I will have you know, Travis Sawyer, that my intentions were honorable all along. I knew the minute I met you that I was going to marry you."

"Is that right? Well, maybe you should have warned me. We could have avoided this whole silly, embarrassing scene. Because I have no plans to marry anyone at the moment."

"I see." She drew herself up proudly. "You were just using me, is that it?"

"No, I was not using you and you damn well know it. We are two adults who happened to be very attracted to each other physically. We have

23

professional interests in common, we're both single and we're working together because you hired me as a consultant. It was perfectly natural that we start an affair. But right now, that's as far as things have gotten. We're involved in an affair, nothing more."

"You're not prepared to make a commitment yet, is that it?" she challenged.

"Do you always hit your dates over the head like this the morning after?"

"As we've already discussed, there haven't been that many morning afters and no, I don't generally hit them over the head like this. But, then, I haven't wanted to marry very many of my dates."

"How many have bothered to ask you to marry them?" Travis asked sarcastically.

"Lots of men have asked me to marry them. I get asked all the time, as a matter of fact. Usually at The Treasure House. That's how I know it's a perfect setting."

"If you've had all those opportunities, why haven't you taken one of the poor jerks up on his proposal?"

Juliana was furious. "Because none of them have been the right man. I've turned them all down. Except for one that didn't work out, anyway."

"So I'm one of the lucky two you've considered suitable, huh? What happened to the other sucker?"

Juliana felt hot tears gathering. She blinked

furiously to get rid of them. "There's no need to be so rude. He wasn't a sucker. He was a wonderful, charming person. A caring person. He was also a real hunk. Beautiful hazel eyes and tawny blond hair. He was a handsome, golden god. And he was a lot taller than you."

"I don't care what he looked like. I just want to know how the guy escaped."

"Why? So you can escape the same way? All right, I'll tell you how he escaped my clutches. He turned tail and ran, that's what he did. Straight into the arms of someone else. Someone who happened to be very close to me. Someone petite and blond and sweet natured. Someone who never argued with him. Someone who never presumed to question his judgment. Someone who didn't overwhelm him the way I did. There. Satisfied?"

"Hell, Juliana, I didn't mean to rake up old memories." Travis rubbed the back of his neck again. "I was just trying to make a point."

"Consider it made. Go ahead, take the same way out my fiancé took three years ago. Run away if you're that skittish. But I have to tell you, Travis, I expected more from you. I didn't think you were the kind of man who was intimidated by a woman like me. I thought you had some guts."

"I am not going to run anywhere," Travis bit out. "But I am also not about to let you pressure me into marriage. Do I make myself perfectly clear?"

Juliana dashed her hands across her eyes, nodding sadly. "Perfectly. It's obvious there's been a terrible misunderstanding here. I guess I misread all the signals." She sniffed back the incipient tears. "I apologize."

Travis's hard face softened. He stepped closer and lifted a hand to stroke the side of her cheek. "Hey, there's no need to upset yourself over this. I've known you long enough to realize you're the impulsive type."

"My cousin Elly says I'm very spontaneous."

Travis smiled, and the rough edge of his thumb traced the line of her jaw. "That, too."

"This is so embarrassing."

"Forget it," Travis said magnanimously. "Last night was very good. I can see where you might have read more into it than . . . uh . . . ."

"More than you intended?"

"Let's just say more than either of us intended."

"Speak for yourself." She turned away, ducking out of the range of his hand. "It's getting late. You'd better get dressed. I'm sure you've got a lot to do today."

"Nothing that can't wait until Monday." Travis watched her closely. "What do you say we spend the rest of Sunday at the beach?"

"No, thanks." She bent down to take one last swipe at the carpet. Then she picked up the empty red cup. "I've got a million chores to do today," she added grimly. "You know how it is. I think I'll start with washing my hair and move on to the laundry. These sheets definitely need to be cleaned."

Travis didn't move. "Are you going to sulk?"

"I never sulk," she assured him grandly.

"Then why the excuses? Last night we both made plans to spend today together."

"Everything's changed now. I'm sure you can comprehend that." She moved into the white-tiled bathroom and replaced the carpet-cleaning materials in the cupboard. "I wish you would hurry up and get dressed, Travis. It's disconcerting having you standing around in my bedroom wearing nothing but a towel."

"I could take the towel off."

She glared at him through the open doorway. "You're not suggesting we go back to bed together, are you?"

"Why not? We both agreed last night was very good."

"I don't believe this." She braced one hand against the door frame. "Travis, you surely do not expect me to jump right back into bed with you now that I know your intentions are not honorable?"

"Will you stop talking like a nineteenth-century heroine who thinks she's been compromised?"

"You don't seem to understand," Juliana said with seething patience. "I am throwing you out. Now. Get dressed and get out of my condo. Everyone makes a mistake now and then, but I do not cast pearls before swine twice. And I

27

have no intention of wasting any more of my valuable time with a man who is as muddle-brained and stubborn as you are."

"Is that a polite way of calling me stupid?"

"You got it. I'm the right woman for you, Travis Sawyer. I was made for you. And you were made for me. If you're too dumb to see that, then there's no point taking this relationship one step further. Get out of here."

His eyes were narrow slits as he snagged his trousers from off the dresser. "Does this mean you're going to try to terminate your contract with Sawyer Management Systems?"

Juliana was startled. If she actually fired him she might never see him again. The thought was too terrible to contemplate. "No, it does not. SMS is the best business consulting firm in this part of California and the future of Charisma Espresso is too important to jeopardize by getting rid of you. Unfortunately."

"Is that right?" Travis yanked on his pants and reached for his white shirt. "Nice to know I'm still appreciated in some respects. But aren't you afraid of mixing business with pleasure? That's how we got into this situation in the first place."

Juliana lifted her chin. "No, I am not the least worried about it. I am quite capable of separating my business from your pleasure."

"Yeah? Well, we'll see just how good you are at it, won't we?" He finished buttoning his shirt and grabbed his wing-tips.

"Are you threatening me, Travis?"

"Wouldn't think of it." He tied his shoe strings with short, savage movements of his fingers and stuffed his tie into a shirt pocket. "But we both know you're the emotional one in this equation. And you want me. Hell, this morning you woke up convinced you were in love with me."

"I never said that."

"Yes, you did," Travis countered coldly. "Last night when you were lying under me, hanging on to me as if I was the only man left on earth. I heard every word."

Juliana felt herself grow very warm with humiliation. Her chin lifted defiantly. "All right, so I said it. I won't deny it. I wouldn't have wanted to marry you if I wasn't in love with you. But you brought me back to reality this morning. Love is probably no worse than the flu. I'll recover just like I did three years ago when my engagement ended. Now go on and get out of here before I lose my temper. You are becoming very annoying."

Travis stalked toward the door. "You'll be sorry you're kicking me out like this."

"Hah. Not a chance. Life is too short for foolish regrets. Like I said, I'll recover. But I warn you, Travis, someday you're going to look back on this whole thing and call yourself a fool."

"Is that right?" He was in her turquoise and apricot living room now, his hand on the doorknob.

Juliana hurried down the hall behind him. He was practically out the door she thought with horror; practically gone. "Yes, that's right. I'm the perfect woman for you, and one of these days you're going to realize it."

He swung around to confront her, the door open behind him. "I already realize we're good in bed together. What more do you want from me?"

She skidded to a halt a couple of feet away from him, breathless. "I want you to realize you love me. And then I want you to ask me to marry you."

"You don't ask much, do you?"

"I never do anything by half measures. You should know me well enough by now to realize that. But —" she paused, gathering her courage "— maybe I should make allowances for the fact that you're a man and therefore not as in touch with your own needs and emotions as you should be."

"Gosh, thanks for all the deep psychological analysis and understanding, lady."

"I'll tell you what, Travis. I'll give you one month. One month and that's all. If you haven't come to your senses by then, I won't give you any more chances."

His brows rose in an intimidating fashion. "One month to do what?"

"One month to figure out you're in love with me and ask me to marry you."

"One month, hmm? I'm surprised at you,

Juliana. You should know me well enough by now to realize that I don't react well to ultimatums."

"Don't think of it as an ultimatum," she urged. "Think of it as a breathing space in which you can sort out your options."

He shook his head, amazed. "You never give up, do you?"

"People who give up don't often get what they want."

Travis went through the door. "I don't need a breathing space. I already know what I want. I've known all along."

"Just exactly what did you want from me?" she demanded, moving to stand in the doorway. "A night in bed?"

"No, Juliana. Getting you into bed wasn't the important thing. Believe it or not, it wasn't even part of my original plan. Just icing on the cake, I guess you could say."

He walked away into the bright, sunny morning. Juliana stood on the brick steps of her Spanish-style white stucco and red-tiled condominium. She watched with dismay as the man she loved climbed into his tan, three-year-old, nondescript Buick.

How could she have been such a nitwit as to lose her heart to a man who drove such a wretchedly dull car and wore such old-fashioned ties, she wondered.

Two

The whole thing had gotten far too complicated, unbelievably complicated, disastrously complicated.

Revenge should have been a simple, straightforward matter filled with strong, clear, uncomplicated emotions. There were just two sides to this thing, Travis reminded himself — his side and the other side. And anybody with the last name of Grant was on the other side.

He sat behind the wheel of his car and stared broodingly out over the Pacific. From the top of this bluff the view was postcard perfect. The town of Jewel Harbor sparkled down below, an artistically charming mixture of Spanish-Colonial-style homes and the latest in California Coast modern architecture. There was an air of trendy prosperity to the whole place that was a bit unreal at times. It would make a good setting for the headquarters of Sawyer Management Systems.

The streets were lined with swaying palms and every yard was lush and green. A lot of the backyards had sapphire-blue swimming pools and orange trees. The cars parked in the wide drives tended to be of German manufacture with the occasional Italian or classic British model thrown in for variety.

The downtown business section of Jewel Harbor looked as casually upscale as the rest of the town. Strict ordinances kept the shops and office buildings low in height and architecturally reminiscent of the Spanish look. White stucco and red tile predominated, just as they did in Juliana's condominium complex. Travis narrowed his eyes momentarily, searching the vista. From here he could just make out the busy shopping plaza where Juliana had opened Charisma Espresso.

He thought about the fateful day nearly four weeks ago when he had walked into the trendy watermelon-red and gray interior of Charisma. He had told himself at the time it was a simple reconnaissance move. He was like a general with a carefully arranged battle plan and he wanted to be sure he had covered all the angles. He had timed everything else involved very carefully, right down to making sure that his trap would close while he was here in Jewel Harbor setting up the newest office of Sawyer Management Systems.

Juliana Grant was the one member of the Grant family he hadn't met five years ago, the one unknown quantity in his equation. She had not lived in Jewel Harbor back then. He vaguely remembered being told that she was working in San Francisco at the time.

Travis wasn't quite certain what he'd been expecting when he pushed open the glass doors of Charisma Espresso, but he had been immediately struck by two powerful forces. The first was the heady aroma of freshly ground coffee and the second, far stronger force, was the vivid, red-haired, incredibly dressed, six-foot-tall goddess behind the counter. The electric-blue jumpsuit she'd been wearing should have looked tacky or at least overpowering, and on anyone else it probably would have. But on Juliana it looked just right. It was as bold and animated as she was.

Juliana Grant was unlike any of the other Grants, and that was probably why he had allowed the situation to get so complicated. Travis remembered the men of the family as being of average height, the women petite and delicate.

Juliana, by contrast, was almost as tall as Travis. Hell, he thought wryly, when she put on heels she was as tall or taller than he was. Her flaming red hair might have been inherited from her father, but it was difficult to be certain because Travis remembered Roy Grant as being gray. The same with Tony Grant, Roy's brother. She had probably gotten her eyes from her mother, Beth. But there was no one in the family with quite the same combination of coloring and height that Juliana had. In her, the genes had obviously come together in a whole new, exotic mixture.

But it wasn't Juliana's looks that had caught

him off balance; it was Juliana, herself. She was different. Not only different from the other members of her family, but different from every other woman Travis had met in his life.

Too much, he thought, seeking the right words to describe her. That was it. Juliana was a little too much of everything. Too colorful, too tall, too emotional, too dynamic, too assertive, too smart. The kind of woman who, a thousand years ago, would have carried a spear and ridden into battle beside her chosen mate. The kind who would give everything and demand just as much in return.

She was, to put it bluntly, the kind of woman most men found overwhelming except in very small doses. Travis knew his own sex well enough to realize that the average male would find Juliana riveting for about fifteen minutes. Shortly thereafter, that same man would be frantically searching for an exit, running as fast as he could in the opposite direction.

No question about it. Women like Juliana could be downright intimidating to the average male.

Travis didn't consider himself average and he refused to be intimidated by Juliana, but that didn't mean he was prepared to jump through hoops for her, either. The lady was a handful but he had no real doubt he could handle her. That was not the problem.

The problem was that he wanted her and, given his current situation vis-à-vis the Grant family, he had no business getting any more involved with her. It had gone far enough. How in hell had he let himself get hired by her as a consultant? He must have been temporarily out of his mind. He hadn't taken on a tiny client like Charisma Espresso for over ten years.

Travis exhaled deeply, trying to think his way through the mess. In the beginning it had seemed simple enough. Juliana was a Grant and he had vowed revenge against the entire clan of Grants. He had told himself that seducing Juliana would add a nice fillip to the masterpiece of vengeance he had concocted. And it was obvious Juliana wasn't going to object to being seduced.

But looking back on it now, Travis wasn't quite sure who had seduced whom. It was Juliana who had paved the way for the affair when she had talked him into taking her on as a client. The moment she had discovered what he did for a living, she'd started bombarding him with questions about how to successfully expand Charisma Espresso.

Travis had taken what seemed an obvious opening and proceeded to play it for all it was worth. One more Grant scalp for his belt.

His mouth twisted grimly at the thought. He didn't need Juliana's scalp. She'd had nothing to do with what had happened five years ago. She didn't even know who he was. It was her misfortune, however, to be related to all the other Grants, and three and a half weeks ago when he had first met her, he'd told himself he might as well make use of her.

Last night he'd stopped thinking in terms of using her for revenge and started thinking in terms of satisfying the craving that had been building up inside him.

This morning he had been too bemused by events to think at all until Juliana had breezed into the bedroom with her chic little art deco teapot and started making marriage plans. That had brought Travis back to reality with a thud.

Juliana was taking over, threatening his plans for vengeance. After all the planning and hunting and patience he'd been obliged to exert during the past few years, Travis was not about to lose control of the situation now.

He rubbed the back of his neck and switched on the Buick's ignition. He should have seen it coming, he told himself. Juliana had fallen in love with him sometime during the past three weeks. He'd known that for certain last night when she had given herself freely, without any reservations. That was the way she did things. And if he was honest with himself, he had to admit he had taken everything she had to give.

He had to keep reminding himself that she was a Grant, Travis thought as he headed back toward his apartment. And he would be damned if he would let any Grant give him an ultimatum.

One month to figure out he loved her? One month to come to his senses? Who did she think she was? Before the month was out he was going to have reaped his revenge on all the Grants.

He would be lucky to have a week with her, at the most.

Because when the manure hit the fan, as it surely would very soon, everyone would have to choose sides. Travis didn't need to be told which side Juliana would choose. Her choice was preordained by the fact that she was a Grant. Travis faced this reality with stoic acceptance. He was used to being the outsider, to being the one not chosen.

But all he could think about as he drove back down the winding road into town was that last night had been something else. The memory of it would haunt him as long as he lived and he knew it. He could still feel the imprint of her nails in his back. Juliana Grant was the kind of woman who left her mark on a man.

Suddenly the prospect of stealing even one more week with her was more than he could resist.

"Hey, how was the big date, Juliana?"

"Saw you having dinner down at The Treasure House last night. You two looked so involved I didn't want to interrupt. Thought maybe the man was about to pop the big question."

"How about it, boss, you wearing a ring this morning?"

Juliana glowered fiercely at the expectant faces of her staff as she came to a halt in front of the long gray counter. "Don't the two of you have anything better to do this morning than stand around asking personal questions?"

"Uh-oh." Sandy Oakes, her gelled hair sleeked back behind her ears to show off the three sets of earrings she was wearing, eyed her co-worker. "Looks like all is not well with our supreme leader this morning. Best go grind a little coffee, Matt."

Matt Linton, whose hair was even shorter than Sandy's and who wore only one earring, frowned in sudden concern. "Hey, we were just teasing. Is everything okay, Juliana?"

"Everything is just fine. Absolutely peachy. Fabulous." Juliana hurled her oversized leather tote into her cubbyhole of an office and then reached for one of the watermelon red aprons that bore the Charisma logo. "I couldn't be happier if I had just found out I had won the lottery. Satisfied? Now get busy. The morning rush will be starting in a few minutes. Sandy, why aren't those *biscotti* in the display case?"

"The bakery just delivered them five minutes ago," Sandy explained in soothing tones. "I'll have them out in a sec." She slid a speculative glance at her boss as she arranged the *biscotti* in a glass case near the cash register.

"Matt, try to look useful. The counter needs straightening. And where's the cinnamon shaker?"

"Ouch." Matt shook his hand as if it had just been bitten by a savage dog.

Juliana groaned. "Look, I apologize for being snappish this morning. But the truth is, I am not in a good mood."

"Funnily enough we could tell that right off," Sandy said. "I take it the turkey didn't ask you to marry him as planned?"

"Not only did the turkey not ask me to marry him, he was apparently stunned to find out I expected him to do so," Juliana informed her. "The whole thing was a complete misunderstanding. I made an absolute fool out of myself. If I ever show any signs of wanting to get involved with anyone else of the opposite sex ever again, I want you to promise to remind me of what happened this time. I refuse to repeat my mistakes."

Matt grinned. "You're going to swear off men forever just because Mr. Right turned out to be Mr. Wrong?"

"He isn't Mr. Wrong. He just doesn't know he's Mr. Right." Juliana turned her back to the door and busied herself with grinding an aromatic blend of beautifully roasted Costa Rican beans. She raised her voice to be heard over the roar of the machine. "But I guess if he doesn't have enough sense to know he's Mr. Right, then he really is Mr. Wrong, isn't he? I mean, the real Mr. Right wouldn't be that dumb, would he?"

She was so busy working through that train of logic that she failed to hear the shop door swing inward.

"Uh, Juliana," Matt began nervously, only to be cut off by Juliana's diatribe.

"But I have to tell you, I think my heart is broken. And what does that say about my intelligence, I ask you? How could I let Mr. Wrong break my heart? I'm too smart to do that."

"Juliana, uh, maybe you'd better . . ."

"What's more," Juliana plowed on forcefully, "if Travis Sawyer is so stupid he doesn't even realize I'm the right woman for him, then he's probably too stupid to be planning the future of Charisma Espresso. This morning I told him I didn't intend to fire him, but now I'm not so sure. I've had time to think about it and I really don't believe I want to put the future of my company in Sawyer's hands . . ."

"Juliana," Sandy broke in hurriedly. "We've got a customer."

"What?" Juliana finished grinding the last of the beans and the machine stopped.

"I said," Sandy repeated very clearly, "we've got a customer."

"Oh. Well? Why make a big deal out of it. Go ahead and see what he wants."

"What he wants," Travis Sawyer said calmly from the other side of the counter, "is the one month you promised him in which to come to his senses."

"*Travis.*" Juliana couldn't believe her ears. Relief and happiness rushed through her. She swung around to confront him, knowing she was smiling like the village idiot, but not caring in the least. Her heart was not broken after all. "You came back."

41

"I never left. At least, not willingly. You're the one who kicked me out."

"I knew you'd see the light. I knew you just needed a little time to get your head screwed on straight." Juliana tossed the sack full of ground coffee toward Matt and dashed around the counter to hurl herself into Travis's arms.

Travis braced himself as she landed against him with an audible thud. He only staggered back a step. "I'm touched by your faith in my intelligence." Travis looked straight into her glowing eyes. She was wearing two-inch heels today. "Does this mean you're not going to try to find a way out of our business contract, after all?"

Sandy spoke up firmly from the other side of the counter before Juliana could respond. "I don't think he's groveled enough yet, Juliana."

"Give the man a break, ladies," Matt growled. "He's here, isn't he? How much more can you ask?"

"Thank you," Travis said gravely, nodding at Matt. "I agree completely. How much more can you ask?" He turned his attention back to Juliana who was smiling with delight, her arms around his neck. "Would you mind very much if we conducted the rest of this grand reconciliation scene in private? I like Matt and Sandy, but once in a while I find I like to operate on a one-to-one basis with you."

"Don't mind us," Sandy said quickly. "We're only too happy to help out."

"Right," said Matt. "We're just like family."

"Not quite," Travis said, taking a firm grip on Juliana's arm and leading her toward the door.

Juliana was bubbling over with laughter by the time Travis had led her to the outside seating area. Morning sun poured warmly through the decorative wooden lattice overhead, dappling the white tables and French café-style chairs.

"They mean well, you know," Juliana said easily as she sat down across from Travis.

"I know they do but I feel like I'm in a gold-fish bowl every time I go into the shop. Do you tell them everything?"

"No, of course not," Juliana assured him quickly. "But they've been sort of monitoring our relationship right from the start. They were there that first day when you walked in and ordered a cup of coffee, remember? They knew how I felt about you at the beginning. They guessed right away this morning that something terrible had happened."

Travis sighed and leaned back in the small chair. It creaked under his weight. "You're a full-grown woman, Juliana. Not a starry-eyed teenager. You'd think by now you would have learned to be a little less, well, less obvious about your personal feelings."

"I'm a very straightforward person, Travis." Juliana grew more serious as some of her initial euphoria subsided. "People always know where they stand with me and I like to know where I stand with them. Life is easier that way. Keeps the stress level down a little."

43

"You're an odd combination of ingredients, you know that?"

"You mean, for a woman?" she asked dryly.

"For anyone, male or female. When it comes to business you're as shrewd a small-businessperson as I've ever met. The success you've made out of Charisma speaks for itself."

"But?"

Travis's mouth kicked up at the corner and his eyes glinted. "But when it comes to a lot of other things, you're a little outrageous. No, that's putting it mildly. You're more like a keg of dynamite. I can't always predict when and how you'll go off. And you always do it loudly."

She shrugged. "You don't know me as well as you think. And I obviously don't know you as well as I thought I did or I wouldn't have put my foot in my mouth the way I did this morning. But that's okay. We've got plenty of time to learn all we need to know about each other, don't we?"

Travis studied her for a long moment. "I'm not going to make any promises, Juliana. I want that clear this time right from the start."

"Are you one of those men who can't make a commitment? If so, just say it straight out because I really don't want to waste any time messing around with a male who's uneducable."

"Damn it, I'm one of those men who won't be rushed into anything, including a commitment. And I just want that fact on the table before we try this relationship again. Knowing that, are you still willing to give me my month?"

Juliana thought about it but not for long. "Sure. Why not? I'm willing to take a risk or two if the prize is worth it."

Travis shook his head in silent wonder. "So reckless."

"Only when I'm going after something important."

"I guess I should be flattered that you consider me a worthwhile prize."

"That remains to be seen. At this point, you're just a potentially worthwhile prize."

"Yeah. Well, as long as we're dealing in warnings, I guess I ought to inform you that you've had yours. No promises from me, Juliana, implied or otherwise. No commitments. We take things a day at a time. I won't be pushed into anything."

"I've had my warning," she agreed smoothly. "But you haven't had yours."

His brows rose. "What warning is that?"

"Since you are unable to see your way clear to make a commitment, I am unable to see my way clear to go to bed with you until we've resolved all the issues between us."

Travis's eyes narrowed coldly. "I didn't think you were the kind of woman who used sex to get what you want."

"I'm not. Just as you're not the kind of man who could be manipulated with sex." She smiled brilliantly. "Therefore, I wouldn't dream of trying to hold you that way."

"Very thoughtful of you," he muttered.

45

"By not going to bed with you I will leave your brain free of hormonal clutter," she added. "You'll be able to think much more clearly about our future."

"Juliana," Travis said with elaborate patience, "last night we discovered we happen to be very good together in bed. Remember?"

"Of course I remember. So what?"

"So why deny ourselves that element of the relationship?" he asked gently. He reached across the table and covered her long, copper tinted nails with his big palm.

"Simple. I happen to view the act of going to bed with a man as an act of commitment. And when it comes to this relationship of ours, I'm not making my commitment again until you've made yours and that's final. I have no intentions of sticking my neck out twice. Still want to use your month's grace period?"

He stared at her for a long, charged moment. "What the hell. Why not? Maybe it's better this way. This relationship doesn't stand a snowball's chance, anyway. I must have been crazy to think I could have my cake and eat it, too."

"What are you talking about? What's all this about a cake?"

"Nothing."

"But, Travis . . ."

He got to his feet. "I'd better get back to the office."

Juliana looked up at him anxiously. She reached out to catch hold of his arm. "Travis,

46

wait. I don't understand what's going on. Do you want to see me again or not?"

He touched her hand as it rested on his sleeve. His gaze was diamond hard in the sunlight. "Yes, Juliana. I want to see you again."

She relaxed. "Even under my terms? You don't look too thrilled about the prospect."

He looked down at the bright copper-colored nails on his sleeve. Then his eyes met hers. "I thought you knew what I wanted better than I did."

Juliana gnawed on her lower lip. "Once in a while I guess wrong, just like everyone else. I can make mistakes. It's happened before."

"With the fiancé who ran off with the petite blonde a few years ago?"

"Like I said, I'm not infallible. Up until last night, I was very sure about you and me. As you said, things clicked between us and not just in bed. But if I'm wrong, I'd rather call it quits right now."

"Would you?"

Juliana drew a deep breath. "You're a very hard man, aren't you?"

"And you are a very volatile woman."

"Maybe that's not such a good combination after all. Maybe all we'll ever succeed in doing is striking sparks off each other. That's not enough, Travis."

"Getting cold feet already, Juliana?"

She reacted to that instinctively. "No. I'll give you your month."

"Thanks." Travis leaned down to brush her mouth with his own. "I'll pick you up for dinner tonight. Six o'clock okay?"

"Yes. Fine." She smiled again, pushing aside the dark second thoughts that had crept into her mind. "I'll be ready. How about the new Thai place on Paloma Street?"

"It's a date." He walked out to where the tan Buick was parked and got inside.

Juliana sprang to her feet and hurried after him. "And do you still want to go with me to my cousin's birthday party next Saturday?" she asked anxiously. "It will mean meeting a lot of my family, including Uncle Tony."

Travis looked up at her through the open window, his expression so startlingly, unexpectedly harsh that Juliana instinctively stepped back a pace.

"Wouldn't miss it," Travis said and turned the key in the ignition.

Juliana smiled uncertainly and waved as he drove off. It was nice to know Travis was the kind of man who didn't mind getting involved with family, she told herself. And immediately wondered why she was not particularly reassured by that information.

The following Saturday evening Juliana sat in the passenger seat of her snappy, fire-engine-red two-seater sports coupé and reveled in the balmy sea air coming through the open sun roof. Travis was at the wheel, and under his

48

guidance the little car hugged the twists and turns of the coast road with easy grace. The blackness of the ocean filled the horizon, merging with the night. Far below the highway, moonlit breakers seethed against the rocks. It was a perfect Southern California evening, Juliana reflected, feeling happy and content. The past few days had been good with Travis, even if there were a lot of uncertainties hovering in the air between them.

"You're wasting your talents behind the wheel of that Buick," Juliana declared as Travis accelerated cleanly out of a turn. "One of these days you'll have to get yourself a real car."

"The Buick suits me. We understand each other."

"Don't you like driving?"

"Not particularly."

"But you do it well," Juliana observed.

"It's just something that has to be done and I try to do it efficiently so that I don't get myself or anyone else killed in the process. That's the extent of my interest in the matter."

Juliana sighed in exasperation. "You've been in a rather strange mood ever since you appeared at Charisma to tell me you wanted your month. And tonight you're acting downright weird. Are you sure you want to go to Elly's birthday party?"

"I've been planning on it for weeks." He braked gently for another curve.

"Yes, I know, but I don't want to force you

49

into this. I mean, a lot of men are not real big on family get-togethers."

"It's a little late to change my mind now, isn't it? We'll be at the resort in fifteen minutes."

"True. You'll like my family, Travis. I would have introduced you to Elly and David before this but Elly's been out of town for the past three weeks. She's been visiting other resorts to get some ideas for Flame Valley. My folks flew into San Diego earlier today to pick up Uncle Tony and drive up the coast to the resort. They should be there by the time we arrive and I know they'll want to meet you. David is —"

"Juliana?"

It worried her more and more lately when he spoke in that particular tone, she was discovering. She did not understand him when he was in this mood. "Yes, Travis?"

"You don't have to sell your family to me."

"Okay, okay. Not another word on the subject. I promise."

He smiled fleetingly, with visible reluctance. "And if I believe that, I ought to have my head examined."

"Hey. I can keep a promise."

"Yeah, but I'm not sure you can keep your mouth shut."

"You got any serious objections to my mouth?" she demanded.

"No, ma'am," he said fervently. "None." He paused. "Has your cousin been married to this David Kirkwood long?"

"Almost three years. They make a wonderful couple. Perfect together. Elly was involved once with someone else about five years ago. I never met the man and she refuses to talk about him, but I know he traumatized her. For a while I was worried she wouldn't let herself love anyone again. And then along came David."

"And they're happily running this resort?"

Juliana smiled. "Flame Valley Inn. One of the most posh on the coast. Wait until you see it, Travis. It's beautiful and it's got everything. Golf course, tennis courts, spa, fantastic ocean view, first-class luxury rooms and a wonderful restaurant. My father and my Uncle Tony, that's Elly's father, built it over twenty years ago. They wanted to cash in on the spa craze."

"And now your cousin and her husband run it." It was a statement, not a question.

Juliana slanted Travis a quizzical glance. "That's right. My father sold a lot of his stock in it to Uncle Tony four years ago but he still holds a minority interest. Uncle Tony was supposed to take over running the place full-time but about three years ago he developed some heart problems and the doctors insisted he start taking it easy. Elly and David took over and they've been running the place ever since. They love it."

"So your cousin's husband has been making most of the decisions about Flame Valley?"

"For the past couple of years, yes. David has a lot of big plans for Flame Valley." Juliana propped her elbow on the door sill and lodged her chin in her hand. "Unfortunately I think he moved a little too quickly on some of those plans, though."

"Too quickly?"

The small show of interest was all Juliana needed. Her brows snapped together as she frowned intently. "David and Elly have a lot of ambitious plans for Flame Valley. If they work out, the place will be one of the premier resorts in the whole world. But if they don't, Flame Valley could be in real financial trouble."

"I see."

"Travis, I've been meaning to ask you something. I know your company consults for a wide variety of businesses. Do you know anything about the resort business?"

There was a beat of silence. Then Travis said softly, "Yes. I know a little something about resorts and hotels."

"Hmmm." Juliana mulled that over. "I wonder if I could get David and Elly to talk to you. I've been a little worried about them lately."

"How much trouble are they in?"

Juliana drew a breath and settled back in her seat. "I really shouldn't say anything more until I've talked to them first. David is very touchy on the subject of his business ability, and Elly gets defensive. But if I could talk them into hiring you for some consultation, would you be willing to take them on as clients?"

"I've got my hands full at the moment, Juliana. I managed to squeeze Charisma into my schedule but I'm afraid that's the limit. Taking on a project the size of Flame Valley would be impossible."

"Oh." Juliana swallowed her disappointment. "Well, in that case, I guess I'd better not say anything to David or Elly."

"That would probably be best."

Juliana brightened. "But maybe you'd have room in your schedule in another month or two?"

Travis gave her a brief, sharp glance. "You never give up, do you?"

She grinned. "Only when the situation is clearly hopeless."

"The question is, would you recognize a hopeless situation when you saw it?"

"Of course I would. I'm not an idiot. Slow down. There's the sign for Flame Valley. Turn right toward the ocean at the next intersection."

Travis obeyed. He was silent as he navigated the narrow strip of road that wound its way toward a glittering array of lights perched on a hill overlooking the ocean.

He continued to say nothing as he parked the red coupé in the lot below the resort. Then he switched off the ignition and sat quietly as Juliana unbuckled her seat belt. He watched as she turned quickly, kneeling on the seat to reach into the back of the car to retrieve a bundle of brightly wrapped gifts.

"Juliana?"

"Yes, Travis?"

"I want you to know something."

"What's that?" She was bent over the back of the seat, fumbling with the biggest of the presents and wondering if she'd made a mistake buying Elly the huge Italian flower vase. Not everyone liked two-foot-tall pillars of aerodynamically shaped black glass.

"When it's all over tonight, try to remember that I never meant to hurt you."

Juliana froze, the packages forgotten. She whipped around in the seat, eyes widening quickly. "My fiancé said exactly those words three years ago just before he announced his engagement to someone else. What are you trying to tell me, Travis?"

"Forget it. Some things cannot be changed once they've been set in motion." He cupped her face quickly between his strong hands and kissed her with a fierce possessiveness. Then he released her. "Let's go." He opened the car door and got out.

"Travis, wait a minute. What's going on here?" Juliana scrambled out of the car, clutching the gifts. The glass beads that trimmed the scoop-necked black velvet chemise she was wearing sparkled in the parking lot lights. "You owe me an explanation. You can't just go around making bizarre statements like that and expect me to overlook them."

"You're going to drop that if you're not careful." He put out a hand and took the biggest package, the one containing the vase, out of her arms. Then he turned and started walking resolutely toward the main entrance to the resort.

Juliana hurried after him, hampered by the remaining presents, the tight chemise skirt and the two-inch heels of her black and fuchsia evening sandals. "You can't get away with this sort of behavior, Travis. I want to know what you meant. If you're seeing someone else, you'd damn well better tell me up-front. I won't be two-timed. Do you hear me?"

"There is no one else." He walked under the dazzling lights that illuminated the entrance.

The massive glass doors were opened by a young man uniformed in buff and gold. "You must be here for the owners' private party," the doorman said with an engaging grin. "Right straight through the main lobby and out to the swimming pool terrace. Can't miss it." He nodded at Juliana. "Good evening, Miss Grant."

"Hi, Rick. How's everything going?"

"Just fine. You should enjoy yourself tonight. The kitchen's been working overtime for the past three days getting ready. A real blowout."

"I believe it. See you later." Juliana smiled distractedly and dashed ahead to catch up with Travis who was still moving forward with the purposeful air of a man heading into battle.

"Honestly, Travis, you're getting weirder by the minute."

He stopped at the doors at the far end of the elegant lobby and paused to hold one open for her with a mocking gallantry.

Juliana scowled at him and then peered through the glass at the throng of people gathered around the turquoise swimming pool. She caught sight of her mother and father, her Uncle Tony and then she spotted her cousin, Elly.

As Juliana looked at her cousin, a tall, fair-haired, good-looking man moved up behind Elly and draped an arm around her shoulders. They made a handsome couple, no doubt about it. Elly, petite, blond and delicate looking, was a perfect foil for her tall, charming husband. When Elly glanced up at David and smiled, it was easy to see the love in her eyes.

Juliana jerked her gaze away from the sight of Elly and David just in time to catch Travis staring intently at the couple. There was something in his expression that sent a frisson of genuine alarm through her veins.

"Travis?"

"We'd better go out and join the others, hadn't we? Wouldn't want to keep them waiting. It's been long enough as it is."

Confused, Juliana walked through the door, conscious of Travis right behind her. Several faces in the crowd turned to smile in a friendly fashion. Juliana paused to say hello a few times before she reached the small group composed of Elly and David and three of their acquaintances.

Elly turned, David's arm still resting affectionately on her shoulders. She smiled with genuine pleasure when she saw her cousin. Her short, silvery hair gleamed in the light.

"Juliana, you're here at last. Uncle Roy and Aunt Beth got here with my father a couple of hours ago. We've been waiting for you."

"How did the spa survey go?" Juliana asked.

"Great. I picked up all sorts of terrific ideas. Now, who is this mysterious date you told me you were bringing tonight?" Elly's gaze switched to the man standing behind Juliana. Her blue eyes widened with shock and the words of greeting died on her lips.

She looks as if she's just seen a ghost, Juliana thought. She watched in sick fascination as her cousin struggled to conceal the panic that had so clearly blossomed at the sight of Travis Sawyer.

Travis did not move but Juliana felt the tension in the atmosphere between him and her cousin. It was the kind of tension that signals powerful emotions and dangerous secrets.

In that moment Juliana's parents came forward with her Uncle Tony, and Juliana saw Elly's shock mirrored on the faces of the other three Grants.

And suddenly Juliana understood it all. Travis was the man from Elly's past, the one she'd been engaged to five years ago, the one no one talked about.

Three

"The most amazing thing is how calm, cool and collected everyone is behaving," Juliana muttered to Elly twenty minutes later when she finally managed to corner her cousin in a remote section of the terrace. "I thought you were going to faint from shock when you first saw him, but then, two seconds later, there you were, greeting him as if he were just another casual acquaintance. Uncle Tony was just as cool. Just an old business associate. And Mom and Dad acted as if they could barely remember him."

"Well? What did you expect us to do?" Elly demanded. "Scream hysterically and fling ourselves over the balcony? It's been five years, after all."

"Yeah, but we both know his showing up here at this point is not just one of those strange little coincidences that sometimes happen in life. Nothing that man does is a coincidence. Believe me."

"You know him that well, do you?" Elly gripped the railing and faced the sea. The evening breeze ruffled her graceful white skirts.

"Let's say I'm getting to know him better by the minute. He's the one, isn't he? The one you

58

were going to marry five years ago. The one who saved Flame Valley Inn from bankruptcy."

Elly bowed her head. "Yes. He wanted me and he wanted the resort, and Dad and Uncle Roy needed his help desperately."

"Nobody ever told me the whole story. All I knew was that you'd been shaken to the core by the whole incident. I thought he'd seduced and abandoned you or something, but it wasn't like that, was it?"

"No." Elly sounded thoroughly miserable. "I'm the one who broke the engagement."

"But not before you'd played the role of Judas Goat, right? You led him on, making him think that you were going to marry him and that he would get a share of the inn that way? But first he had to do everyone the little favor of saving Flame Valley from its creditors."

Elly turned to Juliana, her expression anguished. "It wasn't like that. I honestly thought I was in love with him at first. And everyone encouraged me to think that way, including Travis. It was only as time went on that I began to realize I was not in love with him, that what I had felt was just a sort of fascination, a crush. You ought to understand crushes. Men are always having them on you."

"Sure. And they last all of a day, if that. Within forty-eight hours they always come to their senses and back off as fast as they can. But this was more than a crush, Elly. You got engaged to the man."

"It was a mistake," Elly cried.

"What brought about that great realization?"

"You don't know what it was like. He . . . he frightened me in some ways, Juliana. He was always one step ahead of everyone else. Always plotting and scheming. Always had his eye on the main goal. Always willing to do whatever it took to reach that goal. I decided he was just using me to get a share of Flame Valley. I didn't think he could really be in love with me. A man like that never really falls in love with a woman."

"So you used him? You didn't tell him the engagement was off until after he'd accomplished the job of saving Flame Valley, did you?"

"I didn't use him," Elly protested. "Or if I did, you'd have to say I was evenly matched, because he certainly used me, too. In any event, the reason I didn't tell him the engagement was off was because Dad wouldn't let me."

"Oh, come on, Elly." But Juliana believed her. Anthony Grant had his faults but he had been a good father and utterly devoted to Elly, having raised her alone since the death of his beloved wife years ago. Elly returned that devotion in full measure. She was fiercely loyal to Tony Grant. It would have taken a great deal to make her go against her father's wishes. Obviously whatever she had felt for Travis had not been enough to counter the loyalty she felt toward her father.

Elly swallowed heavily, slanting an uneasy sideways glance at Juliana's angry expression. "It was wrong, I know it was, but I was afraid. I couldn't bring myself to tell Travis the truth and wind up being the reason the inn went into bankruptcy. And make no mistake about it, Travis would have walked out and let the whole place go under if he'd found out the wedding wasn't going to take place. Dad explained to me that Travis had only taken the job in the first place because he wanted a piece of Flame Valley Inn. They had an understanding. A sort of gentleman's agreement. Travis would collect his fee — a partnership in Flame Valley — the day he married me."

"So you continued to wear Travis's ring until he'd pulled Flame Valley out of the red."

"I had to wait until Dad and Uncle Roy were sure everything was stable financially. I felt a responsibility to the family. You should understand that. They'd all worked so hard to build this place. It was a part of them. It still is. I couldn't let them down in the crunch."

"You must have felt something for Travis. You've never wanted to talk about the time you were engaged to him. You've never even mentioned his name to me since I moved here four years ago."

"No one in the family talks about him. No one wants to remember the whole mess, least of all me. It was embarrassing and traumatic and painful. And rather scary, to be perfectly honest. Especially at the end when Travis found out I was calling off the engagement."

"How did he react?"

"Like a chunk of stone. I'd never seen anyone look so cold. It was terrifying, Juliana. I'd expected him to yell or fly into a rage or threaten a lawsuit. But instead he was utterly quiet. No emotion at all. He stood there in the middle of the lobby and just looked at me for what had to be the longest minute of my life. Finally he said that someday he'd be back to collect his fee. Only next time, he said, he wouldn't take just a share in the inn, he'd take the whole damned place. Those were his exact words. Then he turned around and walked out the door. We never saw him again."

"Until tonight. Oh, Elly, what a mess."

"I know." Elly closed her eyes. Teardrops squeezed through her lashes.

"Why do you think you were in love with him in the first place? Why did you develop the crush, as you call it? He's not your usual type."

"I know. But please try to remember that I was a lot younger then. Only twenty-four years old. Travis was older, successful, powerful. It was exciting to realize he was attracted to me. And everyone kept saying what a wonderful match it would be. Dad and Uncle Roy and Aunt Beth all wanted him in the family because they knew they could safely turn Flame Valley over to him. He'd know how to run it. It was easy to think I was in love with him at first and by the time I realized I wasn't, it was too late. Everything was very complicated."

"I'll bet." Juliana went to stand beside her

cousin. "Too bad I wasn't here at the time. I could have told you instantly that you and Travis were all wrong for each other." She hesitated and then added, "Does David know who he is?"

"No." Elly shook her head quickly, her eyes shadowed with worry. "I've never told him about Travis. I'd rather he didn't know. He might not understand. Juliana, what are we going to do?"

"We?"

"Don't torment me. You have to help us."

Juliana shrugged. "I don't see that there's anything we can do. At least not until we find out why Travis has gone to all this effort to stage his big return scene."

"There's only one reason why he'd appear like this out of the blue," Elly hissed. "He's here to get his revenge. He must have found a way to take Flame Valley away from us. Or ruin the place."

"Elly, be reasonable. How's he going to do that?"

The soft sound of shoe leather on flagstone made both women turn their heads.

"I'll tell you how I'm going to do it," Travis said, materializing out of the pool of darkness cast by a clump of oleander bushes. "In the simplest way possible. I'm going to watch Flame Valley Inn go straight into the hands of its biggest creditor."

Elly's hand flew to her throat. "Oh, my God."

Juliana glared at Travis. "Must you creep up

on people like that?" But she didn't think he even heard her. His whole attention was on Elly. Juliana felt very much a fifth wheel and she did not like the sensation at all. She'd felt this way once before when she'd stood near her beautiful cousin and another man.

Elly stared at Travis's harshly shadowed features. "You know, don't you? You know everything."

He moved his hand, and ice clinked in the glass he was holding. "I know. I've known from the beginning."

Juliana's gaze flicked from one face to the other. "What are you two talking about?"

"Don't you see, Juliana?" Elly's voice was thick with unshed tears. "He's here because he knows Flame Valley is having financial problems again. He's here to watch everything fall apart the way it would have five years ago if he hadn't rescued it."

"You and your husband have certainly managed to make a hash of things with your big expansion plans, haven't you?" Travis observed. "If Tony and Roy Grant had kept their hands on the reins, they might have been able to pull it off. But once they turned things over to you and David Kirkwood, it was just a matter of time."

Juliana was incensed. "What do you plan to do, Travis? Sit around like a vulture watching the inn collapse under a mountain of debt? Is that your idea of revenge? If you can't have it, nobody else can, either?"

He glanced at her. "Who says I can't have it?"

"What do you mean?" Elly demanded, sounding even more panicked than she had a moment ago.

Travis took a slow swallow of his drink and then smiled a grim, humorless smile. "I'll spell it out for you, Elly. Five years ago I was promised a one-third interest in Flame Valley Inn. I was cheated out of my share, as I'm sure you recall. So this time around, I'm going to take it all."

"But how?" Elly's voice was no more than a faint whisper of distress.

"You and Kirkwood, in your eagerness to get on with your grand plans for remodeling and expanding Flame Valley, have been borrowing from a lot of sources, but you're in debt most heavily to a consortium of investors called Fast Forward Properties, Inc."

"But what does that have to do with you?" Elly asked.

"I am Fast Forward," Travis said softly.

"You." Elly looked stricken.

"I put together that group of investors and I make all the major investment decisions for them," Travis explained coldly. "When Flame Valley falls, as it surely will sometime during the next six months, it will fall right into my hands. I timed things so that I'd be here to step in and manage everything from the new Jewel Harbor headquarters of Sawyer Management Systems."

"Travis, you can't do this," Elly pleaded.

"You're wrong, Elly. It's as good as done. Everything is in place and all the fuses have been lit. It's too late for anyone to do anything about it. Nothing can stop what's going to happen to Flame Valley."

"Oh, my God." Elly burst into tears. They streamed down her face, sparkling like jewels in the moonlight. She made no effort to brush them away. "I should have known that sooner or later you would come back. I should have known."

"You knew. I told you I'd be back, remember?" Travis took another swallow of his drink.

Elly's tears flowed more heavily.

Juliana had had enough. She scowled at her beautiful cousin. "For pity's sake, Elly, are you just going to stand there crying? Haven't you got any backbone? Don't let him bully you like this."

Travis snapped another quick look at Juliana. "Stay out of this," he said. "It's got nothing to do with you now."

"I've already played my part, right? I helped the big bad wolf stage his grand entrance. Now I know what you meant that morning you left my apartment saying I was just icing on the cake. You wanted the entire Grant family to pay for what happened five years ago, didn't you? Even me, the one member who hadn't been around when you got aced out of your share of the Inn."

"That's enough, Juliana," Travis said, so cold, so quiet.

"Enough?" she yelped furiously. "I've got

news for you, Travis. I haven't even started. If you think you're going to get away with destroying Flame Valley as an act of revenge, you're crazy. I'll fight you tooth and claw."

Travis's eyes glinted. "The inn belongs to the other members of the Grant family, not to you. Stay out of it."

"The hell I will. This is a family matter and I'm family. We'll find a way to fight you, won't we, Elly?"

Elly shook her head mournfully, the tears still flowing copiously. "It's hopeless," she whispered.

"Don't say that, Elly." Exasperated, Juliana caught her cousin by the shoulders and shook her gently. "This is your inheritance we're talking about. It belongs to you and David. Surely you're not going to just give up and surrender like this. You've got to fight back."

Travis wandered over to the edge of the terrace and leaned one elbow on the teak railing. "You're wasting your time, Juliana. Elly's not like you. She doesn't know how to fight for what she wants. She's used to having someone else hand it to her on a silver platter."

Elly's head came up abruptly. "You . . . you bastard. Thank heaven I didn't marry you five years ago. I would have been married to a . . . an inhuman monster." Elly choked audibly on her tears and broke free of Juliana's grip. Without a backward glance she ran toward the relative safety of the crowd near the pool.

Juliana gritted her teeth in disgust as she watched her cousin dash away like a graceful gazelle fleeing the hunter. Then she whirled to confront Travis. "Satisfied? Are you proud of yourself? Do you enjoy hurting things that are softer and weaker than you are?"

"There's nothing all that soft or weak about your cousin. She may not know how to fight a fair fight, but that doesn't mean she's not very good at getting what she wants. She uses her softness, trades on it."

"She's a gentle, loving creature by nature. And you've got a lot of nerve talking about a fair fight. Is that what you call this nasty bit of vengeance you're after? A fair fight? So far all I've seen is a lot of low-down, sneaky, underhanded, manipulative, backstabbing tactics."

"Believe me, everything I know about that kind of fighting, I learned from your family."

"Don't you dare use them as an excuse. My guess is you were born knowing about that kind of fighting."

"It's not an excuse. After what they did to me five years ago, I've got every right to fight as dirty as I have to in order to get even."

"You make it sound like you're involved in some sort of vendetta."

"I guess you could call it that," Travis agreed. "I told Elly the day she called off the engagement that I would come back and when I did, I'd take the whole damned place out of Grant hands. I always follow through on my promises, Juliana."

"Really?" She lifted her chin. "Is that why you were so very careful not to make any promises of marriage to me? You wanted to feel you'd maintained your high standard of business ethics even in bed? How very noble of you."

"I know you won't believe this, but I'm sorry that you got caught in the cross-fire. You're different from the others. I should have left you out of it. I can see that now."

"You couldn't have left me out of it. I'm family, remember? I'm a Grant."

"I remember. And that automatically puts you on the other side. I've known how it would be from the beginning. But, like I said, I'm sorry, all the same."

"Stop saying you're sorry. I don't believe that for a minute."

He swirled the contents of his glass. "No, I don't suppose you do." His gaze was on the darkness of the sea.

"If you were truly sorry," Juliana said suddenly, knowing she was clutching at straws, "you'd call the whole thing off, walk out of here tonight and never bother my cousin and her husband and the rest of the family again."

His teeth were revealed in a brief, macabre grin that vanished as quickly as it had appeared. "Not a chance in hell of that happening, Juliana. I've come too far, waited too long and planned too carefully. Nobody gets away with making a fool out of me the way your family did five years ago."

"Sounds to me like you made a fool out of yourself."

"It's true I made a few mistakes. I let the personal side of my life get mixed up with the business side. I don't make that kind of mistake these days."

"Yes, you made a few mistakes." For an instant Juliana was overwhelmed by the sense of loss that swept through her. "Damn it, Travis, how could you be so blind? We could have had something wonderful together, you and I. I was so sure of us. So certain we were meant for each other. But you're going to throw all the possibilities away for the sake of your revenge. You are a fool."

His face was a taut mask of controlled anger. "You know something, Juliana? I don't expect you to accept the situation and I don't expect you to forgive me. I knew from the start that you'd have to end up on your family's side of this thing. But I think that, given your own talent for straightforward action, you might at least understand my side."

"If you're talking about your desire for revenge," she said, impatient with him now, "I might be able to understand it. But I don't approve of it. How can I? This is my family you're going to hurt."

"Yeah, I know," he said, watching her with what looked like resigned regret. "That's the way the chips had to fall."

"Just for the record," Juliana said, "it isn't your notion of vengeance that has convinced me you're a fool and a muddle-brained, stubborn idiot." She turned on her heel and started toward the crowd near the pool.

"Juliana, wait." The words came roughly through the darkness, as if they had been wrenched out of him.

She refused to turn around. Head high, two-inch heels clicking furiously on the terrace stones, she walked swiftly away from the man she loved.

"Juliana." Travis came up behind her. "Look, I know you're angry and I know you probably won't ever be able to forgive me. But I had my reasons and I did what I had to do. Maybe if I were confronted with the same situation today, I'd handle it differently."

"Don't give me that bull." Juliana didn't pause. She was nearing the long buffet table.

"Okay, so I'd probably do it the same way a second time if I had to do it all over again. That doesn't mean I'm not sorry about involving you the way I did."

"Stop whining. I don't want your apologies. Things are bad enough as it is."

"Damn it, I am not whining. And I'm not apologizing. I'm just trying to explain —" He broke off with a muttered oath, moving quickly to keep up with her. "Juliana, what did you mean a minute ago when you said it wasn't my plans for revenge that convinced you I'm a fool?"

Juliana paused beside the buffet table long enough to pick up a huge glass bowl full of guacamole. She swung around, the bowl held in both hands. "Revenge I can comprehend. Being the magnanimous, liberal person that I am, I could even understand your using me to further your scheme. I don't condone it, mind you, but I can understand your doing it. Given a similar situation, I might have been tempted to do something very much like what you're doing."

"I knew you'd be able to see my side of it." Travis looked bleakly satisfied. "I realize you can't side with me, but at least you understand why I'm doing it."

"What I cannot and will not forgive," Juliana continued fiercely, "is the fact that this whole mess occurred because you thought you were in love with my cousin."

"Now, Juliana, listen to me. That was a long time ago. I was a lot younger then."

"Oh, shut up. I don't intend to ever listen to you again. Why should I listen to a fool? How could you have been so dumb, Travis? She's too young for you. Too soft. You would have run roughshod over her and then gotten angry because she didn't stand her ground. You would have been climbing the walls after six months of marriage. Can't you see that?"

"Uh, Juliana, why don't you put down that bowl?" Travis eyed the guacamole uneasily.

"She's not your type. I'm your type."

"Juliana," Travis said very firmly. "The bowl. Put it down."

"I'll put down the bowl when I'm good and

ready. You asked me why I thought you were a fool. I'm telling you. You're a fool because you fell in love with the wrong woman. You still don't even realize that I'm the right one. I could forgive just about anything but that kind of sheer, unadulterated masculine stupidity."

"Juliana." He put out his hand, as if to catch hold of her.

"Don't touch me."

"Damn it, Juliana. *Juliana.*"

Travis released her and leaped back but it was too late. The avocado-green contents of the bowl were already sailing through the air toward him. He held up a hand and instinctively ducked but the guacamole was faster than he was. It spattered across his white shirt and jacket, a good deal of it hitting him squarely on his striped tie.

Juliana surveyed her handiwork with some satisfaction and put the bowl back down on the table. "Now, at least, you're the right color for a frog."

She swept away from the buffet table, moving through the stunned onlookers like a queen through a crowd of stupefied courtiers. She had never been this infuriated, this hurt before in her life. Not even when her fiancé had told her he wanted to marry another woman.

She had been so sure of herself this time; so sure of Travis. How could she have been so certain and yet be so wrong? It wasn't fair.

Her mother, petite, silver-haired and quite

73

lovely at fifty-nine, was suddenly hovering anxiously in front of her. "Juliana, dear, what's going on? Are you all right?"

"I'm fine, Mom."

"Your father and Tony are worried. That man . . ." Beth Grant's voice trailed off as she glanced helplessly past her daughter to where Travis stood covered in guacamole.

"I know all about it, Mom. Excuse me, please. I'm going home."

"But, Juliana . . ."

Juliana patted her mother reassuringly on the shoulder and hurried around her. She took several deep breaths as she went through the lobby and out into the parking lot. She would have to calm down a little before she got behind the wheel of her car.

Travis stood staring after her, as riveted in place as everyone else around him. He could not quite bring himself to believe what had happened. No one he knew did things like this. Not in front of a hundred people.

"Welcome to the dangerous hobby of escorting Juliana Grant." Like the good host he was, David Kirkwood waded through the shocked crowd toward Travis. He grinned wryly and picked up a napkin off the buffet table. "If you plan to hang around her very long, you'd better get used to being taken by surprise," he advised as he handed Travis the napkin. "She's got a way of keeping a man off balance."

Travis snatched the napkin from his host and

took several furious swipes at the guacamole. He could tell instantly that the tie was ruined. "You sound as if you speak from experience. I take it you've seen other men buried under guacamole?"

"Oh, no. Juliana rarely repeats herself. She's too creative for that. But I've sure seen more than one man with that particular expression on his face."

"What expression?"

"That of a poleaxed steer. If it's any consolation, I know exactly how you feel."

"How would you know?" Travis looked with disgust at the avocado stained napkin.

"I was engaged to Juliana for a short but very memorable month almost four years ago."

Travis crushed the napkin in his fist, his gaze slamming into David Kirkwood's amused eyes. Sympathetic, lively hazel eyes. Six foot three, at least, Travis estimated. Fair-haired. Smile like something out of a toothpaste commercial. An expensive Italian pullover sweater over a designer sport shirt and pleated slacks. A massive gold ring on one hand that went with the massive gold watch on his wrist. Travis's instincts warned him of the truth. No doubt about it, this was Juliana's golden god, the one who'd left her for the petite blonde. The blonde was Elly.

"You were engaged to her?" Travis wiped off more guacamole.

"I know what you mean. Hard to believe, isn't it? She's definitely not my type. I'm not quite sure how it happened, myself. She sure dazzled me for a while, though. Something about her makes a man look twice. Then run."

"What do you mean, she dazzled you?"

"Hey, give me a break. You've been around her long enough to know what I mean. She's a lot of woman. But maybe a little too much woman for most of us poor mortal males. Who wants to marry Diana the Huntress? You always get the feeling you're one step behind Juliana. Say, you want to wash the rest of that stuff off in the men's room?"

Travis shook his head, feeling oddly numb. Before he could think of what to do next, Juliana's father and uncle hove into view, planting themselves squarely in front of him. Just what he needed, he thought. They'd been coolly ignoring him for the most part until now. Apparently Juliana's exit scene had prodded them into staging a confrontation.

"What the hell do you think you're doing, Sawyer?" Roy Grant demanded. He glared hostilely at Travis through the lenses of his bifocals. "What did you say to my daughter to upset her that way?"

"Why don't you ask her? I'm the one wearing the guacamole," Travis muttered. Roy Grant hadn't changed much, he reflected briefly. The younger of the two Grant brothers by two years, he was still in good shape for his age. He'd always been the quieter, calmer one but he wasn't playing that role tonight.

"The real question," Tony Grant interrupted, his florid face turning red with anger, "is what are you doing here tonight, Sawyer? I know you. You've got something up your sleeve. Whatever it is, you'd better watch your step, mister. Roy and I are going to keep an eye on you."

"I'll bear that in mind." Travis wondered if his tie was salvageable. Then a sudden thought struck him, galvanizing him back into action. Juliana was in a mood to walk out of here tonight and effectively strand him thirty miles from his apartment. "Oh, hell. The car."

"If she's got the keys, she's probably long gone by now," David assured him cheerfully.

"I've got the keys." Travis shoved his hand into his pocket and found them. "But I don't know if she's one of those women who keeps a spare in her purse."

"She always did when I dated her," David remarked. "Like I said, she's usually one step ahead of a man. That's our Juliana. Gets on the nerves sometimes, doesn't it?"

"The *car*." At the thought of being left without transport, Travis launched himself toward the lobby doors.

"Now hold on just one damned minute," Roy Grant called after him.

Travis ignored the warning. He flew through the lobby doors, raced past a startled desk clerk and dashed outside to the resort's brightly lit entrance.

The roar of an angry sports car engine was already reverberating through the night air. Travis came to an abrupt halt in the drive and watched in furious dismay as Juliana's red coupé hurtled toward the main road.

"That blasted, redheaded witch." Travis stood with his hands bunched into fists on his hips and watched the coupé headlights disappear around a curve in the highway. "She did it. She went and left me here."

"The thing you have to keep in mind about Juliana," David said as he ambled casually over to stand beside Travis, "is that she tends to be a mite impulsive even under normal circumstances. The only thing she's ever really level-headed about is business. On the other hand, she's generous to a fault. Can't stand to see someone suffering. There's a chance she'll get ten miles, remember you don't have a ride home and come racing back to rescue you."

"Sure there's a chance of that happening. A fat chance."

David nodded. "You're probably right. She did look a little annoyed with you. Hey, but not to worry, mate." He clapped Travis familiarly on the shoulder. "I'll run you back to Jewel Harbor. No problem."

Travis cursed softly, fluently, and with great depth of feeling. The thought of accepting a friendly lift from the man he fully intended to ruin was enough to make him a little crazy. And on top of everything else, Kirkwood was Juliana's ex-fiancé — her golden god.

Travis wondered if Juliana had deliberately abandoned him just so he'd be faced with this bizarre social situation. He was certain the awkward irony of the whole thing would appeal to her warped sense of humor. The woman was a menace.

"Maybe one of your other guests is going to be returning to Jewel Harbor later this evening," Travis suggested tightly.

David frowned consideringly. "Can't think of anyone right off."

"Maybe I can rent a car from your front desk."

"We don't have any car rental service."

"An airport van service?" Travis tried, feeling desperate.

"The kid who drives the airport run has gone home for the night." David grinned. "Afraid you're stuck with me."

Travis flipped another puddle of guacamole off his sleeve and said very carefully, "I think I'd better tell you who I am."

"Why don't you do that?" David said quietly. "I've been wondering just who you are since you walked onto the terrace and caused my wife to nearly faint. And Roy and Tony have been looking as if they'd seen a ghost."

Travis looked at him with a tinge of reluctant respect as it sank in that there was more to David Kirkwood than appeared on the surface. He should have been prepared to find the man had hidden depths, Travis reminded himself. After all,

79

Juliana had once thought herself in love with him.

"I'm the head of Fast Forward Properties, Inc.," Travis said.

David exhaled slowly. "So my time has run out, has it?"

"That's about it. I'm running a business, Kirkwood, not a charitable foundation."

David ran his fingers through his hair. "No, you don't look like the charitable type. But that doesn't answer my question. What else are you besides the big, bad wolf come to blow the house down?"

Travis sighed, not particularly enthusiastic about telling him the rest but knowing it would all be coming out eventually. "I'm the guy who saved this place five years ago when Tony and Roy Grant were teetering on the brink of bankruptcy."

David's gaze sharpened. "The consultant they stiffed? The one who thought he was going to get a third of the resort as his fee for services rendered?"

"You know about that deal?"

"Not all of it. No one talks about that incident. No one talks about you or what happened five years ago. But there were some papers and records left in the office when I took over after my marriage. I think I pieced most of the story together. Including the fact that you were once engaged to my wife."

"Yeah, well, if it's any consolation, I'm not here to claim her. All I want is Flame Valley."

"You don't just want your fee. You're here to get revenge, aren't you?"

Travis shrugged. "I just want what's due me."

"Juliana wasn't a part of what happened five years ago, was she?"

"No." Travis stared out into the darkness. "I shouldn't have involved her."

"No way you could avoid it." David fished a set of keys out of his pocket. "Come on, Sawyer. I'll drive you back to Jewel Harbor."

"Why would you want to do me any favors?"

"I don't see it as a favor," David said as he led the way toward a white Mercedes. "I see it as a final, desperate struggle of a drowning man. I intend to use the time between here and Jewel Harbor to get you to listen to my side of this thing."

"You might as well save your breath, Kirkwood. I've already told you, I'm not running a charity." But Travis reluctantly walked toward the Mercedes. He wasn't going to argue with a free ride at this time of night.

"And you've already told me you wished you hadn't involved Juliana." David opened the door of his car.

"That's right. As far as I'm concerned, she's out of it now. Financially, at least."

David smiled coolly. "I've got news for you, wolf. Juliana is in this up to her big, beautiful eyes."

Travis looked at him sharply over the roof of the car. "How?"

"A year ago she loaned Flame Valley Inn a sizable sum of money."

"She did *what?*" Travis felt as if he'd been punched in the stomach.

"You heard me. It was a very large loan. I was planning to repay her at the end of this year. That won't be possible if you foreclose, of course. If the inn goes under, all Juliana's big plans for Charisma Espresso will probably go with it."

Four

Juliana was in bed, but she was far from being asleep when someone began leaning on her doorbell and did not let up. The famous notes of the opening passage of the William Tell Overture clamored throughout the condo, over and over, endlessly repeating themselves until Juliana's head was ringing. She wondered when the Lone Ranger would arrive.

It didn't take her intuitive mind long to figure out who among her acquaintances was most likely to be standing on her doorstep making a nuisance of himself tonight. It also did not require great mental endowment to figure out that the offender was probably not going to give up and go away anytime soon.

Travis Sawyer was not the type who gave up easily, as he had already demonstrated by waiting five years for his vengeance.

Juliana got out of bed and reached for her peach satin robe. She ignored the dueling chimes long enough to pause beside her dressing table mirror and gloss her lips with a shade of lipstick she thought went well with the robe. She was considering adding blusher when the endless chimes finally got to her. Unable to stand the torture any longer, she stepped into a pair of silver high-heeled lounge slippers and stalked into the living room and threw open the door.

"Whatever you're selling, I'm not buying," she announced. "But you can try."

Travis, who was leaning against the chime button, straightened slowly, his eyes chips of ice. "Why didn't you tell me you had dropped a chunk of your personal savings into Flame Valley Inn?"

"Why should I have told you? It was family business and I only hired you to consult on Charisma. Besides, I had no idea of your deep, personal concern for the inn until a few hours ago."

His gaze swept over her and his cold expression grew even harsher. "You can't stand around out here dressed like that. Someone's likely to drive past any minute. Let's go inside."

"I don't think I want you inside my apartment. Maybe you secretly hold the mortgage on this whole building or something. Maybe you're getting ready to foreclose on the condominium association and kick all us poor owners out into the street."

"Stop talking nonsense. I am not in the mood for any more of your warped humor tonight. Stranding me at Flame Valley was the last straw." He shouldered his way past her, heading for the living room.

"How did you get back to Jewel Harbor?" Juliana closed the door slowly and followed Travis. She noticed that he had removed his tie and there were several green splotches on his white shirt. A niggling sense of guilt shot through her. She quelled it quickly.

By the time she caught up with Travis he had already lodged himself on the salmon-colored leather sofa. He sprawled there with negligent ease, one foot prodded on the black and chrome coffee table.

"What do you care how I got back?"

"It wasn't a question of caring exactly," Juliana explained as she sank into a turquoise chair and crossed her legs. The silver slippers gleamed in the light from a nearby Italian style lamp. "I asked out of simple curiosity."

"I flew."

"On your broomstick?"

"No, you used the broomstick, remember?"

"Ah," said Juliana. "Feeling peevish, are we? I'll bet David gave you a lift home. He would. David's always the perfect host. Did you tell him who you were before or after he went out of his way for you?"

"Before."

"How very upright of you. Your sense of business ethics is certainly an inspiration to the rest of us." The silver sandals winked again as Juliana recrossed her legs and smiled blandly.

"Don't give me that superior look, Juliana. My temper is hanging by a thread tonight."

"Shall I fetch a pair of scissors?"

Travis closed his eyes and leaned his head back against the leather cushion. "No, you can fetch me a shot of brandy. Lord knows I need it."

"Brandy is expensive. Why should I waste any of my precious supply on you?" Juliana asked.

Travis's eyes opened and he looked straight at her. "One of these days you're going to learn when to stop pushing," he said very softly.

"Who's going to teach me?"

"It's beginning to look like I'm stuck with the job. It's obvious there aren't a lot of other candidates and one can certainly understand why. Go and get the brandy, Juliana. I want to talk to you."

She hesitated a few seconds and then, hiding a smile, got to her feet and went into the kitchen. She opened a cupboard, found the expensive French brandy she saved for special occasions and carefully poured two glasses.

"Thank you," Travis said with mocking courtesy as she returned to the living room with the balloon glasses balanced on a clear acrylic tray. He picked up one of the glasses.

"Okay," Juliana said as she sat down again. "What do you and I have to talk about?"

"How the hell could you get yourself engaged to David Kirkwood?"

Juliana had thought she was prepared for whatever bomb Travis dropped, but this salvo took her by surprise. "My, you and David certainly got chummy on the drive back to Jewel Harbor, didn't you? The old male bonding routine, I suppose."

Travis scowled. "Answer me, Juliana."

"How did I get engaged to him? Well, let me think a minute. It's been nearly four years now and my memory isn't totally clear on the subject, but as I recall it was in the usual manner. He took me out to dinner at The Treasure House, the same place you and I ate the night you tricked me into going to bed with you, as a matter of fact, and . . ."

"I didn't trick you into bed and you know it. Stop trying to sidetrack me with a guilt trip. It won't work. And I don't give a damn about where Kirkwood put an engagement ring on your finger. I want to know why in the world you manipulated the guy into asking you to marry him in the first place. Anyone with half a brain can see he's all wrong for you."

Juliana's temper flared at the accusation. "I didn't manipulate him. He asked me and I accepted for all the usual reasons."

"Don't give me that. Nothing happens around you for the *usual reasons*. Damn it, Juliana. How could you even think of marrying a guy like that?"

"David is a very nice man as you witnessed tonight. There aren't a lot of other people in his position who'd give you a lift knowing you planned to ruin them."

"You don't need a very nice man," Travis said

through his teeth. "You need someone who can hold his own with you. Someone who won't let you get away with murder. Someone as strong or stronger than you are."

"You may be right." Juliana smiled and gave him a pointed look. "But a woman can't afford to be too choosy these days. Sometimes we have to lower our standards a trifle."

Travis's smile was as well chilled as hers. "Meaning you lowered your standards when you picked me for your second fiancé?"

"You said it, not me. But now that you mention it —"

"Juliana, that's enough. I've already told you I'm not in the mood for your backchat. All I want tonight are a few straightforward answers. I always thought straightforwardness was your specialty. You once told me you like everyone to know exactly where he stands with you and you like to know where you stand with others."

"I've answered your questions. I got engaged to David because we had a lot in common and because I thought I loved him and he thought he loved me. We both realized our mistake within a month. Actually, to be perfectly truthful, I realized it within a couple of days. At any rate we ended the engagement. No hard feelings."

"You mean David ended things when he realized he would drive himself crazy trying to keep up with you. Elly probably looked like a restful, sweet, golden-haired angel to him after a month of being engaged to you."

"Are you implying I'm a witch?"

"No, ma'am, although I'll admit I called you that earlier tonight when I watched you drive out of that resort parking lot. You're what Kirkwood said you are, a lot of woman. Not every man wants that much woman. Kirkwood sensed he couldn't handle you. You'd have run him ragged and then gotten irritated with him for letting you do it."

"Rather like you and Elly." Juliana felt goaded and warned herself not to let Travis trap her into losing her self-control.

"Like Elly and me? Possible." Travis took a swallow of the brandy and gazed broodingly at the black stone gas fireplace. "You're probably right, as a matter of fact. It doesn't matter anymore, though. Whatever was between Elly and me five years ago is nothing but ashes now."

"I'm not so sure of that," Juliana said coolly. "I saw the way you two looked at each other tonight. There was a lot of old emotion in the air."

Travis waved that observation aside with an impatient movement of his hand. "Old anger on my part. Old fear on hers, probably. Five years have passed and she had convinced herself I'd forgotten all about Flame Valley Inn. She practically had an anxiety attack tonight when she realized that I hadn't forgotten anything."

"Revenge is powerful stuff, isn't it, Travis?"

"It'll keep you warm when you haven't got much else," he agreed. His gaze switched back to her face. "Tell me what happened when you found out Kirkwood was calling off your engagement because he was in love with your cousin."

"That's personal. Why should I tell you?"

"I'll bet there was a hell of a scene. Fireworks and mayhem. Blood and guts everywhere. You wouldn't easily surrender something you wanted. Did you fight like a she-cat to hold on to your handsome tin god?"

Juliana wrinkled her nose. "Golden god, not tin god."

"Let's compromise on papier-mâché god. Did you fight for him, Juliana?"

"Why do you care?"

"I want to know, damn it. I want to know if you fought for him. I want to know how hard you fought for him. Maybe I want to know if you're still fighting for him. Just answer the question."

"I don't owe you any answers. Did you fight for Elly? How hard? Are you still fighting for her? Is taking over Flame Valley your way of trying to reclaim her?"

That appeared to startle him for an instant. Travis looked honestly taken aback. "I don't do things the way you do."

"You just turned around and walked out, huh? Vowing revenge, of course."

"I think," Travis said, "that we had better change the subject."

"You started this conversation."

"Your logic is irrefutable, madam." He saluted her with his glass. "You're right. I started it."

"I love it when you use big words. So professionally macho."

He shook his head. "You're in a real prickly mood tonight, aren't you?"

"I have good cause."

"You're not the one who found himself stranded in the middle of nowhere without a ride home." Travis's crystal cold gaze locked on hers. "Why in hell did you sink your personal cash into Flame Valley, Juliana? You're a better businesswoman than that. You must have realized it was a high-risk investment."

She blinked at the quick shift of topic. "We're back to that, are we?"

"Yes, we're back to that and we're not letting go of it until I get some answers."

Juliana sighed and sipped her brandy. "I've already given you the answer. It was a family thing. I knew David and Elly were in trouble. I made them a loan. So did my parents, for that matter. And Uncle Tony. Everyone in the family has tried to help David and Elly save the inn. We've all got a stake in it."

Travis set the brandy glass down on the table with a snap. "Damn it, Juliana, you're too smart to have let yourself get sucked into that mess. I've been working with you long enough to know you're anything but a fool when it comes to business. You must have known how bad things were with the inn." He slanted her a quick, assessing glance. "Or did Kirkwood lie to you about how precarious the situation was?"

"No, he did not lie. I knew exactly how bad things were."

"And you loaned him the money, anyway."

"He and Elly are fighting to save the resort. I wanted to give them a chance. Of course, the one thing I didn't know at the time was that you and Fast Forward Properties, Inc., were waiting to pounce on the wounded victim at the first sign of blood. We all assumed Fast Forward would be reasonable when David approached them in a month or so and asked for an extension. Obviously that was a false assumption."

"Obviously," Travis agreed dryly. "You're going to lose a hell of a lot of money because of David and Elly Kirkwood."

"If I lose my money, Travis, it will be because of you, not David and Elly."

Travis swore and surged to his feet. "Oh, no, you don't. You're not going to blame this on me." He began to pace the length of the gray carpet. "Don't you dare try to blame this on me," he repeated huskily. "You should never have put money into Flame Valley and you know it."

In a way he was right and Juliana did know it. She lifted one shoulder carelessly. "Win some, lose some."

Travis swung around at the far end of the room and leveled his finger at her. "You can't afford to lose that much cash, Juliana. Not if you want to expand Charisma Espresso. No one knows your financial situation better than me, and I'm telling you that you can't take this kind of loss. All your plans for Charisma will be set back three or four years at the very least."

"Okay, okay. You've made your point. I can't afford the loss. Not much I can do about it now."

Travis shoved his hands into his back pockets. "Is that all you can say?"

"No point crying over spilt espresso."

"You're looking at the end of all your plans to finance Charisma's expansion this year, and that's your reaction?" Travis asked, staring at her in disbelief. " 'No point crying over spilt espresso'? I don't believe I'm hearing this."

"Be reasonable, Travis. There's not much I can do now, is there? The damage is done." She took a swallow of brandy and gazed forlornly into the dark fireplace. "I'm ruined."

"Let's not get melodramatic. We've got enough problems as it is."

"You don't have any problems. I have the problems. And frankly, Travis, if you can't offer anything helpful to the discussion, I'd just as soon stop talking about my dismal business future. It's depressing."

"Helpful?" he snarled. "What kind of help do you expect from me?"

She slanted him a sidelong glance. "Well, Travis, you are the problem, in case you've forgotten. That means you're also the solution."

"I didn't get Flame Valley Inn into this mess," he growled furiously. "Kirkwood and Elly did it all on their own. With a little help from you and your parents and good old Uncle Tony, of course. Don't expect me to solve everything by paying you back after I take over the resort. I've got investors to pay off first, remember? A half dozen of them, and they'll all be standing in line waiting for their money. I've made commitments to them. And I can personally guarantee that none of them are going to be feeling charitable."

"You're right. I don't expect you to pay me off when you and your investors take over the inn."

"Well, what do you expect out of me?" Travis roared.

She pursed her lips primly. "You're my personal business consultant," she reminded him. "You are paid to keep me out of hot financial waters. You are paid to guide and advise me. I'm putting my complete faith, total trust and all my hopes for the future in your hands. I feel certain you'll save my hide."

"Juliana, what the hell am I supposed to do?"

"Save Flame Valley Inn from the clutches of Fast Forward Properties, Inc."

Travis looked thunderstruck. For a moment he just stood there in towering silence, staring at her as if she'd lost her mind. Juliana held her breath.

"Save the inn?" Travis finally repeated blankly. "From myself?"

"From yourself and that pack of hungry wolves you're leading."

"Am I supposed to do this for you?" He looked as if he was having trouble following the conversation.

"I'm one of your clients, aren't I? I hired you to help me put together a workable plan to expand Charisma. As far as I'm concerned, we still have a valid consulting contract. And the bottom line is that we can't expand my firm unless you rescue Flame Valley from Fast Forward Properties. Therefore, yes, I'm expecting you to do this for me."

"You," Travis announced softly, dangerously, "are crazy."

"No. Desperate." If only he knew how desperate, Juliana thought. She wasn't fighting to save the inn or Charisma's future. She was fighting to save the love of her life. But she did not think now was the time to point that out to him. "Will you help me, Travis? Will you salvage the resort from bankruptcy?"

"I already did that once, remember? And I didn't get paid for my efforts. What makes you think I'd repeat that mistake?"

"This time I'll be paying your fee. And you already know I pay my debts."

"Lady, you can't afford my fee."

She frowned. "I'm paying you already, remember? For the consulting work on Charisma. And so far I haven't had any trouble meeting the tab. I paid your retainer right on time and I'll

95

meet the monthly fees, no sweat."

"Oh, Lord. What an innocent." Travis rubbed the back of his neck in a gesture of pure exasperation. He stalked to the end of the room and back again. "Juliana, let me make this crystal clear. The fee you are paying me for consulting work on Charisma is only a fraction of what I would normally charge for that kind of job. In fact, under normal circumstances, I would never even agree to take on a job the size of Charisma. Your operation is far too small for Sawyer Management Systems. Do I make myself plain?"

"You're telling me I got a special deal?"

"You got a hell of a deal."

She caught her lower lip thoughtfully between her teeth. "Why? Because I was a Grant and you were out to punish all the Grants? You wanted to get close to me so that you could hurt me by destroying Charisma?"

"No, damn it, I never intended to destroy Charisma. I don't know what happened. Not exactly. All I know is that I walked into your shop that day last month because I was curious to meet the one member of the Grant family who hadn't been around five years ago." Travis's palms came up in a gesture of total incomprehension. "The next thing I knew I was having a cup of coffee and agreeing to consult on your expansion plans."

"You mean you took on my project out of the kindness of your heart?"

"I don't do business out of the kindness of my heart," he said grimly.

"Well, that's neither here nor there now, is it? We've got a contract. You're my business consultant and I am facing a financial disaster. It seems to me you're duty bound to save me."

"Is that right?" His gaze was unreadable.

"Looks that way to me."

"Since you've got all the answers," he said, "maybe you'll be kind enough to tell me just how I'm supposed to save you?"

"I don't know. That's your business, isn't it?"

Travis shook his head. "You are really something else, you know that?"

"So I've been told. Well? Will you do it?"

He sighed. "It can't be done. I'm too good at what I do, Juliana, and I've set this up so that there aren't any loopholes. Even if I thought I might be able to save the inn and even if I was idiotic enough to agree to try, there's still the little matter of my fee for the job."

"Name it. I'll tell you whether or not I can afford it," she challenged softly.

He looked at her through dangerously narrowed eyes. His gaze moved from the toe of her silver lounge sandals to the top of her red mane. "What if I told you my fee was an all-out affair with you? What if I said that the price of my services, win, lose or draw, no guarantees on saving the resort, was access rights to your bed?"

Juliana stopped breathing for a few seconds. Then she sucked in a dose of air and sat very still to conceal the fact that she was trembling. "My fabulous body in exchange for your services as a consultant? Don't play games with me, Travis. If you don't do business out of the kindness of your heart, you certainly wouldn't do it for sex."

He didn't move. When he spoke his voice was very, very soft. "But what if I said that was the fee, Juliana?"

She shivered, aware that he was deliberately tormenting her. "I'd tell you to take a long walk off a short pier."

He turned away and stood looking out the window into the darkness. "Yeah, I figured that's what you'd say. All right, smart lady, I'll tell you what my fee for trying to salvage Flame Valley Inn and Charisma's future would be. It would be the same as it was five years ago, a piece of the action."

Juliana gasped, not in shock this time but in outrage. "You expect David and Elly to make you a partner in the business if you save it from your own investors? Talk about gall." She thought quickly. "Still, I suppose something along those lines might be worked out. It would have to be a very small piece of the action, however. You simply can't expect David and Elly to give you a half interest or anything like that."

Travis slanted her an assessing look over his shoulder. "No. You don't understand, Juliana. I don't want a piece of Flame Valley Inn. If, by some miracle, I do manage to save the resort, it's going to need a lot of cash and a lot of work and even then there's a good chance it will still go under within a couple of years. I want a piece of a sure thing this time."

She watched him closely, the way she would a predator, and felt her stomach tighten. "What sure thing?"

"Charisma."

Juliana nearly dropped her brandy glass. Her mouth fell open in amazement. She sat staring at him, unable to comprehend what he was saying. "A piece of Charisma? You want a piece of my business? But, Travis, Charisma is mine. All mine. I built it from scratch. No one else in the family is even involved with it. You can't be serious."

"Welcome to the real world. I am very serious. I told you my usual fees are always very high, Juliana. And if you agree to the deal, you'd better understand going in that there's a big possibility I won't be able to save the resort. But I'll expect to be paid, even if I fail to salvage Flame Valley."

Juliana was reeling. She struggled frantically to collect her thoughts. She hadn't been expecting this, although looking back on things, she probably ought to have expected it or something equally outrageous. Travis Sawyer was a formidable opponent. "But Charisma is mine," she repeated, dazed. "I made it what it is all by myself. I learned the ropes managing an espresso chain in San Francisco for years while I saved my money to buy my first machines and find a good location. I've done it all, all by myself."

"And all by yourself you put Charisma's future at risk by making that loan to Kirkwood and your cousin." Travis started for the door. "Think about it, Juliana. I'll drop by your shop Monday morning to get your decision."

"Travis, wait, let's talk about this. I'm sure we can find some reasonable compromise if we just —"

"I don't make compromises when it comes to insuring my fee. I learned my lesson on that five years ago. And I learned it from dealing with people named Grant."

"But, Travis . . ." Juliana leaped to her feet but the door was already closing behind him. She raced to the window in time to see him get into the Buick he'd left parked in front of the apartment earlier in the evening. The headlights came on, blinding her. In another moment he was gone.

"You sneaky, conniving, hard-hearted, son of a . . . Charisma is mine, damn it. *Mine*. And I'm not going to let you or anyone else have a piece of it." Juliana halted in the middle of her tirade as a thought struck her with the force of a lightning bolt.

If she gave him a chunk of Charisma Espresso Travis would be financially bound to her for what might be years. Their lives would inevitably be deeply entangled. It was easier to get

100

rid of a spouse these days than it was a business partner.

And if she had him around to work on all that time, Juliana told herself, spirits soaring, she just knew she could convince him to see that they were meant for each other.

Travis was aware of the tension that gripped his insides as he parked the Buick in front of Charisma Espresso on Monday. He should be used to the unpleasant sensation, he told himself. He'd been awake two nights in a row because of it.

He could see the lights on inside the espresso bar, and he thought he caught a glimpse of Juliana's red hair as she ducked into the back room. His hands tightened on the steering wheel. He still could not quite believe he was doing this. But he supposed he shouldn't be too surprised. This whole deal had been skewed from the moment he had met Juliana. Nothing had happened the way he had planned it since then, so why should anything straighten out now?

He forced himself to get out of the car and walk toward the shop. He had to be prepared for whatever answer he got. Juliana was nothing if not unpredictable. And Charisma was very, very important to her.

He wondered if she would realize what he was doing and why. The whole, crazy scheme had come to him in a flash in the heat of the argument Saturday night and he still was not certain where it would all lead. He had been acting on instinct, trying to buy himself some time.

Would Juliana guess that he was using this bizarre deal to hold on to her for a while longer? And if she did, would she be furious or secretly glad?

Even if she did agree, he would have no way of knowing for certain just why she was doing so. She might be doing it because she was desperate to help her cousin and David Kirkwood. Or she might agree because he had made her aware of just how precarious Charisma's future was because of that foolish loan she had made to the Kirkwoods.

Or she might agree because she still cared a little for him and was willing to take the risk of being snared in financial bonds until they could work out the rest of their muddled relationship.

Then, again, she might simply hurl a cup of espresso in his face and tell him to get lost.

The lady or the tiger? Travis wondered as he pushed open the door. Life with Juliana was never dull.

"He's here, Juliana." Sandy Oakes, her multiple earrings clashing merrily, turned toward the door as Travis entered. Her hair was gelled into a slick 1950s ducktail and her eyes were bright with speculative interest. "Come and get him, boss."

Matt Linton looked up from where he was stacking espresso cups. "Yup. There he is. Poor devil. Look at him standing there, so naive and unsuspecting."

"Kinda breaks your heart, doesn't it?" Sandy observed. "I mean, the poor man doesn't even know what he's in for yet."

"If I were him, I'd run for it," Matt added with a grin.

Travis groaned silently, wishing Juliana would show a little more discretion at times. "You know, the plans for the expansion of Charisma Espresso don't necessarily have to include you two," he muttered.

"Oooh," said Sandy. "I think I detect a threat."

"Juliana," Matt called more loudly. "Come and rescue us. We're being threatened out here."

Travis lifted his eyes beseechingly toward the ceiling and set his back teeth. An instant later Juliana came out of the back room, wiping her hand on a towel. She was wearing a black jumpsuit trimmed with fringe and a pair of black leather high-heeled boots. Her hair was pulled back with two silver combs.

"What's going on out here?" she demanded, brows snapping together. "Oh, it's you, Travis. It's about time you got here."

He shoved his hands into his back pockets. "I didn't realize you'd be waiting so eagerly for my arrival."

"Of course I've been eager for you to get here," she informed him waspishly. "We've got a ton of things to do before this evening, and I won't have time to do them all. You'll have to help out." She ducked into her tiny, cluttered office and reemerged with a piece of paper. "Here. This is what you have to pick up from the grocery store. And on the back are some things I want from the cheese shop three doors down. You might as well pick those up, too. And while you're at it, get some champagne, will you?"

Travis stared blankly at the list in his hand, a conviction growing within him that he had once again lost the upper hand. "Why am I buying all this stuff, Juliana?"

"For the party we're having tonight, of course. Now run along. I'm busy. Got a lot to do here. It's Monday morning, you know. Melvin is late with my shipment of Colombian and I just got word one of the coffee roasters has broken down. I'm swamped."

Travis refused to be budged. "Why are we having a party?" he asked with forced patience.

"Don't worry, it's not a big deal. I've just invited David and Elly over tonight to announce that you'll be going to work to save Flame Valley Inn." She looked straight into his eyes and silently defied him to argue.

Travis returned the challenging look, his fingers tightening on the paper in his hand. His gamble had paid off. "We have a deal, I take it?"

"We have a deal."

"You understand I can't give you any guarantees? I can't promise to save the resort. Things have gone too far for me to make any promises."

She smiled fleetingly. "If anyone can save it, you can."

"But if I can't, Juliana, I'll still expect to be paid."

"Stop carping. You'll get your fee."

Travis glanced around the interior of Charisma, aware of an almost light-headed sensation of relief. "I've always wanted to own part interest in an espresso bar."

"Since when?"

"Since Saturday night. See you later, *partner*."

Five

"There's just one more thing about tonight," Juliana said, tossing a handful of chillies into the Asian-style noodles she was stir-frying. "I don't want anyone to know about our little, uh, arrangement."

The knife Travis had been using to slice mushrooms paused in midstroke. He glanced sideways at Juliana. "Why not?"

"Why not?" she echoed, exasperated. "Isn't it obvious? Because David and Elly will feel terrible if they think I'm paying your fee by making you a partner in Charisma. I don't want to put any more of a guilt trip on them than they're already under."

"Why should it bother them to know you're paying the fee? They didn't have any qualms about taking your money in the form of a loan."

"That was before things got so bad." Juliana concentrated on the noodles. "At the time it looked as if David really was going to be able to pull it off. Travis, have things really gone too far?"

"Probably. I told you, I set the whole thing up so that there wouldn't be any room for Kirkwood to maneuver. I won't know for certain until I get a good look at the books, but if I were you, I wouldn't hold my breath."

Juliana was undaunted. "It'll all work out. I know it will. Hand me those little tart shells, will you?"

"What kind of meal is this going to be, anyway? Just about everything here qualifies as an hors d'oeuvre in my book."

"Stop grumbling. Everyone knows the hors d'oeuvres are always the best part of a meal. I say why bother with the entrée? We'll just graze on the good stuff. When you've finished with the mushrooms, you can help me fill these shells with the cheese mixture."

"You like giving me orders, don't you?"

"It's just that you look cute in that apron. There's a glass platter in that cupboard to your right. Use it for the sliced vegetables. Hurry, David and Elly will be here any minute."

"How do I get myself into these situations with you?" Travis asked so softly Juliana barely heard him. "Every time I think I've got things back under control, I find out I'm barely keeping up with the new detour you're taking." He finished slicing mushrooms with sharp, fast strokes of the knife.

"About our partnership, Travis . . ."

"What about it?" His head came up warily.

"Well, I think we ought to establish right from the outset that you're more or less a silent partner. Know what I mean?" She smiled brilliantly. "I'll consult you, of course, before making major decisions. And heaven knows you're already involved in the planning process for Charisma's future. But in the final analysis, I'm the senior partner in Charisma. I just want to be sure we understand that little point."

"You can call yourself anything you like," Travis said easily, "just so long as you understand you can't make any major decisions without consulting me."

"Just what is your definition of a major decision?" Juliana retorted.

"We'll figure that out when we come across one. Where do you want this tray of vegetables?"

"On the end table near the sofa." Juliana glanced out the window and saw a familiar white Mercedes sliding into the space next to her coupé. "Good grief, they're here already. I'll bet David and Elly are dying of curiosity."

"Why? What did you tell them?"

"That you're going to save everyone's bacon, naturally." Juliana carried the tray of tiny cheese tarts into the living room.

"Don't make too many rash promises, Juliana. I've said I'll do what I can. I can't guarantee anything."

"I'm not worried, Travis. I have complete faith in you."

"You're too smart a businesswoman to have blind faith in any consultant."

"You're not the usual sort of consultant."

Juliana swept toward the door, the skirts of her liquid silk hostess pajamas shimmering around her. "Open the champagne, Travis. David and Elly are going to be thrilled."

"I'll just bet they are."

David and Elly were not precisely thrilled. They regarded Travis with great caution as they walked into the living room and sat down. And they greeted Juliana's grand announcement initially with stunned surprise. The astonishment turned only gradually to careful, cautious concern, and all the questions soon developed a single focus.

"Why are you doing this?" David finally asked point-blank.

"Believe me, I've been asking myself the same question," Travis said dryly, helping himself to one of the mushrooms he'd sliced earlier.

"I mean, what's in it for you?" David persisted, brows drawn together in one of his rare, serious frowns. "Don't think we're not grateful, but I'd like to know what we're getting into here."

Juliana felt Travis's eyes switch to her face, obviously leaving the answer to that one up to her. She smiled at David. "Don't you see, David? He's doing it as a favor to me."

"A favor," Elly repeated, her eyes widening.

"A business favor," Juliana clarified. "Would anyone like another cheese tart?"

"But, Juliana," Elly persisted, "why would he do you any favors?"

"Yeah, why?" David echoed, but he looked thoughtful as he regarded Travis's unreadable expression.

"An interesting and timely question," Travis said. "Why would I want to do you any favors, Juliana?"

Juliana smiled fondly at him. "Why, to save Charisma's future, of course." She turned to David and Elly. "You have to understand something here. Travis had already committed himself to doing consulting work for Charisma before he realized the extent of my financial involvement with Flame Valley Inn. It set up a conflict of interest for him, you see. Once he discovered how much money Charisma stood to lose, he immediately insisted on trying to salvage the Flame Valley situation. He felt ethically bound to help Charisma. Travis is very big on business ethics, you know."

"Is that right?" David murmured. "I believe I will have another cheese tart, Juliana. And another glass of champagne."

"Certainly." Juliana reached for the tray of tarts. It was all going to work out just fine, she told herself as David plied Travis with further questions. Things were falling into place nicely. She ignored Elly's anxious glances.

Twenty minutes later, however, David cornered her in the kitchen where Juliana had gone to whip up another batch of stir-fried noodles.

"Just what the devil is going on?" David demanded in low tones.

Juliana glanced up from the wok. "What do you mean?"

"You know damn well what I mean. Juliana, I don't for one minute believe all that nonsense about Travis Sawyer suddenly being stricken with a severe case of business ethics. It's just you and me out here in the kitchen, kiddo, and we've been friends a long time. Now tell me the truth. Why is he galloping to Flame Valley's rescue? The real reason this time. What's in it for him?"

"David, really." Juliana switched off the electric wok and busied herself piling noodles onto a platter. "Keep your voice down. Elly and Travis will hear you."

"Are you the fee this time around, Juliana?"

David took a step closer, towering over her in the way Juliana used to find so exciting. She realized now she found it annoying.

"Don't be silly, David."

"I'm serious, Juliana. What's going on between the two of you? Are you sleeping with him in exchange for his help? Is that the deal? Because if so, I won't let you do it and that's final. God knows I'll take all the help I can get up to a point. But I sure as hell won't let you sell yourself to Sawyer as payment for saving Flame Valley."

Juliana opened her mouth to order him out of her kitchen but before the words could be uttered, Travis interrupted from the doorway. His voice was cold and hard.

"Leave her alone, Kirkwood. Juliana doesn't

111

need your protection. The deal she and I have made is a private one, and you don't have any need to know the details."

"He's right," Juliana said briskly. "The arrangement is strictly between Travis and myself. We're both satisfied with the details so you need not concern yourself. Here, take these extra napkins out to Elly. We'll need them for the next round of noodles."

"But, Juliana . . ."

"Go. Shoo. Get out of my kitchen. The noodles are getting cold. And stop worrying about me, David. You know perfectly well I can take care of myself. I always have."

"I know, but I'm not sure I like the setup here."

"You may not like it but you'll go along with it, won't you?" Travis asked softly. "Because it's the only chance you've got to save the resort."

David met his eyes for a few seconds and then, without another word, walked back into the living room with the napkins.

Travis turned back to Juliana. "You do realize that's how everyone's going to interpret this crazy deal we've made? Your folks, your Uncle Tony, David and Elly, they'll all come to the conclusion you're sleeping with me in exchange for my help."

"But you and I know the truth, don't we? This is a business arrangement. Nothing more."

"Oh, sure. Right. A business arrangement. Get real, Juliana."

"Look at it this way," Juliana said as she shoved a plate of noodles into his hands. "Since we know that the business side of this deal is just that, strictly business, we're free to continue working on our relationship without worrying about each other's motives." She started briskly for the door, heedless of the shock in Travis's eyes.

"Juliana. Damn it, come back here. What did you mean by that?"

Juliana ignored him, sailing into the living room with a warm smile for David and Elly. "Everyone ready for another round of noodles? Don't be shy. I've got plenty of hot sauce."

Travis closed in on Juliana the instant the Kirkwoods' Mercedes pulled out of the lot.

"About our relationship," he began darkly.

Juliana had been feeling very confident earlier, but now that she was alone with Travis she wasn't quite so sure of herself. So she compensated by acting more sure of herself than ever. It was an old habit. When people assumed you could take care of yourself, you learned to do it. "What about it?" she asked as she began picking up napkins and plates.

Travis paced toward her very deliberately. "The last time I inquired into the subject, you had sworn off sleeping with me until I had made a commitment to marry you. As I recall, I was given one month to come to my senses. Then you said very specifically that all we had was a business arrangement. Now you're talking about a relationship. Are you changing the rules on me again, Juliana?"

"You want the truth? All right, I'll give you the truth. We're going to be seeing a lot of each other from now on, and I've been in love with you since the day I met you. To be perfectly blunt, I don't think I'll be able to resist you for much longer. Not if you put your mind to seducing me, that is." She picked up an empty tray and carried it into the kitchen.

"What the hell does that mean?" Travis stormed into the kitchen behind her and caught her in his arms as she set down her burden. "Tell me exactly what it means. No more cute games, Juliana."

"It means just what it sounds like. Are you going to put your mind to seducing me?" She twined her arms around his neck and brushed his hard mouth lightly with her own soft lips.

"Juliana." Travis breathed her name on a hoarse, urgent groan. He caught her head between his hands and crushed her mouth hungrily beneath his own.

Juliana responded instinctively, willingly and with all the pent-up eagerness that she had been fighting to keep in check.

"Hold it," Travis ordered, although his arms were already tight around her waist. "I've got to figure out what's going on here. You say our business arrangement is separate. That it has nothing to do with this end of things. And you claim you still love me. So what is all this? You think that if you let me back into your bed I'll suddenly decide I have to offer marriage, after all?"

"Who knows? You might, you being such an ethical type and all."

He shook his head in slow wonder. "You never give up, do you?"

"No. But I don't want to talk about marriage right now."

"Good. Neither do I."

Travis kissed her hard and then took her hand firmly in his and started toward the bedroom. He turned out lights en route and by the time they reached the end of the hall they were deep in the never-never land of night and shadow.

"Travis?" She smiled up at him as he drew her to a halt near the bed. She trailed her fingertips along his shoulder, enjoying the strength she found there.

"I've been going out of my mind wondering when it would happen with you again." Travis caught her earlobe between his teeth as he began to unbutton the top of Juliana's silky hostess outfit.

"I should never have set that month deadline the first time around," Juliana confided, leaning into his vital warmth. "I should have known I'd never last that long."

"We agreed after that first night together that what we have between us is very special." Travis finished unbuttoning the yellow silk top and eased it off her shoulders.

Juliana felt the whisper softness of the fine material on her skin and shivered with anticipation. Travis looked down at her bare breasts and his crystal eyes glinted in the darkness. His palm closed gently over one budding nipple and Juliana inhaled deeply.

"You feel so good." Travis's mouth found the curve of her throat. "So right."

"So do you." She tugged at his tie and tossed it aside. Then she unbuttoned his shirt with fingers that trembled. As soon as she had freed him of the shirt, she locked her arms around his waist and leaned her head against his bare shoulder. The crisp mat of hair on his chest grazed her nipples. She wriggled against him experimentally, enjoying the teasing sensation.

Travis laughed softly, the sound husky and very sexy in the darkness. He slid his palms beneath the waistband of Juliana's silk pants and pushed them down over her hips. A moment later she stood nude in his arms.

"This is how it's supposed to be between us," Travis muttered as he picked her up and carried her to the bed. He settled her on the pillows and stood back to rid himself of the rest of his own clothing.

"I know." Juliana lay gazing up at him, glorying in the solid, hard shape of him. The strength in his shoulders and thighs appealed to all her senses and the boldness of his arousal pleased her on a deep, primitive level. But it was not simply the physical aspect of Travis that captivated her and made her so aware of him on a sensual plane, she realized. There were other men just as solidly masculine and far better looking.

"You look very serious all of a sudden," Travis observed as he came down beside her on the bed, a small foil packet in one hand. "What are you thinking?"

She looked up at him as he loomed over her. "I was just wondering what it is about you that makes you so sexy."

He grinned, teeth flashing in the darkness as he dealt with the contents of the packet. "You're trying to figure out why you couldn't resist me any longer?"

"Um-hm." She trailed her fingers through his hair and tangled one leg between his.

"Don't think about it too hard," Travis advised, leaning down to kiss the valley between her breasts. "Some questions don't have any clear answers. Just accept it and go with it."

"Whatever you say, partner." Juliana arched invitingly as his hand stroked down the length of her to her hip. When his fingers squeezed the curve of her thigh she sighed, feeling her own response deep within her body. He had such good, strong, knowing hands, she thought. The barest touch of his rough fingertips aroused her, and when he kissed her breast, she almost cried out. She could feel herself growing warm and damp already.

Turning slightly in his arms, Juliana snuggled

more closely against Travis. She began to explore him intimately, rediscovering the secrets she had first learned during their one previous night together.

She could feel the sexual tension in him as her hand slipped over his body. The muscles of his back and shoulders were taut with the desire he held in check. Travis's obvious need elicited a fierce reaction within Juliana. She stirred restlessly against him. When his fingers slid between her thighs into the dampening heat of her she parted her legs and whispered his name softly. Her nails grazed his shoulders.

"The kind of woman who leaves her mark," Travis muttered, rolling her onto her back.

"What?"

"Never mind. Wrap your legs around me, sweetheart. Tight. Take me inside and hold me close."

Juliana gathered him to her, aware of his hard shaft pressing against her softness. She was aching for him, her body throbbing with need.

"Now," Juliana whispered, her eyes closed as she clung to him and tightened her legs around his lean hips. "Come to me now. I love you, Travis."

"I couldn't wait any longer if I tried."

Travis eased himself into her carefully, filling her completely, stretching her gently until she felt the spiraling tension begin to build to unbearable heights.

She lifted her hips and Travis responded by driving more deeply into her and then withdrawing with excruciating slowness. Juliana clung more fiercely, urging him closer. He stroked into her again and again withdrew.

"Travis," she hissed, clutching at him.

"I'm not going anywhere without you."

He surged into her again, more deeply than before and Juliana gasped as the delicious tension within her went out of control. She gave herself up to the release, Travis's name on her lips as she felt him plunge into her one last time. The muscles of his back went rigid beneath her palms and she was crushed into the bedding beneath his solid weight.

For a time there was nothing else of importance in the entire universe.

A long while later Travis stirred on top of Juliana and looked down at her, his gaze intent in the shadows. She smiled sleepily, her fingers sliding slickly through the traces of perspiration on his shoulders.

"Don't worry," she said, yawning delicately. "I'm not going to bring up the subject."

"What subject?"

"Marriage."

"You just brought it up," Travis pointed out, kissing the tip of her nose.

"Well, I'll change it to a more pressing issue."

"Such as?"

"Such as which of us gets up first in the morning to make the tea."

"I'll do it this time," Travis said.

"Such an accommodating man." Juliana yawned again and drifted off to sleep.

Three days later, at ten-thirty in the morning, Travis sat at his desk in the new Jewel Harbor offices of Sawyer Management Systems and tried to decide if he really needed to take a coffee break. Charisma Espresso was only a few blocks away. He could walk over, say hello to Juliana, grab some coffee to go and be back at the office within twenty minutes.

Or he could save fifteen minutes by helping himself to the office pot. Lord knew he needed every spare minute he could salvage to work on the Flame Valley Inn problem.

Travis eyed the stack of papers on his desk with a brooding eye. He and Fast Forward Properties had certainly done a very thorough job of setting the inn up for a takeover. The grim truth was that he was not at all certain he could save it now. A lot of creditors, including his own group of investors would have to be put on indefinite hold and a big infusion of cash would be required to get the inn back on its feet financially. It was a nightmare of a task.

Every time he thought of Juliana's blithe faith in his ability to pull the fat back out of the fire, he swore silently. Lately he found himself trying very hard not to think about what would happen if he failed.

The intercom on his desk murmured gently. "Mr. Sawyer, there's a Mrs. Kirkwood here to see you."

Travis gritted his teeth. The last thing he needed right now was a visit from Elly, but there wasn't much he could do about it. "Send her in, Mrs. Bannerman."

A few seconds later Elly came through the door, her fragile features accentuated by her short gamin haircut, her delicate figure set off by close-fitting white pants and a white silk blouse. The big blue eyes that had once seemed so clear and guileless to Travis were filled with an irritating apprehension. Had he ever really been in love with this woman, he wondered fleetingly. She was a sweet, pale azalea next to the vibrant orchid that was Juliana.

"Hello, Elly. Have a seat." Travis got reluctantly to his feet and waved her casually to a chair.

"Thank you." Elly sat down, never taking her eyes off him. She crossed her feet at the ankles and folded her hands in her lap. She looked very nervous but very determined. "I had to come here today, Travis. I had to talk to you."

"Fine. Talk." Travis sat down again and leaned back in his chair. He wished he'd made up his mind about going over to Charisma five minutes earlier. He would have missed this little heart-to-heart chat. It went to show that he who hesitated was definitely lost.

Elly took a deep breath. "You're sleeping with

Juliana. You're having an affair with her."

Travis glanced out the window. He had a pleasant view of the harbor from here. He could even see The Treasure House restaurant down at the marina. "That's none of your business and you know it, Elly."

"It is if you're using her. She's my cousin, Travis. I won't have her hurt by you."

Travis looked at her. "Did you worry a lot about hurting Juliana when you lured Kirkwood away from her?"

"That's not fair," Elly cried. "I didn't lure him away. We fought our feelings for each other until . . . until . . ."

"Until Juliana finally noticed and let Kirkwood go?"

"Damn it, Travis. You don't know what you're talking about. Juliana didn't just let David go."

"Did she fight for him?" He waited tensely for the answer. He wanted to know how important Kirkwood had once been to Juliana.

Elly looked shocked at the question. "There was no fighting involved. They both realized they had made a mistake and they ended the engagement by mutual agreement. It was all very calm and civilized. Ask her. She'll tell you the truth. Travis, I don't want to talk about the past."

"Not even our own past?" he asked with mild interest.

"Especially not our own past." Her gaze slid away from his.

He nodded and tossed his pen down onto the

desk. "You're right. It's a rather boring subject, isn't it?"

Elly leaned forward anxiously. "Travis, I must know if you're blackmailing Juliana into sleeping with you."

"Why?"

"Because if that's the price you've put on saving Flame Valley for us, it's too high. I won't let Juliana pay it."

"You think your cousin would let herself get blackmailed into sleeping with a man, regardless of the reason?"

Elly frowned, looking momentarily confused. "Not under normal circumstances, but this is different."

"How is it different?"

"She's very attracted to you. In fact, I think she's in love with you. I don't think it would be very difficult for her to convince herself that it was all right to sleep with you in exchange for your help. But she would be hurt when the end came, and I don't want her to pay that kind of price."

Travis was suddenly impatient. "Don't worry about it, Elly."

"She's my cousin. I can't help but worry. You've thrown everything into such chaos by coming back this way, Travis. You're a very dangerous man and I'm the only one who seems to fully comprehend that fact. David refuses to look below the surface to see what you're really up to, and Juliana is blinded by her emotions. My father and aunt and uncle are totally confused by the whole situation. They want desperately to believe the inn can be saved, and although they all love Juliana, they're used to letting her take care of herself."

"So you've decided you're the only one left to sound the warning? Save your breath, Elly. Juliana won't listen to you."

Elly got to her feet and walked nervously over to the window. She stood there, clutching her leather purse, her back toward Travis. "You're doing this to punish me, aren't you?" she whispered.

Travis thought about that for perhaps twenty seconds. "No," he said.

"You didn't come back just because you felt you hadn't been paid for saving the inn five years ago. You came back because of what happened between us."

Travis studied Elly's slender back and shook his head. Juliana had been right: this fragile, delicate, easily startled doe of a woman was not for him, had never been right for him. "I came back because I'd been cheated out of my fee, Elly. That's the only reason I came back."

Elly swallowed a sob. "I'm sorry about what happened five years ago. I was sorry at the time. I didn't want to do it. I wanted to tell you the truth, about how I wanted to end the engagement, and I wanted to tell you just as soon as I realized we weren't right for each other. But Dad was sure you would walk out, and he said if you did, we would lose Flame Valley. He said things had gone too far and that you were the only one who could save it. He said I had to make you think I was going to marry you, and that you would get the partnership in the inn as a wedding present."

"So you went along with the lie. I'll have to admit you were very convincing, Elly."

She swung around, an anguished expression in her damp eyes. "What was I supposed to do? I had to make a choice between my family and you."

"And your family won."

"Yes, damn you. There was no other option. But I won't have Juliana hurt because of what I had to do five years ago."

"You make a charming martyr, Elly. But you can relax. I am not hurting Juliana."

"You're using her. She's a stand-in for me. I know that's what it is. If you can't have one Grant woman, you'll take another, is that it? Your revenge wouldn't be complete unless you got one of us into bed."

Travis stared at her, real anger gathering in him for the first time. He stood up. "That's enough, Elly. You don't know what you're talking about. Just for the record, I will tell you that there is no way in hell Juliana could ever be a stand-in for you. Juliana is totally unique. She's not a stand-in for anybody. She could never be that. Now, I suggest you leave before you say another word."

"I haven't finished." Elly lifted her chin, defying him with all the boldness of a traumatized deer. "I want you to know I'm on to your tricks. I don't trust you one inch, regardless of what the others think."

"Is that right? Well, I'll certainly have to watch my step, won't I?"

Travis took a menacing step forward and Elly stifled a small shriek. "Don't touch me, you big brute." She leaped back, turned and ran for the door.

Travis watched with mild disgust as his former fiancée fled his evil clutches. Juliana would never have run like that, he reflected. She would have stood her ground.

Travis sat back down at his desk and picked up the stack of papers he'd been wading through a few minutes earlier. Elly's words flickered through his head.

I had to make a choice between my family and you.

Travis knew he'd been wrong when he had told Elly that Juliana was totally unique. She had one very important thing in common with Elly. She was fiercely loyal to her family.

Travis prayed he would never be in the same situation with Juliana that he'd been in five years ago with Elly — one where the lady was forced to choose between him and her family. He always seemed to lose in those situations.

That realization made him wonder again what would happen if he failed to save Flame Valley Inn.

The door burst open in the middle of that thought, and Juliana swept into the room bearing a paper sack with the Charisma logo emblazoned on the side.

"All right, what did you do to my cousin, you big brute?" She smiled sunnily as she set the sack down on the desk and removed an extra-large Styrofoam cup of coffee.

Travis studied her, a smile playing around the edge of his mouth. She looked very chic today in a full, pleated menswear style skirt, a tight fitting vest and a wide-sleeved shirt.

"Big brute?" Travis repeated, eyeing the cup of coffee with enthusiasm.

"I believe those were her very words." Juliana uncapped a second cup and sat down in the chair her cousin had just vacated. "The two of you had a little scene up here, I take it?"

"It's hard to have more than a small scene with Elly. She always takes off running just when things start getting interesting." Travis popped the lid on his cup and inhaled gratefully. "Just what I needed."

"I figured you'd be ready for your mid-morning coffee break about now. Thought I'd save you the hike over to Charisma. Besides, I needed a break myself. Melvin and I are feuding again. So what did you and Elly argue about?"

"She thinks I'm blackmailing you into sleeping with me." Travis watched Juliana over the rim of the cup as he took a swallow of the rich, dark-roasted coffee.

"Her, too, huh?"

"I wouldn't be surprised if Sandy and Matt, your parents and Tony Grant suspected the same thing. Which would make the opinion practically unanimous."

"Too bad it's not true," Juliana remarked. "It's certainly an exciting thought."

Travis was irritated. "You weren't very excited the other night when I came up with the suggestion. As I recall you turned me down flat when I offered to take your fair body as my fee for saving Flame Valley."

"Well, of course I turned you down. I had to, for your sake."

"My sake?"

"Sure. If I'd agreed to let you blackmail me into sleeping with you, you would have worried yourself to a frazzle in no time."

Travis set down his cup cautiously. "About what?"

"About the real reason I was having an affair with you. Every time I kissed you or told you I loved you, a part of you would have wondered if I really meant it or if I was just saying it to keep you working like a dog on behalf of the resort. This way we keep business completely separate from our personal relationship."

Travis swore wryly. "Always one step ahead, aren't you?"

"I can't help it. It's my nature. I confess it was

for your own good that I had to put my foot down on your thrilling blackmail attempt." Juliana got to her feet and came around to his side of the desk. Her eyes were filled with seductive, mischievous humor. "But never think for a moment that I didn't find the offer wonderfully exciting."

Travis grinned, set aside his cup and reached for her. He tumbled her down across his thighs, cradling her in his arms. "For your information, if you had accepted my blackmail offer, I wouldn't have worried about it all that much."

"No?" She kissed his throat.

"No. Next time you do something for my own good, consult with me first, okay?"

"Okay." She nuzzled his neck invitingly. "When's your next appointment?"

"I have no scheduled appointments this morning." Travis slid his hand up along her thigh, aware of the growing tautness in his lower body. Everything about this woman aroused him, he thought in wonder. He crushed a thick handful of red curls in his fist.

Juliana giggled. "Have you ever done it on the job, so to speak?"

"No." He grinned. "You?"

"Never."

"Sounds like we have some catching up to do."

He gently eased her up onto the desk so that she perched on the edge, facing him. Then he pushed the gray flannel skirt slowly up to her thighs. His fingers closed around her knees. Juliana had very nice knees, he thought. Beautifully rounded knees. Funny, he had never realized how attractive a woman's knees could be.

"Travis?" Her hand rested on his shoulders. Her topaz eyes gleamed.

"Yeah?"

"It wouldn't hurt once in a while if we sort of pretended you were blackmailing me into this, would it?"

He grinned wickedly and slowly spread her gorgeous knees apart. "I'm always willing to oblige a fantasy for you, sweetheart."

Six

"Juliana, this is a very serious situation. We are all gravely concerned about the implications. That's why we all decided to hold this family conference. We need to understand just what is going on."

"Sure, Mom. I understand. But everything's under control. Trust me." Juliana confronted the three serious faces around the table, sighed inwardly and then dug a spoon into her bowl of zesty black bean soup. Not a lot of trust in this little family group; just a lot of anxiety, she decided. The excellent California-style Mexican food provided here at the colorful shopping mall restaurant would have to serve as compensation for the doleful company she was obliged to endure.

When the invitation to lunch had come that morning, Juliana had tried to think of a polite excuse to refuse but there simply was no acceptable way to turn down her parents and Uncle Tony. Lunch this afternoon had been more or less a command performance.

"Tony and I have been over this whole thing a number of times," Roy Grant informed his daughter. He watched her closely through his bifocals, concern etched in the lines of his face. "We can't figure out what Sawyer is up to but there's no question but that it's got to be bad news for all of us."

"Damn right," Tony Grant chimed in, scowling at his plate, which was overflowing with a sour cream and chicken enchilada. "Sawyer is nothing but trouble. We all know that. He's after Flame Valley and he's going to take it one way or another. Spent five years setting us up. Can you believe it? *Five years.* Christ, the man holds a mean grudge, doesn't he?"

"We knew five years ago the situation could get dicey when he found out Elly didn't want to marry him," Roy said. "Tried to pay him off then, but he wouldn't have any of it."

Juliana looked up. "As I understand it, the fee was supposed to have been a one-third interest in Flame Valley. Did you offer him that after Elly turned him down?"

"Hell, no," Tony exploded. "Couldn't go givin' away one-third of the place to someone outside of the family. For pity's sake, girl, where's your common sense? Flame Valley is a family thing. When the marriage was called off, that meant the deal was off, too. But Roy and I tried to compensate Sawyer. Wasn't like we just stiffed him completely."

"Uh-huh." Juliana helped herself to a handful of tortilla chips, dipping them into the salsa. "Did you actually offer him the cash equivalent of one-third of Flame Valley?"

Her mother's mouth tightened. "Of course not, dear. That would have meant giving him something in the neighborhood of close to a million dollars. We couldn't possibly have paid him or any other consultant that much in cash. But we did offer him a very reasonable consulting fee."

"The only thing wrong with the fee you offered was that it wasn't the fee that had been agreed to when he took the job."

Tony Grant turned red in the face. "It was just an understanding, not a signed contract, and it all hinged on his marrying Elly. That didn't work out, thank the good Lord. Whenever I think of how unhappy she would have been married to Sawyer, I get sick to my stomach. Hell, my stomach's upset right now."

Juliana rolled her eyes and munched chips. "I think that's the green chili sauce on your enchilada. You know it always upsets your stomach, Uncle Tony."

"Look, this is getting us nowhere," her father said brusquely. "Honey, we're not here to rehash the past. Right or wrong, what's done is done. It's the present we've got to worry about."

"But there's nothing to worry about, Dad." Juliana smiled soothingly. "I told you, it's all under control. Travis is working for me now. Because of his obligation to Charisma, he's agreed to try to salvage Flame Valley. He knows that if the resort goes under, I stand to lose a chunk of dough."

"Darling, it makes no sense," Beth Grant said anxiously. "Why should Travis Sawyer do you any favors?"

"I told you why. Charisma is a client of Sawyer Management Systems. It's not a favor, it's a business arrangement."

"All those bloody investors in Fast Forward Properties are clients of his, too," Uncle Tony rumbled. "Bigger clients than Charisma, believe me. They'll come first and don't you forget it."

Juliana blushed. "All right, so maybe there's more to it than just Travis's sense of commitment to Charisma. Maybe he's doing it for me."

Her parents and Uncle Tony all stared at her, their mouths open. Beth Grant recovered first.

"For you? What do you mean, Juliana? What is going on here?"

"Well, if you must know, Mom, I'm hoping to marry the man one of these days. And I think he's kind of fond of me, too, although he's still shy about admitting it."

Her father looked appalled and then outraged. "Has that bastard led you down the garden path?"

"We're getting there," Juliana assured him.

"Damn it, girl, has he proposed to you?" Tony Grant blustered. "Is that what this is all about? Is he trying to make you a substitute for Elly in this blasted revenge business?"

"Well, no, Uncle Tony, he hasn't actually proposed yet, but I have great hopes."

Her mother leaned forward, topaz eyes deeply concerned. "Darling, surely you wouldn't be taken in by a proposal of marriage from that man? You're an old hand at dealing with proposals. Men propose to you all the time. They're never serious for long."

Juliana winced. "Thanks, Mom."

"Oh, dear, I didn't mean to hurt your feelings, but you know it's the truth. You've been collecting proposals since you were in college. Men are always asking you to marry them within about twenty-four hours of meeting you and then, about twenty-four hours after that, they change their minds."

"And start looking for the nearest exit. Yeah, I know, Mom. A sad but true phenomenon. No doubt about it, I have a curious effect on men. I overpower them. But Travis is different. Nothing overpowers him." Juliana brightened. "To tell you the truth, I consider it a good sign that he hasn't rushed into a proposal."

"Why?" Her father looked suspicious.

"Well, as Mom said, the others all do it within about twenty-four hours of meeting me. Travis is apparently giving the matter much thought, which means that when he does ask me, he'll be sure of what he's doing." Juliana smiled. "But, then, Travis always knows what he's doing. He's always one step ahead."

Uncle Tony pointed his fork at her from the

other side of the table. "You remember that, my girl. That man is always one step ahead of everyone. That's what makes him so dangerous."

Juliana shook her head in exasperation. "Just because he's fast and smart doesn't mean you can't trust him. I, for one, trust him completely. If anyone can save Flame Valley, he will."

A sudden thrill of awareness caused Juliana to glance around. Travis was there behind her. She smiled in welcome as his large hand settled on her shoulder.

"Thanks for the vote of confidence, honey," Travis said.

"*Travis*. What are you doing here?"

"I called Charisma to see if you could get away for lunch. Sandy said you'd already been kidnapped and were being held for ransom here at the mall." Travis surveyed the glowering expressions of the other three people at the table. "Thought I'd stop by and rescue you."

"She don't need rescuing," Tony Grant muttered.

"Damn right," Roy Grant added grimly. "My daughter can take care of herself."

"She certainly can," Beth Grant said with maternal pride.

"Everybody needs rescuing once in a while," Travis said easily as he sat down next to Juliana and picked up a menu. "I'm starving. Working on Flame Valley finances all morning is enough to give a man an appetite. Either that or make him slightly nauseous, depending on his point of view. Fortunately I've got a strong stomach. The resort's in a hell of a mess again, isn't it?"

Tony Grant blanched and then turned red. "Pardon me," he muttered, pushing his chair back from the table. "Got to get going if we're going to make it to San Diego this evening. Roy, you and Beth have a plane to catch."

Beth rose to her feet and nodded at her husband. "Yes, dear, we must run along." She frowned at Juliana. "You will remember what we talked about?"

"Sure, Mom. Have a good trip back to San Francisco."

With one last uncertain glance at Travis, Beth turned to follow her husband and brother-in-law out of the restaurant.

"Don't look now," Juliana said, "but I think I just got stuck with the tab."

"You can afford it. You've been making money hand over fist since the day you opened Charisma."

"Does that mean I'm going to get to pay for your lunch, too?"

Travis closed the menu, looking thoughtful. "It's customary for the client to pick up the consultant's expenses."

"Oh."

"Just how many marriage proposals have you collected since college, Juliana?"

Juliana blinked. "Did a bit of eavesdropping, did you?"

"Couldn't help it. Everyone was so wrapped

up in that little intimate family conversation, I hated to interrupt."

"Forget my long and sordid history of collecting marriage proposals. None of them meant anything."

"If you say so."

Two days later Juliana again sat down to eat in a restaurant. This time she was alone with Travis.

"So how did the meeting with David go today?" Juliana speared one of the pan-fried oysters on her plate and chewed with enthusiasm. The Treasure House always did this dish particularly well, she thought. When there was no immediate response from the other side of the table, her brows came together in a firm line. "Not good?"

"Let's just say that Kirkwood is not a happy camper at the moment." Travis ate his swordfish in a methodical fashion that did not indicate great enjoyment.

"You know, Travis, you've been in a rather difficult mood for the past couple of days," Juliana pointed out.

"I'm working eighteen hours a day trying to save that bloody resort, get my new office up and running, keep a lot of important clients happy and find some time to spend with you. What kind of mood do you expect me to be in?"

"Maybe going out to dinner tonight wasn't such a hot idea."

"It wasn't. I've got a pile of papers back at the office I should be going through even as we speak. But since eating out was my idea, I suppose I ought to keep my mouth shut."

"So how did the meeting with David go?"

"Like I said, he's not happy. I pointed out one possible way out of the mess and he didn't like it."

"What was that?"

"Find a buyer for Flame Valley. Maybe one of the big hotel chains. An outfit that will agree to pay off the resort's creditors and agree to let Kirkwood stay on as manager."

Juliana winced. "I can see why he didn't jump at that. The last thing he wants to do is sell the place. The whole point is to hang on to it." She paused, thinking of what her Uncle Roy had said at lunch two days earlier. "It's a family thing."

"He reminded me of that. Juliana, I have to tell you, there's a chance, a very real chance, that I won't be able to pull this off."

"You'll do it." She smiled at him with all the confidence she felt.

"Damn it, I wish you weren't so irrationally sure I can save the resort." Travis's impatience blazed in his eyes. "Oh, hell. Look, I don't really want to talk about Flame Valley tonight."

"All right. Want to talk about Charisma instead?" Juliana helped herself to the last oyster on her plate. "I've been thinking about adding a new line of mugs with the store logo on them.

Subliminal advertising, you see. Every time a customer uses one of them at home, he'd think of Charisma."

"It's probably not a bad idea but right now I don't want to talk about it or anything else to do with Charisma."

"Well, what do you want to talk about?" she asked patiently as she forked up the remainder of her anchovy and garlic spiked salad.

"Us."

"Us?" She paused in midchew and eyed him intently. He definitely was in a strange mood tonight. "What about us?"

She watched Travis glance around the casually chic dining room. The place was filled with casually chic diners, casually chic ferns and a lot of waitpersons who could have modeled for magazine covers. Classic California restaurant style.

"Did you say Kirkwood brought you here the night he asked you to marry him?" Travis asked.

"Yup. So did most of the others who have proposed since I arrived in Jewel Harbor."

"How many would that be?"

She scowled at him. "Still after a number, hmm? Well, there really weren't that many. Two or three, at the most. I mean, you can hardly count the real estate agent who got me the lease for Charisma or the hunk who sold me my first espresso machine. They were very nice men, but extremely superficial. Salesmen types."

"Uh-huh."

140

"Look, Travis, I'm sorry about the fact that several men have asked me to marry them. I've explained before that none of them were serious for long."

Travis smiled wryly. "You terrorized them all, didn't you?" He reached for his wallet. "Come on, let's get out of here."

"Where are we going?"

"For a walk."

"At this time of night?"

"This isn't downtown L.A. It's Jewel Harbor, remember? I want to talk to you and I don't want to do it in here." Travis caught the waiter's eye and the young man came hurrying over to present the check.

Five minutes later Juliana allowed herself to be led outside the restaurant and down to the marina. For a while Travis strolled beside her in silence. It was a lovely evening, filled with soft breezes and the scent of the sea. Beneath Juliana's feet water slapped the wooden slats of the docks. The boats bobbed in their slips, and here and there cabin lights indicated owners who lived on board.

Travis took a seemingly aimless path that led them to the farthest row of slips. He paused finally and stood in brooding silence, staring out over the water.

Juliana tolerated the silence for a moment or two before her curiosity overcame her. "Why did you bring me out here, Travis?"

"To ask you to marry me." He didn't look at

her. He seemed mesmerized by the dark horizon.

Juliana couldn't believe her ears. *"What?"*

"You said I had a month to come to my senses. I don't need a month. I've known for quite a while that I want to marry you. It was just that everything was so damned complicated. It still is, for that matter. Nothing has changed. But I'm tired of waiting for the right time. At the rate things are going with the resort, there may not be a right time."

"Travis, turn around and look at me. Are you serious? You want to marry me?"

He turned his head slowly, a faint smile curving his mouth. "I'm serious. I haven't had time to buy a ring, but I'm very serious."

"You're not going to change your mind within twenty-four hours like the others, are you?" In spite of her confidence in him, old habits died hard, Juliana discovered. She was instinctively cautious when it came to receiving marriage proposals.

"Juliana, I guarantee I'm not going to change my mind about wanting to marry you. Trust me."

She smiled tremulously. "I do."

"Do I get an answer tonight or are you going to make me suffer awhile?"

"Oh, Travis, how can you even ask such a silly question? Of course I'll marry you. I practically asked you first, remember?"

"I remember."

Elation seized Juliana as she studied his face in the soft light. She couldn't recall a happier moment in her whole life.

"I'm sorry about the anchovy and garlic on the salad," she said as she hurtled toward him, her face raised for his kiss.

"Juliana, no, wait . . ."

It was too late. Normally there would have been no problem. Travis was getting used to catching Juliana's full weight against his body. But tonight the dock under his feet was bobbing precariously and he was caught off balance when she went flying into his arms.

At the last possible instant Juliana realized that disaster loomed. She clutched at Travis, her eyes widening in startled dismay as she felt him stagger back a step. She tried to catch her own balance but one of her two-inch heels got caught between the dock slats.

Travis groaned in resignation as they both went over the edge of the walkway, landing with a splash in the waters of the marina.

Juliana surfaced a few seconds later, spitting salt water out of her mouth. Her hair had been instantly transformed into a wet, tangled mop. She could feel the weight of her clothing dragging at her.

"Travis? Where are you?" She turned quickly, searching for him.

"Right here," he said from behind her, splashing softly as he made his way back toward the dock.

Juliana whipped around in time to see him plant both hands on the edge of the dock and haul himself out of the water. Juliana smiled up at him, vastly relieved. "Thank goodness. Are you all right?"

"I'll survive." Sitting on the edge of the dock he reached down to grab her hand. "I should have known that the simple task of asking you to marry me would turn into an adventure in Wonderland. By rights I ought to charge you hazardous duty pay."

"Just add it on to your usual fee, Mr. Sawyer."

His usual fee. Much later that night as he lay awake in bed beside Juliana, Travis reflected that for his usual fee, he usually produced results. This time he was not at all certain he could satisfy the client.

And he wondered if an engagement ring would be strong enough to hold Juliana if he failed to save Flame Valley Inn.

He couldn't seem to escape the premonition of disaster that hovered over him these days. In an effort to fight it off he turned on his side and gathered Juliana more tightly into his arms. She came to him willingly, fitting herself instinctively against him. After a while Travis was able to find sleep.

Juliana lounged back in her squeaking desk chair, her booted feet propped on the edge of her desk and cheerfully chewed out her supplier who was late on a delivery.

"No, Melvin, I do not want a double shipment of the regular Sumatra. I want the aged stuff. I can tell the difference, so don't try to con me. I've got standards to uphold, remember?"

"You don't even like coffee," the man on the other side of the line complained good-naturedly.

"That doesn't mean I don't know how to taste it. By the way, how are you doing getting me another batch of those good Guatemalan beans? I'm using them in my new house blend."

"I'm trying, Juliana, I'm trying. The two estates I usually buy from have cut back shipments for a while. Weather problems. How are you doing with the decaffeinated blends?"

"Going like hotcakes. Although why anyone would want to drink decaffeinated coffee defeats me. Seems sort of pointless. I mean, why drink the stuff at all if you're not going for the caffeine jolt? Say, Melvin?"

"Yeah?"

"You know anything about buying tea?"

"Sure. Tea is a staple sideline in my business. Why? You interested in adding a line of tea there at the shop?"

"I'm thinking about it. Actually, I was thinking about opening a whole shop devoted to tea."

"Forget it. There aren't enough tea drinkers around here to keep you in business. Try adding tea there at Charisma as a sideline before you go off the deep end."

"I'll discuss it with my new business partner."

"Partner? You've got a partner now? That's a surprise. I thought you liked owning Charisma lock, stock and barrel."

"The new partner is my fiancé," Juliana confided, feeling smug. She studied the toe of her lizard skin boots. The exotic footwear went nicely with the yoked and pleated pastel jeans and the snappy little bolero jacket she had on today.

There was silence on the other end of the line. "I don't know, Juliana. I'm not so sure it's a good idea to mix business and marriage. Just look at me. I've been through three wives. Gave them all a piece of my business. Every time I got a divorce I got wiped out financially and had to start over again."

"You should have paid as much attention to your wives as you do to your coffee-importing business," Juliana chided. She glanced out the door of her office and spotted a familiar dark-haired woman entering the shop. "Look, I've got to run. See what you can do about the aged Sumatra, okay? And the Guatemalan stuff. As a favor to me, Melvin."

"Juliana, if I do you any more favors I'll probably go out of business."

"Just be sure you leave me a list of other coffee importers I can go to if you go under."

"You're a hard-hearted woman, Juliana Grant."

"I'm a businesswoman, Melvin. Just trying to make a living and keep the customer satisfied. Talk to you later."

Juliana slid her feet off the desk and recradled the phone as she stood up. She hurried out of the office and hailed the woman who had just come into the shop.

"Angelina. Just the person I want to see."

Angelina Cavanaugh smiled from the other side of the counter. Her aristocratic Spanish ancestry was evident in her fine dark eyes and the sleek brown hair she wore in a classic chignon. "Good morning, Juliana. How are you today?"

"Great."

Sandy grinned. "Be careful, Angelina. She's finally cornered her man. She's brought him to his knees and she's still wallowing in her victory."

Angelina laughed in delight as Sandy handed her a small cup of intense, dark espresso. "Is that true, Juliana? You got a proposal out of your business consultant?"

Matt leaned over the counter conspiratorially. "It wasn't easy. Sawyer told me this morning just how it happened. She tripped the poor guy, got him off balance and threw him into the marina. He said that by the time he surfaced, he knew he was finished. He decided to surrender before she tried something more drastic."

"That," proclaimed Juliana, "is a gross distortion of events."

"Hey," said Matt, "I got the story from the victim, himself."

"Never mind him, Angelina. Come over here and sit down. I want to talk to you about the engagement party and wedding plans. Have you got room for me on your client list?"

Angelina's bright red lips curved in a smile. "Angelina's Perfect Weddings always has room for one more client, not that we aren't quite busy, what with repeat clients."

"This wedding will be a one-time event," Juliana declared.

"That's what they all say until the divorce. Have you set a date?"

Juliana frowned. "Not for the wedding. Travis is very busy right now with, uh, other matters. But I don't see any reason why I can't go ahead and schedule the engagement party on my own. Travis won't mind."

"Are you thinking of a major event?"

"Are you kidding?" Juliana chuckled in anticipation. "I'm pulling out all the stops. I want Travis to have the wedding of his dreams, and that includes the perfect engagement party."

"I see," Angelina drawled. "And have you checked with Travis to find out just exactly what his dreams entail?"

Juliana waved that aside. "I told you, he's very busy these days. I'll take care of the wedding and engagement party details for him."

Matt, eavesdropping unabashedly, nearly choked. "Poor Travis. And he thought getting dunked in the marina was the end of his problems."

"Ignore him, Angelina. How do you like the espresso?"

"It's wonderful. Full-bodied and distinctive flavor. Very rich and strong."

"Yeah, it'll put hair on your chest, all right," Juliana agreed. "Have another cup while I start making some notes on the engagement party. Sandy," she called across the room, "fix me a cup of tea, will you, please? I hid a tin of English Breakfast behind the counter this morning."

Travis, his sleeves rolled up to the elbows, his tie loosened and his shirt rumpled, regarded the man who sat across the overflowing desk.

"There just aren't a lot of options, Kirkwood. You've been teetering on the edge of bankruptcy for months and you know it. I'm telling you, the best I can do is try to find a buyer for the inn."

"No, damn it." David leaped to his feet and paced to the window. His expression was haggard. "I told you selling out is not an option. I can't sell the inn. Just call off your wolves and buy me some breathing space."

"Breathing space isn't going to do you any good." Travis flicked a pile of papers that all spelled impending disaster. "I might be able to stall my investors but that still leaves the banks you've been dealing with. It also leaves you needing cash. A lot of it. Even if I can get my group to hold off for a few months, which is unlikely, there's no way in hell I can ask them to pour more money into your operation."

"Fast Forward is your company. You told me

yourself, you make the investment decisions."

"I do. But I've got obligations to my backers. I've made certain commitments that have to be met."

David looked back over his shoulder, his eyes intent. "I can't sell the inn, Sawyer. Even if you can dig up a buyer at this late date, I just can't sell out."

Travis studied him in silence for a minute. "Because of Elly?" he finally asked quietly.

David turned back to the view of the harbor. A deep sigh escaped his chest. "Yes. Because of Elly. I've made a lot of commitments, too, Sawyer. Told her I was going to make the resort the biggest and best on the coast. Told her I'd keep it in the family, just like her Daddy wanted. Told her she'd always be proud of it. I don't think she'll ever forgive me if I lose it."

"What will Elly do if you can't keep your promises?"

"I don't know."

"You think she'll leave you? Is that what you're afraid of?"

"Shut up, Sawyer. You worry about saving the inn. I'll worry about my marriage, okay?"

"Whatever you say. But you'd better get it through your head that I may not be able to save the inn."

David hesitated and then said under his breath. "And I may not be able to save my marriage."

There was another moment of silence. "It

looks like we'd better get back to work," Travis said eventually.

Fifteen minutes later he looked up again. "Did I tell you I'm engaged to Juliana?"

David reluctantly dragged his gaze away from an accounts journal he had been studying. "What's that?"

"I said, I've asked Juliana to marry me."

David smiled slowly. "Are you sure that's the way it happened? You asked her? She didn't ask you?"

"As I recall, she told me I had a month to ask her properly. I did so last night and she rewarded me by pushing me into the water at the marina."

"Very romantic. I hope you know what you're getting into."

"So do I." Travis smiled to himself as he remembered Juliana sending them both into the harbor the previous night.

"You're doing this for Juliana, aren't you?" David asked. "Not because you're her business consultant, but because you want her to marry you and you know that if you destroy the inn, you'll lose her."

Travis shrugged and went back to his papers.

"Be a little weird if this turns out to be an instant replay of five years ago. Maybe Juliana is leading you on, getting you to save the inn and planning to dump you once Flame Valley is in the clear."

"That's enough, Kirkwood."

"You know, it's obvious to me that you would have been all wrong for Elly and she would have been wrong for you," David said conversationally.

Travis put down his pencil and folded his elbows on the desk. "Yeah?"

"Yeah." David waited, his eyes full of challenge.

"I'll tell you something, Kirkwood. You're right. Elly and I would have been all wrong for each other. And I'll tell you something else."

"What's that?"

"You and Juliana would have been a damned bad mismatch, too." Travis picked up his pencil and then reached out for the phone.

"Who are you calling?"

"An eccentric old venture capitalist I know. Tough as nails. Got money coming out his ears and no one to spend it on. Thrives on a challenge. Sometimes he'll go for something off the wall that no one else will touch."

"I told you, I don't want to sell the inn," David said angrily.

"I'm not going to try to sell this to him," Travis explained. "I'm going to see if I can talk him into paying off your biggest creditors and pour some cash into the resort."

David's expression lightened. "Think he'll go for it?"

"No, but it can't hurt to ask. We're running out of options." Travis concentrated on the phone. "This is Travis Sawyer," he said when a pleasant voice came on the line. "Tell Sam Bickerstaff I'd like to talk to him for a few minutes, please. . . . Yeah, I'll wait."

Elly flew through the door of Charisma Espresso shortly after lunch. She plowed through the standing-room-only crowd of coffee aficionados, searching the place until she spotted Juliana.

Juliana saw her cousin approaching and knew immediately she had heard about the engagement. Elly looked stricken.

"*Juliana*. Juliana, I just heard. Oh, my God, how could you do it? You don't know what you're doing."

"I always know what I'm doing, Elly, you know that. Everybody knows that. Now calm down. Here, I'll have Sandy make you a nice latte. You like her lattes. You can drink it in my office."

Five minutes later with Elly cradling the cup of steamed milk and coffee in her hands, the two women shut the door of Juliana's office and sat down.

"All right," Elly said tensely. "Tell me first of all if it's true. Are you engaged to Travis?"

"It's true," Juliana said cheerfully. "How did you hear about it? I was going to call you this evening. In fact, what are you doing here in Jewel Harbor? Surely you didn't feel compelled to race all the way into town just because you got the word about my engagement."

153

"David has an appointment with Travis today. I came with him into town. I just saw him at lunch a few minutes ago and he told me Travis is claiming he's engaged to you. Juliana, how did that happen?"

"Well, it wasn't easy, I can tell you that. Travis has been so busy what with all this business with the resort. But last night . . ."

"I knew it, he coerced you into sleeping with him but that wasn't enough to satisfy his ego, was it? Oh, no. He had to go all the way and trick you into an engagement." Elly shook her head sadly. "He really is trying to duplicate the past except that this time around he's using you instead of me."

Juliana smiled a little grimly. "You know me well enough to realize I wouldn't have gotten engaged to a man unless I was in love with him. Look how much practice I've had gracefully declining marriage proposals. Now stop carrying on about how I'm being suckered in a revenge plot and let's get on to a more interesting subject."

"Like what?"

"My wedding." Juliana scrabbled around on her desk, uncovering two hefty tomes she had picked up at the Jewel Harbor Library.

Elly blanched at the sight of the books on wedding etiquette. "You can't. Juliana, please. Think about this. Don't rush into anything. You mustn't set yourself up like this. He's only using you. He has no intention of marrying you."

"I talked to Angelina Cavanaugh this morning. Remember her? She has that wedding business in town. She gave me a booklet to read and recommended these books. The first step is the engagement party. I was thinking of something spiffy at The Treasure House. They have a special room they rent out for catered events."

"Juliana, this is insane. Listen to me, he won't go through with it. He just wants revenge and he'll get it by leading you on and then dumping you when he finally takes over Flame Valley."

Juliana opened one of the etiquette guides. "He's working to save the inn, remember?"

"I don't believe it," Elly whispered. "I know David believes him, but I don't. It's all a game with Travis. A game of vengeance. Neither of you know him. Ask your folks or my father. They know Travis for what he is. They know how dangerous he can be. They heard him vow revenge five years ago."

"People change," Juliana said easily.

"God. I feel like Cassandra calling out a warning that no one will heed."

Juliana nodded in commiseration. "Always a frustrating role."

"It's not funny," Elly flared. "This is serious. Very serious. Right now Travis is feeding David all sorts of nonsense about getting another investor involved in the inn. Somebody named Bickerstaff. That's all we need. Another creditor. Oh, Juliana, what are we going to do? It's such a mess."

"Travis will straighten things out. Now about my engagement party. I think I want a buffet affair with lots of yummy goodies rather than a sit-down dinner. And it would be fun to have a band, don't you think? I wonder if Travis knows how to dance."

"I can't stand it. Nobody will listen to me." Elly put aside her cup of latte, covered her eyes with her fingers and wept. Juliana sighed and reached for a tissue. She handed the tissue to her cousin as she got to her feet.

"Here, Elly. Dry your eyes. I'll be right back."

"Where are you going?" Elly asked, lifting her tear-stained face.

"To get you a cup of tea. When the chips are down, a good cup of tea is infinitely better for the nerves than a cup of coffee."

Juliana walked back into the office a few minutes later, tea in hand and found that Elly had, indeed, managed to stop the flow of tears.

"Thank you," Elly mumbled as she took the tea.

"Feel better?"

Elly nodded, sipping daintily at the brew. "I'm sorry to be so emotional but I'm frightened, Juliana."

"I can see that. But you're worrying yourself sick over nothing. Everything's going to be all right. Travis is going to save the inn and give you and David a second chance with it. You'll see."

"But what if he doesn't? Even if he's not deliberately plotting against us, he might not be able to save it. David hinted at that much today. Juliana, I'm scared about what will happen if we lose Flame Valley."

Juliana drummed her nails on the desk. "It would be unfortunate, but it would not be the end of the world, Elly."

"It might be the end of my marriage."

"Oh, come on now."

"I mean it, Juliana. David's been acting tense lately. Not like his usual self at all."

"He's worried about the inn. We all are."

"It's more than that." Elly looked up from the steaming tea. "If we lose the Flame Valley, I might lose David."

Juliana sat very still. "That's ridiculous. Why do you say that?"

"He wanted the resort very badly back in the beginning. You know that, Juliana." Elly's voice was a mere thread of sound.

"I know he's been interested in it right from the start. He's thoroughly enjoyed running the place and planning for its future," Juliana agreed carefully. "But . . ."

"Sometimes I think that he married me to get Flame Valley. . . ."

"Elly. How can you say such a thing? It's not true. It's absolutely not possible. I was there when you two met, and I was there when you realized you were in love, remember? In fact, I realized the two of you were in love before either one of you admitted it to each other."

"We tried so hard to hide our feelings, didn't we? Even from ourselves," Elly recalled wistfully. "We didn't want to hurt you, Juliana."

"I'm well aware of that. Now pay attention. David cares very deeply about you. One of the reasons he cares so much about Flame Valley is because he knows how important it is to you. It's the legacy your father wants you to have, and David feels obliged to hold on to it for you at all costs."

"I tell myself that over and over again, but lately I've begun to wonder. And now I realize that a part of me has always wondered. Ever since . . ."

"Ever since what, Elly?"

"You have to remember that I just narrowly escaped being married once before because of the inn. To Travis Sawyer. I guess I'm sensitive on the subject."

Juliana narrowed her eyes and studied her cousin. "It's no wonder women occasionally question the motives of the male of the species," she observed. "We've all been burned a few times. But women are born to take risks. You know what they say — no guts, no glory."

Elly smiled mistily. "Juliana, you're incredible."

Juliana smiled and picked up another piece of paper she'd been studying earlier. "Now, about the engagement party menu. What do you think about having those little rounds of marinated goat cheese wrapped in grape leaves?"

"Nobody actually likes goat cheese, Juliana. They just eat it because it's trendy."

"Being trendy is an excellent reason for including goat cheese. Besides, believe it or not, I like it."

Elly's brows rose. "And it is your party, isn't it?"

"Right."

Seven

That evening Travis turned the key in the lock of Juliana's front door and was amazed at the quiet sense of pleasure he experienced in the small, mundane act. Another day was over and he felt as if he were home. It was true he had not yet officially moved in with Juliana, but he was spending so many nights here, he might as well do so. He could smell something savory in the oven, and he knew his redheaded lady would be waiting on the other side of the door with a glass of wine in her hand.

What more could a man ask, he wondered as he stepped into the white-tiled hall and set down his heavy briefcase. If only he didn't have to worry about how long he would be able to claim these small, vital treasures.

"I'm home." Travis listened carefully to the words as he said them. Not quite the whole truth and they might never be the whole truth, but he liked the sound of them anyway.

"Be right there," Juliana called from the kitchen.

Travis walked through the living room, picking up the evening paper that was lying on the

160

coffee table. He scanned the headlines and then glanced up as Juliana appeared in the kitchen doorway. He smiled slowly.

She was holding a glass of wine in one hand and a spatula in the other. Her hair was caught up in a high shower of curls and she was wearing a Charisma Espresso apron over her pastel jeans. He saw that she had removed the lizard-skin boots she had put on that morning and replaced them with a pair of fluffy pink slippers. The slippers had bunny faces in front and a pouf of a bunny tale at the ankles. Juliana had once carefully explained to him that the silly looking slippers had something called "a charming wittiness." Whatever it was, they still looked like dead rabbits to Travis.

"You seem exhausted," Juliana announced. She came forward, and because she wasn't wearing heels, had to go up on her toes to give him a kiss as she put the wineglass in his hand.

Travis felt her tongue tease his bottom lip and he groaned as his body reacted immediately. "I am exhausted. But I know my duty and I'm sure that, with the right stimulus, I can manage to get in a quickie with you before dinner."

"Absolutely not," Juliana said with mock horror. "Married people save it until after dinner."

"We're not married yet," he complained, following her into the warm, fragrant kitchen.

"We have to start practicing. Besides, I don't dare leave this walnut and blue-cheese sauce just now, and the corn bread would certainly burn if I surrendered to your lecherous ways."

"I take it back. Maybe I am too exhausted. What a day." Travis, newspaper still in hand, sat down at the cozy little breakfast table and took a sip of wine.

"Another meeting with David, right?"

"If you can call it that. I have to tell you, Juliana, I can see now why Flame Valley is on the brink of collapse. Even I didn't realize how easy it was going to be to take over the inn. Kirkwood is pigheaded stubborn in some areas. No wonder he's in trouble."

"He's stubborn in areas that have to do with Elly, and the resort has a lot to do with Elly."

"Yeah, I'm beginning to see the problem. Any man who lets his business decisions be dictated by his need to please a woman is setting himself up for a —" Travis broke off abruptly as he realized what he was saying.

Juliana batted her eyelashes outrageously. "Yes, dear? What was that about basing one's business decisions on the need to please a woman?"

Travis felt a rueful smile tug at his lips. "All right, so the poor jerk and I have something in common besides the fact that we both proposed to you."

"You'd better not have too much in common. I don't want to catch you running off with a petite blonde."

"Not a chance. I'm only interested in redheads these days. Tall redheads who know how to cook." Travis paused, eyeing her thoughtfully as he remembered Kirkwood's fear of losing his wife as well as the inn. "Did it hurt a lot?"

"Did what hurt a lot?" Juliana asked, concentrating on the sauce.

"When Kirkwood left you for Elly?"

"Well, it wasn't the high point of my emotional life, I'll say that much. But by the time it happened, I was more or less prepared for it. I'd seen it coming before either of them did. Whenever they were in the same room together there was a certain electrical charge in the air. They both just sort of hummed with it. I envied the force of the attraction but I knew from the beginning I couldn't duplicate it. Not with David, at any rate."

"So you just let him go and wished them well?"

"That's me. Gracious, even in defeat," she agreed brightly.

"Is that so? Funny, the word *gracious* never came to mind the night you threw guacamole dip all over me." Travis smiled at the memory.

"That was different," she retorted.

"Was it?"

"Darn right. I could see early on I'd made a mistake thinking David was the right man for me. But I learn from my mistakes, and this time around I was sure I had picked the right man. It really annoyed me when I discovered you hadn't had the same blinding realization."

"Oh, I'd had it. Sort of," Travis mused, thinking about it. "But there was the business with Flame Valley and a big chunk of the past in the way. Blinding realizations sometimes take a while to clarify themselves. Juliana?"

"Umm?" She frowned over the cheese sauce and stirred ferociously.

"If things turned sour again between us, would you fight for me a little harder than you fought for Kirkwood? I don't think I want you being gracious in defeat in my case."

She didn't look up from the thickening sauce. As she stirred frantically with one hand, she tossed a huge handful of pasta into a pan of boiling water with the other. "If I ever catch you hanging out with petite blondes, I'll nail your hide to the office door. There. How's that for feminine machismo?"

"Very reassuring," Travis murmured and wished the situation with Flame Valley Inn was as simple and straightforward as dealing with a petite blonde would have been.

"Did you and David make any progress today?" Juliana asked, changing the subject as she scooped the pot of cheese sauce off the burner.

"Not much. He's in worse shape than I thought and that's saying something. I played a long shot and called a guy named Bickerstaff. Offered him the wonderful opportunity of paying off the inn's creditors and pouring a ton of cash into the place. In exchange, I promised to guide the restructuring of Flame Valley and personally guarantee to get the present owner back on his feet."

Juliana smiled, looked very pleased. "Good idea. I knew you were the resourceful type. Did this Bickerstaff go for it?"

"He said he'd consider it and get back to me."

"What does that mean?"

"Knowing Bickerstaff, it means he'll consider it and get back to me."

"Oh." Juliana chewed on her lower lip as she emptied the cheese sauce into a bowl. "It sounds like a good idea to me. And you can be very convincing. I'll bet he goes for it."

Travis shook his head, sighing to himself over Juliana's irrepressible faith in his business acumen. "Don't hold your breath. I'm not. Bickerstaff likes a calculated risk, but he didn't get where he is by playing real long shots."

"We'll see." Juliana opened the oven door and a tantalizing aroma wafted through the kitchen. She bent over to get the pan of golden corn bread out of the oven. "Ready to eat?"

Travis took another swallow of wine and studied the sight of Juliana's pastel jeans pulled taut over the full curve of her derriere.

"Starved," he said. "Let me know if you want me to do anything."

"No, not tonight. You've been working too hard lately as it is." She straightened with the pan of steaming corn bread in hand.

"You know, Juliana, I have to tell you that you really have a terrific gluteus maximus. World class, in fact."

"Why, thank you. That only goes to show you can find something nice to say about anyone if you try hard enough. To be honest, I think yours is rather cute, too." She busied herself getting dinner on the table. "Did I tell you I had a nice chat with the lady who's going to help me plan our engagement party and wedding?"

The wine slopped precariously in Travis's glass, and his stomach, which had been relaxing nicely, tightened abruptly. "No, you didn't tell me. Isn't that moving a little fast? You just agreed to marry me last night."

"No point waiting, is there?"

"Uh, no. I guess not." He felt dazed. What would happen if she actually married him before he found out if he could save the inn? Travis wondered.

"We'll need to start putting together a guest list. Start jotting down names as you think of them, okay?"

Travis reflected briefly, trying to catch up with her. "I don't have any names to jot down."

"Don't be silly. Of course you do."

"Can't think of anyone to invite. Well, maybe the staff here at the new office. That's about it."

"Just your staff?" Juliana gave him a severe glance as she sat down across from him and began slicing corn bread. "What about your parents, for heaven's sake?"

He shrugged, his mind on the corn bread as he watched her transfer a chunk to his plate. He hadn't had homemade corn bread in years. "I don't see any point in you going out of your way to invite my folks. They didn't bother to come to my last wedding."

Juliana's hand froze over the corn bread. Her gaze collided with his, her topaz eyes full of demanding questions. "You've been married before?" she got out in a throaty whisper.

"It was a long time ago." Travis massaged the back of his neck, aware he hadn't handled the announcement very well. He certainly hadn't meant to drop it on her like this. He wasn't thinking clearly tonight. Too tired, probably. Lord, he was exhausted. "Back in my early twenties. It was a mistake. Didn't last long."

"What happened? Who was she? Where is she now? Are there any kids? Exactly how long were you married? And why didn't your parents come to the wedding?"

Travis wondered why he hadn't kept his mouth shut. He really didn't feel like going into all this tonight. But it was too late now. "Jeannie and I were married less than a year. She was a secretary at the firm where I got my first job. There was a slight misunderstanding on both our parts. I thought she wanted to build a successful future with me. She thought I could make her forget her first husband."

"What happened?"

Travis took a big bite of corn bread. "She

went back to her first husband. I left the firm to go out on my own. Turned out we both made the right decision."

"Kids?"

"No kids. There's nothing more to the story than that, Juliana. It was over a long time ago."

"Why didn't your parents come to the wedding? Did you elope?"

"No."

"Didn't your folks approve of your intended bride?"

"They never even met her. Approval wasn't the problem. Neither Mom nor Dad showed up at the wedding because each knew I had invited the other."

Juliana frowned. "I don't get it."

Travis helped himself to more corn bread. "They divorced when I was fourteen. It was a very bitter separation and neither one has been able to say a civil word to or about the other since the legal proceedings were final. Not that they had too much to say to each other before the divorce, either. When I invited them to my wedding, each wanted to know if the other had been invited."

"Oh, dear." Juliana's eyes filled with sympathy.

"When I said yes, my mother made it clear she would not attend unless I promised to disinvite my father. And my father made the same stipulation. I refused to be put in the middle like that and they both got even by not attending the wedding."

"Oh, Travis, that's terrible. They put you in a terrible position. Of course you couldn't not invite one or the other. Didn't they realize that? Didn't they realize how that would make you feel?"

"I don't think my feelings came into the matter," Travis said dryly. "They were both too wrapped up in their own emotions. Always were. I hated the weekends I spent with Dad because he always told me what a lousy mother I had and when I got back to Mom she always grilled me on what my father was doing and who he was dating."

Juliana grimaced. "How awful."

"Not that uncommon in this day and age, and we both know it. Frankly, it was easier on everyone not to have either of my parents show up at my first wedding, and I think I can safely promise the same thing this time around. Don't bother inviting my side of the family, honey."

"Do you have any brothers or sisters?"

"A couple of stepbrothers and two stepsisters. I don't know them very well. Mom and Dad both remarried soon after the divorce and started new families. But I left for college three and a half years after they split up, so I didn't spend much time with my new baby brothers and sisters. The only time I was invited to get involved in their lives was when each of my parents asked me to make a contribution toward their college funds."

Juliana wrinkled her nose. "Which you did, I bet."

"Sure. Why not? I can afford it and they all know it. I'm still picking up the tab on three of them. The oldest graduated last year and got a job at a company where I know a few people."

"And everyone just lets you do it? Lets you finance your stepbrothers' and sisters' educations and lets you help find them all jobs? Yet they can't be bothered to come to your weddings?"

"I believe picking up the college costs is seen as my contribution to the family. Everybody definitely agrees I'm the one in a position to handle it financially. And what the hell, they're right. Look, don't go wasting a lot of sympathy on this, okay? It's not worth it. Are you going to let that cheese sauce congeal or are you going to serve it over the pasta?"

Juliana gasped and leaped to her feet. "Good grief, the pasta. I forgot about it. It's going to be soggy mush. I can't stand overcooked pasta. It's no good if it isn't *al dente*."

"The cheese sauce isn't too bad over the corn bread," Travis said, running an experiment on his plate.

"Travis," Juliana said from the sink where she was dumping the pasta into a colander, "I want you to give me the addresses of your parents. I really think we should invite them, regardless of what they did the last time you got married. After all, that was a long time ago. They've probably mellowed by now."

"I doubt it, but do what you want."

"Do you ever see your parents?"

"Once in a while. I call them on their birthdays and they call me on mine, and I've managed a few short visits over the years. Which is sufficient for all concerned."

"Why do you say that?"

Travis helped himself to more corn bread and conducted another experiment with the cheese sauce. "After they remarried, my folks concentrated most of their attention on their new families. They both wanted to make fresh starts, I think."

Juliana's eyes widened as realization struck. "But you were a reminder of the past, weren't you? You were the living evidence of their failure. They probably felt guilty about the way they'd torn your world apart and made you witness their battles for so many years. It's easier not to have to face people who make you feel guilty."

"I think it was definitely a relief to all concerned when I left for college. The interesting thing," Travis said thoughtfully as he poured more cheese sauce over the corn bread, "was that they turned out to be fairly good parents the second time around, at least as far as I can tell. My stepbrothers and stepsisters seem happy and well adjusted. And my parents' second marriages seem to have worked out."

"Ouch. Darn it." Juliana turned on the cold water faucet and held a finger in the spray.

Travis glanced at her with concern. "What happened? Burn yourself?"

"Just a little. It's all right," Juliana said quickly. "I'll be over there in a minute. I think we lucked out. The pasta isn't too squishy after all."

"After they remarried, they concentrated most of their attention on their new families."

Travis's laconic explanation of why he wasn't very close to his parents echoed through Juliana's mind that night as she sat in bed waiting for him to emerge from the bathroom. It was clear to her that after the divorce and remarriage of his parents, Travis had been left out in the cold. He had not become a real part of either of the new families.

A few years later, his first marriage had ended in divorce when his wife had gone back to her first husband.

Five years ago his engagement had ended when his fiancée had used him to help her family and then called off the engagement.

All things considered, Juliana decided, Travis had not had a particularly good experience with family life. He'd always been the one left out. He was never the one chosen when choices had to be made — never a real member of a family. But people were quite willing to use him when it suited their purposes.

The bathroom door opened and Travis strolled out with a towel around his hips. He yawned and Juliana decided he was enormously sexy, even when he was yawning. He looked like a big, sleek wild animal that had somehow wandered into her very civilized white-on-white bedroom.

Travis saw her looking at him. His eyes glinted. "Still think I'm too short?"

"There are compensations," Juliana declared loftily. She put down the book she had been reading. "I called Melvin today."

"The guy who supplies your coffee?"

"Right. I asked him if he could supply the shop with tea, too. He said yes."

"Uh-huh." Travis did not look overly interested in the conversation. He rubbed the back of his neck as he walked toward the bed.

Juliana plunged ahead with enthusiasm. "I've been thinking about getting into tea in a big way. Travis, have you ever noticed that there are zillions of espresso shops and coffee houses opening all along the coast clear to the state of Washington, but no tea shops?"

"There's a reason for that," Travis explained as he slung the towel over a chair. "The money's in coffee, not tea. Nobody drinks tea."

"That's not true. I drink tea. Lots of people drink tea." Juliana was momentarily sidetracked by the sight of Travis's nude body. "And I'll bet there are thousands of people just like me."

"Closet tea drinkers? I doubt it." He pulled back the covers and slid into bed beside her.

"We aren't closet tea drinkers, we just tend

not to drink a lot of tea in public because it's so hard to get it properly made in restaurants. Wimpy tea bags plopped into lukewarm water don't cut it for a real tea enthusiast. Most tea drinkers drink coffee when they're out rather than pay for bad tea."

"What's your point?" Travis plumped up his pillows and leaned back against them. He reached for a stack of papers he'd left on the nightstand.

"My point is that if tea drinkers knew they could get properly made tea at a certain place, they would go there and order it. Tons of it. They'd buy it in bulk to take home with them. They would experiment with different teas of the world and enjoy them the way coffee drinkers enjoy coffee."

Travis scanned the figures in front of him, frowning intently. "You want to add a tea option at Charisma? No problem. Go ahead and do it."

"Travis, you don't understand. I don't want to just put in a line of tea at Charisma. I'm going for the whole enchilada. I've been thinking about this for several months now and I've decided to open a shop devoted entirely to tea. The first in a chain."

"No, you're not." Travis didn't even bother to look up from his paperwork.

"Now, Travis, I'm serious about this. The tea shops would be a first around here."

"And a last because everyone around here drinks coffee. You'd lose your shirt, and right now, Juliana, you cannot afford to lose a single dime. Believe me, I'm in a position to know."

"We'd push for the upscale crowd, the same way we do at Charisma. We'd make tea drinking trendy. We'd serve power teas to business people and their clients, and we'd package a whole line of tea under our own label."

Travis grunted and finally raised his eyes from his paperwork as the enthusiastic determination in her voice finally sank in. He glanced at her nightstand. "You've been reading that book again, haven't you?"

Juliana looked at him innocently. "What book?"

"That book about the Boston tea heiress whose ancestor was a witch. Leaves of something-or-other by Linda What's-her-face. I saw it on your nightstand earlier."

"*Leaves of Fortune* by Linda Barlow," Juliana corrected automatically. "It's a great story and it's given me all sorts of ideas, Travis."

"It's given you delusions of grandeur. You are not going to make your fortune in tea, Juliana. Fancy coffee is where the future is and you're perfectly positioned with Charisma to take advantage of it. As your consultant and partner I'm not about to let you fritter away your money and energy chasing an unrealistic business goal."

"I've been giving this a lot of thought," Juliana persisted. She broke off as Travis began rubbing the back of his neck again. "What's wrong?"

"Nothing. I'm just a little stiff and sore from

175

sitting hunched over four years' worth of Kirkwood's income tax forms all day." Travis eased himself into a slightly different position on the pillows.

Juliana pushed back the covers. "Turn over on your stomach and I'll rub your back for you."

He hesitated and then shrugged. "It's a deal."

Travis rolled onto his stomach, sighing heavily as Juliana straddled his thighs. The long skirts of her frothy French nightgown flowed around his lean hips.

The muscles of his back were sleek, strong and well defined, she thought as she leaned forward to begin working on his shoulders. Very sexy, very masculine. She could feel the rough hair on his legs against her soft inner thighs. She adjusted her position, settling herself more firmly.

"No fair wriggling." Travis's voice was muffled in the pillow.

"Sorry." Juliana applied herself to the massage, easing the tension in Travis's shoulders and neck with smooth, deep movements.

"Lord, that feels good. If you'd told me you could give a massage like this I probably would have asked you to marry me the first day I met you."

"I wanted you to admire my brain, not my brawn. Now, about my tea shop idea. I've been going over a lot of different aspects of the project lately and I've come up with a basic plan. When you've got some free time I'll lay it out for you. It's going to work, Travis. I know it is."

Travis said nothing. Encouraged by the lack of a negative response, Juliana kept talking as she massaged. Slowly the hard muscles beneath her palms relaxed. As she stroked the strong contours of Travis's back she began to think of other things beside the future of her tea shops. She grew increasingly aware of the hardness of Travis's buttocks beneath her much softer shape and wondered just how tired he really was. Perhaps the massage would prove invigorating as well as relaxing. In the meantime, she kept talking about tea.

Fifteen minutes later when she finally came to a momentary halt in the middle of her monologue, Juliana realized that Travis had fallen asleep somewhere along the line. She groaned.

She was ruefully aware that while her efforts had apparently thoroughly relaxed Travis, they had had the opposite effect on her. Smiling wryly, Juliana dismounted from her sleeping stallion and crawled over to her side of the bed.

"Don't think that you can escape every discussion of my tea shop plans this easily," she whispered as she turned out the light.

Travis did not respond.

Two days later Sandy poked her head into Juliana's office. "Don't forget the coffee tasting at noon," she said. Then she frowned at the array of crumpled papers on the floor. "Hey, what's going on in here? You writing out a resignation or something? Going to turn Charisma over to me and Matt? I knew you'd see the light one of these days. We've already made plans to put in an ice-cream parlor out front."

Juliana didn't look up from where she was busily penning still another version of the letter she had been trying to write all morning. "The problem with owning your own business is that it's tough to resign. Forget the ice-cream empire. As it happens, I'm writing to Travis's parents."

"Introducing yourself?"

"Yes and inviting them to the wedding."

Sandy looked at all the aborted efforts. "What's so hard about writing a simple note telling them Travis is making a brilliant marriage to you?"

"I want to get just the right tone. They didn't come to Travis's first wedding and he doesn't think they'll come to this one. They're divorced and remarried, and apparently there was a lot of bitterness between them after they split. One won't be found dead in the same room with the other, not even at their son's wedding."

"Hmm. A messy situation, socially speaking."

Juliana sighed and leaned back in her chair. She tapped the tip of the pen on the desk. "Such childish behavior for adults. It's incredible. It's disgusting."

"It's also fairly common these days. Not much you can do about it."

"Travis's folks have been divorced for years and have raised second families. It's time they remembered their first-born son."

Sandy shrugged. "They'll probably remember him fast enough if and when he gives them a grandchild. It's been my observation that the older people get, the more interested they are in their descendants."

Juliana stared at her. "You know something, Sandy, you have just made a brilliant observation."

"I've been telling you since the day you hired me that I'm brilliant." Sandy folded her arms and leaned against the doorjamb. "How are you going to get Travis's parents to come to the wedding?"

"Well, I have been working on a pleasant, conciliatory approach." Juliana waved a hand at all the crumpled notes on the floor. "Something along the lines of what a really terrific daughter-in-law I'll be and how I want to get to know my husband's parents, et cetera, et cetera. But after talking to you, I think I'll try a different tactic."

"What's that?"

"Threats."

Sandy raised an eyebrow. "What sort of threats?"

"I don't know yet. I'll have to think about it." Juliana got to her feet. "Everything all set for the coffee tasting?"

"Yes. You wanted to do a comparison of Indonesian, Hawaiian and Mexican coffee today, right?"

Juliana wrinkled her nose. "Right."

Sandy laughed. "Your enthusiasm is overwhelming. You should be looking forward to going out front in a few minutes. This series of comparative tastings you've been running the past month has really increased sales. The shop is already crowded."

"The problem with tasting days is that I have to actually drink the stuff." Juliana groaned. "The sacrifices I make for my business."

The phone rang just as Juliana was about to follow Sandy out of the office. She reached over and grabbed the receiver, hoping Travis would be on the other end of the line.

"Oh, hello, Melvin."

"You sound disappointed. And after all I've done for you."

Juliana chuckled. "You caught me on the way to my lunch-hour coffee tasting."

"I won't keep you. Just wanted to let you know I've got that aged Sumatra for you. I'll deliver it this afternoon."

"Great. I've got 'em standing in line for it. Customers seem to go for the word *aged*."

"Makes 'em think of fine wine, I guess," Melvin said absently. "Although there's no comparison between aged coffee and aged wine. Most aged coffee is kind of flat tasting. This Sumatra's not bad, though. Nice, heavy body. Should blend well. How are the plans for the tea shops going?"

"I'm still discussing the concept with my partner."

"Which, translated, means you still haven't sold him on the idea? I'm surprised at you, Juliana. What is it with this fiancé of yours? He must have a will of iron if you haven't managed to whip him into shape by now. Can't imagine any man holding out this long in an argument with you."

"We're not arguing about it, we're discussing the possibilities," Juliana snapped, irritated. "You make me sound like a shrew, Melvin. One of those tough, hard-edged, aggressive business-women men always dislike."

"Hey, don't put words in my mouth," Melvin said hastily. "I only meant that you're a very forceful lady and when you go after something, you usually get it, that's all. I'm just surprised this guy you're engaged to hasn't thrown in the towel and acknowledged the brilliance of your tea shop concept yet, that's all."

"Goodbye, Melvin," Juliana muttered. "Make sure that Sumatra gets here by three o'clock, or I'll find myself another supplier." She tossed the receiver into the cradle and stood glowering at Sandy.

"Something wrong?" Sandy asked politely.

"Tell me the truth. Do you find me forceful? Even a tad aggressive, perhaps? The sort of female who usually gets what she wants, no matter how many hapless males get in her way?"

Sandy grinned. "Definitely. And I want you to know I admire you tremendously. I consider you my mentor. When I grow up I want to be just like you."

Juliana smiled brilliantly. "Good. Glad I'm not losing my touch. For a while there I worried that being engaged might have softened my brain a bit. Let's go drink some coffee."

Eight

Travis was poring over the papers he had spread out on the kitchen table when he heard the refrigerator door open and close in a stealthy fashion.

Out of the corner of his eye he watched Juliana pry the lid off the container she had just taken out of the freezer compartment. She had been working quietly at the kitchen counter for several minutes now, her back to him so that he could not see precisely what she was doing. All he knew was that it had something to do with a banana.

"I give up," he said, tossing down his pencil. "What are you doing over there?"

"Fixing you a little something special. I think it's about time you took a break. You've been working there since we finished dinner." She did not turn around but it was obvious she was very busy.

Travis exhaled heavily. "I think it's about time I took a break, too."

"Any word from Bickerstaff today?"

"No."

"Does that mean yes or no?"

"It means," said Travis, "that he's still considering it."

"Good. I'm sure he'll go for it."

Travis shook his head, awed, as usual, by her boundless faith in him. He could guess what would happen when that faith was shattered. Juliana was a businesswoman. She would expect him to live up to his end of the deal they had made, and if he didn't . . . "How are the engagement party plans going?"

"Great." Juliana opened a cupboard and removed a package of nuts. "Everything's all set for the fourteenth at The Treasure House. Be there or be square."

Travis groaned. "Why The Treasure House?"

"For sentimental reasons, of course. That's where you proposed."

"Not precisely. Unlike every other male who proposed to you in the restaurant, I showed some creativity. I took you down to the docks, remember?"

"Details, details. You're just irritable on the subject because you fell in the water that night." She reached for a jar of chocolate sauce.

"I did not fall in the water, I was thrown in. What are you making?"

"I told you, it's a surprise. Just be patient. I talked to Melvin again today."

"Yeah?" Travis mentally girded his loins for battle. He knew what was coming next.

"He wanted to know how the plans for the tea shops were going. I explained you and I were still discussing the concept but that things are moving forward rapidly."

"The hell they are," Travis said mildly. "They haven't moved forward one inch and you know it. You are not opening a tea shop, Juliana, and that's final."

"You're just feeling a bit negative because you've got so many other things on your mind," she assured him. "We'll get down to details after this business with the resort is settled."

"We will never get down to details because there are no details to get down to. There will be no tea shops. I would be worse than a fool, I would be criminally negligent in my responsibilities if I allowed you to go ahead with your bizarre plan."

"You've said, yourself, I'm a very good businesswoman, Travis."

"You are. Very savvy and very realistic. A natural entrepreneur. Except when you get emotionally involved with something the way you did when you loaned money to Kirkwood and the way you're doing lately with the idea of a tea shop. When it comes to things like that, you let your personal feelings and emotions take over. That's a bad way to do business and we both know it." Just look at the situation he was in because of personal feelings and emotions, Travis thought as he glanced bleakly at the paperwork in front of him.

"I really think there's a wonderful potential for the tea shops," Juliana said resolutely, opening a jar of maraschino cherries.

"There is no potential for the tea shops."

"I can make them work."

"Nobody could make them work. You can add a line of tea at Charisma but that's the end of the tea business for you."

"I appreciate your consulting expertise," Juliana said, an edge on her words. She opened a drawer with a jerk and snatched a spoon out of the silverware tray. "I assure you I will bear your comments on the subject in mind as I make my decision."

"You can't make any decisions of this magnitude without me," Travis reminded her quietly. "I'm not just your business consultant, I'm your partner. Remember?"

"I remember. Believe me, I remember." Juliana turned toward him, her culinary masterpiece held in both hands. Her eyes sparkled militantly. "But you're going to have to realize that I got where I am with Charisma all on my own. I know what I'm doing."

"Most of the time. But everyone's got a blind spot. Tea happens to be yours. Along with Flame Valley, of course."

"My, you are in a grouchy mood tonight. Maybe this will perk you up."

Travis's attention wavered from the argument as he studied the most spectacular banana split he'd ever seen. Three giant scoops of ice cream resided between halves of a plump banana in a large glass bowl. All three scoops and the ba-

nana were lavishly glazed with chocolate top-
ping, nuts, whipped cream and three cherries.
Travis's mouth watered.

"Maybe," he agreed.

"How does it look?" Juliana asked expec-
tantly. She nudged aside some papers and put
the concoction down on the table in front of
him.

"I haven't seen anything like this since I was
eight years old and the one I ordered then
wasn't nearly this big." Travis picked up the
spoon and wondered where to begin.

"I decided you needed some quick energy."
Juliana reached across the table and dipped the
edge of her own spoon into one of his scoops of
ice cream. She popped the bite into her mouth.
"Now, about the tea shops."

The lady was as tenacious as a terrier, Travis
thought, not without a sense of grudging admi-
ration. "I've told you before, forget the tea
shops. You're fated to get rich as a coffee mer-
chant." He carefully chose his first mouthful of
ice cream and nuts.

"I want to try the tea shops, Travis."

"Look. When I've got some time, I'll sit down
with you and show you just why you won't make
any money with tea shops, okay? Right now I've
got my hands full trying to save Kirkwood's
rear."

"Damn it." Juliana jumped to her feet, eyes
suddenly ablaze. "Talk about a blind spot. You
won't even listen to me."

Travis scooped up another bite of ice cream. "I've listened to you. Your idea is lousy. As your partner, I'm not going to agree to allow you to go ahead with the plans. That's all there is to it."

"Well, I am going through with my plans and that's final," she hissed, her hands on her hips.

"You're not going to do a thing without my approval."

"You can't start giving me orders, Travis. As far as I'm concerned you haven't earned your fee yet. You haven't saved Flame Valley and until you do, you're not a real partner in Charisma."

"Yes, I am. We already agreed that the partnership was my fee and I would collect my fee regardless of how successful my efforts were with Flame Valley."

Juliana folded her arms under her breasts and stood defiantly, feet braced slightly apart, in the middle of the kitchen. "Charisma is mine. I created it and I made it what it is today. Even if you're a partner in it, you're the junior partner. Don't ever forget that, Travis. I make the decisions about the future of my business and that's the end of it. Don't think that just because you're going to marry me you can start telling me what to do."

Travis sighed. He had known this showdown was coming. He just wished it hadn't arrived tonight. He had too many other things on his mind. "And don't think that just because you got me to propose to you that you can lead me around like a bull with a ring through its nose," he said evenly.

"Why, you mule-headed, stubborn, hard-nosed son of a . . . You do remind me of a bull. A very thick-headed one." Juliana turned on her heel and stalked out of the kitchen.

A moment later the bedroom door slammed shut.

A fine example of high dudgeon, Travis decided as he reluctantly went back to work. He wished he'd gotten in a few more bites of the ice cream before Juliana had exploded. He hadn't even started on the banana. A slow grin edged his mouth. Juliana was the kind of woman who would keep a man young or wear him out. Either way, he would never be bored.

Juliana reappeared an hour later. Travis's eyes narrowed as he slowly became aware of her presence behind him. He turned his head and saw her lounging with sultry insouciance in the kitchen doorway. She was wearing what he privately considered her sexiest nightie, the black see-through one with the small lace flowers strategically placed over the relevant portions of her anatomy. Her mass of red hair frothed around her shoulders. Her feet were bare and her eyes were luminous.

"I've decided to forgive you," she said, her voice husky.

Travis felt desire seize his insides. "This must be my lucky day."

"I shouldn't have tried to discuss the tea shop idea with you tonight. You're much too involved with Flame Valley right now to be bothered with other business decisions."

Travis decided this was not the time to tell her that his opinion on the tea shops was not likely to alter regardless of how busy he was. "You're sure you're not just trying to use sex to get me to see the brilliance of your plan?"

She smiled with glowing innocence. "I would never stoop to that sort of tacky behavior."

"Too bad. I've always wondered what it would be like to be the victim of that sort of behavior."

Juliana held out her hand invitingly. "We could always pretend."

"Yeah. We could. We're pretty good at creating fantasies together." Travis got to his feet and went toward her. She was fantastic, he thought. He'd never before met anyone quite like her, and he knew deep inside he never would again. *He must not lose her.* He would cling to this fantasy with all his strength.

"Travis?" She was still looking at him with smoky sensuality, but there was a trace of concern in her gaze as she studied his face. "Is something wrong?"

"No," he muttered as he came to a halt directly in front of her. "Nothing's wrong."

But there was and he knew it. Every day that passed without a response from Bickerstaff or any of his other contacts meant losing one more piece of the small chunk of whatever hope he had of saving the inn. But he couldn't think about that tonight, Travis decided, not with Juliana standing here, inviting him to make love to her.

He kissed her, startling her a little with his sudden urgency. She hesitated a split second and then responded, as she always did, with everything that was in her. There was nothing like being wanted by this woman, Travis reflected, his hunger for her soaring.

"Do you feel yourself changing your mind about the tea shops yet?" Juliana whispered teasingly against his throat.

"No, but I definitely feel lucky." He was taut and heavy with his desire. He began nibbling on her ear. "Is this the way you're going to forgive me every time we argue?"

"Probably. I'm not the type to hold a grudge." She unbuttoned his shirt slowly until it hung open to reveal his chest. Then she ran her fingers through the crisp hair. Her eyes were soft with a woman's sweet need as her nails lightly circled his flat nipples.

"No," he acknowledged softly, "you won't hold grudges, will you? That's not your way. You'll yell at me for a while, slam a few doors and then put on a sexy nightgown and seduce me. I won't stand a chance."

"Putty in my hands," she agreed, pressing her breasts against him. "Does the thought make you nervous?" She unzipped his pants and unbuckled his belt.

"I'll take my chances." His hand slid down over her soft, curving belly. He found the small lace flower that barely concealed the triangle of red curls at the apex of her thighs. His palm closed over the flower and a glorious sense of satisfaction roared through him when Juliana moaned and melted against him.

He caressed her intimately through the filmy material of the gown until he felt her growing hot and damp. Her fingers slipped inside his pants and he groaned as she touched him. When he could take the subtle torture no longer, he picked Juliana up and carried her into the bedroom and put her down onto the bed.

A moment later, his own clothes in a heap on the floor, Travis slid into bed, gathered Juliana into his arms and rolled onto his back. She sat astride him taking him deep within her warmth. Her legs pressed demandingly against his thighs.

Travis reached up to cup her breasts, coaxing the nipples into tight, sensitive buds. When he grazed the delicate peaks with his palms Juliana caught her breath and stiffened. Travis could feel her tightening around him and it was all he could do to muster some remnants of self-control.

Juliana began to breathe more quickly and her head tipped back. Her hair flowed in a silken wave around her shoulders. Travis waited until he couldn't stand it any longer and then he moved, easing Juliana down onto her back. When she reached for him, pulling him close once more, he slid all the way into her heat and surrendered to the fabulous oblivion.

A long time later Travis felt Juliana stir in the shadows beside him.

"Are you awake?" she asked softly.

"Umm." He had been unable to sleep, wondering if he should put in another call to Bickerstaff's office in the morning. But he didn't want to look too anxious, he told himself. Bickerstaff would get skittish.

"I shouldn't have lost my temper tonight just because you don't agree with me about opening a tea shop. The thing is, Charisma's always been mine. I've always made the decisions, all the decisions, about its future."

"I know." He let his hand drift over her curving thigh.

"I guess it's like having raised a kid all by yourself and then marrying and having to let someone else have a say in the kid's future."

Travis said nothing. As usual, he felt himself edging away from any topic that was even remotely concerned with children. One of these days he would have to deal with the subject, he told himself. But he intended to put it off as long as possible.

"Travis?"

"Yeah?"

"I know you mean well and I know you think you know what's best for Charisma but . . ."

"But you don't like having me tell you what's

best for Charisma when it conflicts with what you want, right?"

"Right."

"Don't take it personally, Juliana. Don't let emotion enter into your decision-making process. Charisma is business. Keep it that way."

"Sometimes the two get mixed up, don't they?"

Travis thought about his current situation. Revenge, business and desire were irrevocably entwined into a knot he was not at all certain he could unravel.

"Yes," he said quietly. "Sometimes they get mixed up."

Juliana waited another two days before she introduced a subject that had not yet been discussed and which she had hoped Travis would bring up first. So far he had not done so and she was, as usual, too impatient to wait for things to happen in their own time.

She decided to do it very casually. Craftily she waited until she had coaxed him away from his desk for a walk on the beach. She led up to her topic slowly.

"You don't mind living in my condo until we decide where we're going to live permanently?" she asked.

"Your condo's fine." Travis's hand tightened around hers as he paced barefoot beside her on the damp, packed sand at the water's edge. "As far as I'm concerned, we can live there permanently."

"It's a little small."

"Plenty of space for two people." He sounded unconcerned. "I'm just about moved in now as it is and everything's working out fine."

That much was true. Juliana still found it something of a novelty to open the closet and discover a row of conservatively tailored men's white shirts hanging inside, but she was adjusting. She had complained briefly about Travis using up all the hot water during his morning showers but he had resolved the issue by making her share the shower with him. So far none of the problems of living together had been anything more than a minor challenge.

Obviously commenting on the small size of the condo was not going to open up the area of discussion that was foremost on her mind today. She would have to find another approach. *Subtle,* she told herself. *Keep it subtle.*

"Everything is all set for the engagement party," she reported. "Seven o'clock this Friday night. Just about everyone who was invited is coming. Even my parents are coming down from San Francisco. I've had several talks with the chef at The Treasure House, and the food is going to be fabulous."

"Fine."

The neutral tone of his voice disturbed her. Lately it seemed to Juliana she had been hearing that tone more and more from Travis. She sought for a way to ease the conversation from engagement party plans to the more important subject and gave up. So much for being subtle.

She couldn't wait any longer to bring up the one undiscussed subject that remained. She would have to take the bull by the horns.

"So," Juliana said boldly, "what would you say if I told you I was pregnant?"

It didn't take long for her to sense she had made a mistake.

Travis came to an abrupt halt and spun halfway around to face her, his face rigid with anger. "You're *what?*"

It dawned on Juliana that she had obviously thrown the poor man into shock. "I was just wondering how you would feel if it turned out I was . . ."

"You're not pregnant," he cut in swiftly. "You can't be pregnant. We've been taking precautions. There haven't been any accidents."

"I know, but . . ."

"Are you telling me you are pregnant?" he asked through set teeth.

"No, no, it's okay, Travis. You're right. I'm not pregnant. It was just a hypothetical question."

"A hypothetical question? Are you nuts? You don't throw hypothetical questions like that around. What the hell got into you?"

"All right, so I didn't phrase it very well."

"No, you did not."

He stood looking at her with an expression that rocked her as nothing else had done since the moment she had watched him confront her cousin Elly that night on the terrace. Juliana pulled herself together quickly.

"I'm sorry, Travis," she said quietly. "I didn't mean to alarm you. I just thought it was time we talked about children. It's something we haven't discussed yet."

His unreadable gaze searched her face and then shifted to the ocean horizon over her left shoulder. "No, we haven't talked about children, have we? I somehow got the impression you weren't particularly interested in having kids. Ever since I've known you, you've seemed wrapped up with your plans for Charisma. You never said anything about wanting to have babies."

"I hadn't thought about babies very much before I met you," she admitted, realizing for the first time the truth of that statement. "There was never a time or a man that made me think about having kids. But now there's you and we're getting married and neither one of us is exactly young and, well . . ." The sentence trailed off.

"And you've decided you want children." Travis closed his eyes wearily and then opened them again. His crystal gaze was more unreadable than ever.

Juliana drew a deep breath. "Are you trying to tell me you don't?"

He began to massage the back of his neck. "This is a hell of a time to bring up the subject."

"What better time?" She studied him anxiously. "If you don't want children you should tell me now, Travis."

"Kids complicate things."

"Living is a complicated business. What sort of complications are you worried about?"

"Damn it, Juliana, you know what the complications are. Don't act naive. If things don't work out between us, we don't want to hurt anyone else, do we?"

She sucked in her breath. "You're already looking ahead to a divorce?"

"No, of course not. It's just that these days people have to be realistic. Half of all marriages fail, and there are probably a lot of others that would collapse if given a slight push."

"So what are you suggesting? That people stop having babies?" she snapped.

"I'm suggesting they give the matter a lot of thought before they go ahead with something as irrevocable as the decision to have a child," he muttered, resuming his pace.

Juliana hurried to catch up with him. "I agree with you, Travis. Babies should be planned and wanted. No question about it. But if two people are sure of their commitment to each other and if they both want children, then they shouldn't be afraid to go ahead and have them."

"Do we have to discuss this now, Juliana?"

Her palms were damp, Juliana realized vaguely. A sick feeling slashed through her. For the first time since she had met Travis she questioned her own judgment. Had she chosen the wrong man, after all?

"No," she said. "We don't have to discuss this now."

"Good." He glanced at his watch. "Because I've got to get back to the office. Bickerstaff still hasn't called and there are one or two other people I want to contact this afternoon."

"Sure. I understand. I'd better get back to Charisma, too. Lots to do today. I want to talk to my staff about trying a darker roast on some new beans I bought." She tried a bright smile on for size and thought it stayed put fairly well.

Travis slanted her a brief glance, nodded and turned back up the beach to where Juliana had parked her car. Little was said on the drive back into town.

Juliana dropped Travis off at his office and drove very slowly back to her apartment. She would think about trying a darker roast on the new coffee beans some other time.

She parked the red coupé in front of her apartment and went inside. The first thing she did was put a kettle of water on the stove. The second thing she did was not answer the phone when it rang two minutes later.

When the tea was ready she carried it over to the kitchen table and sat down. She was still sitting there, staring out the window, when she saw Elly's car pull into the parking space beside the coupé.

The phone had been easy to ignore but there was no point ignoring the knock on the front door. Juliana knew there was no way she could pretend she wasn't inside.

"There you are," Elly said as Juliana opened

the door. "I stopped at Charisma but your staff said you hadn't returned, and when you didn't answer the phone I thought I'd just stop by on my way back to Flame Valley. What's wrong?"

"Nothing's wrong. Why do you ask?"

"Don't give me that. You never come home in the middle of the day." Elly stepped past her and went straight into the kitchen. "And you're having a cup of tea all by yourself. What's going on around here?"

"Elly, please, I'm just a little tired. I'm not feeling very sociable."

Elly peered at her. "Something's wrong, isn't it? Don't bother trying to lie, Juliana. We've known each other too long."

"I've had a rough day." Juliana sat down at the table and picked up her teacup.

"So have I. That's one of the reasons I wanted to talk to you today. I'm getting very nervous about David and what's going to happen if Travis doesn't pull off the deal with Bickerstaff."

Juliana nodded without much interest. "I know you're worried, Elly."

"But," Elly continued quietly as she sat down across from her cousin, "at the moment I am a lot more worried about you. You're not acting like yourself, Juliana."

"How can you tell? You just walked in the door."

"I can tell. You're normally as vivid as a neon sign. Right now you look as if someone has just unplugged you."

In spite of her morose mood, Juliana managed a flicker of a smile. "Not bad, Elly. A good analogy."

"It's Travis, isn't it? Tell me."

"There's not much to tell. I'm just wondering if I'm making a mistake. That's all."

Elly's brows rose. "I don't believe it. After all my ranting and raving failed to deter you, after the lectures from your parents and my father went unheeded, *now* you're suddenly wondering if you've made a mistake? That's a shock. All right, let's have it. What went wrong today?"

"I asked Travis how he'd feel if I got pregnant. He was furious."

"Are you?" Elly asked sharply.

"Pregnant? No. It was just a hypothetical question. I wanted to find a subtle way to introduce the subject of babies."

"Men aren't big on hypothetical questions," Elly observed with unexpected insight. "Or subtlety, either. You probably gave the guy the shock of his life. There he was, not thinking about anything except trying to salvage Flame Valley and you hit him up with something like that."

"He didn't calm down when I explained I just wanted to discuss the possibility of having children." Juliana met her cousin's eyes. "I don't think he wants kids, Elly. I think he wants to hedge his bets."

"What do you mean?"

"I get the feeling a part of him doesn't really

201

expect our marriage to work. I think that deep down he doesn't want to have any loose ends around if things collapse. Probably because, as a kid, he was a loose end, himself."

"I think I'm beginning to see the problem. But, frankly, I'm surprised you're letting it get you down. You usually rise to a challenge the way a fish rises to bait. I've always envied your talent for self-confidence. Nothing really shakes it, not even when you lose once in a while. You just reorganize and bounce back. You're always so strong, Juliana. Everybody in the family says that, you know. We all see you as the strong one."

"I don't feel strong now. If you must know the truth, I feel scared. I was so sure of him, Elly. So sure he was the right one for me. I knew it the day he walked through the front door of Charisma. I practically jumped on him then and there and told him he was going to marry me. I knew I'd been waiting for him all my life. It was all I could do to control myself until the night he . . . we . . . the night we went to bed together for the first time."

Elly studied the table for a moment. "You were still sure of him even after he and I had staged that scene out on the terrace at Flame Valley, weren't you? You were mad at him, but still certain he was the right man for you."

Juliana nodded. "It's true. I was very annoyed with him that night. Furious that he'd tricked me about his past relationship with my family

202

and even more upset to discover he'd once wanted to marry you. I mean, it's obvious you and he are all wrong for each other."

"Obvious."

"But I figure everyone's entitled to a mistake or two. Heck, I've made a few small ones, myself."

"Very understanding of you."

"And he soon saw the error of his ways," Juliana continued. "Didn't he turn right around and agree to try to save the inn and didn't he ask me to marry him?"

"True. That's just what he did."

"But this thing about the babies has shaken me, Elly. This is a different matter entirely."

"Not everyone wants to have children, Juliana. You've never shown much interest in them yourself until today."

"But I've always known that when the right man came along I would want to have them. There was never any question in my mind."

"It takes two to make an important decision like that."

"I know." Juliana sighed. "If it were a simple matter of Travis not wanting to be a father, I might be able to understand. But there's more to it than that. He talked about not wanting to see kids hurt in a divorce. He talked as though one went into a marriage planning for the worst possible case."

Elly sat back in her chair, frowning. "Whereas you, with your boundless certainty and enthusiasm are going into it prepared to give it your all."

Juliana looked up, feeling raw and very vulnerable. "Exactly. Elly, I can't marry a man who isn't as committed to making the marriage work as I am. I won't marry a man who feels he has to hedge his bets just in case things don't work out."

"Be reasonable. What do you expect from Travis? He's a businessman, Juliana. And I've told you from the very beginning there's a cold-blooded streak in him. He's not an emotional creature like you. If you ask me, it's perfectly in character for Travis to hedge his bets. Be grateful he's got enough integrity not to want to leave you holding the baby, so to speak."

Cold-blooded? Travis? Never. But it was true he was a businessman and he could be incredibly stubborn. Juliana reminded herself of how difficult he became every time she brought up the subject of the tea shops. Perhaps he was looking at marriage the same way he would look at a potential business investment.

The thought was enough to make her nauseous.

"Juliana? Would you like another cup of tea?" Elly got quickly to her feet. "Here, I'll make it for you."

At any other time the notion of Elly reversing roles with her and becoming the reassuring, bracing one would have made Juliana laugh. But when Elly put the fresh cup of tea down in front of her, she could only feel wanly grateful.

"Thank you," Juliana murmured.

Elly sighed. "I suppose I should be encouraging you to have these second thoughts. After all, I'm the one who's been warning you not to marry the man. But for some insane reason I can't bear to have you think you've made such a horrendous mistake. It's just not normal for you to be acting like this. You've got to get a grip on yourself, Juliana. Depression doesn't look good on you."

"I know." Juliana sipped her tea. It was too weak but that didn't seem to matter today. Nothing really mattered today except that she was staring at the possibility of having been totally wrong in her estimation of Travis Sawyer. He was afraid to have kids because he was afraid the marriage would end. Which meant that he was not really committed to making it work.

"Juliana? Feeling any better?"

"No."

"Oh, Juliana, I'm so sorry."

Juliana stared unseeingly out the kitchen window. "What am I going to do, Elly?"

"I suppose you may have to consider calling off the wedding, if you're feeling this uncertain of the future."

Juliana gripped the teacup. "Heaven help me, Elly. I don't think I have the courage to do that."

Nine

"Juliana? I'm sorry, but it looks like I'm not going to be able to get away from here in time for dinner tonight."

"That's two nights in a row, Travis. Is something happening with Flame Valley?"

"I'm not sure yet. Maybe."

"You don't sound encouraged."

Probably because I'm not, Travis thought. "I don't want anyone to get his hopes up. Look, I'd better get back to it. I don't know how long I'll be." He waited, praying she would tell him that it didn't matter how late he was, she would be expecting him when he was finished at the office.

"You'll probably be exhausted when you're done for the evening."

"Yeah. Probably." Travis's fingers tightened on the phone as he prepared himself for what he sensed was coming next. Juliana was going to say the same thing she had said last night when he'd called her to tell her he'd be late. *"You'll probably want to go straight home to your place and fall into bed."*

"Don't bother stopping by here. I know you're

tired. You'll want to go straight home to your own apartment and collapse into bed," Juliana said with far too much calm understanding for Travis's taste.

"Yeah. I was just thinking that might be best. I'll see you tomorrow, honey."

"Fine."

"Everything on line for the party Friday night?"

There was a slight pause before Juliana spoke. "Yes. Everything's fine."

Travis could feel her sliding away. He gripped the phone harder, frantically searching for a way to keep her on the other end for just a bit longer. "Did you find a dress yet?"

"No. Elly insists I give it one more try tomorrow. She's going with me. I told her not to worry about it. If I don't find something new, I can always dig something out of my closet."

Travis closed his eyes in bleak despair as he heard the lack of enthusiasm in her voice. He knew that under normal circumstances Juliana would never have said such a thing. She would have been searching California from one end to the other for a new gown to wear to her engagement party.

"Good luck shopping," Travis finally said, knowing he was the one whose luck was running out faster than water through a sieve. "I'll try to stop by Charisma tomorrow for a cup of coffee."

"All right. See you tomorrow." Juliana hesitated. "Good night, Travis."

"Good night."

Travis slowly hung up the phone and watched as night enveloped Jewel Harbor on the other side of the floor-to-ceiling windows. The darkness looked cool and comforting — velvety soft. A place to hide. Inside his office everything was fluorescent bright and there was no place to hide from the failure he saw looming on the horizon.

That failure had been crouching there all along, of course. He had caught glimpses of it right from the start when he had first agreed to try to save the inn. But Juliana's indestructible faith in him had somehow obscured reality for a while. Even as her breezy confidence in him had irritated him, it had buoyed him.

Hell, for a while there, he had almost believed he could pull it off.

But Juliana hadn't expressed any of the familiar, serene assurance in his abilities for the past two days. There had been no bright, bracing lectures on how everything was going to work out.

There had also been no call from Bickerstaff.

But Travis knew the cold feeling in his gut tonight didn't come from facing the harsh business reality; it came from having to face the fact that Juliana was distancing herself from him before disaster even hit.

She had been growing cool and remote since that day they had walked on the beach, when she had asked him what he would do if she told him she was pregnant.

Since then, Travis reflected, she had stopped trying to pin him down to a wedding date. There had been no talk of the tea shops. And now Juliana seemed to have lost interest in shopping for a new dress for the engagement party. The signs couldn't get more ominous than that.

His bright, vibrant, enthusiastic Juliana was slipping out of his grasp even though nothing had happened to Flame Valley yet.

Travis reran the conversation on the beach in his mind for what had to be the hundredth time, trying to figure out what had happened to make Juliana grow cold.

His first thought this morning was that he had made a terrible mistake. Perhaps she had gotten pregnant accidentally and had tried to tell him, and his anger and refusal to believe her had hurt her deeply.

But that couldn't be it. He remembered he'd asked her point-blank if she was pregnant and she had denied it. She wouldn't have lied to him about a thing like that.

His second thought was that sometime during the past couple of days she had finally realized that the chances for saving the inn and the money she had in it, weren't good after all. Deep down, under all that flash and optimism, Juliana was still a realistic businesswoman. Maybe she was finally getting realistic about Flame Valley and its future.

And the bottom line was that if he couldn't rescue the inn, Travis knew he would automatically revert to the role of the bad guy. He had recognized that from the start. He was faced with only two options. He would be either the problem or the solution. There was no middle ground. If he didn't save Flame Valley, he would become the one who destroyed it.

Travis had tried to force himself to face that prospect from the beginning but somehow he had let himself believe some of the glowing press Juliana had insisted on giving him.

When he was with Juliana it was difficult not to get caught up in her enthusiasms, Travis reflected. But he had never had to deal with a Juliana who had lost her effervescent assurance.

It was beginning to look as if she had lost her faith in him.

Travis reminded himself that he had known from the start that if the chips were down, Juliana would side with her family. She would blame him for destroying Flame Valley and probably her cousin's marriage in the bargain.

The raw truth was that she would be right.

When it was all over, Travis knew he would be the outsider again. It was a role he had played often enough in the past and he recognized it immediately.

When choices had to be made, he got left out.

He tried to tell himself that maybe it was better this way. His relationship with Juliana had been on borrowed time from the start. Maybe it would be easier if she began to withdraw from him now.

But the idea of losing her before the final roll of the dice was more than Travis could endure. He would face the end when there was absolutely, positively no hope left of saving Flame Valley. Until then he was determined to grab what he could of Juliana's fire.

He loosened his tie and went back to work searching for the loophole that he knew didn't exist. It was ironic that he was going to walk away from this mess with the resort. After five years of being obsessed with the damned place, he now discovered he never wanted to see it again.

But one way or another, he decided, he would see that Juliana got her money back. It might take a while, but he would find a way to see she got paid off. He knew that wouldn't buy her back but it was the least he could do for her.

She had given him a great deal during the past few weeks and Travis prided himself on always paying his debts.

"Good grief, Juliana, you can't be serious about that outfit." Elly stared in shock as her cousin paraded forth from the dressing room.

"What's wrong with it?" Juliana glanced down the length of the demure winter-white crepe two-piece suit she was wearing. The long sleeves, high neckline and modest below-the-knee skirt were totally inoffensive as far as she could tell.

"What's wrong with it?" Elly's delicate brows

snapped together in a severe frown. "Are you out of your mind? That dress is not you at all. It's got no spark, no sizzle, no color. It's plain, plain, plain. It might look fine on the sweet angelic type or a traditional preppy type, but it's definitely not you."

Juliana felt a momentary flash of annoyance. "Well, you suggest something, then. I'm getting tired of trying things on."

"You never get tired of shopping and trying on new clothes."

"I'm tired of it today, okay?"

"All right, calm down. You're not yourself today, Juliana. Just settle down and listen to me. Go back into the dressing room and try that green and gold number, the one with the V-back cut to the waist."

Juliana heaved a sigh as her irritation died and was replaced by the now familiar sensation of disinterest. She trooped back into the changing room and reached for the racy green evening dress Elly had selected earlier.

As she smoothed the slender skirt down over her hips a part of her realized that the green and gold gown was done in the sort of dashing style that normally appealed to her. The deep V-back was at once elegant and daring. It ended in an outrageous bow at the small of her back. The snug outline of the skirt emphasized the curve of her hips and her long legs. For a moment or two Juliana almost got enthusiastic as she considered how a pair of rhinestone shoes she had recently spotted in a shop window would look with the dress.

But apathy set in again as she recalled exactly why she was buying the gown.

"Much better," Elly decreed as Juliana emerged from the dressing room. "In fact, perfect." She glanced at the hovering saleswoman. "She'll take it."

Juliana started to protest and then shrugged, not feeling up to arguing.

Twenty minutes later Elly led her out into the parking lot of the huge shopping center. The green dress was in a bag under Juliana's arm. A pair of rhinestone studded heels were in another sack.

"I've never seen you like this before, Juliana. You just aren't yourself today." Elly slid into the front seat of the Mercedes and turned the key in the ignition. "Things are really going bad in a hurry between you and Travis, aren't they?"

"How would I know? I haven't seen Travis in nearly three days. He's spent the last two nights at his old apartment."

"But he hasn't told you to cancel the engagement party, has he? He's a very assertive individual," Elly pointed out as she pulled away from the parking slot. "If he wanted to call off the party, he would do it. You, I presume, haven't changed your mind?"

Juliana stared out the window. "No. I've been telling myself things are terribly wrong and I should call it all off while I still can but I just can't get up the nerve to do it. I love him, Elly. What am I going to do if it turns out he doesn't love me enough?"

"I don't know." Elly eased the Mercedes onto the freeway, her expression sober. "I spend a lot of time asking myself the same question lately."

Juliana was instantly contrite. "You're still worrying about what David will do if you lose the inn, aren't you?"

"David and I haven't been communicating very well lately, to put it mildly. Not much better than you and Travis, as a matter of fact. He spends all his time either locked in his office or closeted with Travis. When he comes to bed at night, he falls asleep before I get out of the bathroom. The next morning he's gone before I get out of bed. I can't tell what he's thinking, but I know he's depressed and worried. I'm scared, Juliana."

"Join the club."

Travis was surprised at how hard he had to work to psych himself up to stop at Charisma Espresso on Friday. The fact that he had to work at the task at all alarmed him. He was accustomed to facing problems head-on but the problem of Juliana seemed to be unique.

It was noon, he reminded himself as he got out of his car. Only a few more hours to go until the engagement party. And no word yet from Bickerstaff. Time was running out on him fast. The closer Travis got to the bitter end, the more he perversely tried to believe Bickerstaff would call at the last minute and say he wanted in on the deal.

Talk about the irrational hope of the doomed.

Charisma was filled with people who were standing around with small cups and little notepads in their hands. Belatedly Travis remembered that this was one of the coffee-tasting days Juliana had inaugurated last month. Through the glass doors he could hear her behind the counter giving her lecture on coffee while Matt and Sandy poured sample cups.

Travis pushed open the door and stood quietly, listening to the windup of Juliana's talk.

"Always keep in mind that most of the coffee in your cup is water so you must pay attention to water quality. There's no point brewing a pot of coffee using water that doesn't taste fresh and good. Now, let's run through the three blends we tasted today. The first was the dark roasted Colombian. Remember that when you drink dark roasted coffees, you're tasting mostly the effect of the roasting process, not the specific characteristics of the beans used. The coffees taste stronger, but the caffeine level is actually about the same, sometimes even less than in lighter roasts."

Juliana looked a little wan today, Travis thought, frowning. As if she weren't feeling up to par. He wondered if she was coming down with a cold.

"The second cup we tasted was the Kona

blend. The coffee grown in the Kona district of Hawaii is the only coffee grown in the United States. Production levels are small but the coffee, at its best, can be outstanding. Medium acidity, smooth, clear flavor."

She not only looked a little wan today, Travis decided; she looked a little preoccupied. Usually when she played to a crowd of customers, she was like a good actress onstage, full of presence and definitely *on*. Today she seemed to be just going through the motions. She was a professional, however, and she conducted the coffee tasting with all the panache of a wine-tasting event.

"The third sample was a blend using chiefly Tanzanian arabica beans from the slopes of Mount Kilimanjaro. Tanzanian coffee is known for its excellent balance. I hope you noted the intense flavor and full body." Juliana smiled at her customers. "And that wraps it up for today, folks. I hope to see you all next week when we'll be trying out several coffees brewed by a variety of methods. We'll also be discussing some more coffee history."

Travis watched Juliana smile one last time at her audience, a pleasant enough smile, but it lacked that extra measure of brilliance he was accustomed to seeing in it. Then she spotted him at the back of the room, and for just an instant he thought her smile bounced up to its usual dazzling wattage. He wasn't certain because the effect didn't last long. The smile

216

slipped right back into the pleasant, polite level and stayed there as she came around from behind the counter.

"Hello, Travis. Taking a lunch break?"

"I want to talk to you."

Something that might have been fear flashed in her eyes but it disappeared instantly. "All right. I'm finished with the tasting. Let's go outside and sit at one of the courtyard tables."

He followed her as she wove her way through the milling crowd of people ordering freshly ground coffee at the counter. A minute later they emerged into the relative calm of the courtyard.

"Well? What is it, Travis? Having second thoughts about tonight?" Juliana asked with typical bluntness as they sat down.

"No. But I thought you might be having a few." Travis faced her across the table, willing the truth from her. He felt as if he were standing on the edge of a cliff.

"It's not as if we're getting married tonight," Juliana pointed out coolly. "It's just an engagement party. Nothing permanent. No reason to panic."

"Right. Are you sure you're not panicking?"

"I'm a little nervous, but I'm not panicked," she retorted with a burst of anger.

Travis nodded. "All right, calm down. I was just asking."

"Why?"

"Because you've been behaving a little

strangely for the past few days," he said quietly. "Since that day we took a walk on the beach, in fact."

"Oh. Maybe it's nerves."

He waited but when there was no further explanation forthcoming, he tried again. "Juliana, did I say something to upset you that day? If you're hurt because I came down on you like a ton of bricks when you implied you might be pregnant, I'm sorry. It was just that I was so sure you couldn't be pregnant, I was stunned to hear that you might be and I . . ." He let the sentence wind down into nothing. "I overreacted, I guess."

"Don't worry about it. I didn't bring up the subject in a diplomatic fashion, did I?"

"Someday we'll talk about kids," Travis promised.

"Will we?"

He nodded, hastening to change the subject. "Is there anything else that's worrying you?"

She looked straight at him. "No."

"I thought maybe you were concerned about the situation with Flame Valley."

"No."

Of course she wouldn't come right out and tell him she was wavering in her faith in him, not after all the buildup she had given him for the past few weeks. She would keep her growing uncertainties to herself.

"It's not looking good, Juliana," Travis felt obliged to say one last time.

"You've already told me that several times," she said impatiently.

Travis felt his temper fray. He got to his feet abruptly. "Yeah, I have, haven't I? Maybe it's finally sinking in. See you tonight, Juliana. Shall I pick you up?"

"No. I'll drive myself. I want to go to the restaurant a couple of hours early to make certain everything's in order." She jumped to her feet. "Travis, I didn't mean to snap at you. It's just that I'm a little tense."

"Yeah. Me, too."

He stalked out of the sidewalk café into the sunlight. When he reached the Buick he glanced back and saw that she was still staring after him. He thought he saw pain in her eyes and he almost went back to her. But even as he hesitated, uncertain of what to do or how to handle her, she turned and walked back into Charisma without a backward glance. He saw her dab at her eyes with a napkin she had picked up off the table and his stomach twisted.

Travis looked down into the yawning chasm below the edge of the cliff and wondered how it was going to feel when he went over.

He found out exactly how it was going to feel when he hung up the phone after the final conversation with Bickerstaff. It felt rather as he'd expected it to feel — as if the ground had just dropped out from under his feet. There was nothing to hang on to, nothing he could use to save himself.

It was over.

Travis wondered at the unnatural sense of calm he was feeling. He rubbed the back of his neck and glanced at the clock. A few minutes past seven. The engagement party was already underway and he was late. He wondered if Juliana would guess the reason why.

Feeling more weary than he could ever remember feeling in his life, he got to his feet, went around the desk and retrieved his jacket. No point going back to the apartment to dress for the occasion. He wouldn't be staying long at his engagement party.

"If this is any indication of the future, Juliana, you'd better be prepared to find yourself standing alone at the altar."

"Not a good sign, friend, when the future groom is late to his own engagement party."

"I can't believe it, Juliana. How could you plan everything right down to the shrimp dip and then forget to make sure your fiancé got here on time? Not like you, pal. The prospect of marriage must have addled your brain."

Juliana managed a smile as she endured another round of good-natured teasing. It had been like this since shortly after seven when guests had begun arriving and discovered that Travis was not yet there. Most were treating it as a joke, fully expecting Travis to walk through the door at any moment.

The only ones who showed any real evidence of concern were Elly and the other members of Juliana's family. The Grants, as a group, looked decidedly grim.

"Do you think you should call his office? Or his apartment? Something may have happened, Juliana." Elly spoke from right behind her cousin.

"He'll be here when he's ready," Juliana said, wondering at the odd sense of resignation she was feeling. She felt almost anesthetized, she realized. It was something of a relief after all the painful anxiety and uncertainty she'd been experiencing most of the week.

She took another look around the room and saw that everything was running smoothly, if one overlooked the minor fact that the future groom was not present.

The large room The Treasure House rented out for special occasions was festively decorated with silver balloons, colorful streamers and a wealth of exotic hothouse flowers.

The centerpiece was a magnificent buffet table that stretched almost the entire length of the room. It was laden with a staggering array of delicacies that included everything from garlic toast to shrimp brochettes. In a moment of nostalgia, Juliana had even ordered a bowl of guacamole and had stipulated it be set in a place of honor in the middle of the table.

The room was filled with laughing, talking

people decked out in typical California style, which meant that every conceivable variety of fashionable attire from silver jeans to elegant kimonos were represented.

David and Elly had arrived at six, volunteering last-minute services. It was the first time Juliana had seen David since the night she had invited him and Elly to dinner to tell them Travis was going to save the resort. One glance at his handsome face had warned her Elly was right; he did look worried. He tried to hide it behind his familiar genial smile, but Juliana knew David was deeply concerned. Elly was struggling just as hard to maintain a cheerful front.

Her parents and Uncle Tony were putting on a good front chatting with other guests but every so often one of them glanced toward the front door and scowled darkly.

Thank heavens she had decided to invite Travis's parents only to the wedding, not to the engagement party, Juliana thought.

She sighed to herself as she studied the crowd. How had she let things get this far? She ought to have called the whole thing off days ago, right after that fateful walk on the beach. She glanced at the clock for the fiftieth time. Seven-thirty. She wondered if Travis would bother to show at all.

Juliana was seriously considering the possibility of disappearing out the back door of the restaurant when she heard a murmur of awareness go through the crowded room. She swung around instantly, knowing Travis must have arrived. As she looked toward the door, her spirits lifted briefly. Hope died hard, she was discovering.

A roar of approval went up as Travis walked into the room. There was another round of teasing comments and congratulations and much laughter.

Travis ignored it all. He walked straight toward Juliana without glancing at anyone else in the room. He was wearing his familiar working uniform, a white shirt with the sleeves rolled up, a conservatively striped tie and dark trousers. He had his jacket hooked over his shoulder.

Juliana took one look at his grim, implacable face and knew that everything was lost.

She stood very still in the middle of the room as Travis paced toward her. She realized her hands were trembling. She folded them together in front of her. The crowd began to realize that something was wrong. The teasing became more muted and gradually disappeared. People stepped out of Travis's way and a hush descended.

Travis walked the last few steps in a charged silence. He seemed unaware of anyone else in the room except Juliana. His eyes never left her face as he came to a halt in front of her.

"I just got off the phone with Bickerstaff," Travis said in a terrifying cold, quiet voice. "It's all over. He doesn't want to get involved with Flame Valley. Too big a risk, he said. And he's right."

Juliana's mouth went dry. "Travis?"

"Sorry it went down to the wire like this. For a while it was close. You even had me thinking there was a chance, and I, of all people, should have known better. Bickerstaff was the last shot and he's out of it now."

"What are you trying to say?" Juliana demanded tightly.

"That I can't save Flame Valley from the wolf — can't save it from myself. Looks like I'll have my revenge, whether I want it or not. I just came here tonight to let you know that you don't have to call off our engagement. I'll do it for you and save you the trouble."

Travis turned on his heel and walked swiftly back out of the room.

Juliana stared after him, feeling as if she had just been kicked in the stomach. The layer of anesthetizing numbness that had been protecting her for the past few days began to crack. Underneath it lay a world of hurt.

Travis was walking out of her life.

"Juliana?" Elly came toward her quickly, keeping her voice low. "What's wrong? What did Travis say to you?"

"That Flame Valley's last chance just went down the tubes so he's calling off the engagement. Saving me the trouble of doing it myself, he said."

"Oh, my God." Elly closed her eyes. "What will David do?" Then the rest of Juliana's words hit her. Elly's eyes flew open. "Travis is calling off the engagement? Now? Tonight? Just like that? In front of all these people?"

"You've got to hand it to the man, once in a while he displays a real flair for the dramatic."

"Juliana, I'm so sorry. So very sorry. I didn't think it would end like this. I really didn't. Do you know, during the last couple of weeks I'd decided he really did love you, that he wasn't using you for revenge. I'd actually decided we'd all been wrong about him and you'd been right."

"He never once told me he loved me, you know," Juliana said wistfully. "I thought he was working up to it, though. I really did."

"What are you going to do, Juliana? All these people. All this food. The music. What will you tell everyone?"

The last of the numb feeling fractured and disintegrated. The pain was there, just as she had known it would be. But so were a lot of other emotions, including anger.

"How dare he do this to me?" Juliana said through her teeth. "Who the hell does he think he is? He's engaged to me, by heaven. And if he thinks he can walk out on me like this, he's got another think coming."

She started through the crowd toward the door.

"Juliana," Elly hissed. "Where are you going? What shall I tell the guests?"

"Tell them to enjoy the food. It's paid for."

Juliana dashed through the startled crowd of

guests and out the front door of the restaurant. She came to a brief halt on the sidewalk, scanning the parking lot for the familiar tan-colored Buick.

She heard the engine before she spotted the car. Travis was just pulling away from the curb.

"Come back here, you bastard . . . I said come back here." Juliana hiked up her skirt and ran at top speed across the restaurant driveway, no easy feat in her glittering high heels.

She cut through two rows of parked cars and reached the Buick just as Travis paused to glance over his shoulder to check the traffic behind him.

He did not see her when she threw herself on the hood of the Buick but he certainly heard the resulting thud. His head came around very quickly, and he stared at the woman in green sprawled across the engine compartment as if he had just seen a ghost.

"Juliana!"

"You're engaged to me, you bastard," she yelled back through the windshield. "You can't walk out on me like this. I deserve an explanation and I'm warning you right now, whatever that explanation is, it won't be good enough. Because we aren't just engaged, we're partners, remember? You might be able to end an engagement like this, but you can't end a business relationship so easily."

Travis switched off the engine and opened the car door. "I don't believe this," he muttered as he got out. "On the other hand, maybe I do. Get down from there, Juliana."

She ignored the order and stood up on the tan-colored hood, balancing a little precariously. She paid no attention to the marks her heels were leaving in the paint. She folded her arms and gazed down at him with fire in her eyes. "I'm not going anywhere until I choose to do so. I want an explanation for the way you're trying to end our engagement. You owe me that much, Travis Sawyer."

He looked up at her, the lines of his face harshly etched in the glare of the parking lot lights. "I gave you your explanation, Juliana."

"What? That business about not being able to save Flame Valley? That's no explanation, that's an excuse."

"Didn't you hear me? I can't salvage the damned inn for your precious cousin and your ex-fiancé and the rest of your family. Flame Valley is going to go under and there's nothing I can do to save it."

"Stop talking about that stupid resort. I don't care about it right now. Our engagement party is a hell of a lot more important."

"Is that right?" he demanded roughly. "Are you really trying to tell me you want to go through with marriage to the man who's going to be single-handedly responsible for destroying Flame Valley?"

"Yes!" she yelled back.

Ten

The only thing that made it possible for Travis to hang on to his self-control was the sure and certain knowledge that if he lost it now, he would never be able to regain it. He looked up at the magnificent creature standing on the hood of his car and felt the blood pounding through his veins. Her hair was a wild, crazy shade of orange in the glare of the parking lot lights. Her shoes sparkled garishly as if they'd been coated with some sort of cheap glitter dust, and the huge satin bow at the back of her green dress had come undone.

Travis knew he had never in his life wanted a woman as badly as he wanted this one.

"Juliana, listen to me. I'm the Big Bad Wolf in this story, remember? That resort has been in the Grant family for over twenty years. I'm going to tear apart everything your father and his brother built. I'm going to ruin your cousin and Kirkwood. And you're going to lose a big chunk of your savings in the process. This is bottom-line time. You have to choose sides whether you like it or not and *I'm on the wrong side.*"

"So you decided you'd make the choice for me? Forget it, Travis. I make my own decision."

His hand clenched into a fist. "You're going to hate my guts when you watch Elly and Kirkwood lose the resort."

"I could never hate you, although I might get madder than hell at you from time to time."

"Juliana, sometimes you have to make choices. You can't be on your family's side and my side, too, not in this. Don't you understand? You'll have to choose. I've already told you, I'm on the wrong side."

"I don't care which side you're on. That's the side I'm on and that's final. You can't get rid of me by telling me I have to choose between family and you. That's not how it is. Besides, I've already made my choices. I made them the day I met you. I chose you, Travis."

Travis took a step forward, coming up against the hard metal of the Buick's fender. He could have reached out and touched Juliana but he didn't dare. Not yet.

"Are you trying to tell me that you still want to get engaged to me? That you want to marry me? Even though I can't stop what's going to happen to Flame Valley?" he demanded. He could hear the rasp in his voice. His mouth was dry.

"Travis, for a reasonably intelligent man, you are sometimes awfully slow to catch on. Yes, that's what I'm trying to tell you. For crying out loud, I didn't fall in love with you because I thought you could save Flame Valley. I fell in love with you weeks before I knew anything about your connection to the inn."

"But after you found out about my connection to the inn things changed, didn't they?"

"I got mad but I didn't stop loving you. I've never stopped loving you. Besides, you offered to try to fix the damage. That was good enough for me. You made all the amends you needed to make."

"I didn't do a very good job of fixing things, did I?"

Her smile glowed. "That doesn't matter," she said, her voice suddenly husky. "You tried. If anyone could have saved Flame Valley, it would have been you."

Travis swore. "Trying isn't always good enough, Juliana."

"Yes it is. Most of the time, at any rate. And certainly this time it is."

"What makes it good enough this time?"

"Because you did it for me." She threw her arms open wide, and her smile was even more dazzling than usual. "And you did your best. You worked night and day to try to save the inn."

"But I didn't pull it off. Don't you understand?"

"You're the one who doesn't understand. You don't understand what your efforts meant to me. Nobody has ever even tried to do something on that scale for me before. They all think I can take care of myself. But you went to the wall for me, Travis. For *me,* not for Elly or David or my parents or Uncle Tony. You did it for me. It was for me, wasn't it?"

"Hell, yes, it was for you. If you hadn't been involved everything would have been a damned sight simpler, that's for sure. I'd have taken over Flame Valley without a second's hesitation and I would never have looked back."

"That's true. And you would have had every right. But you didn't do it because of me. Nobody does stuff like that for me, because I'm the strong one. Do you know how wonderful it is to have someone step in and try to save me?"

Travis was at a loss for words for a few seconds. All he could think about during that brief moment of charged silence was that until now no one had ever chosen him when a choice had to be made.

"Are you sure you want me?" was all he could manage to get out. "Your parents, Elly and David, Tony, they're all going to blame me for not being able to pull the fat out of the fire."

"They can blame anyone they want. We both know you did everything that could be done," she retorted vehemently.

"You're overlooking the fact that the fat was in the fire in the first place because of me," Travis felt obliged to point out.

"That doesn't matter. You had your reasons for doing what you did."

"Revenge is a good reason?"

"Well, certainly it's a good reason. You had a right to get even for what happened five years ago. One can hardly hold that against you."

"Your logic is incredible. But who am I to argue with it?"

Her smile was brighter than the parking lot lights. "Does this mean you're going to come back inside The Treasure House and celebrate your engagement to me in front of all those people?"

He touched the toe of one of her glittering high-heeled shoes. "Yes, ma'am, that's exactly what it means."

"Then what are we waiting for?" She held out her arms.

Travis felt the joyous laughter well up from somewhere deep in his gut. For a soul-shattering instant he knew the meaning of pure happiness. He reached out and scooped his lady off the hood of the car, paying no attention to the small paint scars left by her heels.

"You know something, Juliana? You make one hell of a hood ornament."

She laughed up at him as he set her on her feet. With great care he retied the huge satin bow at the small of her back. When he was fin-ished he traced the elegantly bare line of her spine with his finger. She was warm and silky and so magnificently feminine that he ached for her. But there were a lot of people waiting inside The Treasure House, he reminded himself.

He reparked the car. Juliana reached for his

hand as they started back across the parking lot but Travis forestalled her. He'd never felt more swashbuckling in his life. So he picked her up and carried her into the restaurant.

A cheer went up as Travis strode into the crowded room with Juliana in his arms. The band immediately struck up a waltz. Travis set Juliana on her feet and took her into his arms. He whirled her out onto the empty dance floor before she had quite realized what was happening.

"I didn't know you could waltz," Juliana murmured as the applause rose around them.

"Neither did I. But I think that tonight I could do just about anything." *Except save Flame Valley from myself.*

Out of the corner of his eye Travis caught glimpses of Elly and David and the Grants watching with anxious concern. By now they must have realized that their precious resort was history, he thought, but no one made a move to stop Juliana from dancing with him.

Then a sense of exultant satisfaction swept through him. Of course no one was going to try to come between him and Juliana. No one in his or her right mind got in Juliana's way when she wanted something, and she was making it very clear tonight that she wanted him.

Juliana had made her choice.

Several hours later Juliana was still humming a waltz to herself as Travis took her key and opened the front door of her condominium. He looked at her, amusement and something far more intense gleaming in his crystal eyes.

"Enjoy your engagement party?" he asked as he followed her into the hall.

"Had a lovely time. It was a perfect engagement party. The engagement party to end all engagement parties." She did one or two twirls on the carpet, enjoying the way her rhinestone shoes glittered beneath the green skirt of her gown. "What about you?"

Travis folded his arms and leaned one shoulder against the wall. He watched her dance around the living room. "It was a hell of a party, all in all."

"I thought so." She came to a halt in the center of the room and studied her long-nailed fingers. A diamond ring sparkled in the lamplight. "You even remembered the ring."

"I bought it right after I asked you to marry me. I've been carrying it around ever since."

"And you brought it to the party with you tonight even though you were only planning to stay long enough to say the engagement was over." Juliana smiled, feeling deliciously smug.

"I'd put it in my pocket earlier before I went to the office to call Bickerstaff one last time," Travis explained.

"Maybe it was your good-luck charm."

Travis's smile came and went. "If it was, it didn't do me much good when it came to dealing with Bickerstaff."

"Forget Bickerstaff. The business with the

inn is over." Juliana walked toward him. "Now that we've settled the little matter of our engagement, I think it's time we ironed out a few small details of our relationship."

Travis's brows rose. "Such as?"

She looped her arms around his neck and looked straight into his eyes. "I love you, Travis. Do you love me?"

Travis unfolded his arms and put his hands around her waist. His eyes were startlingly serious now. "I love you."

"Is this a forever kind of love or the kind that lasts until the divorce?"

He pulled her hips tightly against his thighs and kissed her hard on the mouth. "It's the forever kind."

She relaxed, believing him. "You've never said, you know. I got a little nervous there for a while."

"That day on the beach. The day we talked about babies. That's when you started getting nervous, didn't you? I could feel you retreating from me, pulling back emotionally. I thought you were finally beginning to realize I might not be able to save the resort."

"I got scared, all right, but not because of Flame Valley. It occurred to me for the first time since I had met you that I might be making a mistake. You didn't seem ready to make a complete commitment. You were afraid to talk about anything as permanent as a baby. That's what made me so nervous."

He lifted his hands from her waist to spear his

fingers through her thick hair. "I'll be honest, honey. The thought of having kids makes me uneasy."

"Perfectly understandable, given your background. That I can handle. We can work on your fears together. But I was worried that maybe you weren't sure of your commitment to me. And that terrified me."

"There was never any doubt in my mind about my feelings for you. But tonight when I talked to Bickerstaff and realized Flame Valley was going under, I didn't want to hear you tell me the engagement was off. So I decided to tell you first. I should have known you wouldn't let me get away that easily."

Juliana brushed her mouth against his. "Yes. You should have known. How could you do it, Travis? Would you really have walked away from me tonight and never looked back?"

"I figured I didn't have a prayer of marrying you, at least not anytime soon, but I sure as hell didn't intend to walk out of your life."

"Because you knew I'd come after you?"

His smile was slow and tantalizingly wicked. "No, not because I knew you were going to throw yourself on the hood of my car. I knew I'd be seeing you again because I'm your partner in Charisma, remember? We have a deal, you and I. I am supposed to collect my fee regardless of whether or not I saved the inn. There's nothing like business to make sure two people see a lot of each other."

She laughed up at him, delighted. "Very clever."

"A man has to be clever to stay one step ahead of you."

"Who says you're one step ahead of me?" she purred, liking the way her new ring flashed when she stroked her fingers along his shoulders.

"Right now I don't care which one of us is ahead of the other. I just want to get together. It's time we really celebrated our engagement."

His mouth closed once more over hers, and Juliana felt herself being lifted up into his arms for a second time that night.

"I'm a little big to be carried around like this," she murmured against his mouth.

"You're just the right size for me," he said as he carried her into the kitchen.

"Funny, I was just thinking the same thing. What are we doing in here? The bedroom is the other direction."

"Open the refrigerator," he ordered.

She did so and saw the bottle of chilled champagne inside. "Ah-hah. This beats crackers in bed anytime."

"Don't forget the glasses."

Juliana plucked two glasses off the counter and cradled them, along with the champagne, as Travis carried her down the hall to the bedroom. Inside the white-on-white room he set Juliana lightly onto the bed. She pulled her legs up under her green skirt, and the rhinestone shoes glittered in the shadows. Travis watched her as he stripped off his clothing. Then he reached out and took the bottle of champagne and the glasses out of her hands and put them on the nightstand.

"Definitely the world's most stunning hood ornament," he muttered as he came down beside her on the bed.

"You don't think I was a bit gaudy? For a Buick, I mean?"

"Juliana, my sweet, you're always in the best of taste."

He kissed her shoulder and simultaneously found the bare skin of her back. Juliana trembled as his rough fingertips traveled down the length of her spine to her waist. Then he slowly undid the satin bow.

She stretched languorously, her pulse throbbing with anticipation as Travis slowly lowered the bodice of the green gown. "Your hands feel so good," she whispered.

"You're the one who feels good. Soft and silky and sleek." He leaned down to kiss the slope of her breasts as he bared them.

A moment later the dress was on the floor along with the rhinestone heels and Juliana's filmy underwear. Her breath was coming more quickly now in soft little gasps as her senses reeled with the gathering excitement.

"You always go wild in my arms," Travis said, sounding thoroughly pleased and deeply awed. "You make me crazy with the way you want me. You know that, sweetheart? No one's ever wanted me the way you do."

"I've never wanted anyone this way before," she confessed, clinging to him as he slid his hand down toward her moist heat.

"We agreed the first time that what we have is special. We were right." He parted her gently and found her sensual secrets.

Juliana clutched his arms and arched herself against him. *"Travis."*

He raised his head to look down at her. "I'll never get tired of watching you when you're with me like this." He stroked his fingers into her and withdrew them with calculated slowness.

"Oh, Travis, I can't seem to wait, I want . . ." Juliana felt the small convulsive contractions seize her with little warning. She cried out.

"No need to wait," Travis assured her, pulling her hips close to his thighs again. "There's more where that came from and we've got all the time in the world tonight. Go wild again for me, sweetheart."

"Not without you." She reached for him, sliding one of her legs between his, searching out the heavy, waiting length of him.

When she touched him intimately, cupping him and caressing him, Travis sucked in his breath. After that there were no words, only soft sounds of growing need and spiraling desire.

Juliana was lost in the wonderland of passion when she felt Travis open her legs with his strong hands. Her nails sank into his shoulders as he thrust boldly into her, and the delicious shock set off another ripple of release that seemed to travel throughout her whole body.

She heard him mutter her name as he surged into her again and again, felt his body tighten under her hands and then there was only the marvelous sense of free-fall that always followed the peak of their lovemaking. Juliana welcomed it, losing herself in her lover's arms even as he lost himself in hers.

The rhinestone shoes lying on the white carpet glittered in the shadows, the small stones sparkling as if they were diamonds.

"Juliana?" Travis said her name softly in the darkness a long time later. He was sitting naked on the side of the bed, pouring champagne into the two glasses they had brought from the kitchen.

"Ummm?" She was feeling deliciously relaxed and content. She studied his broad shoulders and strong back with a loving eye. He really was just the right size, she told herself happily.

"If you did happen to get pregnant, hypothetically speaking, I'd be the happiest man alive."

"It's just too bad you couldn't appreciate the expressions on everyone's face last night when you went racing out the door chasing Travis and then returned ten minutes later in his arms. It was an absolutely priceless scene. A scene of legendary proportions, as far as the management of The Treasure House is concerned."

Elly sipped at her coffee latte and shook her head in wonderment.

"Sometimes a woman has to go after what she wants." Juliana savored the rich color of the Keemun tea in her cup and then automatically glanced around Charisma's pleasantly crowded serving area with an appraising eye.

Saturday mornings had typically been light until three months ago when she had put in a full range of newspapers and breakfast pastries. Customers had surged into the shop ever since on Saturdays to take their morning coffee and a croissant while reading something exotic and foreign like the *New York Times*.

"Your parents and my father were a bit stunned, I'll have to admit," Elly continued. "Especially after I told them that Travis was not going to save Flame Valley after all. You know what your father said?"

"What?"

"He told Dad to wait and see. 'It wasn't over until the fat lady sings,' I believe were his exact words."

"I hope the rest of you aren't holding any false expectations," Juliana said gently. "Travis said there really was nothing more he could do except find a buyer for the resort. That way, at least, we won't all lose our money. But you and David aren't going to own and operate Flame Valley, and the inn will definitely go into new hands."

"I know. David and I had a long talk last night

about our future," Elly said. "We settled a lot of things we probably should have settled much sooner."

Juliana frowned. "Well? How did it go? Are you still worried about him leaving now that the resort is dead in the water?"

Elly smiled sweetly, her eyes clear. "Oh, no. He never was thinking of leaving me. The poor man was scared I might leave him. That's why he was so uptight these past few weeks. Can you believe it?"

"Yup. I always knew the two of you were meant for each other." Juliana sat back. "So what are you going to do?"

"Well, one possibility, if Travis finds us a buyer, is to take the money and try the resort business again, this time on a much smaller scale. Maybe a little bed-and-breakfast place on the coast. Another possibility, according to David, is to see if Fast Forward Properties can negotiate us a contract with the new owner, whoever it is, to run Flame Valley. Under Travis's guidance, of course. We don't want to make the same mistakes we made last time."

"Would that bother you, Elly? To stay on at the resort after it goes to a new owner?"

"I think I could handle it and so can David. I'm not sure Dad or your parents will like it. It will probably gall them to see family members reduced to being just the managers of Flame Valley."

"On the other hand," Juliana pointed out,

"Uncle Tony and my folks aren't the ones who have to make the decision are they? It's you and David who have to decide what to do with your future."

"That's exactly what David and I told each other last night. I feel amazingly relaxed about the whole thing now that it's over. It's as if because of this mess David and I have finally stepped out from under Dad's shadow. Whatever happens now, our marriage will be the stronger for it."

"Uncle Tony always means well," Juliana said, "just like my parents always mean well."

"True. And we know they love us. That's the most important thing. But there's no denying they can be a little overbearing at times."

"Look at it this way, as irritating as the situation can be, it's better than winding up with parents who couldn't care less about what you do with your life."

"Who's got parents like that?" Elly asked in amazement.

"Travis."

Elly looked at her. "Oh. That's right. I suppose you have plans to fix the problem?"

Juliana smiled with cheerful confidence. "Let's just say I've got plans to give Travis's parents one more chance to dance at this wedding."

"And if they don't show up?"

"They'll show up."

"How can you be so sure?"

"Because thanks to a talk with Sandy I decided no more Ms. Nice Guy. I'm going to blackmail Travis's folks into showing up and behaving themselves."

Elly's eyes widened. "You're going to blackmail them? I should have guessed. You don't lack the nerve, Juliana. You always go after what you want, no holds barred."

"It's one of my best features," Juliana agreed. "Just ask Travis. Say, I was planning to start the big hunt for my wedding gown this afternoon after Charisma closes for the day. Want to come with me? I guarantee this is going to be one special dress."

"Now I know for certain you're back to normal," Elly said.

"Strange how some of the things you fear most in life aren't the ones you have to worry about after all, isn't it?" David took a long swallow of his beer and looked out over the harbor.

"Life's funny that way sometimes." Travis was sitting next to David at one of the outdoor tables of the Golden Keel, a trendy pub near the marina. Both men had repaired to the bar by mutual agreement to discuss the future of Flame Valley. But so far all they'd talked about was their relationships with the women in their lives. "I take it you and Elly have come to an understanding."

"She was worried I'd leave if we lost the inn. I think some part of her had never been completely certain I hadn't married her because of Flame Valley."

"Juliana was always sure the two of you were in love right from the start."

David chuckled. "Juliana is always sure of everything."

"Yeah." Travis grinned fleetingly. "And sometimes she's right."

"I've got to hand it to you, Sawyer. I don't know of any other man who could handle Juliana Grant."

"If any other man ever tries, I'll break his neck," Travis said calmly.

"That's assuming Juliana doesn't do it first."

"True. So what now, Kirkwood? You want me to try to find a buyer and see if I can get you a contract to manage the inn?"

David lounged back in his chair. "I've got a proposition for Fast Forward Properties."

"What's that?"

"How about letting me and Elly run the place after you take possession of the property? Hell, nobody knows the resort as well as we do."

Travis studied the condensation on his beer glass. "It wouldn't be the same as owning it," he warned. "I'll have a responsibility to my investors. It would be my job to make sure the resort got back on its feet. I'd be on you all the time. Looking over your shoulder. Watching every move. I'm good at what I do, Kirkwood, but the fact is, I'm hell to work for."

"I think I could deal with it." David looked

unperturbed. "Who knows? I might learn something about financial management from you."

"I'll think about it. Run it past my investors."

"Fair enough," David agreed. "So when's the wedding?"

"Juliana's got it scheduled for the end of the month."

"The end of the month? Why so soon? You're engaged and you're practically living together. Why is Juliana rushing the wedding?"

"It was my idea," Travis said. "I'm not taking any chances. When I sew up a business deal, I sew it up tight. No loopholes."

David grinned. "I know. You sewed up the deal on Flame Valley so tight even you couldn't find a way out. Juliana doesn't stand a chance."

"That's the whole idea."

Eleven

"He's going to stand her up at the altar, I just know he is." Beth Grant, looking every inch the mother of the bride in mauve lace and silk, paced back and forth in the small church anteroom.

"Uncle Roy and Dad will get a shotgun if Travis tries to duck out now," Elly declared with a small smile. She was fussing with the satin train of Juliana's gown.

"Relax, both of you. Travis isn't going to stand me up." Juliana scrutinized her image in the mirror. The sweetheart neckline had been the right choice, after all. She'd had a few second thoughts yesterday when she'd tried the gown on one last time. But today it looked perfect. The wedding dress was everything a wedding dress was supposed to be, spectacular, frothy and extravagant. She'd spent a fortune on it and didn't regret one dime.

"I'm not so sure Travis wouldn't walk out at the last minute," her mother said. "It would be the ultimate revenge on the Grant family, wouldn't it? First Flame Valley falls into his clutches and then he leaves you at the altar. Where is that man?"

"He'll get here on time if he knows what's good for him," Elly murmured, giving the train another small twitch. "He knows that if he doesn't show there'd be no telling what Juliana would do. He's still complaining about the scratches her heels left on the hood of his car."

"I'm not worried about Travis getting here," Juliana said, feeling perfectly calm as she bent closer to the mirror to check her lipstick. She wondered if she should have opted for a more vivid shade. The coral looked a tad pale. Then again, brides weren't supposed to go to the altar looking as if they'd just walked out of a makeup ad, she reminded herself. Angelina Cavanaugh had made that very clear to her a week ago. *"Tone it down, Juliana. Brides are supposed to look sweet and demure, not like an empress claiming her empire."*

"You don't look concerned about anything except your makeup," Beth sighed, watching Juliana. . . .

"Well, to tell you the truth, I am a little concerned about one thing. I was wondering if Travis's parents had arrived. Any sign of them?"

"I'll check with the usher," Elly said, heading for the door. She disappeared out into the hall, obviously glad to have something useful to do.

Beth came toward Juliana with a misty, maternal gleam in her eyes. She hugged her daughter briefly. "You look beautiful, dear. I'm so proud of you."

"Thanks, Mom."

"I will personally throttle that man if he doesn't show up."

"He'll show up." Juliana spoke with complete confidence. Nevertheless, she was touched by her mother's unusual protective instincts. "The thing about Travis is that you can always count on him."

Beth shook her head wonderingly. "I cannot understand how you are always so certain of him. As far as I can tell he's done absolutely nothing to warrant your total confidence in him. He didn't even save Flame Valley."

"He did everything that could be done. At least with him in charge of Fast Forward Properties the transfer to new ownership will be as smooth as possible. And David says he and Elly are going to get the contract to run it."

"I suppose that's better than nothing. Maybe someday we'll find a way to get the resort back into Grant hands. Your father still thinks there's every possibility this will all work out in the end. Even your Uncle Tony seems amazingly optimistic. But I still don't see how you could have had such faith in Sawyer right from the start."

"You don't really know him. I do."

"What makes you think you're such an authority on Travis Sawyer?"

"We have a lot in common," Juliana said simply, adjusting a straying curl beneath her veil. "We trust each other and we love each other. It's all really very simple."

"I wish I could be sure of that. I hope you know what you're doing, Juliana."

"I always know what I'm doing, you know that, Mom. Would you hand me my flowers, please? It's almost time."

Beth looked more anxious than ever as she handed the bouquet of exotic flame-colored orchids to her daughter. "You can hardly walk down the aisle in front of all those people if Travis isn't waiting at the other end. I won't allow it. The humiliation would be unbearable for the entire family."

"He'll be there."

Beth eyed her daughter's serene expression and smiled reluctantly. "Your complete faith in him is getting contagious."

"You'd better go take your seat, Mom. Dad and Uncle Tony will be getting nervous."

Beth looked at her. "You're absolutely certain Travis will show up today?"

"Absolutely certain."

The door burst open and Elly stuck her head around the corner. She was excited and amused. "They're here, all right. The usher says he just seated the groom's parents. All four of them. And a bunch of stepbrothers and stepsisters."

Juliana nodded, content. "Good. Another example of winning by intimidation. Mom, run along now. It's time."

Elly bit her lip. "Uh, Travis's folks are here, but Travis isn't here yet, Juliana."

"He'll be here."

Beth cast her daughter one last worried look. "You're sure?"

"Yes, Mom, I'm sure."

"You do look lovely today, dear." Beth smiled tremulously and went out of the room.

Elly gave Juliana a hard stare. "*Are* you sure?"

"Of course I'm sure. Would I be standing here in this dress if I thought I'd get stood up at the altar? Travis would never do that to me."

"Well, there was that little incident at the engagement party," Elly reminded her delicately.

"Travis showed, didn't he? He didn't stand me up. He just didn't plan to hang around very long after he got there, that's all."

"That's a very charitable view of the situation. If you hadn't run after him and thrown yourself on the hood of his car, I don't know what would have happened that night."

"A woman has to go after what she wants."

"And you really want Travis Sawyer, don't you?" Elly said with soft understanding.

A knock on the door interrupted Juliana's reply. "We're ready, Miss Grant," said a muffled voice on the other side.

Juliana nodded with satisfaction and pulled the waist-length veil down over her face. "Right on time. Get your flowers, Elly."

Juliana opened the door and walked confidently down the hall to the point where she would make her entrance with her father. She peeked down the aisle toward the altar and was not at all surprised to see Travis standing there,

251

waiting for her. He looked incredible in formal clothes, she thought fondly. She'd have to find a way to get him dressed up more often.

"He just drove up two minutes ago," Roy Grant muttered, shaking his head as he took his daughter's arm. "Tony and I were just about to go after him. Thought for sure you'd been stood up."

Everyone was feeling protective of her today, Juliana thought happily. It must have had something to do with her role as a bride. Parental instincts coming out, no doubt.

"There was no need to worry, Dad. Travis said he'd be here."

The music swelled. Juliana smiled her most brilliant smile and started down the aisle on her father's arm.

Travis never took his eyes off her as she came toward him. Her gaze met his through the veil. When she reached the altar he accepted her hand as Roy Grant released her.

"Sorry I'm a little late," Travis murmured very softly. "Got held up at the office. Bickerstaff changed his mind."

"He *what?*" Juliana hastily lifted the gossamer veil so she could get a better look at Travis's laughing eyes.

"You heard me. Flame Valley is technically again in Grant hands as of about fifteen minutes ago. Got a hell of a load of debt hanging over it, but David and Elly are the official owners."

Juliana threw her arms around him, laughing

with surprise and delight. "I knew you'd pull it off, Travis. There's nobody like you in the whole world."

A ripple of astonishment went through the crowd as Juliana hugged Travis. The minister coughed to get everyone's attention. "I believe we're ready to begin," he said with a touch of severity.

Juliana released Travis, grinning. "First you have to make an announcement," she informed the man of the cloth.

The minister's brows rose in amused curiosity. "What sort of announcement would that be?"

"Just say that Bickerstaff changed his mind."

The minister looked out over the crowded church. "Bickerstaff," he intoned solemnly, "has changed his mind."

Juliana thought she heard a small gasp from Elly and then the bride's side of the church broke out in wild applause. Anxious not to offend, the rest of the guests quickly followed suit.

When the applause finally faded the minister looked sternly at an unrepentant Juliana. "Now may we begin the wedding service?"

"You bet," Juliana said.

"Hang on a second," Travis said and reached out to lower the veil back down over Juliana's dazzling smile. When he was finished arranging the filmy stuff he nodded, satisfied with the old-fashioned, demure effect. "A man's got a right to insist on a little tradition once in a while."

"I'm thinking of buying an interest in this res-taurant," Travis muttered some time later as he stood, champagne glass in hand, surveying the throng of guests at the reception. "At the rate we're using this place, we might as well own a share of it. One of these days you're going to have to arrange a party at one of the other res-taurants in town. Just for variety."

"Now don't grumble, Travis. The Treasure House always does a wonderful job with wed-ding receptions."

"Uh-huh." He sipped his champagne. "At least I'm finally putting an end to your hobby of collecting proposals here."

She batted her lashes at him. "Yours was the only proposal that counted."

"Damn right."

Juliana beamed. "Do you realize this is the first moment I've had you to myself since the wedding? I thought I'd never get you away from David and Uncle Tony and Dad."

"They wanted all the details about the Bickerstaff deal."

"I'll bet they did. Why did Bickerstaff change his mind?"

"It's complicated and I don't really feel like going into it now, but to sum it up, I called in an old favor from a banker friend of mine. When he found out Bickerstaff was interested in the re-sort, he agreed to restructure some of Flame Valley's debt. That, in turn, tipped the scales as far as Bickerstaff was concerned. He decided to go ahead with the deal."

Juliana whistled faintly in appreciation. "Sounds tricky. What about your investors?"

"They'll be paid off the same way they would have been in a buy-out. It's complicated, but I think it's going to work. Assuming Kirkwood cooperates."

"He will."

"Yeah, I think he will." Travis glanced over Juliana's shoulder and his eyes hardened faintly. "Here comes Mom and her second husband."

"I like your mother. And your father. We all had a nice chat earlier." Juliana turned to smile at the attractive, champagne-blond woman who was approaching with a slightly portly, well-dressed man in tow.

"Hello, Mrs. Riley. Mr. Riley. Enjoying yourselves, I hope?"

Linda Riley returned the smile and so did her husband. "Very much, dear." She looked at her eldest son. "You've chosen a very lovely bride, Travis."

"She chose me," Travis said, not bothering to conceal his satisfaction. "Glad you and George and the kids could get here today," he added a little gruffly.

"Wouldn't have missed it for the world," Mrs. Riley said dryly as she slanted an amused glance at Juliana. "We never see enough of you, Travis. You really ought to come visit more often. The kids are always curious about their mysterious big brother, you know. They admire you. I believe Jeremy wants to talk to you about going into land development."

"Is that so?" Travis looked wary but interested.

Mrs. Riley's smile deepened with understanding as she turned back to Juliana. "Thank you very much for inviting all of us, Juliana. You know, sometimes families drift apart without really meaning to. People lose perspective in the heat of selfish emotion. Pride becomes far more important than it should. But that doesn't mean any of us want it that way or that we can't see the light eventually."

"I know, Mrs. Riley," Juliana said, returning her mother-in-law's smile. "As I told Travis, people change. Weddings are great opportunities for getting families together, aren't they?"

"Better than funerals," Travis remarked.

Juliana wrinkled her nose at him and then helped herself to a canapé from the buffet table while he talked to his mother and stepfather. She was quite pleased with the way things had turned out, she decided. Everyone had been well behaved at the church and seemed to be acting like adults here at the reception.

Travis's father, a tall, distinguished-looking man who had introduced himself earlier, was at the other end of the room with his second wife. Both were in deep conversation with Roy and Tony Grant.

Travis's stepbrothers and sisters, ranging in age from the late teens to early twenties, were a lively, talkative crew who appeared to regard the brother they shared with some awe and fascination.

"So how did you do it?" Travis asked as his mother and stepfather drifted off to join another group.

"Do what?"

"Don't play the innocent with me. I know you too well. How did you get both my mother and my father to attend the wedding?"

"Travis, I think you should make allowances for the fact that reasonable people are quite capable of change over a period of time. It's been years since they refused to attend your first wedding. They were probably still very bitter toward each other back then. Now they've had a chance to mellow and mature. Intelligent people grow up sooner or later."

Travis picked up a cucumber and salmon canapé and popped it into his mouth. He considered Juliana's words carefully and then dismissed them. "I'm not buying it. What you say about their maturing may be true but I don't see you just sending out invitations and hoping for the best. You wouldn't take any chances. You wanted them here, so what did you do to make certain they showed up today?"

"Blackmailed them."

Travis grinned. "With what?"

"I made it very clear that neither your mother nor your father would be invited to see their first grandchild if they didn't have enough courtesy to attend the wedding."

"I should have guessed." He put down his champagne glass. "Care to dance, Mrs. Sawyer?"

"I would love to dance, Mr. Sawyer."

Juliana went into his arms, the heavy skirts of her wedding gown whirling around her low-heeled satin slippers.

"I see you decided not to wear high heels today," Travis observed as he looked down slightly to meet her eyes.

"I figured a bride should be able to look up to her husband on her wedding day," Juliana explained demurely. "Tradition, you know."

Travis laughed and the sound of his uninhibited masculine pleasure turned the head of everyone in the room. "Are you sure you didn't wear the low heels because you wanted to be prepared to run after me in case I didn't show up at the church?"

She looked up at him with all her love in her eyes. "I knew you'd show. I never doubted it for a moment."

Travis's gaze grew suddenly, fiercely intent. "You were right. Nothing on earth could have kept me from being at that church today."

"I love you, Travis."

He smiled. "I know. I've never been loved by anyone the way you love me. Just for the record, I love you, too."

"I know," she said, pleased. "Hey, you want to sneak out of here early and start on our honeymoon?"

"That depends. What, exactly, are you planning to do on our honeymoon? Toss me in the marina? Chase me through parking lots? Dump guacamole over my head?"

"Gracious, no. I was thinking we could spend the time going over the plans for my chain of Charisma tea shops."

"You never give up, do you?"

"Never."

"I've got a better idea," Travis said. "What do you say we go someplace private and talk about babies?"

"While it's true I never give up," Juliana responded smoothly, "I can be temporarily distracted. I would love to go someplace private and talk about babies."

"It's a deal," Travis said.

He came to a halt in the middle of the dance floor, took Juliana's hand in his and led her toward the door — and their future.